Algorithms for Intelligent Systems

This book series publishes research on the analysis and development of algorithms for intelligent systems with their applications to various real world problems. It covers research related to autonomous agents, multi-agent systems, behavioral modeling, reinforcement learning, game theory, mechanism design, machine learning, meta-heuristic search, optimization, planning and scheduling, artificial neural networks, evolutionary computation, swarm intelligence and other algorithms for intelligent systems.

The book series includes recent advancements, modification and applications of the artificial neural networks, evolutionary computation, swarm intelligence, artificial immune systems, fuzzy system, autonomous and multi agent systems, machine learning and other intelligent systems related areas. The material will be beneficial for the graduate students, post-graduate students as well as the researchers who want a broader view of advances in algorithms for intelligent systems. The contents will also be useful to the researchers from other fields who have no knowledge of the power of intelligent systems, e.g. the researchers in the field of bioinformatics, biochemists, mechanical and chemical engineers, economists, musicians and medical practitioners.

The series publishes monographs, edited volumes, advanced textbooks and selected proceedings.

More information about this series at http://www.springer.com/series/16171

Subhojit Dawn · Kedar Nath Das ·
Rammohan Mallipeddi · Debi Prasanna Acharjya
Editors

Smart and Intelligent Systems

Proceedings of SIS 2021

Springer

Editors
Subhojit Dawn
Department of Electrical and Electronics Engineering
Velagapudi Ramakrishna Siddhartha Engineering College
Vijayawada, Andhra Pradesh, India

Rammohan Mallipeddi
School of Electronics Engineering
Kyungpook National University
Daegu, Korea (Republic of)

Kedar Nath Das
Department of Mathematics
National Institute of Technology Silchar
Silchar, Assam, India

Debi Prasanna Acharjya
School of Computer Science and Engineering
Vellore Institute of Technology
Vellore, Tamil Nadu, India

ISSN 2524-7565 ISSN 2524-7573 (electronic)
Algorithms for Intelligent Systems
ISBN 978-981-16-2108-6 ISBN 978-981-16-2109-3 (eBook)
https://doi.org/10.1007/978-981-16-2109-3

This Springer imprint is published by the registered company Springer Nature Singapore Pte Ltd.
The registered company address is: 152 Beach Road, #21-01/04 Gateway East, Singapore 189721, Singapore

Preface

The invention is the successful exploitation of a new indication. Through invention, we achieve MORE from LESS for MORE. The invention of smart and intelligent systems will not only help mankind but also contribute to the economic and technological advancement of the country as well as to the globe. The book provides advanced ideas for sustainable development through smart and intelligent systems to conserve our daily life.

It is essential that the global problems of environmental degradation are addressed immediately; thus, we need to consider the way of expansion and evolve the concept of sustainable development through a smart and intelligent concept. Indeed, the new environmentally friendly technologies are fundamental to attain sustainable development. The book provides a large number of innovative green technological ideas with a smart and intelligent system to maintain and improve the quality of the environment and will help us to achieve a resource-efficient and sustainable thinking society of the future.

The book *Smart and Intelligent Systems* will provide an inter-disciplinary innovative approach to address various technical issues and advancements in smart and intelligent systems for Scientific and Technological Development, Smart Information, Communication, bio-monitoring, smart cities, environmental aspects, alternative energy, Sustainable Infrastructure Development, etc.

The book will provide useful information to the people from academia, budding engineers, research scholars, upcoming young minds, and industries in bringing awareness about the recent advances in various fields mentioned above. The book will help the readers to open up new ideas wherein the shortcomings of the presented ideas can be addressed.

Vijayawada, India — Subhojit Dawn
Silchar, India — Kedar Nath Das
Daegu, Korea (Republic of) — Rammohan Mallipeddi
Vellore, India — Debi Prasanna Acharjya

Contents

About the Editors

Dr. Subhojit Dawn has completed his Ph.D. in Electrical Engineering from National Institute of Technology Silchar, India; Master of Technology (M. Tech) in Power and Energy Systems Engineering from National Institute of Technology Silchar, India; and Bachelor of Technology (B. Tech) in Electrical Engineering from West Bengal University of Technology, India. Dr. Dawn has published more than 25 research papers in reputed international journals/conferences in the field of power system. His current research includes power system economics, renewable energy integration, power system planning, congestion management, and smart grid and energy management. Dr. Dawn is a member of IEEE. He also serves the IEEE Student Branch, NIT Silchar, India, as Chair during 2016–2018. Dr. Dawn is Associate Editor of Journal of Electrical Engineering & Technology (JEET), Springer. He is a continuous reviewer of many reputed international (SCI/SCIE/ESCI) journals including IET Renewable Power Generation, IET Generation, Transmission & Distribution, Renewable Energy (Elsevier), Applied Energy (Elsevier), etc., and various IEEE/Springer/Elsevier hosted/sponsored conferences. Dr. Dawn is an editorial member of several international journals including American Journal of Electrical Power and Energy Systems, SCIREA Journal of Electrical Engineering, International Journal of Energy Policy and Management, etc. He was participating in many international conferences as Organizing Chair, Session Chair and member in Technical Program Committee. He is Editor of "Intelligent Techniques and Applications in Science and Technology" book published by Springer.

Dr. Kedar Nath Das is currently working as Assistant Professor in the Department of Mathematics, NIT Silchar, Assam. He qualified the GATE examination and was awarded his Ph.D. degree from IIT Roorkee in the year 2008. His area of research interest includes operations research, evolutionary optimization techniques, networking optimization, and multi-objective optimization. As of now, Dr. Das has over 50 research papers to his credit including papers in international journals of repute, which has around 400 Google Scholar Citations till date. There are 1 book and 3 book chapters to his credit. He has already guided 5 Ph.D. scholars and recently 4 are working under him. Also, he has successfully guided about 6 M.Sc. projects and 2 undergraduate internships. Dr. Das has visited around 11 foreign countries (including Las Vegas, USA; London, UK; Amstradem, Netherlands; Bankok, Thailand; Singapore; Dubai; Kuching, Malaysia; Bratislava; Budapest, Vienna, Prague, Liverpool, UK; Wales, UK) and delivered about 25 expert lectures at different institutions in India (like IIT Roorkee, VIT Vellore, GMIT Guwahati, NIT Silchar, JIET Jodhpur, GIST Udaipur, SAU Delhi, NIT Nagaland, S. S. College Hailakandi, Prativa College Chatrapur Odisha, RTU Kota, IIIT Guwalior, BJB College Bhubaneswar, GC College Silchar). As coordinator, he has organized 11 events including STTP, international conference, workshops, and self-financed courses. He has already reviewed more than 100 research papers. He is the life member of ORSI (Operational Research Society of India), ISTE (The Indian Society for Technical Education), OMS (Odisha Mathematical Society) and member of IEEE. He has been recently awarded "Slovak National Scholarship" in the year 2019.

Rammohan Mallipeddi is Associate Professor working in the School of Electronics Engineering, Kyungpook National University (Daegu, South Korea). He received Master's and Ph.D. degrees in computer control and automation from the School of Electrical and Electronics Engineering, Nanyang Technological University, Singapore, in 2007 and 2010, respectively. His research interests include evolutionary computing, artificial intelligence, image processing, digital signal processing, robotics, and control engineering. He co-authored papers published IEEE TEVC, etc. Currently, he serves as Associate Editor for "Swarm and Evolutionary Computation," an international journal from Elsevier and a regular reviewer for journals including IEEE TEVC and IEEE TCYB.

Debi Prasanna Acharjya received his Ph.D. in Computer Science from Berhampur University, India; M.Tech. degree in Computer Science from Utkal University, India, in 2002; M.Phil. from Berhampur University, India; and M.Sc. from NIT, Rourkela, India. He has been awarded with Gold Medal in M.Sc. Currently, he is working as Professor in the School of Computing Sciences and Engineering, VIT University, Vellore, India. He has authored more than 100 national and international journal and conference papers, and five books; Fundamental Approach to Discrete Mathematics, Computer Based on Mathematics, Theory of Computation; Rough Set in Knowledge Representation and Granular Computing; Introduction to Information and Communication Technology and Computer programming to his credit. He has also edited seven books to his credit. He has also published many chapters in different books published by international publishers. Besides, he has produced seven Ph.D.'s in computer science and engineering. In addition, he is a reviewer of many international journals such as Fuzzy Sets and Systems, Knowledge Based Systems, and Applied Journal of Soft Computing. He has been awarded with Eminent Academician Award from Khallikote Sanskrutika Parisad, Brahmapur, Odisha; Outstanding Educator and Scholar Award from National Foundation for Entrepreneurship Development, Coimbatore, India; The Best Citizens of India Award from The International Publishing House, New Delhi, India; and Bharat Vikas Award from Institute of Self Reliance, Bhubaneswar, Odisha, India. Dr. Acharjya is actively associated with many professional bodies like CSI, ISTE, IMS, AMTI, ISIAM, OITS, IACSIT, CSTA,

IEEE, ACM, and IAENG. He was Founder Secretary of OITS Rourkela chapter. His current research interests include rough sets, formal concept analysis, knowledge representation, data mining, granular computing, and business intelligence.

Chapter 1
Optimal Sizing and Siting of Distributed Generation for Losses Minimization in Distribution System Using Fractional Lévy Flight Bat Algorithm

Aditya N. Koundinya, Galiveeti Hemakumar Reddy, Z. Mohammed Khalander, and Revanasidda

1 Introduction

The performance of the distribution system (DS) is causally linked with that of the entire power system as the DS is responsible for supplying electricity to the end users. The DS operates at high current and low voltage [1], which leads to high power losses in the system. According to the data provided by the ministry of power [2], in India, the occurrence of power losses in the DS is around 22% of the generated power. As the power losses on the distribution side increase, the operational cost of the entire system increases, which in turn increases the cost of electricity supplied to the consumers.

Literature survey evinced that the researchers have used different methods to minimize power losses in the DS, such as network reconfiguration [1], placement of FACTS devices [3], integration of distributed generation (DG) [4], etc. Among all the methods which have been used, integration of DGs has gained a lot of popularity, as it has numerous benefits like reduction of power loss, voltage improvement, reduction of load demands during peak hours [5], reduction of carbon footprint [6], critical load pickup [7], voltage sag mitigation [8], and increase in the system reliability [9, 10]. On the other hand, inappropriate placement of DGs lead to increase in the power losses in the system than that without DG [4]. Therefore, it is vital to select the size and spot of DG optimally before placing it.

Different meta-heuristic optimization techniques have been utilized for the placement of DGs in the DS. The authors in [8] have used PSO for the placement of DG to maximize the critical load pickup. In [11], Simulated Annealing has been used for reducing power losses by placing DG. The authors of [12] have used Ant colony search (ACS) to place DG. The Artificial Bee Colony technique has been used

A. N. Koundinya · G. H. Reddy (✉) · Z. M. Khalander · Revanasidda
Department of Electrical and Electronics Engineering, MVJ College of Engineering, Bengaluru, Karnataka 560067, India

S. Dawn et al. (eds.), *Smart and Intelligent Systems*, Algorithms for Intelligent Systems,
https://doi.org/10.1007/978-981-16-2109-3_1

in [13] for allocation of DG to reduce the losses of real power. In [14], the authors have used Cuckoo Search (CS) algorithm to optimally place DG to reduce active power loss. In [15], Bat Algorithm (BA) has been implemented for the integration of DG in the DS to minimize the power loss and improve voltage stability index.

From the literature, it is perceived that the above-mentioned algorithms have a low convergence rate, low accuracy, and fall easily into local optima solution. In this work, Fractional Lévy Flight Bat Algorithm (FLFBA) has been proposed for optimal placement of DGs for power loss minimization. This algorithm was proposed by Redouane et al. [16]. The ability to elude from local ideal solutions has been improved in this modified version of BA [17], which makes the algorithm robust and helps to untangle the problem considered in this paper.

2 Mathematical Formulation

The problem considered in this paper is to place DGs optimally in the DS to minimize the active power losses in the system. Before placing the DG, load flow analysis must be done. The load flow equations are computed based on the single line diagram of the feeder shown in Fig. 1. It consists of several buses, but two buses, namely 'k' and 'k + 1,' are considered for the computational purpose.

$$P_{k+1} = P_k - P_{Loss(k,\, k+1)} - P_{L(k+1)} = P_k - \frac{R_k\left(P_k^2 + Q_k^2\right)}{|V_k|^2} - P_{L(k+1)} \tag{1}$$

$$Q_{k+1} = Q_k - Q_{Loss(k,\, k+1)} - Q_{L(k+1)} = Q_k - \frac{R_k\left(P_k^2 + Q_k^2\right)}{|V_k|^2} - Q_{L(k+1)} \tag{2}$$

$$|V_{k+1}|^2 = |V_k|^2 + \frac{R_k^2 + X_k^2}{|V_k|^2} * \left(P_k^2 + Q_k^2\right) - 2 * (R_k P_k + X_k Q_k);\ k \in \{1, 2, 3, \ldots, NB\} \tag{3}$$

Using the following equation, real power losses in the branch section connecting kth and $(k + 1)$th bus is computed:

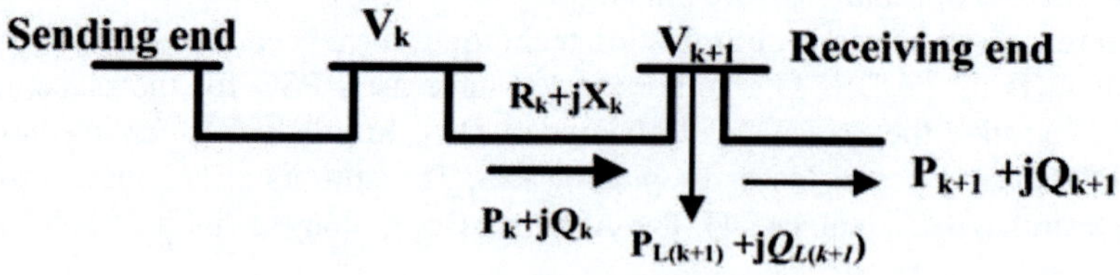

Fig. 1 Typical representation of radial type distribution feeder

$$P_{Loss(k,k+1)} = \frac{R_m\left(P_k^2 + Q_k^2\right)}{|V_k|^2} \tag{4}$$

The sum of the real power losses of the DS having NB buses can be found as shown below.

$$P_{TLoss} = \sum_{k=1}^{NB} P_{Loss(k,k+1)} \tag{5}$$

where P_{k+1} denotes the active power coming out of (k + 1)th bus, P_k is the active power flowing through the branch line, $P_{Loss(k,k+1)}$ is the active power losses with the branch linking buses k and $(k + 1)$, $P_{L(k+1)}$ denotes the active power demand at (k + 1)th bus, Q_{k+1} denotes the reactive power coming out of $(k + 1)$th bus, Q_k denotes the reactive power flowing through the branch line, $Q_{Loss(k,k+1)}$ is the reactive power losses with the branch linking buses k and (k + 1), $Q_{L(k+1)}$ is the reactive power demand at (k + 1)th bus, R_k is the resistance of the branch line, V_k is the voltage at bus k, V_{k+1} is the voltage at bus $(k + 1)$, X_k is the reactance of the branch line, P_{TLoss} is the sum of the active power loss of the feeder [1].

2.1 Objective Function

Reducing the total active power losses in the DS is considered the sole objective in this paper.

$$\text{Obj } f = \min(P_{TLoss}) \tag{6}$$

The objective function must be formulated without violating the operational constraints because it must be ensured that the DS operates safely. The voltage and thermal limits are given by

$$V_{min} \le V_k \le V_{max} \tag{7}$$

The minimum voltage of the bus is V_{min}, and the maximum voltage of the bus is denoted as V_{max}

$$I_k < I_k^{max} \tag{8}$$

where I_k^{max} is the maximum current that can flow through the branch.

The total power generated in terms of real and reactive power must balance the total real and reactive power required to meet the load demand and the total real and reactive power losses in the DS.

$$\sum_{k=1}^{NB} P_{kGen} = \sum_{k=1}^{NB} (P_k + P_{Loss,k}) \tag{9}$$

$$\sum_{k=1}^{NB} Q_{kGen} = \sum_{k=1}^{NB} (Q_k + Q_{Loss,k}) \tag{10}$$

The DG capacity limits are:

$$P_{DGk}^{min} < P_{DGk} < P_{DGk}^{max} \text{and } Q_{DGk}^{min} < Q_{DGk} < Q_{DGk}^{max} \tag{11}$$

The Eq. 11 represents the upper and lower bounds of the size of DG [7].

2.2 *Optimal Placement of DG*

As mentioned in the first section, it is necessary to select the size and location of DG optimally so that it does not create any adverse effects. The selection of the size and location of DGs are performed using optimization techniques. An optimization algorithm follows an iterative procedure and helps find the best solution by comparing the latest solution with the previous one. In this paper, FLFBA has been chosen for the optimal siting and sizing of DGs.

2.3 *Fractional Lévy Flight Bat Algorithm*

The FLFBA is a revamped version of the Bat Algorithm. This algorithm is based on differential evolution, fractional calculus, and Lévy Flight. FLFBA is inspired by the bat's hunting strategy. Bats fly with a velocity v_a, from position x_a while searching for prey, and they emit an ultrasonic sound with a frequency f_{min} that echoes after bouncing back from nearby objects. This helps the bat to find the size as well as the distance of its prey. The emitted sound pulses will be as high as 110 dB when the bat is far from its prey, but when it reaches close to its prey, the loudness of the sound pulses emitted by the bat will be much quieter. This strategy of bats has been incorporated in this algorithm.

The FLFBA begins with generating a 'P' number of random locations in a population and initializing the velocity vector by assigning some values. The generated locations are bounded by a search space of 'Dim' dimensions [16].

$$x_{ab}^{0} = x_{min} + (x_{max} - x_{min}) * rnd \tag{12}$$

where a ∈ [1,…,P], b ∈ [1,…, *Dim*], and rnd is an arbitrary vector with elements distributed consistently produced in the range between 0 and 1.

x_{min} is a vector containing the upper boundary, while x_{max} is a vector containing the upper boundary in all the dimensions *b*, respectively. Generally, the preliminary value of velocity is set to 0, i.e., $v_a^0 = 0$. A new location is produced by calculating the variation in the randomly selected two local best solutions, and this difference is multiplied by 'ε,' which is an arbitrary value. This arbitrary variable is added to a^{th} local best solution.

$$s = \varepsilon\left(x_q^l - x_p^l\right) \tag{13}$$

$$x_n = x_a^l + s \tag{14}$$

where x_n is the new location and x_a^l is the analogous best solution.

The objective function is evaluated for each population after it is initialized, and then the best solution is selected by comparing the current best solution with the previous one. If the new location has produced a better result, then the solution will be updated. This updated solution is obtained using Lévy Flight search. Lévy Flight helps in increasing the search space exploration, while Fractional Lévy Flight enhances the utilization of predominant regions.

$$x_a^{t+1} = x_a^l + 0.01\frac{(u)}{|v|^{1/\beta}}\left|\bar{x}^l - x_a^t\right| \tag{15}$$

where x_a^{t+l} is the updated solution, $\bar{x}^l$ is the average of the local best solution vector, β is an arbitrary vector, and it lies in the range of 0 to 1.

The subsequent part of the population is calculated by applying fractional calculus and Differential Equation (DE) velocity upgrade equation.

$$v_a^t = \sum_{k=0}^{o} s_k v_a^{t-k} \tag{16}$$

where s_k is a coefficient and o ∈ [1, 10] is the fractional derivative's order.

The velocity equation shown in Eq. 16 is updated by the DE approach.

$$v_a^{t+1} = \omega^t v_a^t + f_a^t \xi_1^t\left(x^g - x_a^t\right) + f_a^t \xi_2^t\left(x_a^l - x_q^t\right) \tag{17}$$

where v_a^{t+l} is the updated velocity vector, ω^t is the weight of inertia, f_a^t denotes the frequency, x^g denotes the global best location, ξ_1^t and ξ_2^t are learning factors, x_q^t is an arbitrary set solution obtained from the population where $q \neq a$.

The above equations follow an iterative procedure until $F\left(x_a^1\right) < F(x^g)$ is achieved.

The implementation of FLFBA for the optimal selection of locations and sizes of DGs is shown in Fig. 2.

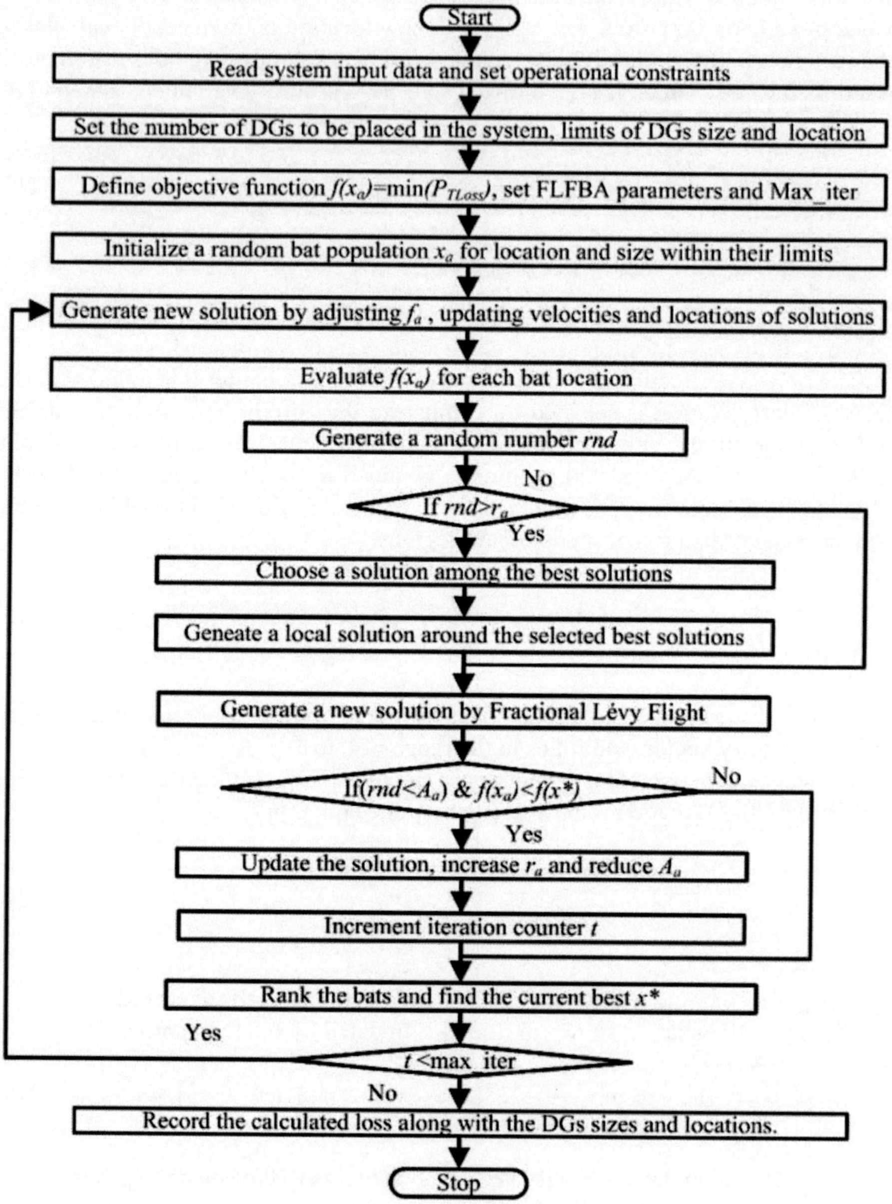

Fig. 2 Flowchart of FLFBA for optimal siting and sizing of DGs

3 Results

The potency of the proposed technique is evaluated on the standard IEEE-33 bus test system (TS). It consists of 32 lines and has a voltage of 11 kV. The TS has a maximum reactive power load of 2.30 MVAr and active power load of 3.72 MW. The one-line representation of the TS is depicted in Fig. 3.

The optimal DG placement problem is untangled using FLFBA. The parameters of FLFBA taken into consideration are $alpha = 0.99$, $gamma = 0.9$, size of population = 30, Maximal number of iterations = 50, maximal frequency (f_{max}) = 2, minimum frequency (f_{min}) = 0, maximum loudness (A_{Max}) = 2, minimal loudness (A_{Min}) = 0, $w_{Max} = 0.9$, $w_{Min} = 0.2$ (maximum and minimum inertia weight). Two DGs are simultaneously installed in the TS by considering minimization of active power losses as an objective function. The performance of FLFBA is compared with the PSO technique, and it is found that FLFBA has a faster convergence rate compared to PSO. PSO converged after 32 iterations, whereas, FLFBA converged after 21 iterations. The convergence curve for optimal placement is shown in Fig. 4.

The DG locations and sizes are shown in Table 1. From Table 1, it is perceived that the objective function value has been reduced considerably. Both the DGs are situated in the same locations while using PSO and FLFBA. The sizes of both DGs are different. FLFBA is giving more size for DG1 and less size for DG2, while keeping the sum of the capacity of both the DGs almost the same. It can be observed that FLFBA has given a better solution than PSO for loss minimization.

The reduction of active and reactive power losses is shown in Table 2. The TS without DG placement is considered the base case for a better comparison of power loss reduction. It is detected that reduction of power losses took place significantly with DG integration. Even though the intention was to minimize real power losses, the reactive power losses have also been reduced drastically. Also, it is noticed that FLFBA has managed to reduce power losses slightly greater than PSO.

The power loss at each branch is shown in Table 3. It is perceived that in the base case, power losses are more in every branch. After DG integration, power loss at most

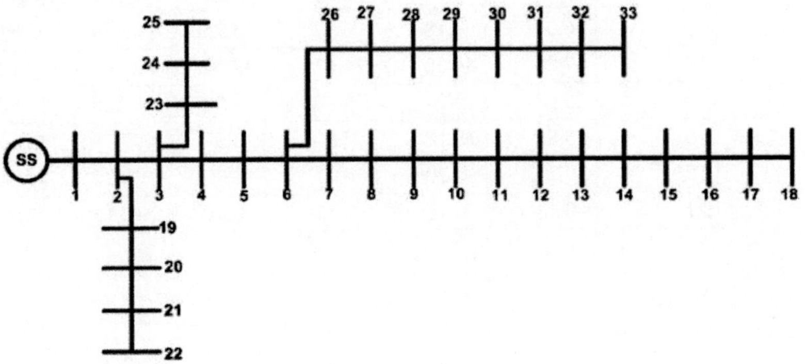

Fig. 3 Test system diagram

Fig. 4 Convergence curve

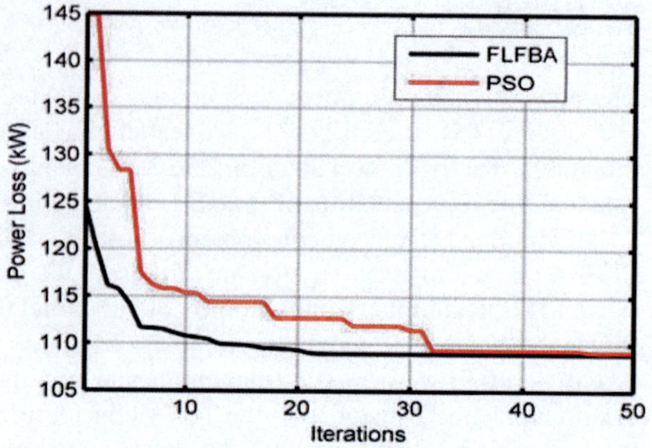

Table 1 DG location and size

Optimization technique	Location		Size (MW)		Objective function (kW)
	DG1	DG2	DG1	DG2	
PSO	30	13	1.4453	1.0803	109.12
Proposed FLFBA	30	13	1.4985	1.0593	108.96

Table 2 Real power and reactive power losses (in kW and kVAr, respectively) with and without DG placement

Instance	P_{Loss}	Q_{Loss}	Reduction in P_{Loss} (%)	Reduction in Q_{Loss} (%)
Base case	275.72	184.87	–	–
PSO	109.12	75.12	60.42	59.37
Proposed FLFBA	108.96	74.98	60.48	59.44

Table 3 Active and reactive power loss (in kW and kVAr, respectively) in each branch before and after integration of DG

From bus	To bus	Base case		PSO		Proposed FLFBA	
		P_{Loss}	Q_{Loss}	P_{Loss}	Q_{Loss}	P_{Loss}	Q_{Loss}
1	2	16.07	8.18	6.37	3.25	5.89	3.01
2	3	67.86	34.56	24.63	12.54	22.66	11.54
3	4	25.98	13.24	8.29	4.22	7.85	3.99
4	5	25.88	13.18	7.80	3.97	7.36	3.75
2	19	0.21	0.20	0.21	0.20	0.21	0.20
20	21	0.13	0.16	0.13	0.16	0.13	0.16

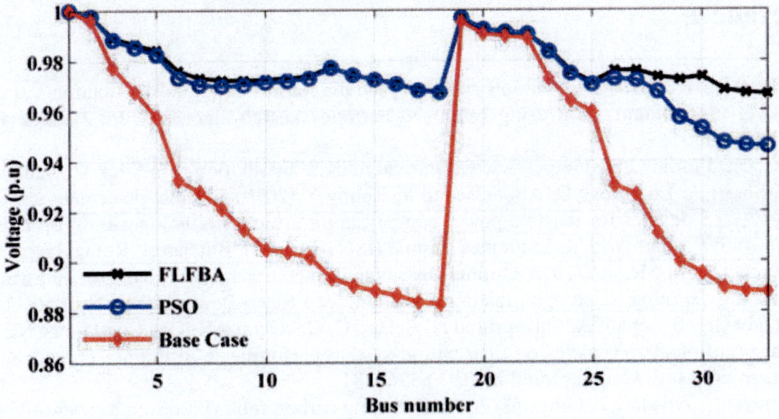

Fig. 5 Bus voltages with and without DG placement

branches has reduced remarkably, while in a few, it has remained the same. The results show that the proposed FLFBA has outperformed PSO in reducing the losses.

Figure 5 depicts the voltage profile with and without DG. For the base case, minimum bus voltage (0.8830 p.u) occurred at bus 18. After DG integration, there is an improvement in the minimum bus voltage of the system, and it is 0.9457 p.u at bus number 33 for PSO and 0.9666 p.u at bus 33 for FLFBA. The proposed FLFBA method has an edge over the PSO for increasing the minimum system voltage. From Fig. 5, it is observed that DG integration greatly improved voltages at all buses. This signifies that the placement of DG has helped in voltage profile improvement along with minimizing the real power losses.

4 Conclusion

A modified optimization technique called FLFBA is proposed for simultaneous placement of multiple DGs for minimization of real power losses in the DS. Two DGs are placed simultaneously in the test DS. This method is validated by weighing up the results with that of the PSO. From the results obtained, it can be concluded that placement of DGs in the test DS using FLFBA succeeded in minimizing the real power loss with a fast convergence rate and gave better results compared to PSO. Besides, it also reduced the total reactive power losses and helped in improving the bus voltages of the DS. The placement of DGs in the test DS using FLFBA proved to be effective. Hence, it can be concluded that FLFBA is a pertinent strategy to solve the optimal selection of the size of DG and placing it in a DS.

References

1. Khetrapal P (2020) Distribution network reconfiguration of radial distribution systems for power loss minimization using improved harmony search algorithm. Int J Electr Eng Inf 12(2):341–358
2. Electrical India (2018) https://www.electricalindia.in/distribution-losses-how-to-reduce/
3. El-Sherif A, Elkobrosy G, Abouelseoud Y, Helmy Y (2019) Optimal placement and settings of FACTS devices for reactive power compensation using a firefly algorithm. In: IEEE PES Innovative Smart Grid Technologies Europe (ISGT-Europe), Bucharest, Romania, pp 1–5
4. García JA, Gil-Mena A (2013) Optimal distributed generation location and size using a modified teaching–learning based optimization algorithm. Int J Electr Power Energy Syst 50(1):65–75
5. Reddy GH, Badepalli A, Shivaprasad N, Behera C, Goswami AK, Dev Choudhury NB (2019) Impact of electric vehicles on distribution system performance in the presence of solar PV integration. Int J Comput Intell IoT 2(1):376–381
6. Monyei C, Aiyelari T, Longe O (2013) Reducing carbon footprint through renewable energy, distributed generation and smart government policies. CISDIARJ, vol 4, pp 17–20
7. Reddy GH, Chakrapani P, Goswami AK, Dev Choudhury NB (2017) Optimal distributed generation placement in distribution system to improve reliability and critical loads pick up after natural disasters. Eng Sci Technol Int J 20(3):825–832
8. Behera C, Debbarma M, Banik A, Reddy GH, Goswami AK (2018) Voltage sag mitigation using distributed generation for an industrial distribution system. In: 2018 IEEE international conference on power electronics, drives and energy systems (PEDES), Chennai, India, pp 1–6
9. Reddy GH, Kiran MK, Kumar PS, Goswami AK, Dev Choudhury NB (2020) Fuzzy reliability assessment of distribution system with wind farms and plug-in electric vehicles. Electric Power Compon Syst 47(19–20):1791–1804
10. Reddy GH, Goswami AK, Dev Choudhury NB (2018) Impact of distributed generation integration on distribution system reliability. Indian J Sci Technol 11(34):1–13
11. Sutthibun T, Bhasaputra P (2010) Multi-objective optimal distributed generation placement using simulated annealing. In: ECTI-CON2010: The 2010 ECTI international conference on electrical engineering/electronics, computer, telecommunications and information technology, Chiang Mai, pp 810–813
12. Nwohu M, Olatomiwa L, Ambafi J, Sadiq AA, Mogaji A (2017) Optimal deployment of distributed generators using ant colony optimization to minimize line losses and improve voltage profiles on distribution network. In: World congress on engineering computer science conference, San Francisco, USA
13. Abu-Mouti FS, El-Hawary ME (2011) Optimal distributed generation allocation and sizing in distribution systems via artificial bee colony algorithm. IEEE Trans Power Deliv 26(4):2090–2101
14. Ghosh M, Kumar S, Mandal S, Mandal K (2019) Optimal sizing and placement of DG units in radial distribution system using cuckoo search algorithm. Int J Appl Eng Res 12(1):362–369
15. Prakash R, Sujatha BC (2016) Optimal placement and sizing of DG for power loss minimization and VSI improvement using bat algorithm. In: 2016 national power systems conference (NPSC), Bhubaneswar, pp 1–6
16. Boudjemaa R, Oliva D, Ouaar F (2020) Fractional Lévy flight bat algorithm for global optimisation. Int J Bio-Inspired Comput 15(2):100–112
17. Yang X-S (2010): A new metaheuristic bat-inspired algorithm. In Gonzalez JR, Pelta DA, Cruz C, Terrazas G, Krasnogor N (eds) Nature inspired cooperative strategies for optimization (NICSO 2010), pp 65–74

Chapter 2
Performance Analysis of a Standalone Inverter System Under Variable Loading Conditions

David Hmingthanmawia, K. Lalmalsawma, Samuel Lalngaihawma, Subir Datta, Subashish Deb, Ksh. Robert Singh, and Sadhan Gope

1 Introduction

In India, there has been a rapid growth of renewable energy sources for electric power generation in the past few decades. The main driving forces are due to the adverse environmental impacts of conventional plants, their huge cost, and losses in long transmission lines [1]. Renewable energy can be defined as an energy that is collected from resources which are naturally generated in short time, such as sunlight, wind, geothermal heat, etc. Therefore, for all practical purposes, these resources can be considered to be inexhaustible, unlike conventional fuels [2]. The implementation of distributed generation (DG) has been highly increasing. Compared to the conventional centralized power generation, DG units have many advantages such as higher energy utilization efficiency, flexibility in installation location, and less power transmission losses. Nowadays, microgrid is one of the most up-to-date and important topics in the scope of power systems [3]. The microgrid concept was first proposed in the USA by the Consortium for Electrical Reliability Technology Solutions [4]. A microgrid is defined as a cluster of DG units and loads, serviced by a distribution system, and can operate in two modes—grid-connected mode and islanded mode. The basic functions of a microgrid are [5] regulating the microgrid's voltage magnitude and frequency within their normal ranges during autonomous mode; controlling active power and reactive power flow from DG units to loads while working in autonomous mode; managing power flow between microgrid and the main grid during grid-connected mode; and providing a smooth transition between islanded mode and grid-connected mode. Most DG units are connected to the microgrid through DC/AC inverter interface. Thus, by proper control of those inverters, microgrid energy management is sufficiently accomplished. The fundamental control variables of a microgrid are active power, reactive power, voltage,

D. Hmingthanmawia · K. Lalmalsawma · S. Lalngaihawma · S. Datta (✉) · S. Deb · Ksh. R. Singh · S. Gope
Department of Electrical Engineering, Mizoram University, Aizawl, Mizoram, India

S. Dawn et al. (eds.), *Smart and Intelligent Systems*, Algorithms for Intelligent Systems,
https://doi.org/10.1007/978-981-16-2109-3_2

and frequency. In grid-connected mode, the microgrid frequency and the voltage at the Point of Common Coupling (PCC) are predominantly dictated by the main grid. In this case, the major function of the microgrid control is to manage both active and reactive powers produced by the DG units and the load requirements [6]. Injecting reactive power into the main power grid can be used to provide ancillary services such as power factor correction, elimination of harmonics, or voltage control. In some cases, the utility may not permit voltage control at PCC by DG units to prevent interfering with similar actions provided by the utility [7]. In [8], a standalone power inverter was used to convert DC into AC and, these types of inverters are mostly used in remote areas where utility grid is not available, and these are powered from solar-PV, wind turbine units, etc. These type of inverters are usually used for residential purposes. A three-phase inverter is an inverter with six IGBT switches and converts a DC input into a three-phase AC output. Its three legs are normally delayed by an angle of 120° so as to produce a three-phase AC supply. In islanded mode, the microgrid works totally independent. Therefore, this situation is more difficult than being connected to the main grid, as maintaining load-supply equilibrium necessitates the application of precise load sharing mechanisms to adjust and equilibrate any unexpected power mismatches. Neither Voltages nor frequency of the microgrid are still determined by the main grid, thus, they must be controlled by the DG units [9]. In case of standalone system, another significant matter is the suitable distribution of the required power to the microgrid loads between generation units. Droop control method is used to retain the rated frequency (fundamental) and the voltage magnitude of microgrid so that the proper powers can be shared. Often used droop control methods to improve energy sharing and synchronization of the voltage/frequency are active power–frequency (P–F)-based droop control method and reactive power–voltage magnitude (Q–V)-based droop control method [10]. Based on the researches, the droop control method is one of the most effective approaches for synchronization of power generation among numerous generators, since the stability can quickly be achieved by using these methods and communication unit is also not required among generation units in case of droop approach. So, using this droop control, the reactive and active power can be controlled [11]. In [12], phase-locked loop (PLL)-based control scheme for the inverter was used to detect frequency and phase along with classical P–f and V–Q droop techniques. The voltage regulator computes and regulates the desired voltage magnitude of the inverter. Lastly, the PWM generator takes the desired voltage magnitude and phase and creates the PWM output signals [12]. We know that the main application is in microgrid. It can also be used as uninterrupted power supplies (UPS) in computer system and motor drives [13].

In this paper, a standalone inverter system is considered, and the overall system is implemented in MATLAB/Simulink environment to study its performances under varying load conditions. In addition, PI controller gain values are optimized using Firefly Algorithm.

1.1 Standalone Inverter and Its Control Scheme

A typical standalone inverter base system is shown in Fig. 1. A voltage source converter converts DC input to AC output voltage. A linear RL load and variable load are connected to the inverter. A capacitor is connected across the load to smoothen its voltage.

The dynamic equations of the system in three-phase abc frame are as follows: Transmission line dynamic equations are [1],

$$\left.\begin{aligned} V_{sa} &= \mathrm{V}_{ma} + R_s I_{sa} + L_s p I_{s1a}) \\ V_{sb} &= \mathrm{V}_{ma} + R_s I_{sb} + L_s p I_{s1b}) \\ V_{sc} &= \mathrm{V}_{mc} + R_s I_{sc} + L_s p I_{s1c}) \end{aligned}\right\} \tag{1}$$

The equations for linear load (L_1) [1] are given as,

$$\left.\begin{aligned} V_{ma} &= R_L I_{La} + L_L p I_{La}) \\ V_{mb} &= R_L I_{Lb} + L_L p I_{Lb}) \\ V_{mc} &= R_L I_{Lc} + L_L p I_{Lc}) \end{aligned}\right\} \tag{2}$$

Here, the linear load phase currents are I_{Labc}.

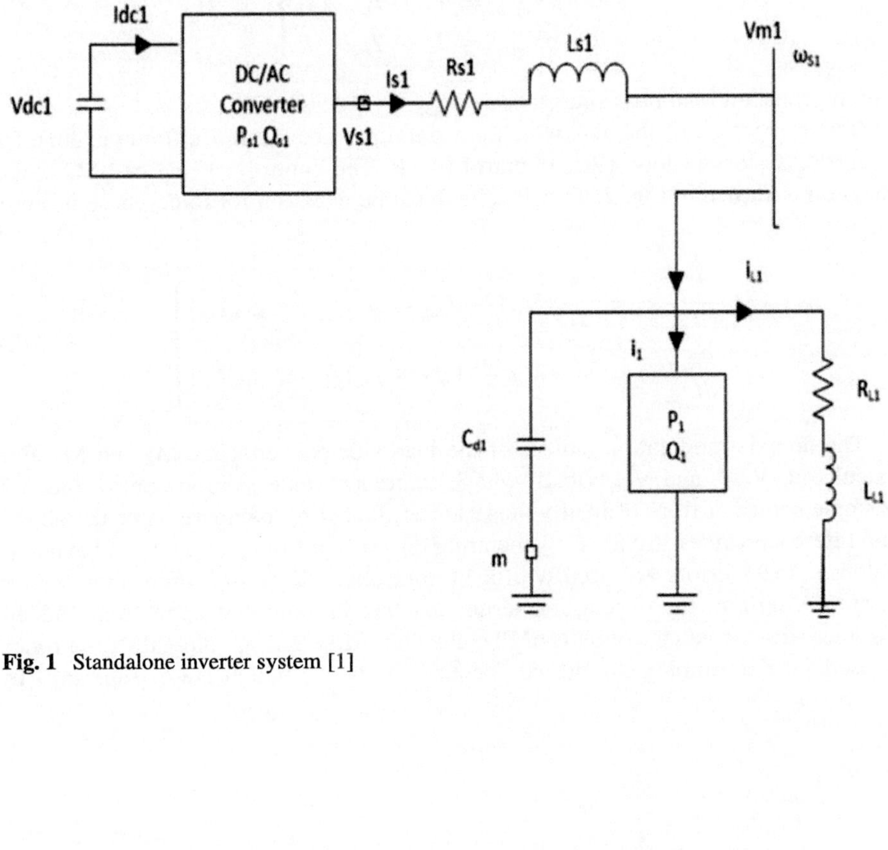

Fig. 1 Standalone inverter system [1]

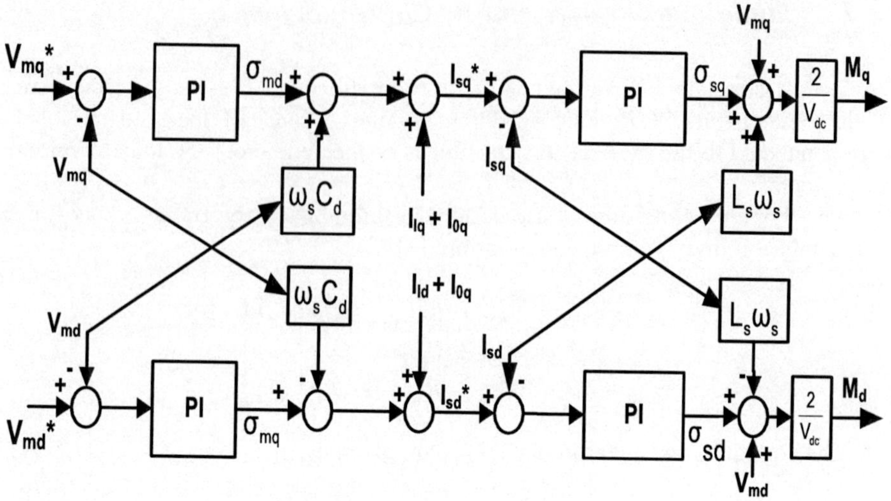

Fig. 2 Control scheme for three-phase inverter [14]

The dynamic equations for the capacitor are given as [1],

$$\left.\begin{aligned} C_{d1}\mathrm{pV}_{m1a} &= \mathrm{I}_{sa} - I_{La} - I_a \\ C_{d1}\mathrm{pV}_{m1b} &= \mathrm{I}_{sb} - I_{Lb} - I_b \\ C_{d1}\mathrm{pV}_{m1c} &= \mathrm{I}_{sc} - I_{Lc} - I_c \end{aligned}\right\} \tag{3}$$

The constant load phase currents are I_{1abc}.

Transformation of the axis which is from dq to abc reference frame is done by using Phase-locked loop (PLL) control block. The control scheme of LSC block diagram is shown in Fig. 2. The dq axis dynamic equation for three-phase inverter is,

$$\left.\begin{aligned} \frac{d}{dt}(I_{sq}) &= \frac{1}{L_s}(M_q\frac{V_{dc}}{2}V_{mq} - R_s I_{sq} - L_s\omega_s I_{sd}) \\ \frac{d}{dt}(I_{sd}) &= \frac{1}{L_s}(M_d\frac{V_{dc}}{2}V_{md} - R_s I_{sd} + L_s\omega_s I_{sq}) \end{aligned}\right\} \tag{3}$$

The dq axis modulation indices of the load side converter are M_d and M_q. For calculating V_{md}* and V_{mq}*(load voltage reference value), droop control method is implemented. Errors of load voltage are evaluated by comparison of the actual and reference values. Again, for generating dq axis controlled currents and external voltages, these errors are used with a PI controller. Then, to acquire the inverter reference current, the decoupling terms and the dq axis load currents are added together with the voltage outputs of PI controller. After that, an internal PI controller is used for determining the dq axis controlled signals which are σ_{sq} and σ_{sd}. To determine the actual controlled signals, the two signals are then added with the cross coupling terms and dq axis components of actual load voltage. By using these signals, the dq axis Modulation index of an inverter can be determined. These 'dq' signals are converted into 'abc' signals using inverse park transformation technique and, thereafter, 'abc' signals are processed through PWM generator to generate triggering pulses of the three-phase inverter. In order to maintain the frequency and voltage, a droop controller is implemented. This method is used in AC power generators that are included in power plants. It is used for sharing the total load and also controlling the frequency along with the voltage magnitude at particular ranges in an autonomous mode of operation of a Distributed generation.

1.2 Firefly Algorithm (FA)-Based PI Controller

It is an optimization technique which is used to determine the PI controller gain values of the inverter control scheme. The optimal controller gain values are given in Table 1. This algorithm works on the concept of the fireflies. The firefly's flashes light to attract the mates. The attractiveness of the fireflies is based on intensity of the flashing light. The less bright firefly move towards the more brighter one, and the brightness is depending on the distance. This algorithm can be applied as to measure the intensity of the flashing light. The distance should be computed between two fireflies. The attractiveness should be measured with respect to distance between two fireflies. The movement of one firefly to the brighter one is an attractiveness function. Figure 3 shows the flow chart of firefly algorithm. Firstly, to generate the variables or parameters say x_i; (i = 1, 2, 3,............n). Then to find the objective function f(x); x = (x_1, x_2,...., x_d). Update the intensity of the light, i.e., formulate the light intensity I so that it is associated with f(x). For example, for maximization problem, I is directly proportional to f(x) or simply I = f(x). Rank and update the position of the function. Loop ('For', 'While' etc.) functions are used to check the conditions for obtaining the optimal values of PI controllers. Until the optimal value is reached,

Table 1 Optimal gain values of the PI controllers obtained using FA

Parameters of the PI controllers	Optimal gain values
K_p (q-axis voltage controller)	0.377
K_I (q-axis voltage controller)	0.001
K_p (d-axis voltage controller)	0.377
K_I (d-axis voltage controller)	0.0051
K_p (q-axis current controller)	25.13
K_I (q-axis current controller)	628.32
K_p (d-axis current controller)	25.13
K_I (d-axis current controller)	628.32

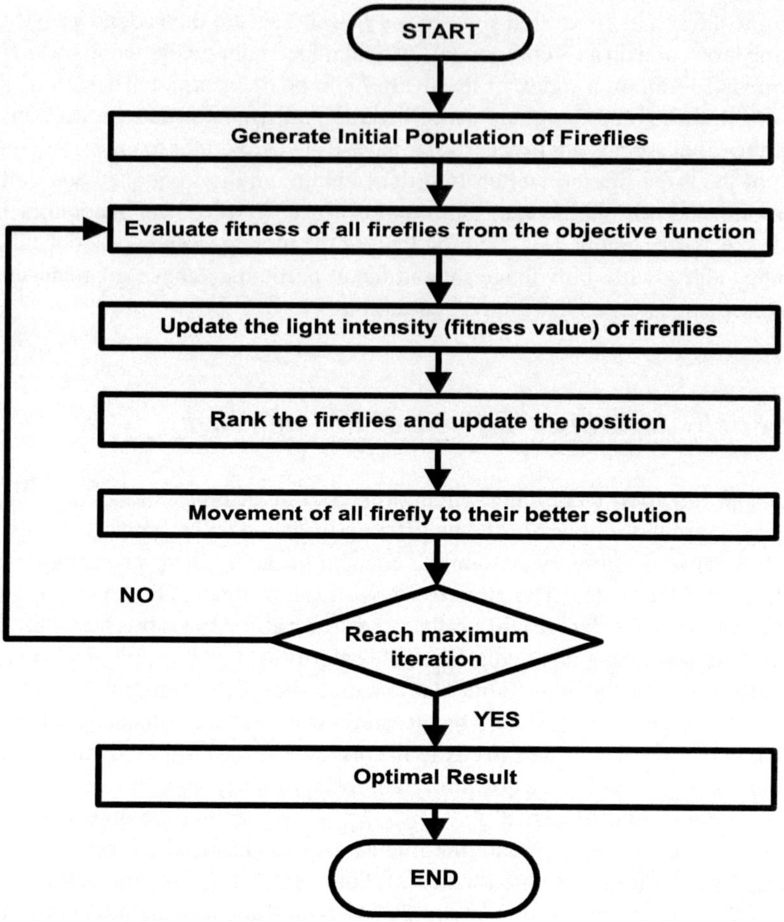

Fig. 3 Flow chart of firefly algorithm

the program will run again and again, and once the optimal value is reached, the program will stop.

2 Simulink Results and Discussion

The standalone microgrid inverter model and its control scheme is implemented in MATLAB/Simulink R2020a software. Time domain responses of the system are taken from the simulink model under variable linear loading conditions to study the performance and effectiveness of the control scheme.

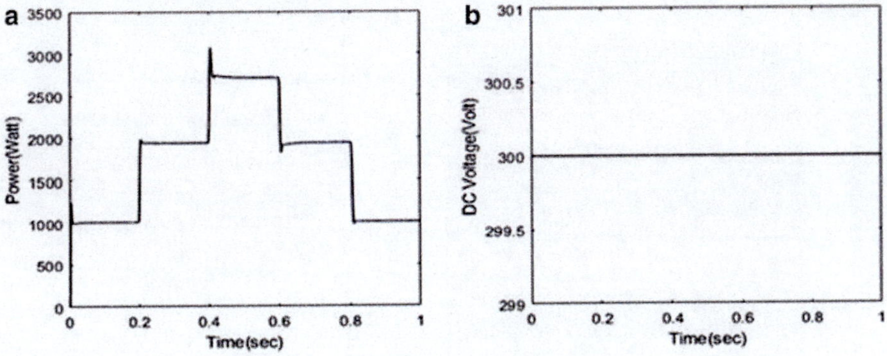

Fig. 4 **a** Variable load power (Watt), and **b** Input voltage of inverter (Volt)

The load power required will be delivered by the DC battery. The DC source is assumed to be constant at 300 V shown in Fig. 4b, but the load is assumed to be varying as shown in Fig. 4a, and the performance of the system is shown below.

For time t = 0–0.2 s: At this instant time, load is around 1 kW. The load voltage is around 120 V and frequency increases to a peak value of 420 Hz and decreases to a value of 376.8 Hz and remains constant till 0.2 s.

For time t = 0.2–0.4 s: At this instant, load is increased to around 1.9 kW. During this instant, the load voltage still attains the same value and frequency also remains the same.

For time t = 0.4–0.6 s: At this instant, load increases and reaches its peak value at around 2.7 kW. The load voltage still maintains its value of around 120 V and frequency also remains the same.

For time t = 0.6–0.8 s: At this instant, the load power decreases to 1.9 kW. The load voltage and frequency still maintain their values and remain constant.

For time t = 0.8–1 s: During this time, load drops to its minimum value again which is 1 kW. Till this point, the load voltage and frequency attain its constant value of 120 V and remain unchanged. Load voltage and current versus time graph is shown in Fig. 5. Waveform of load frequency is shown in Fig. 6. And, inverter voltage and current versus time graph is shown in Fig. 7.

It is noticed, from Simulink results, that the control scheme works properly, and it is capable enough for maintaining load voltage and frequency to their rated value under varying load conditions.

3 Conclusion

The performance of a standalone converter system are studied in this paper. A MATLAB/Simulink software is used for designing the complete system and is simulated under varying load conditions as shown in the simulink results. In order to

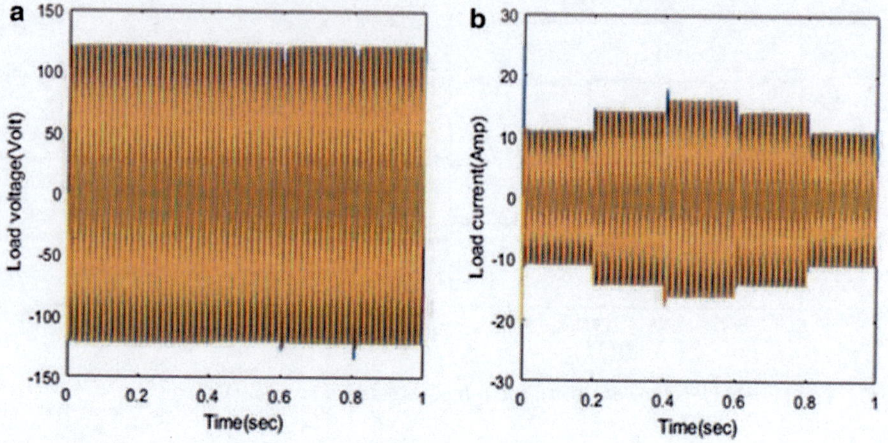

Fig. 5 **a** Load voltage (Volt), and **b** Load current (Amp) under varying loading conditions

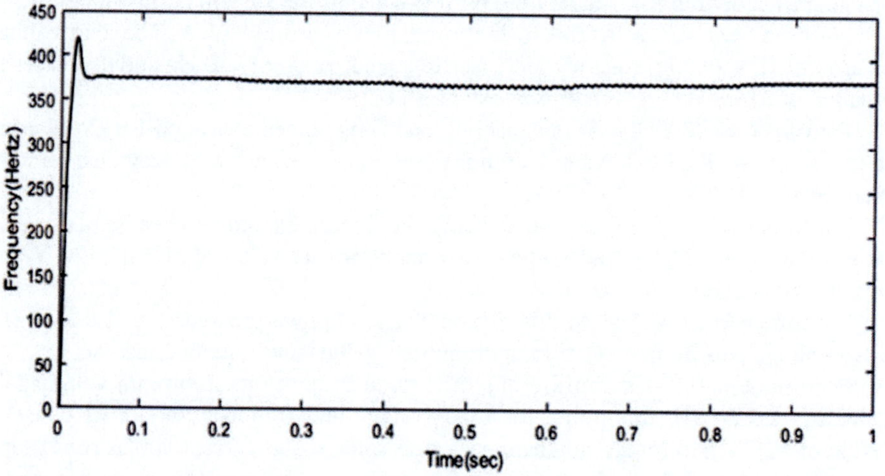

Fig. 6 Load frequency (rad/sec)

maintain frequency and voltages, droop control method is also used. A PLL is incorporated in the system to design the vector control scheme for inverter, and an optimization technique (Firefly Algorithm) is used to determine the PI controller gain values. The Simulink results are analyzed, and results show that the inverter control scheme is properly working, and it is capable enough to maintain rated voltage and frequency at the load bus under varying load conditions.

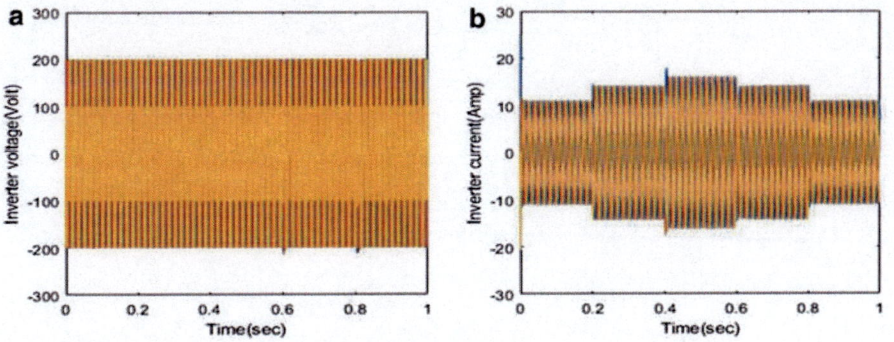

Fig. 7 **a** Output voltage of inverter (Volt), and **b** Output current of inverter (Amp)

References

1. Davijani HK (2012) Analysis and control of a microgrid with converted fed distributed energy sources. Dissertation, Tennessee Technological University
2. Lim U, Kim HW, Cho KY, Bae JH (2018) Stand-alone microgrid inverter controller design for nonlinear, unbalanced load with output transformer. Electronics 7:1–16
3. Moran B, Lorentzen M (2016) Assessing the role of energy efficiency in microgrids. TRC Solutions
4. Jiravan M (2014) Energy storage for stability of microgrids. Electric power, Universite de Grenoble
5. Sun X, Hao Y, Wu Q, Guo X, Wang B (2016) A multifunctional and wireless droop control for distributed energy storage units in islanded AC microgrid applications. IEEE Trans Power Electron 32:736–751
6. Palizban O (2016) Distributed control strategy for energy storage systems in AC microgrids. University of Vaasa, Faculty of Technology, Energy Technology, Finland
7. Zheng L, Zhuang C, Zhang J, Du X (2015) An enhanced droop control scheme for islanded microgrids. Int J Control Autom 8:63–74
8. M-Merza A, Shaheed AH (2013) Design and implementation of three phase inverter based on microcontroller. Dissertation, Electrical Engineering Department, University of Babylon
9. Xia XZ, Wei FH (2012) Impacts of P-f and Q-V droop control on microgrids transient stability. In: International conference on applied physics and industrial engineering, physics procedia, science direct, pp 276–282
10. Vignesh SS, Sundaramoorthy RS, Megallan A (2016) The combined V-F, P-Q and droop control of PV in microgrid. Int J Res Appl Sci Eng Tech 4:989–994
11. Bevrani H, Shokoohi S (2013) An intelligent droop control for simultaneous voltage and frequency regulation in islanded microgrids. IEEE Smart Grid 4:1505–1513
12. Hiskens I, Fleming E (2008) Control of inverter-connected sources in autonomous microgrids. In: American control conference, Seattle, WA, pp 586–590
13. Mukherjee S, Chowdhury VR, Shamsi P, Ferdowsi M (2017) An improved control scheme for standalone inverters in the stationary frame of reference with a zero sequence controller. In: IEEE Power and Energy Conference at Illinois, Champaign, IL, USA
14. Datta S, Deb S, Samanta S, Maity NP, Maity R, Adhikari S (2019) Power management of a solar-battery based stand-alone system using adaptive neuro fuzzy inference system based controller. WSEAS Trans Power Syst 14:145–155

Chapter 3
Performance Study of a Wind-Battery-Based Islanding System

Samuel Lalngaihawma, C. Rohmingtluanga, Rahul Roy, David Hmingthanmawia, Subir Datta, and Nidul Sinha

1 Introduction

Nowadays, the use of Renewable energy resources (RERs)-based microgrid and the research in this field are becoming popular so as to overcome the problem that arises due to the fast depletion of naturally occurring fuels which leads to uprising cost and the scarcity of fuels [1]. The use of renewable resources for the production of electricity not only helps in reducing pollution and preserving the conventional sources but also actually helps in the improvement of reliability in the system which further helps in improving power quality [2].

One of the most popularly used renewable energy is the wind energy. A wind turbine (WT) is used to convert the kinetic energy into mechanical energy and then to the end product, i.e., electricity, by means of rotating generator. There are two types of WTs, namely, Vertical Axis wind turbine (VAWT) and Horizontal-axis wind turbine (HAWT) [3]. The types of WT can also be categorized into two, namely, fixed speed and variable speed turbines. These turbines are then connected to generators through a gearbox or also without gearbox. The different types of WTGs can be given as Squirrel cage induction generators (SCIG), Wound rotor induction generator (WRIG), Permanent magnet synchronous generator (PMSG), Wound rotor generator (WRSG), Doubly fed induction generator (DFIG), and are discussed in [4]. For this project, PMSG is chosen for the generation of electricity. It has many advantages over the other WTGs like the absence of rotor windings which results in low maintenance cost and that it runs in gearless operation which thereby results in minimization of weight. It is efficient because of its higher power density. Even at a low wind speed, excessive amount of torque can be reached [5]. The main drawbacks in WTGs or any other renewable resources are the uncontrollable production of electricity due to

S. Lalngaihawma · C. Rohmingtluanga · R. Roy · D. Hmingthanmawia · S. Datta (✉)
Department of Electrical Engineering, Mizoram University, Aizawl, Mizoram, India

N. Sinha
Department of Electrical Engineering, NIT Silchar, Assam, India

S. Dawn et al. (eds.), *Smart and Intelligent Systems*, Algorithms for Intelligent Systems,
https://doi.org/10.1007/978-981-16-2109-3_3

the complete dependence on weather condition (in our case which is wind) which is not controllable or fully predictable. As inputs are varying, direct connection to the main grid or to the load can cause instability in the grid and further failure of the system too [6].

PMSG-based WT system is interfaced with power electronic converters to get the desired operation under different conditions [7]. The use of passive diode rectifier along with the IGBT inverter and the other configuration in which rectifier is substituted by a PWM dual converter permits variable rotational speed control of the generator [8, 9]. The designing of control scheme is the challenge that arises so as to maintain the frequency and voltage throughout the microgrid system [10]. The vector control and droop control schemes are the most common and popular techniques used to achieve the VSCF operation, and also, to maintain the magnitude and angle of both voltage and currents [11–14].

Moreover, for reliable operation and continuation of power supply, energy storage systems like battery, super capacitor, etc., are provided through a bidirectional boost converter in a microgrid system [12]. At the time of surplus generation, battery gets charged, and at times when load is high, battery can back up the WT system. The output of the converters are then connected to the DC common link [12, 13].

In this paper, a wind-battery-based stand-alone system is implemented in MATLAB/simulink platform along with its control schemes. Then, the performance of the system is studied under different load and source conditions. The Simulink results show that the system is operated according to its coordinate control scheme. This paper has been organized as follows: the detail description of the system and simulink results are presented in Sects. 2 and 3, respectively. Finally, the conclusions are drawn in Sect. 4.

2 System Description

Figure 1 shows a wind energy and battery-based microgrid system. It is designed to generate electric power by means of a WTG from wind energy and a battery system for backup purposes. The turbine is coupled to a PMSG, and the output of the PMSG is linked with a three-phase converter, i.e., Machine side converter (MSC), which is further linked to the DC bus along with the output of the battery. Then, the other side of the DC bus is connected to the Load side converter (LSC) passing through a filter circuit for smooth distribution of power to the load. The MSC is used to control the speed of the generator under varying wind velocity, whereas LSC is used to control the magnitude of load bus voltage and current. A bidirectional boost converter is connected between battery system and DC-link capacitor to maintain the constant voltage across DC-link capacitor irrespective of the direction of battery power.

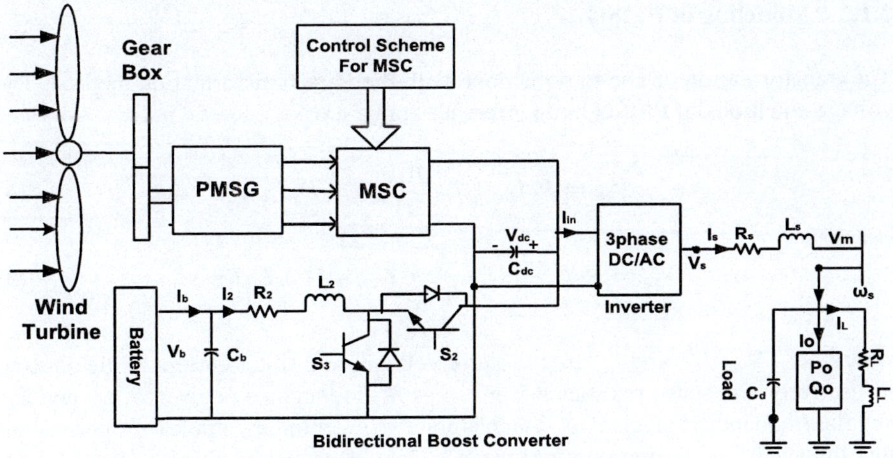

Fig.1 Block diagram of a stand-alone system with wind and battery units

2.1 *Wind Turbine System*

The WT system includes the conversion of the kinetic energy of the wind to mechanical power which is converted into electrical energy. It includes the modeling of turbine, generators, converter, and control scheme.

2.1.1 Modeling of Wind Turbine

The mechanical power P_T captured by the turbine from the wind for a given wind speed (Vω) is computed by the following expression [15]:

$$P_T = \frac{1}{2}\rho A C_p(\lambda, \beta) V_\omega^3 \tag{1}$$

where A is the swept area and is equal to πr^2, r is the radius of the turbine, the air density is ρ (kg/m), the velocity of wind V_W (m/s), and the power coefficient is C_P. The tip ratio is λ and blade pitch angle is β. The power coefficient and tip ratio are calculated by using Eqs. (2) and (3), respectively.

$$C_p = \frac{1}{2}\left(\lambda - 0.22\beta^2 - 0.5\right)e^{(-0.17\lambda)} \tag{2}$$

$$\lambda = \frac{R * \omega_T}{V_\omega} \tag{3}$$

2.1.2 Modeling of PMSG

The transformation of abc to dq is done with the Park transformation method. The voltage equations of PMSG in dq reference frame are;

$$V_{sd} = R_s I_{sd} + L_d \frac{dI_{sd}}{dt} - \omega_e L_q I_{sq} \tag{4}$$

$$V_{sq} = R_s I_{sq} + L_q \frac{dI_{sq}}{dt} - \omega_e \psi_{fl} + \omega_e L_d I_{sd} \tag{5}$$

where, V_{sd}, V_{sq}, I_{sd}, and I_{sq} are the stator voltages and stator current in the dq axis, respectively, the stator resistance is given as R_s, inductance is given as L_d and L_q, angular frequency is given as ω_e which is equal to the number of pole (n_p), permanent flux linkage is ψ_{fl}. Expression for the electrical torque can be given as:

$$T_e = 1.5 n_p \left[\psi_{fl} I_{sq} + (L_d - L_q) I_{sd} I_{sq} \right] \tag{6}$$

If PMSG is surface mounted, then $L_d = L_q$, and the torque can be written as:

$$T_e = 1.5 n_p \psi_{fl} I_{sq} \tag{7}$$

Mechanical equation for Permanent magnet synchronous generator can be written as

$$T_m = T_e + B w_m + J \frac{dw_m}{dt} \tag{8}$$

where, the moment of inertia is J, coefficient for friction is B, T_m and T_e are the mechanical torque and electrical torque, respectively.

For the steady state conditions, both active and reactive powers of the permanent magnet synchronous generator can be given as:

$$P_s = \frac{3}{2}\left(V_{sd} I_{sd} + V_{sq} I_{sq} \right) \tag{9}$$

$$Q_s = \frac{3}{2}\left(V_{sq} I_{sd} - V_{sd} I_{sq} \right) \tag{10}$$

2.1.3 Control Scheme of MSC

The output power of the PMSG is then converted to an acceptable DC power by using power electronic devices which is to be delivered to the DC bus.

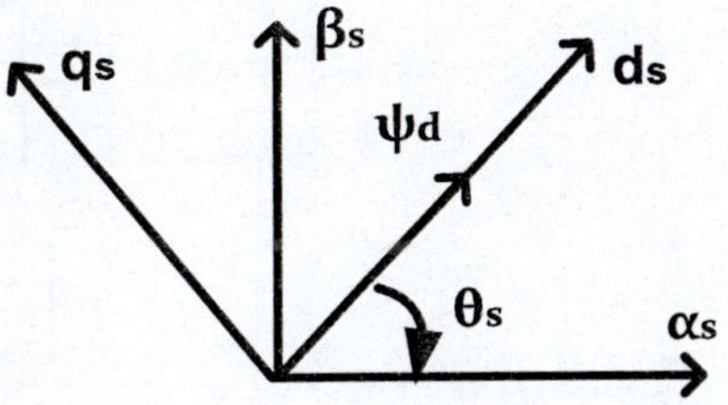

Fig.2 Phasor diagram for MSC [7]

In Fig. 2, the synchronously rotating reference frames are given as d_s and q_s while the stationary reference frames are α_s and β_s. ψ_d and ψ_q are the flux aligning with the d axis and q axis, respectively. Equation for the main flux is,

$$\psi = \psi_d + j\psi_q \tag{11}$$

Then, $\psi_q = 0$, assuming main flux to align in the d axis. So, Eq. (10) becomes,

$$\psi_d = \psi \tag{12}$$

As $\psi_d = \psi$, then, $V_d = 0$ and $V_q = V$. Putting the value of V_d and V_q in Eq. (8),

$$P = \frac{3}{2} V_q I_q \tag{13}$$

Here, V_q is considered constant. So, we can achieve control of the active power by controlling I_q. And also, $V_d = 0$, so Eq. (9) becomes,

$$Q = \frac{3}{2}(V_q I_d) \tag{14}$$

From Eq. (13), it is shown that by controlling $I_{d,}$ the reactive power Q can be controlled. Figure 3 shows the block diagram of the MSC. The Perturb and observe (P & O) algorithm-based MPPT [8] is used to capture maximum power at rated and below rated wind velocities, and the corresponding speed (called as reference speed (ω_m*)) of the maximum power is calculated to design control scheme for the MSC.

2.1.4 Control Scheme of LSC

Figure 4 shows a block diagram for control scheme of LSC. A Phase-locked loop (PLL) is used for axis transformation, i.e., from ‘abc’ to ‘dq’ and vice versa. The dq axis dynamic equation for three-phase inverter is given in Eq. (15).

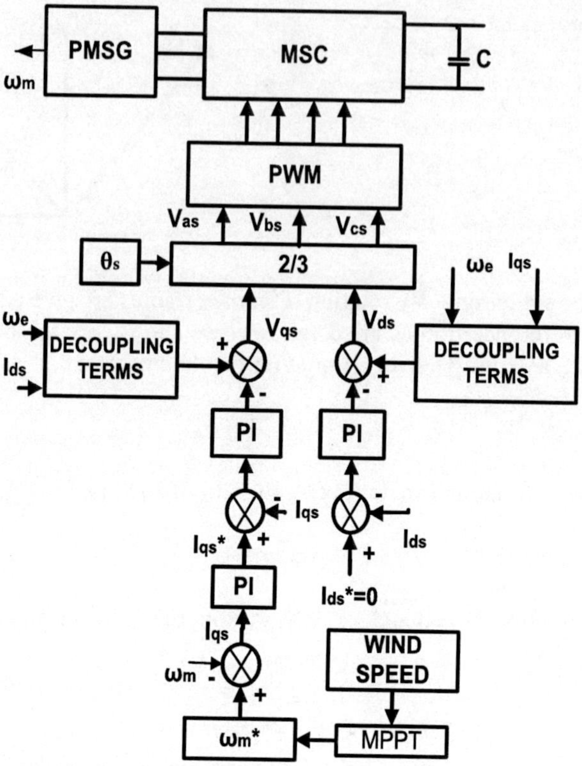

Fig.3 Control block for MSC [7, 8]

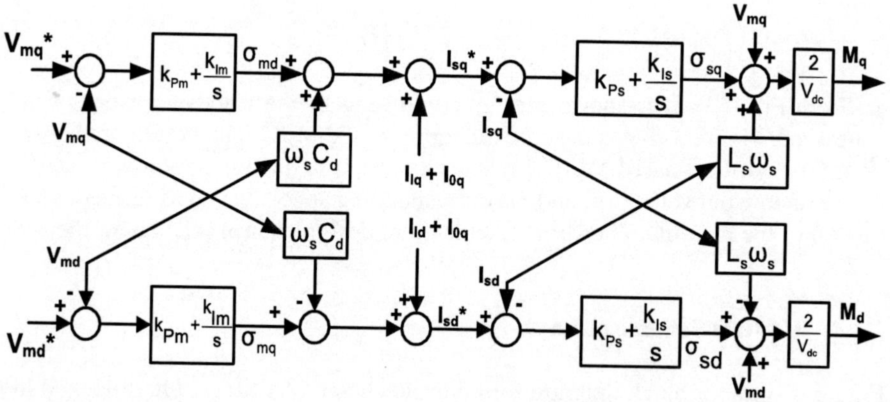

Fig.4 Block diagram of control scheme for DC-AC three-phase converter [12]

$$\left.\begin{aligned}\frac{d}{dt}(I_{sq}) &= \frac{1}{L_s}(M_q\frac{V_{dc}}{2}V_{mq} - R_sI_{sq} - L_s\omega_sI_{sd})\\ \frac{d}{dt}(I_{sd}) &= \frac{1}{L_s}(M_d\frac{V_{dc}}{2}V_{md} - R_sI_{sd} + L_s\omega_sI_{sq})\end{aligned}\right\} \tag{15}$$

where, M_d and M_q are the dq axis modulation indices of LSC. A method called droop control is used for calculating the reference value for load voltage, i.e., $V_{md}{}^*$ and $V_{mq}{}^*$. These values are compared with the actual values for evaluating load voltage errors. These errors are used for generating the dq axis control currents, and external voltage PI controllers are used for it. Thereafter, dq axis load currents and decoupling terms are added with the outputs of the voltage PI controllers for obtaining inverter reference current. Then, internal PI controllers are used to determine the dq axis controlled signals (σ_{sq} and σ_{sd}), and these two signals are added with cross coupling terms and dq axis components of actual load voltage to calculate the actual controlled signals as shown in Fig. 4. Modulation index of an inverter of the dq axis is computed by using these signals. These outputs are then converted into a three-phase component and go through the pulse width modulation generator for generating three-phase inverter pulses.

2.2 Battery System

A battery system is included in the system for backup purposes, i.e., to get reliable operation or for continuous distribution of power from generator side to the load side because output power of renewable energy system can fluctuate due to its dependence on weather condition.

Battery system is equipped with a bidirectional boost converter for efficient conversion of the DC voltage to required voltage. Battery is charged when surplus amount of generation power is produced and discharged when load demand is greater than generation. A block diagram of control scheme for DC/DC bidirectional boost converter for the battery system is shown in Fig. 5 [16].

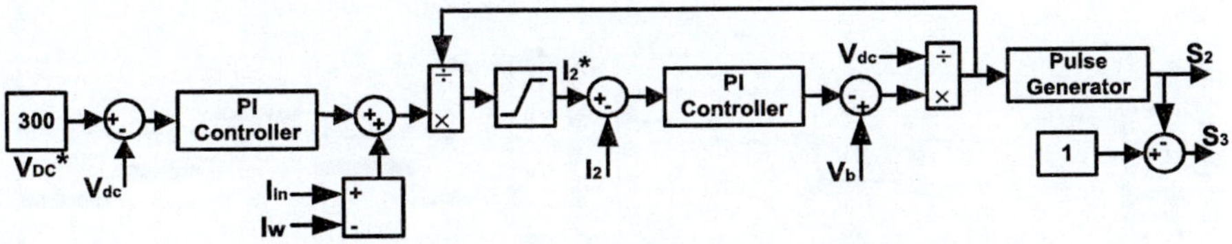

Fig. 5 Control scheme for DC/DC Bi-directional boost converter of battery unit

3 Simulation Results and Discussion

In this section, the performance of the wind-battery-based microgrid system is studied. The simulation is done in different cases, i.e., under different load and source variations.

3.1 Case-1:Fixed Wind Power and Variable Load

In this case, generation of power from the WTG is assumed to be constant, producing 1.3 kw to feed the varying load at different times. Load varies from 1.08 to 1.5 kw and then elevates upto 2.3 kw and then drop backs to 1.5 kw and to 1.08 kw. Wind power, load power, and power of battery is shown in Fig. 6a.

At time t = 0–0.2 s: During this time duration, the load is assumed to increase from 0 to 1.08 kw which means that generated power of wind is greater than the load power. At this time instant, battery charges with the surplus amount of power produced by the use of bidirectional DC/DC converter. So, at the DC-link/bus, voltage is maintained at 300 V to be fed to the load as shown in Fig. 6b. And, the battery charging voltage and current can be observed in Fig. 6c and d, respectively. Speed of PMSG (rad/sec) versus time is shown in Fig. 7d, and it is assumed to be constant, producing constant power throughout the time.

At time t = 0.2–0.4 s: During this period, load continues to increase from 1.08 kw to more than 1.3 kw which is greater than the power generated from wind turbine which further results in discharging of battery system for delivering desired power required by the system.

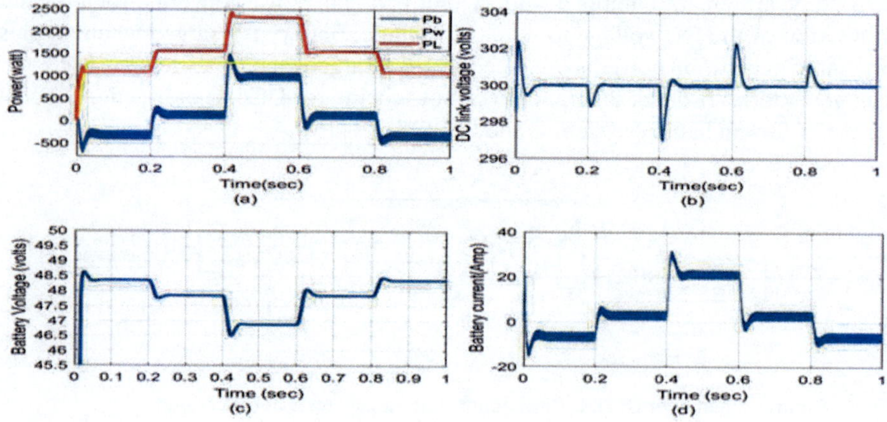

Fig. 6 **a** Power (watt), **b** DC-link voltage (volt), **c** Battery voltage (volt), and **d** Battery current (amp)

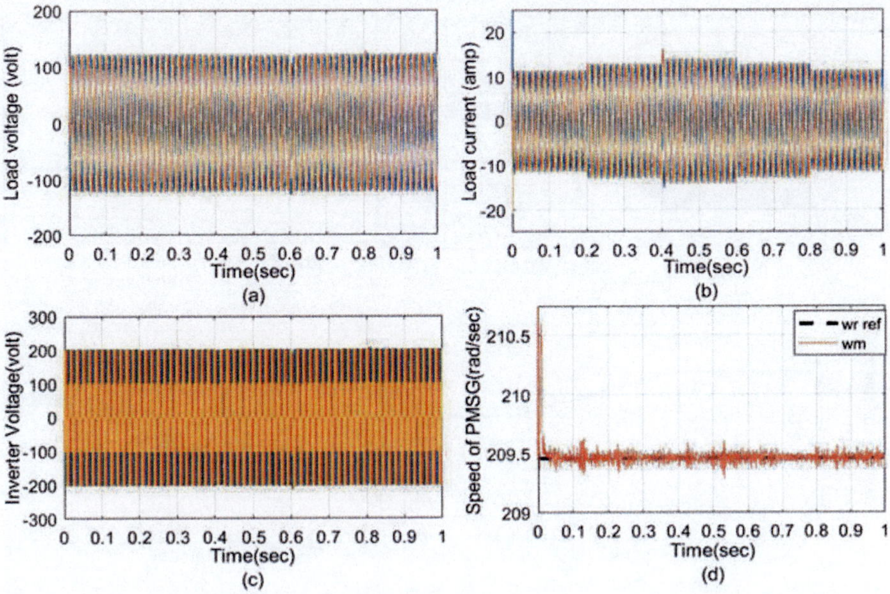

Fig. 7 **a** Load voltage (volt), **b** Load current (amp), **c** Inverter voltage (volt), and **d** Speed of PMSG

At time t = 0.4–0.6 s: In this scenario, the load power is at its highest point which is about 2.3 kw. So the gap between the production of power and the load required is maximum during this period. Battery keeps on discharging power to supply the power needed. Load voltage (in volt) and load current (in amp) versus time graph is given in Fig. 7a and b, respectively.

At time t = 0.6–0.8 s: The system response is same as that of time t = 0.2–0.4 s, since load starts to decrease to 1.5 kw.

At time t = 0.8–1 s: Here also, system response is same to that of time t = 0–0.2 s, since load is decreased to 1.08kw, which means that generation is more than the load power required. Also, at this time duration, battery starts to charge again. Inverter voltage versus time graph is given in Fig. 7c.

3.2 *Case-2: Variable Wind Power and Constant Load*

In this case, the wind power generated is considered to be varying as in real life scenario. The speed of the PMSG gradually decreases from time t = 0.4 until t = 0.8 s which is shown in Fig. 9d. The time when the wind power generation decreases, the amount of power which is used to charge the battery decreases. When the load power is greater than the generated power at time t = 0.6 s onwards, battery stops charging and starts discharging power to the system as required. By this way, power

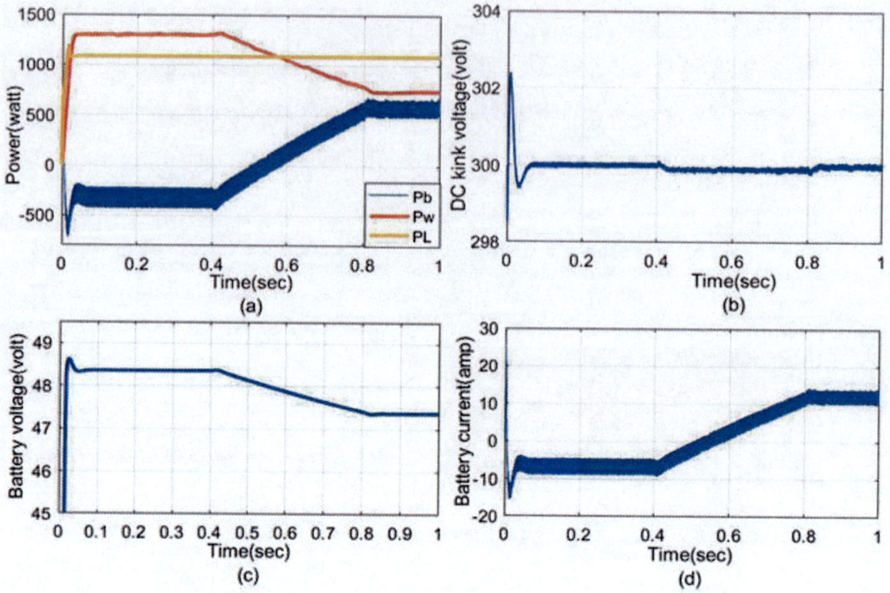

Fig. 8 **a** Power (watt), **b** DC-link voltage (volt), **c** Battery voltage (volt), and **d** Battery current (amp)

is maintained in the DC bus system as shown in Fig. 8b. The load power is set to be constant at 1.08 kw as shown in Fig. 8a. Load voltage and load current with time is shown in Fig. 9a and b, respectively.

At time t = 0–0.4 s: During this time, the load power is at 1.08 kw which is fed by a wind power of 1.3 kw which means that the generation is more than the load power demand, and therefore, the battery is charging during this time.

At time t = 0.4–0.6 s: At this time instant, the generation from wind starts decreasing, and the battery system gets less power from the source. At 0.6 s, the generation is equal to the load.

At time t = 0.6–0.8 s: During this time, the generation is now less than the load power which results in the complete discharging of battery power to the system. Power from wind keeps on decreasing until it reaches point t = 0.8 s. Battery voltage (volt) and battery current (amp) versus time graph can be seen in Fig. 8c and d, respectively.

At time t = 0.8–1 s: During this time, the wind power stops dropping and produces power at constant 0.73 kw until t = 1 s which is lower than the load power. System stability and supply of power is maintained by the battery system. Three-phase inverter voltage with time is shown in Fig. 9c.

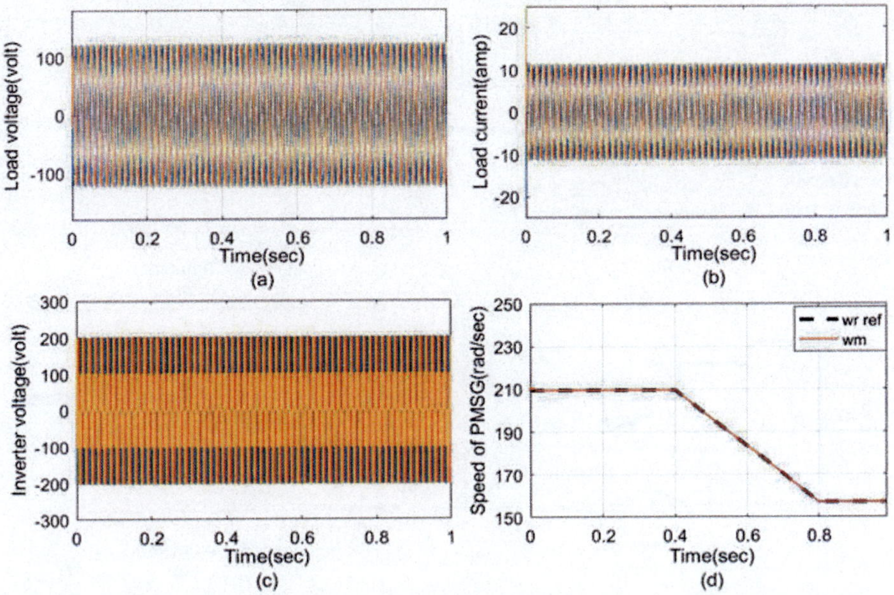

Fig. 9 **a** Load voltage (volt), **b** Load current (amp), **c** Inverter voltage (volt), and **d** Speed of PMSG (rad/s)

3.3 Case-3: Variable Wind and Load Power

A real life scenario of varying source and load condition is studied and mentioned in this case. The wind power, load power with battery power is shown in Fig. 10a. Initially, the load power is about 0.78 kw, and then increases until time t = 0.6 s and starts to fall again. Wind power is about 1.3 kw at the start and begins to decrease gradually till the end. With the help of the battery system, power is delivered and system voltage is maintained as shown in Fig. 10b. Inverter voltage is also shown in Fig. 11c. Speed of PMSG (in rad/sec) versus time graph is shown in Fig. 11d.

At time t = 0–0.2 s: The load power is about 0.78 kw at the instant until it hits next point. The power from wind is about 1.3 kw. So, excessive power is used by the battery system to charge with the help of bidirectional DC/DC converter. Battery voltage is shown in Fig. 10c along with battery current graph in Fig. 10d.

At time t = 0.2–0.4 s: Load increases from 0.78 to 1.24 kw. But it is still lower than the wind power. So, the system remains the same only that battery charging power is less than before. Load voltage and load current versus time is shown in Fig. 11a and b, respectively.

At time t = 0.4–0.6 s: During this time, wind power starts to fall gradually. Load power reaches maximum point where power required is about 2.02 kw which is more than the generated power from wind. So, at this time, battery power gets into the system to help supply the required amount of power.

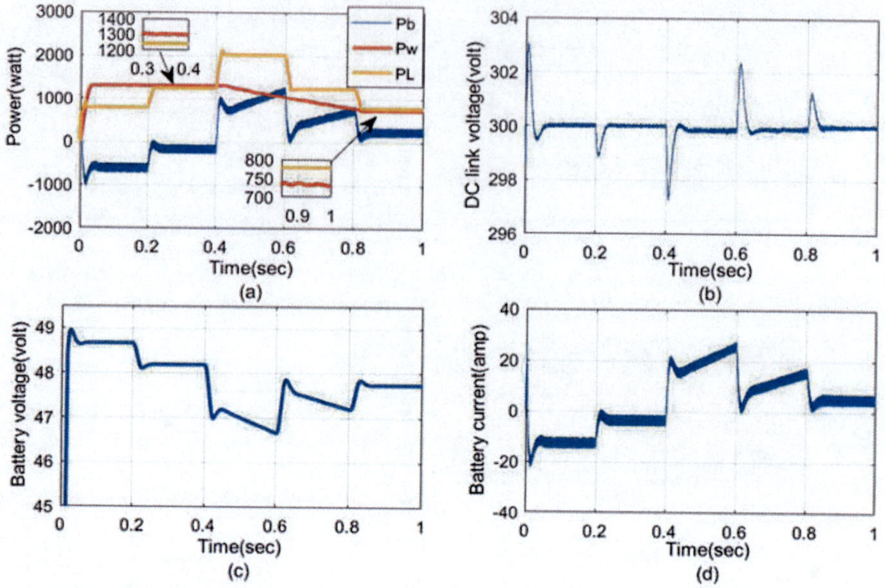

Fig. 10 **a** Power (watt), **b** DC-link Voltage (volt), **c** Battery voltage (volt), and **d** Battery current (amp)

At time t = 0.6–0.8 s: Power generated from wind keeps on decreasing until it hits time t = 0.8 s. The load power demand also starts to fall but still greater than the generated power from wind.

At time t = 0.8–1 s: During this time, the wind power is around 0.73 kw. The load power demand also falls to 0.78 kw which is still greater than the wind power. So, battery system plays an important role in this case to supply power continuously, thus, improving voltage stability and further improving the overall system stability.

4 Conclusion

In this paper, a stand-alone system is implemented in MATLAB/Simulink environment. This system is equipped with PMSG-based WT and battery system. Field and voltage oriented vector control schemes are used to control MSC and GSC, respectively. The P and O algorithm-based MPPT is used to capture the maximum power from a particular wind speed, and also droop control is used to maintain the magnitude of the rated load voltage and its frequency under various scenarios. A battery system is used as a backup to deliver required power when load is greater than generation, and keeps charging when generation is greater than load by its surplus power generated. The considered system is simulated in MATLAB/Simulink under three different conditions to study the performance of the system. The Simulink results

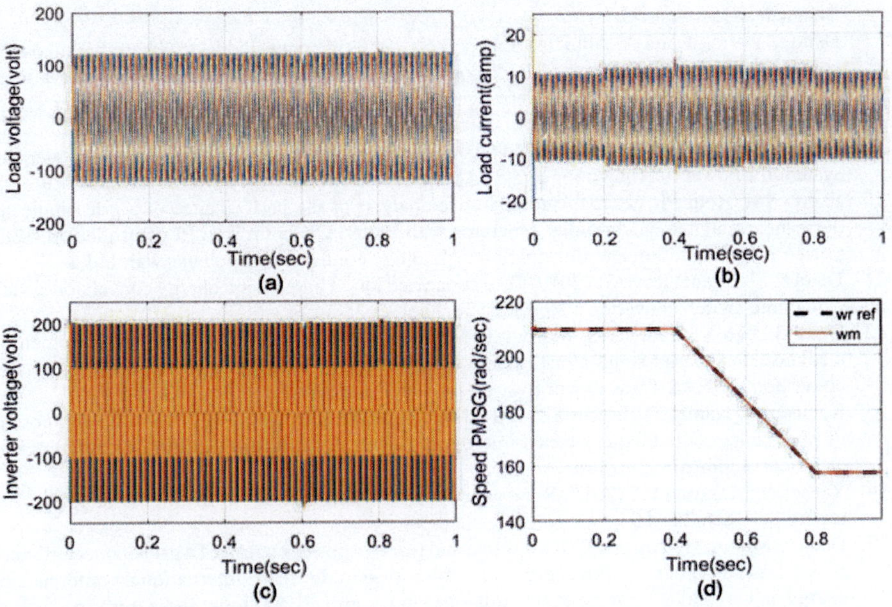

Fig. 11 **a** Load voltage (volt), **b** Load current (amp), **c** Inverter voltage (volt), and **d** Speed of PMSG (rad/s)

are analyzed in detail. It is observed that the wind-battery-based microgrid system is capable enough to deliver continuous power to the load even under varying source and load conditions without any disruption.

References

1. Nayak S, Shwetha MNVR, Nanditha J, Zulfa S (2019) and Prasad DBR: microgrid technologies and its need in distribution system in India: a review. IJERT 8:469–471
2. Zhou X, Guo T, Youjie M (2015) An overview on microgrid technology. In: Proceedings of IEEE, international conference on mechatronics and automation, Beijing, China
3. Dang T, Rashid MH (2009) Introduction, history, and theory of wind power. North American power symposium, Starkville, MS. USA
4. Babu B, Divya S (2017) Comparative study of different types of generators used in wind turbine and reactive power compensation. IOSR J Electr Electron Eng 2:95–99
5. Vijayalaksmi S, Saikumar S, Saravanan S, Sandip RV, Sridhar V (2011) Modelling and control of a wind turbine using permanent magnet synchronous generator. Int J Eng Sci Tech 3:2377–2384
6. Zhou T, François B (2011) Energy management and power control of a hybrid active wind generator for distributed power generation and grid integration. IEEE Trans Ind Electron 58:95–104
7. Datta S, Islam A, Saikia T, Adhikari S (2019) Performance study of a grid connected permanent magnet synchronous generator based wind turbine system. In: IEEE International conference on computer, communication, chemical, materials and electronic engineering, Rajshahi,

Bangladesh
8. Datta S, Deb S, Datta A, Adhikari S, Roy B (2019) Grid connected PMSG based wind energy conversion system using Back-to-Back converter. In: International conference on innovation in modern science and technology, learning and analytics in intelligent systems, Springer, Cham, vol 12, pp 373–382
9. Erramia Y, Ouassaid M, Maaroufia M (2013) Control of a PMSG based wind energy generation system for power maximization and grid fault conditions. Energy procedia 42:220–229
10. Parida SM, Rout PK (2015) A comparative analysis of the performance of a grid connected permanent magnet synchronous generator with PI and DE optimized PI controller. In: IEEE power, communication and information technology conference, Bhubaneswar, India.
11. Datta S, Mishra JP, Roy AK (2019) Grid connected DFIG based wind energy conversion system using nine switch converter. J Appl Res Technol 17:258–271
12. Datta S, Deb S, Samanta S, Maity NP, Maity R, Adhikari S (2019) Power management of a solar-battery based stand-alone system using adaptive neuro fuzzy inference system based controller. WSEAS Trans Power Syst 14:145–155
13. Kucuker A, Kamal T, Hassan SZ, Li H, Mufti GM, Waseem MH (2017) Design and control of photovoltaic/wind/battery based microgrid system. In: IEEE, International conference on electrical engineering, pp 1–6
14. Yousef A, Maksoud SA (2015) Review on Field oriented control of induction motor. Int J Res Emerg Sci Tech 2:5–16
15. Datta S, Mishra JP, Roy AK (2015) Active and reactive power control of a grid connected speed sensor less DFIG based wind energy conversion system. In: IEEE International conference on energy, power and environment: towards sustainable growth, Shillong, India, pp 1–6
16. Davijani HK (2012) Analysis and control of a microgrid with converted fed distributed energy sources. Dissertation, Tennessee Technological University

Chapter 4
Fabric Defect Detection Using Computer Vision

V. Likith Kumar, A. Hari Priya, N. Jahnavi Chakravarthy, and Padarti Vijaya Kumar

1 Introduction

In this fashion-oriented era, the rate of production of the fabrics has been highly increasing daily. The textile field [1, 2] is primarily concerned about quality. However, during the production cycle of fabric, there is an obvious occurrence of the defects or damages in the fabric which is sold at lower prices thereby causing a substantial loss to the organization. Automation of both clothing and textile manufacturing has been developing interest over the decades, and is still a challenging task because of the unpredictable variability of fabric material and their properties. The accurate measurement of texture parameters of the fabric such as weave structure, thread counts and surface roughness has wide applications in other areas like virtual environments, e-commerce and robotic tele-manipulations apart from the textile industry.

The different textures of the fabric depend upon the types of weaves used. Textures are given to all types of fabric like cotton, silk, wool, leather and linen. Defect detection [3–6] or inspection is a practice of identifying and locating defects. A fabric glitch is a result of the manufacturing process. Quality inspection is a crucial aspect of industrial manufacturing. In fabric, the faults may occur due to machine faults, colour bleeding, yarn problems, scratch, poor finishing, dirt spots, excessive stretching and crack point. The traditional fabric inspection is done by manual visual inspection with high labour cost and low efficiency. It is found that only 70% of the fabric deformities could be detected by the most highly trained inspectors. Added to the limitation in the quality [7–13] is increased cost of manpower and time taken for inspection is more. Therefore, an automated system is required for the inspection and evaluation of the quality of the desired fabrics. Different kinds of methods are

V. Likith Kumar · A. Hari Priya · N. Jahnavi Chakravarthy · P. Vijaya Kumar (✉)
Department of ECE, Velagapudi Ramakrishna Siddhartha Engineering College, Vijayawada, Andhra Pradesh, India
e-mail: vijayakumar.padarti@vrsiddhartha.ac.in

S. Dawn et al. (eds.), *Smart and Intelligent Systems*, Algorithms for Intelligent Systems,
https://doi.org/10.1007/978-981-16-2109-3_4

divided into three categories, namely spatial thresholding method in which the normal grey scale is defined as the number of pixels in the statistical sub window exceeds the range and checks the child window to see if there are any defects. The morphological method for detecting the fabric defects [14, 15] using corrosion and extended edge detectors. The neighbourhood association describes the relation of the current point and the neighbourhood point by a trained neural network (Figs. 1, 2, and 3).

There are different types of defects like holes, oil stains, weft, wrap, missing thread and pin mark; some of them are shown in the figures below. It is important to develop a real-time automated fabric defect detection system which is easy to install

Fig. 1 Hole

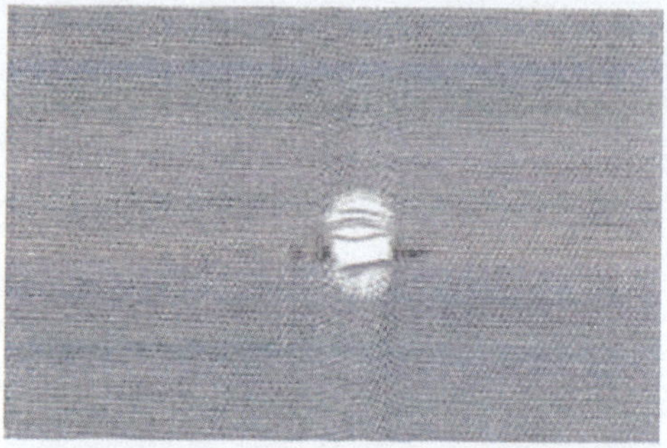

Fig. 2 Oil stain

Fig. 3 Pin mark

with better efficiency than previous approaches and low complexity, so the image processing help us to achieve these requirements with less cost and works, and faster rate. The previous methods are seen in the below section which have disadvantages like less accuracy and requiring more time for processing, so this method is proposed to overcome them. The haar cascades algorithm is one part of Viola Jones algorithm proposed in the year 2001 which is used in various apps like Snapchat and Instagram for face detection. Until 2015, even Apple Company used it.

2 Existing Methods

The fabric detection is very important to the textile industries. In the ancient period, they allocated many workers to detect the defects through manual analysis, but there are still cases of many defects that are not accurately detected. One of the method that has come up is real-time vision-based system which can only detect small defects, later comes the hand prediction with FNN whose complexity is more as the number of inputs and output increases. The computer vision techniques [16, 17] propose different types to detect different defects. Wavelet decomposition is also used to detect defects at the wrap knitting machine that can be applied on plain fabrics. Over a period of time, many image processing techniques have been proposed which have many disadvantages like cost, accuracy and compatibility. The technique we are proposing can be used in the real-time as it is automated and cost is also low compared to other techniques which are existing (Table 1).

Table 1 Comparision of existing methods based on different parameters

Reference papers	Cost	Accuracy	Delay	Noise	Size
[5]	Low	More	High	Less	Complex
[7]	Lower	Higher	Less	Least	Simpler
[9]	Lower	Higher	Less (but considerable)	Least	Simpler
[8]	Low	Reasonable	Less	Considerable	Complex
[4]	High	High	Considerable	Less	Simple
[2]	Low	High	Less	Considerable	Simple
[6]	High	Low	High	Less	Complex
[1]	Low	Reasonable	Less	Less	Complex
[3]	Low	Reasonable	Considerable	Less	Simple

3 Proposed System

The system that is proposed to detect the defect using image processing in which we start by training the defect images, and later the camera captures the video images and compares it to the pre-trained images. Initially, the Cascade trainer GPU is used to create datasets. In order to create datasets, there is a need to collect positive and negative images. The positive images are the ones with the defect since those are the ones which are to be detected. The negative images are the images which do not have defects, that is, plain fabrics are taken. Both of the images are put together as datasets, and training is done after choosing and making changes in the parameters as per the requirement. The below Figs. 4 and 5 show the dataset collection of positive and negative images, respectively. The XML file is created using positive and negative images, along with this cascade file is created. Image processing is the one applicable in many applications which is used in this system to improve the factors like accuracy, speed, and has less complexity and low cost.

In this project, machine learning algorithms like haar cascade methods are used. The installation of web camera is done at the place of the rollers where fabric is rolled up into bundles and transported to sell them in the market. This web camera captures the video which is the input later converted into sample of photos. The procedure

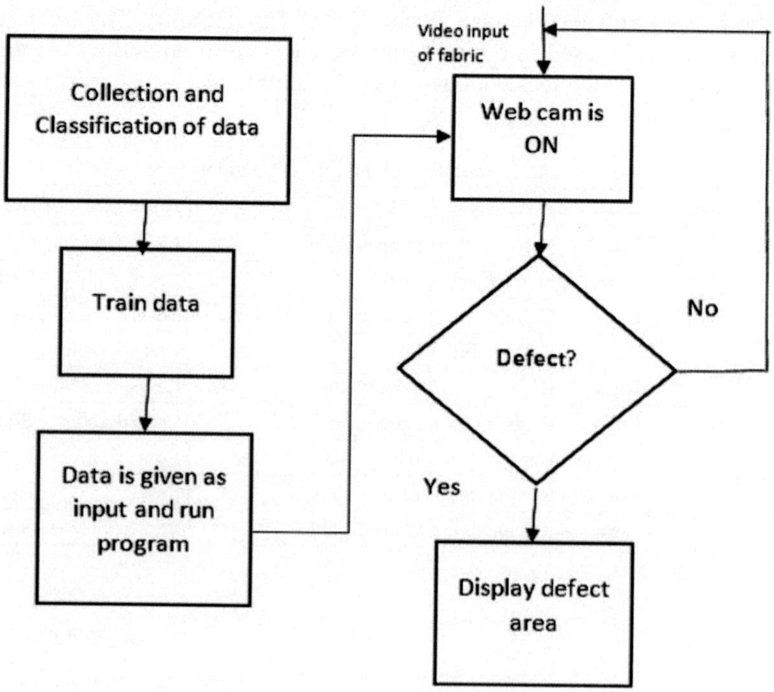

Fig. 4 Proposed system Model

Fig. 5 Positive images

starts by converting the photos into grayscale images. The images that are acquired after conversion are compared with the XML file that is being created as prerequisite. After the comparison of both trained and acquired images, the defect is plotted. In haar cascades method, we use standard libraries like numpy and opencv (Figs. 6 and 7).

The coding is done using the python programming, and the libraries like numpy and opencv are used to programme. The Raspberry Pi is used as an interface between the software and the webcam. VNC software can be used to control the system through another device that is connected to the internet.

Fig. 6 Negative images

Fig. 7 Experimental setup using hardware

4 Experimental Results

There are different types of defects. By training a large amount of data, we can detect any type of data very accurately. Figures 8 and 9 show the hole and oilstain that are detected.

Fig. 8 Hole defect plotted

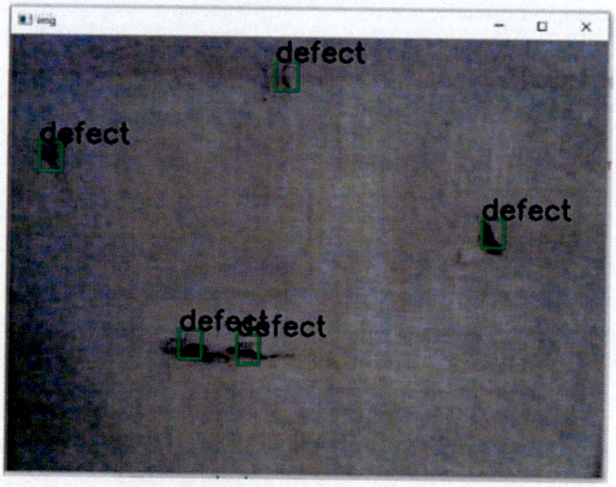

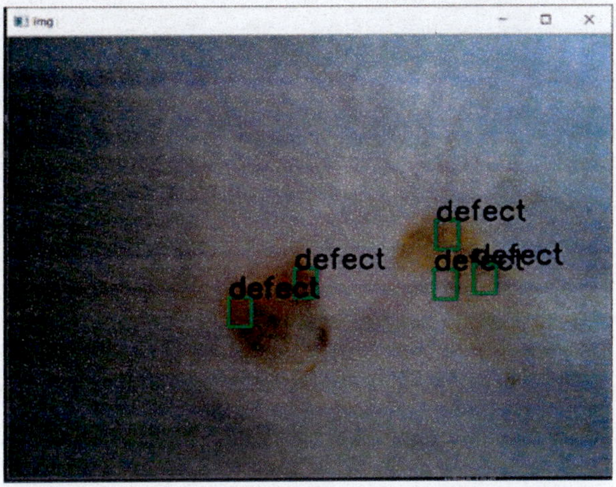

Fig. 9 Oil stain defect plotted

Similarly, any type of defect can be detected by training and creating datasets (Figs. 10 and 11).

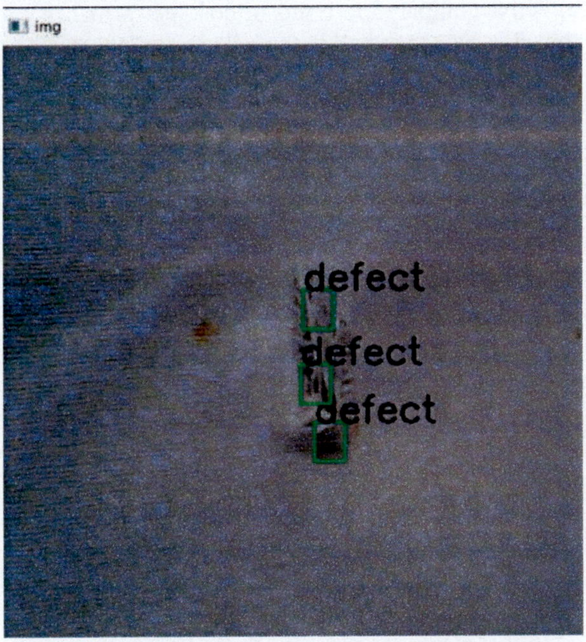

Fig. 10 Warp defect plotted

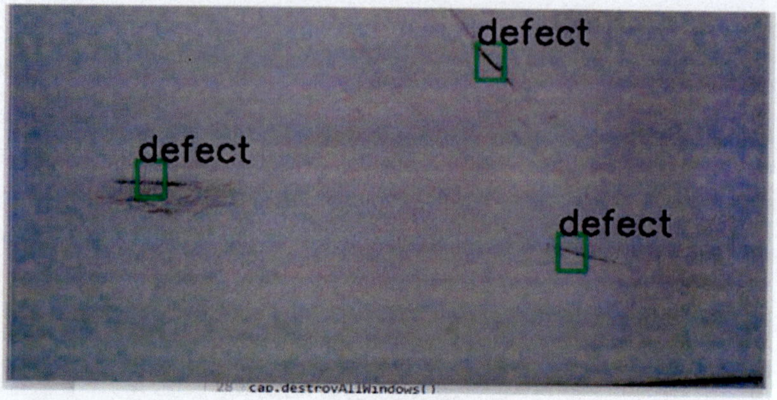

Fig. 11 Weft defect plotted

5 Conclusion

Fabric detection has always been important for the textile industries since ages because the defects cause the industries to lose their profits. This project is more effective compared in terms of cost, performance and processing time. Apart from this, it is easy to install in the industries since python programming is used in development part. And also, Raspberry pi is affordable and a small-size computer board can be employed in any type of environment. In industries, major concern is the accuracy which has increased more significantly. The accuracy improvement depends on the dataset created so it can be easy to increase the accuracy by just making sure of collecting the fabric defects.

References

1. Jiang P, Li N (2017) Fabrics defect detection via visual attention mechanism. IEEE
2. Nadaf FS, Kamble NP (2017) Fabrics fault detection using digital image processing
3. Karlekar VV, Biradar MS (2015) Fabrics defect detection using wavelet filter
4. Jmali M, Zitouni B (2014) Fabrics defects detecting using image processing and neural networks. IEEE
5. Anuja J, Agilandeswari V (2014) Fabric quality testing using image processing
6. Li Y, Di X (2013) Fabric defect detection using wavelet decomposition
7. Wang X, Gerogans ND (2010) Fabric texture analysis using computer vision techniques. IEEE
8. Yu Y, Hui C-L (2010) Intelligent fabric hand prediction system with fuzzy neural network. IEEE
9. Cho C-S, Chung B-M (2005) Development of real time vision-based fabric inspection system. IEEE
10. Weninger L, Kopaczka M, Mwrhof D (2018) Defect detection in plain weave fabrics by yarn tracking and fully convolutional networks. IEEE
11. Jadhav SP, Biradar MS (2014) Fabric defect detection by using neural network technique. IOSR-JECE

12. Li Y, Zhao W, Pan J (2016) Deformable patterned fabric defect detection with fisher criterion-based deep learning. IEEE
13. Tajeripour F, Kabir E, Sheikhi A (2007) Fabric defect detection using modified local binary patterns. EURASIP
14. Raheja JL, Kumar S, Chaudhary A (2013) Fabric defect detection based on GCLM and Gabor filter: a comparison. OPTIK
15. Rathinavel S (2016) The improvement of fault Detection in Different Fabric for Weaving Loom. IRJET
16. Padarti VK, Rao NV (2020) Adaptive SOICAF Algorithm for PAPR Mitigation in OFDM Systems. Wirel Pers Commun 113:927–943
17. Padarti VK, Nandhanavanam VR (2021) An improved ASOICF algorithm for PAPR reduction in OFDM systems. Int J Intell Eng Syst 14(2):352–360. ISSN No: 2185–3118. (April, 2021)

Chapter 5
Energy Audit and Advancement of Solar Installation in SIT: A Case Study

Shreya Shree Das, Subhojit Dawn, and Sadhan Gope

1 Introduction

This practical work has been arranged with a view to simplify our understanding of the power consumption pattern of different buildings of Siliguri Institute of Technology (SIT) at different times of the year. The report describes the energy efficiency measures to enhance the use of non-conventional and renewable energy in such a way that people can maintain or modify their daily energy consumption, and can reduce the use of conventional/non-renewable resources of energy. By using non-conventional way (like solar / PV cell), how we can reduce the overall cost of electricity consumption in domestic or commercial purpose as well as cost of production in industry is also discussed in this work.

An energy audit is the learning process of a plant to decide how the energy is used, where the energy is used, and to detect the methods for energy savings. Now, the status of the daily living of human beings is upgrading, which required much electrical power. In the last few years, researchers are doing their work for making the world techno-economic. As a part of techno-economic model, the deregulated power market concept has been introduced, which makes the electricity market more transparent and economically sustainable. The social welfare, system profit can be maximized for a renewable integrated power system by minimizing the system generation cost.

In the past, many researchers have given their concerns about the energy audit in their publications. Dongellini et al. [1] presents an energy audit report containing the energy consumption and energy savings data of a car manufacturing company in Italy. Kumar et al. [2] report an energy saving method by performing an Intelligent Energy Audit under "MGR vision 10 MW." The paper [3] shows the usefulness

S. S. Das (✉) · S. Gope
Department of Electrical Engineering, Mizoram University, Aizawl, India

S. Dawn
Department of Electrical and Electronics Engineering, Velagapudi Ramakrishna Siddhartha Engineering College, Andhra Pradesh, India

S. Dawn et al. (eds.), *Smart and Intelligent Systems*, Algorithms for Intelligent Systems,
https://doi.org/10.1007/978-981-16-2109-3_5

of energy audit for business organizations. Authors have taken some small manufacturing companies of Germany as samples. Kluczek and Olszewski [4] show the importance of energy audit in the industrial processes. The Ref. [5] presents the application and methodology of an energy audit which is applicable for integrated smart system. The basic concept of Energy Audit with the detailed processes has been discussed in [6]. From the literature, it is revealed that energy audit is very important for the economical running of any industry/organization. Taking the encouragement from the stated literature, we have chosen SIT as the site for energy auditing and shown the benefit (economical) of solar electrification in SIT in this paper.

The energy audit is a continuous process of energy preservation by examination, review, and study of energy flow in any organization/plant without disturbing the output of the process. The energy audit gives an idea of how much energy is being consumed by an enterprise and other organizations. Energy consumed should adhere to the "National Energy Conservation Laws and Regulations." These regulations and laws help govern energy consumption and energy audit management.

Step-by-step approach by audit activities is as follows:

(a) All energy system identification.
(b) System condition evaluation.
(c) Impact analysis for system improvement.
(d) Energy Audit report preparation.
(e) Scope for integration of RES is to be enhanced.

The economic analysis has to be done after auditing based on the collected data. According to the report of several electricity sectors of Govt. of India, energy audits can save over crores every year.

The vision of this work is to make Siliguri Institute of Technology (SIT) as a more energy efficient institute. The SIT consumes a large amount of energy since being a big technical institute, having a huge scope of energy saving by proper energy management through energy auditing. The energy auditing will not only make the campus energy efficient but also will help reduce expenses. This will also enable us to fulfill our moral obligation towards society for proper utilization of scarcely available resources.

These tasks of making SIT energy efficient will be a pioneer approach in igniting inspiration among individuals of the society. It can be stepping stones towards fulfilling our moral obligation towards a greener and efficient energy future.

2 Site Details

Organization name	Siliguri Institute of Technology
Site name and address	Siliguri Institute of Technology Sukna, Siliguri, Dist., Darjeeling, Pin: 734,009

(continued)

(continued)

Building included	First Year Building Main Building Engineering Science and Humanities Building Library Building Workshop Building Canteen area First AID Building Guest house Area
Survey conducted by:	Subhojit Dawn Shreya Shree Das

3 Energy Audit: Objectives and Significance

The objective of this Preliminary Energy Audit, sometimes referred to as a "Walk Through Audit," is to primarily check the viability of energy efficiency upgrade of the facility under consideration. This assessment will help understand the commercial viability before investing in energy performance contract or subsequent investment grade audit.

The purpose of this work is to enhance the use of non-conventional and renewable energy in such a way that we can maintain or modify our daily energy consumption and can reduce the use of conventional, non-conventional resources of energy.

By using non-conventional way (as solar/PV cell), we can reduce the overall cost of electricity consumption in domestic purpose as well as cost of production in industry which is also the objective of this work.

This objective will be achieved by:

- Quantifying potential energy savings by identifying energy performance indicators for existing and target sources. This helps to understand the impact of energy conversion measures and provides a sense-check of calculation.
- To identify and frame suit of measures comprising of savings, implementation budget, and payback period in one package. This helps to create financial viability of project, easier implementation, and quantifying non-energy savings.
- Essential client requirements are to be identified to incorporate in works and providing savings and implementation budget figures.
- Identify and quantify entities which may have reached end of life or resolution of comfort.
- To identify the need for additional metering and recording requirements that are likely to be required should the verification and measurement of savings be necessary. The associated installation budget should be included.
- Identify potential risk to the project, be it technical, financial, or any other.

4 Energy Audit Methodology

The audit is done in the following manner:

- As a first step in this regard, a group of members was formed.
- Visual inspection and data collection are to be done by the audit group.
- Quantification based on the observation of the general condition of the equipment and facility.
- Measuring energy consumption and other parameters, and to verify that data.
- Assumption validation, detailed calculation, and analyses are to be performed based on the collected data.
- Implementation of potential energy saving opportunities.

5 Graphical Representation of Energy Consumption

For better understanding of the energy use pattern in the plant, graphical analyses of hourly/daily/monthly/annual energy use of the various types of energy used in the plant are to be done. This helps in opening up new avenues to modify the way energy is used and save energy based on the energy usage pattern. For example, it is quite unlikely to see a seasonal variation in energy use in heavy process industries like cement factory due to weather changes. If such pattern is observed, then there might be a need to investigate the possible energy losses in the system.

Sometimes, the variation of energy use alone may not be sufficient to have a better visualization on the condition of energy efficiency in a plant. In that case, the graphical analysis of plants' energy intensity (EI) gives better and accurate results.

$$\text{Energy Intensity (GJ/tonne)} = \frac{\text{Energy Consumption (GJ)}}{\text{Production (tonne)}}$$

It might also be observed that EI pattern does not vary over years but the fuel intensity pattern may vary with season. This is due to the fact that fuel is used for heating during cold season as the temperature and moisture of the spinning part must be kept constant.

Energy Audit has been Broadly Classified into Two Categories:

- Preliminary energy Audit
- Detailed Energy Audit

In case of detailed energy audit, we study and measure each and every energy parameters in different sections and sub-sectional areas, and also suggest different cost effective methods which can reduce energy use without reducing the effective output.

6 Solar Energy

Solar energy is inexhaustible and green (non-polluting) source of energy. It is abundant and periodic in nature. This energy can be used to harness energy for various applications like electricity, providing light, maintaining interior temperature, heating of water, etc., be it for personal, commercial, or industrial.

The periodicity and direct utilization of solar radiation is the main cause of increased popularity throughout the world. The GOI is currently promoting solar by providing subsidies. As by 2022, India has a target of reaching 160GW RES energy out of which 100GW is solar [7].

Advantages of Solar Energy

- Solar is abundant and green source of energy.
- Solar energy is devoid of charges.
- Solar energy is inexhaustible in nature.
- Zero maintenance of equipment.
- On a long run, it may have a high rate of return.

Disadvantages of Solar Energy

- Solar energy is available for a limited period of time, i.e., it is unavailable during night and rainy season.
- The intermittency and quirky nature of solar energy make it less reliable.
- Solar panel requires energy converter and energy storage system, thus, high initial investment is required.
- Large areas of land are required for solar farming which is quite a crisis in our country.
- Lower efficiency of solar panel compared to other RES converters.
- The equipment is fragile and requires extra insurance to protect PV investment.

7 Economic Benefits of Solar Installation in SIT

Siliguri has an elevation range 114–140 m (average 400 ft.) from sea level and its latitude is 26.71°N [8]. Due to suitable location, it is very adventurous for implementation of solar PV array on the rooftop of SIT campus buildings.

The determination of the Shadow Profile for installation of PV array for the proposed green campus by MNRE (Ministry of New and Renewable Energy) are as follows:

Initially, rooftop of two departmental buildings of SIT were selected for installation of PV array for making a Grid tied-up Street Lightning System and other associated Renewable Energy-based Projects in order to convert SIT to a Green Campus on the basis of the area of the available rooftops. These two buildings are as follows:

(i) Central office and Main Engineering Building.
(ii) Engineering Science and Humanities Building.

Effective Area (Total Shadow Area Availability):

(1) Main Building:
Total area—872 sq.m
Shadow Area—330.52 sq.m
Effective Area—541.48 sq.m
(2) Engineering Science and Humanities Building:
Total Area—871 sq.m
Shadow area—74.81 sq.m
Effective Area—796.19 sq.m

So, the available effective area (Deducing the average Shadow area) is much greater in the Engineering Science and Humanities Building than the rooftop of Central Office and Main Engineering Building.

According to WEBREDA data, by each 10 sq.m solar PV array, we can generate 1KW power. So, if we utilize the rooftop of Engineering Science and Humanities Building, we can generate = (796.19/10) = 79.619 KW power (max). Due to humidity, foggy, or rainy weather, the average value of generated power will be reduced to 50% of maximum value, i.e., 39.8 KW.

Energy generated each day in effective solar hours (8.30 am to 3.30 pm = 7 h) is $(39.8 \times 7) = 278.6$ KWh.

Energy generated each month = $(278.6 \times 30) = 8358$ KWh.

Total amount saved each month = (8358×7.34) = Rs. 61,347.72 /-

Total amount saved each year = $(61{,}347.72 \times 12)$ = Rs. 736,172.64 /-

Whereas, energy cost is Rs. 7.34/- (average) per KWh.

Installation cost of 800 sq.m rooftop solar PV array is 20 Lakhs INR approx. according to WEBREDA (West Bengal Renewable Energy Development Agency).

Payback Period = 2,000,000 / 736,172.64 = 2.7 years.

How to Install Solar Panels Step-by-Step

Installation of solar panel is a time taking process and requires certain approach for powering up. We have outlined the seven-step solar panel installation guide below:

Step 1: The mounting structure provides the base for the entire solar system, so make sure it is sturdy and properly fastened to the rooftops of your house or commercial establishment. A typical mounting structure is made up of aluminum. The performance of the solar panels depend upon the direction in which these panels are placed. The best direction to face solar panels is south, since here they receive the maximum sunlight. The East and West directions also work well. North is the only direction that we should not want to put our panels on. Since India lies in Northern Hemisphere, south direction works best here.

Step 2: Once the solar structure is fixed accurately, we will connect it with solar modules. We should ensure that all nuts and bolts of solar modules are fixed with solar structure so that it is properly secured and lasts long.

Step 3: MC4 connectors are used to connect solar panels. These are universal connectors and can be connected with any type of solar panels. The solar array wiring becomes simpler and faster using MC4 connectors.

Few modern solar modules come with wire leads that have MC4 connectors on the ends, else they have a built-in junction box at the back with wires jotting out. In a series connection, you will have to connect the positive wire from one module to the negative wire of another module. In a parallel connection, you connect the positive to positive and negative to negative leads. A parallel connection maintains the voltage of each panel, while a series connection increases the voltage in order to match it with the battery bank.

Step 4: The backside of an inverter is connected with solar panel wire. Connect the positive wire from the solar panel with the positive inverter terminal, and the negative wire with negative terminal of the inverter.

There are other connections too, like battery wire connection and output wire connection with the inverter. In all, Solar panel, Solar Battery, and Grid input are connected with the solar inverter to produce electricity. The output of a series string of solar modules is connected to the input of the inverter. Make sure the inverter is turned off while the connections are being done.

Step 5: In an off grid solar system, Battery is mandatory where it is used to store power backup. This battery is connected with solar inverter to recharge it with solar panel and grid. The positive terminal of the battery is connected with the positive of the inverter and vice versa.

Step 6: In order to connect the inverter to the grid, simply plug it in the main power switch board, so that it gets power from the grid. The output wire is also connected with board that is supplying electricity in home or any commercial establishment.

In order to calculate the excess energy generated from the solar system, we need to install a metering device. We need to connect the positive wire from the metering device with the line terminal, and the negative wire to the neutral terminal of the inverter.

Step 7: After all the connections are done, we switch on the mains. There is a digital display which shows the total solar unit generated during the day, what is supply volt and current (amp) from solar panel, etc.

For performing this work, we have collected several practical data from SIT campus. Table 1 shows a sample of the taken power consumption data in SIT. As calculated earlier, payback period of SIT is 2.7 years, which gives a clear indication about the profit maximization in terms of solar installation in SIT.

8 Conclusion

The reserve quantity of coal and fossil fuels is decreasing with a huge rate throughout the world. If we do not want to sacrifice our comfort in terms of using electrical equipment, moving from conventional sources to renewable sources is not the option, it is compulsory. The basic investment cost of renewable energy sources is very high.

Table 1 Load distribution table of Siliguri Institute of Technology

EE Department (Ground Floor, Room no. 001)	No. of fans	6
	No. of lights	12
	No. of 6 amp plug point	1
	No. of 16 amp plug point	2
	No. of Switches	22
	No. of AC	0
	No. of external power point	0
EE Department (Ground Floor, Room no. 002)	No. of fans	6
	No. of lights	15
	No. of 6 amp plug point	1
	No. of 16 amp plug point	2
	No. of switches	15
	No. of AC	0
	No. of external power point	0
EE Department (Ground Floor, Room no. 003)	No. of fans	7
	No. of lights	13
	No. of 6 amp plug point	1
	No. of 16 amp plug point	2
	No. of switches	19
	No. of AC	0
	No. of external power point	0
EE Department (Ground Floor, 1st Wing)	No. of fans	0
	No. of lights	18
	No. of 6 amp plug point	5
	No. of 16 amp plug point	1
	No. of switches	23
	No. of AC	0
	No. of external power point	0
EE Department (Ground Floor, Project Lab)	No. of fans	5
	No. of lights	13
	No. of 6 amp plug point	56
	No. of 16 amp plug point	9
	No. of switches	84
	No. of AC	0
	No. of external power point	0

But, due to their very less running cost, customers get more benefit after some time period from the installation. Before the installation of any renewable energy sources in running plants, it is necessary to perform an energy audit for maximizing the economic sustainability of the system. The detailed concept with several methodology of energy audit has been designated in this work. The integration process of solar energy is also studied here. From the study, it is clear that, after solar power installation, SIT gets more profit after some years. So, this is not a very long term policy from SIT's end. This work can also help the organizations, which are thinking about the installation of renewable energy sources in their campus, in economic conditions.

References

1. Dongellini M, Marinosci C, Morini GL (2014) Energy audit of an industrial site: a case study. Energy Procediam 45:424–433
2. Kumar A, Ranjan S, Singh MBK, Kumari P, Ramesh L (2015) Electrical energy audit in residential house. Procedia Technol 21:625–630
3. Schleich J, Fleiter T (2017) Effectiveness of energy audits in small business organizations. Resour Energy Econ. https://doi.org/10.1016/j.reseneeco.2017.08.002
4. Kluczek A, Olszewski P (2017) Energy audits in industrial processes. J Clean Prod 142(4):3437–3453
5. Belussi L, Danza L, Salamone F, Meroni I, Galli S, Svaldi SD (2017) Integrated smart system for energy audit: methodology and application. Energy Procedia 140:231–239
6. Allouhi A, Boharb A, Saidur R, Kousksou T, Jamil A (2018) Energy auditing. Compr Energy Syst 5:1–44
7. Dawn S, Tiwari P, Goswami AK, Mishra M (2016) Recent developments of solar energy in India: Perspectives, strategies and future goals. Renew Sustain Energy Rev 62:215–235
8. The latitude and longitude gps coordinates of Siliguri (India). http://thegpscoordinates.net/india/siliguri

References

Chapter 6
Random Fault Positioning Based Voltage Sag Assessment for a Large Power Transmission Network

Chinmaya Behera, Arup Kumar Goswami, Galiveeti Hemakumar Reddy, Sadhan Gope, and Chetan M. Bobade

1 Introduction

The definition of voltage sag as per IEEE 1159 standard [1, 2] is that any reduction in the voltage level (i.e., 0.1–0.9 pu) for a duration of 0.5 cycles to 1 min is named as voltage sag. To distinguish voltage sag from other PQ disturbances, it requires the information regarding voltage magnitude and its duration. The occurrences of voltage sag in a network is due to several reasons, like interaction of birds, snakes, rats, cattle, animals, wind gust on conductors and attachments, tree touching, cattle. In [3], they first recognized the occurrence of voltage sag and proposed different computational tools to identify voltage sag parameters (i.e., voltage magnitude and its duration). An analytical method has been proposed in [4] to estimate the probability distribution function (PDF) for the evaluation of voltage sag due to symmetrical faults only for small power networks. To understand the influence of power system fault at different location of a transmission network and how the generators serving the load demand causes voltage sag is given in [4]. An analytical method proposed in [5] helps in stochastic evaluation of unbalanced and balanced voltage sag for large transmission systems. In [6], to evaluate voltage sag, an analytical method is developed to simulate the faults, and it is applicable to both large and small power system networks. In [6], a model is developed to estimate the voltage sag due to the influence of transformer. Martinez et al. [7] have proposed a digital simulator for the prediction of voltage

C. Behera (✉) · C. M. Bobade
G. H. Raisoni University Amravati Maharashtra, Amravati 444701, India
e-mail: chinmaya.behera@ghru.edu.in

A. K. Goswami
National Institute of Technology Silchar, Silchar Assam 788010, India

G. H. Reddy
MVJ College of Engineering, Bangalore 560067, India

S. Gope
Mizoram University, Aizawl, Mizoram 796004, India

S. Dawn et al. (eds.), *Smart and Intelligent Systems*, Algorithms for Intelligent Systems,
https://doi.org/10.1007/978-981-16-2109-3_6

sag in distribution networks, which further deliberates the detailed approach and modeling guidelines for voltage sag simulations. The consideration of the protection system and applicability of MCS technique in time-domain simulation in [8] is used to estimate the number of voltage sag in a redial system. In [9], reliability indices are used for the evaluation of interruption duration so that voltage sag indices can be calculated. And, depending upon these indices, the voltage sag is evaluated for a distribution network.s

A modified stochastic method is proposed in [10, 11] to estimate the number of voltage sag for a transmission network. As voltage sag depends upon the fault rate, therefore, in [12], time-varying fault rate and generator scheduling are taken into consideration for the evaluation of the residual voltage at each bus of a power system network. In [12], sensitive load buses are taken into consideration by assuming unaltered power system operating conditions and network topology for the yearly prediction of voltage sag. The fault rate is assumed to be constant throughout the year. In [13, 14], the Sag Severity Index (SSI) is used to identify the vulnerable buses in a power system network so that assessment of voltage sag can be performed. A table is prepared using a concept of bivariate frequency distribution and voltage sag data, so that voltage sag map can be developed. This voltage sag map helps the industrial personnel to identify the weak buses so that proper action can be implemented for the mitigation of voltage sag. The sensitive equipment connected at load end shows different characteristics towards the voltage sag data. Therefore, [15] shows how the sensitive equipment is affected by the occurrence of voltage sag data at different buses in an industrial distribution network. During the occurrence of voltage sag at load buses, the equipment may malfunction which in turn leads to huge financial losses. Therefore, in [6], the probabilistic method is used to evaluate the number of equipment and process trips, and a C-matrix is used to evaluate the financial losses. Similarly, a straight forward calculation procedure is given in [16] to evaluate the financial losses. Depending upon the previous literature, it is found that the occurrence of voltage sag can trip sensitive equipment and can lead to huge financial losses. Therefore, utility engineers should focus on reducing the number of voltage sag. In [17, 18], DG and FCATS devices scheme is implemented for industrial system to mitigate the number of voltage sag per year. In [19], author shows that the spreading of voltage sag due to the occurrence of power system fault varies from network to network. In this paper, an analytical expression is developed to simulate the fault randomly in a power system network. This analytical expression is valid for both the radial and meshed networks. To simulate different types and locations of faults, uniform and normal probability distribution function is used with MCS technique for the evaluation of the voltage sag. The power system faults can be caused due to many factors; therefore, fuzzy is used to handle the uncertainty as per [20, 21], so that fault rate can be evaluated, and depending upon which the number voltage sag will be obtained. Also, in this paper, IEEE 1159 standard is implemented so that different types of voltage sag can be categorized when power system faults occurred at IEEE 118 bus system.

2 Voltage Sag Assessment Technique

The analytical expressions are formulated using impedance modeling and classical Thevenin equivalent circuit approach for the assessment of voltage sag.

2.1 Voltage Sag Due to Bus and Line Faults

Different types of power system faults are considered in this section for evaluation of magnitude of voltage sag at each bus of IEEE 118 bus system. Let us consider Fig. 1 which is a generic power network where the buses and loads are connected. Let suppose at bus no. *n,* the power system fault occurs, and using Thevenin equivalent theorem, the impact of fault at bus *n* and its neighboring bus *m* can be analyzed.

S_n and S_m are the apparent power of bus *m* and *n,* respectively, the faulted bus impedance and current at bus *n* are Z_{fn} and $I_{nc,}$ respectively.

The prefault bus voltages for the system is expressed as follows:

$$U_{pfbus} = \left[U_{pf1} \cdots U_{pf\,n} \cdots U_{pf\,m}\right]^T \tag{1}$$

The prefault current of the Fig. 1 is neglected as the system is in balanced condition, and the expression is given below.

$$I_{pf\,bus} = \frac{1}{Z_{bus}}.U_{pf\,bus} \tag{2}$$

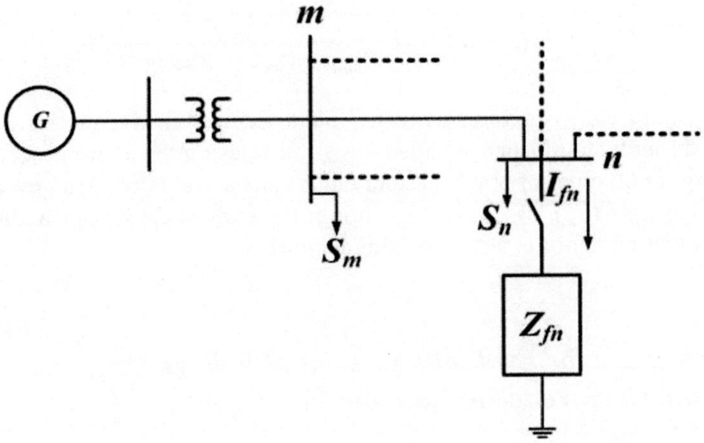

Fig. 1 General n-bus power system

Fig. 2 Sequence impedance network for single line to ground fault

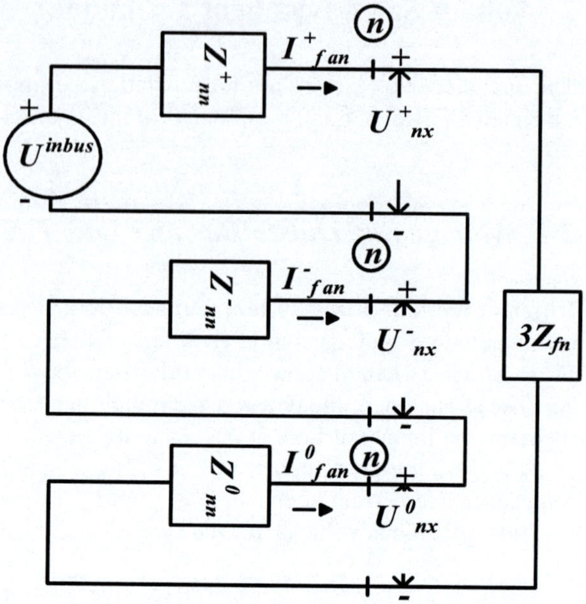

Now, let unsymmetrical fault (i.e., Line to ground fault) be taken into consideration. Now, the sequence impedance at the fault point (i.e., *bus n*) and the representation of the sequence components of faulted bus n are shown in Fig. 2.

The fault current for line to ground fault can be evaluated as per [6] and is given in Eq. (3).

$$I^{+}_{f\,na} = I^{-}_{f\,na} = I^{0}_{f\,na} = \frac{U^{in\,bus}}{Z^{0}_{mm} + Z^{+}_{mm} + Z^{-}_{mm} + 3Z_{f\,na}} \tag{3}$$

Now, the current expression when LG fault occurred is given in Eq. (3), and it will flow through an impedance called $Z_{f\,na}$. Now, using the above procedures, the voltage sag for all types of power system faults can be evaluated. After evaluation of sequence currents $I^{+}_{f\,na}$, $I^{-}_{f\,na}$, and $I^{0}_{f\,na}$, it is easier to evaluate voltage at the point of fault and other parts of the power system network.

2.2 *Procedure for Evaluating Residual Voltage Due to the Occurrence of Random Faults*

In the previous section, the voltage sag can be evaluated when the faults are simulated at a particular bus or line. As power system faults are random, so MCS technique is used to evaluate the voltage at different buses of IEEE 118 bus system. The voltage

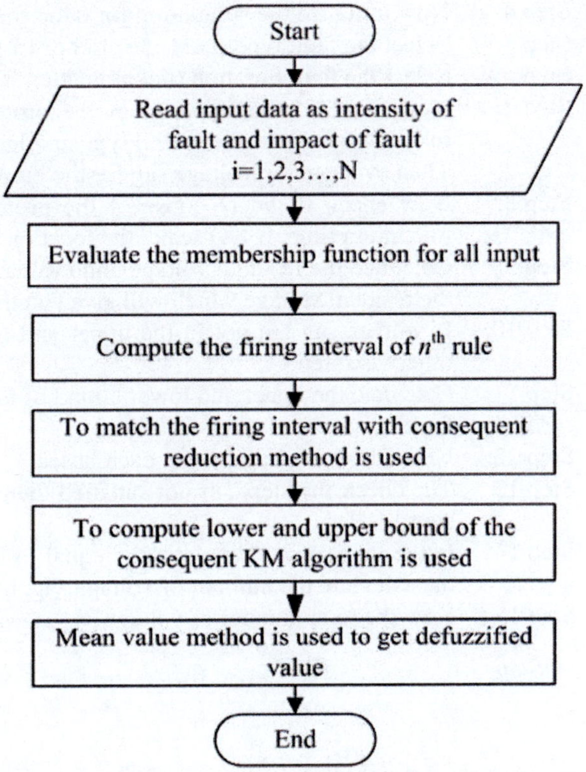

Fig. 3 Type-2 fuzzy fault rate evaluation flow chart [20]

sag evaluation due to random faults is shown in Fig. 3. From the flow chart is found that occurrences of power system faults are kept in the range of 0–1. It means the fault position can move any position from 0–1; if fault occurs near to the fault position, then it is considered to be 1, and if it occurs in the middle of the line, then it is considered to be 0.5.

Then, the voltage sag magnitude is evaluated as per equation given in [6]; if any voltage value is less than 0.9 pu, then that value is considered to be voltage sag at that bus. Similarly, the voltage sag is also taken randomly between 50 and 300 ms. The probability of three phase, single line to ground, line to line, and double line to ground fault considered to be 5%, 70%, 15%, and 10%, respectively, is used for the evaluation of voltage sag. The steps involved for simulating random faults are given below:

Start

Step 1: Collect the system data (i.e. failure rate or fault rate) for each line and buses of power system network.

Step 2: Perform the load flow operation.

Step 3: Using Z-bus building algorithm develop z-bus matrix and evaluate the prefault voltages.

Step 4: Now initialize the simulation for 8760 times.
Step 5: Select the fault type (i.e. LL or LG or LLL or LLG).
Step 6: Select the fault duration (i.e. protection device response time).
Step 7: Now check if the response of the protection device is less than the maximum time. If yes the all type and location of faults are considered. Then evaluate the voltage sag for the bus or line.
Step 8: Now check if the response of the protection device is less than the maximum time. If NO select the location of the fault.
Step 9: Calculate the residual voltage and evaluate the upper and lower limit of the residual voltage which will give you the exact number of voltage sag.
Step 10: If voltage sag are not in the upper and lower bound then go to step 8 otherwise step 11.
Step 11: Calculate the upper and lower bound of fault location and probability of fault occurrence.
Step 12: Now add the voltage sag at each buses.
Step 13: The check the step 7 if not satisfied then go to step 5 and 6, otherwise step 14.
Step 14: Is the iteration of simulation is equal to maximum number of iteration the calculate the number of voltage sag, otherwise go to step 4.
Step 15: Now the total number of voltage sag is evaluated for each buses.

Stop

3 Evaluation of Fault Rate

There are many factors that influence the occurrence of power system faults. Some of the factors are tree leaning, wind, gusts, lightning, equipment failure, etc. These factors need to be modeled properly in terms of fault rate so that the accurate number of voltage sag can be evaluated. Therefore, in this section, the fault rate is evaluated using a technique called type-2 fuzzy system. The intensity and probability of occurrence of fault in a power system are used to calculate the fault rate. In [20], the input to the type-2 fuzzy system are causes and intensity of faults, and the output is failure rate. This fault rate will be used in the Sect. 2 to evaluate the number of voltage sag. The methodology involved in the evaluation of the fault rate is shown in Fig. 3.

4 Case Study and Analysis

To implement the proposed methodology discussed in the previous sections, a four bus power system network is considered as a test case. The impedance value of the four bus system is shown in Fig. 4. The fault rate value is evaluated using flow chart

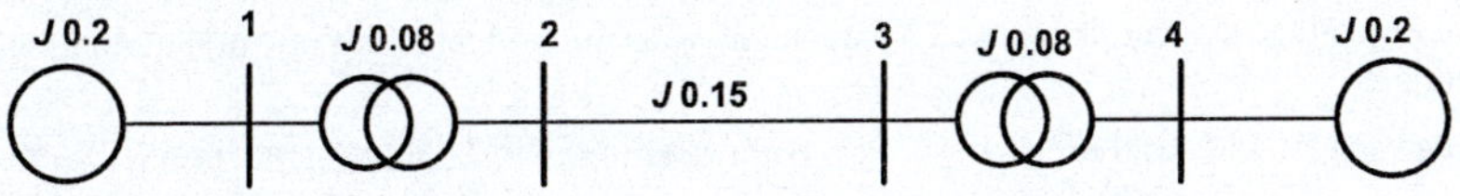

Fig. 4 Four bus system (impedance values are in per unit)

shown in Fig. 3. Fault rate value for the lines of Fig. 4 are $L_{1\text{-}2} = 0.5215$; $L_{2\text{-}3} = 0.6423$; $L_{3\text{-}4} = 0.4932$.

Now, let us consider LG fault that occurred at bus no. 3. The fault current is evaluated using Eq. (3) [6]. Then the voltage sag at bus no. 3 can be evaluated as per the equations given in [6] which is given as follows:

$$U^{a}_{sag\,(bus3)} = 1 - I^{+}_{f\,a(bus3)}\left(Z^{0}_{33} + Z^{+}_{33} + Z^{-}_{33}\right) \tag{4}$$

Similarly, the voltage sag can be evaluated for the other two phases using Eq. (4). Annually, the number of voltage sag at bus no. 3 due to symmetrical and unsymmetrical faults are calculated using proposed method and are given in Table 1. Then the faults were simulated using MCS technique. The advantage of MCS technique is that it creates simulated data considering possible uncertainties using random numbers. The comparison between the MCS-aided fault position method and analytical method is shown in Table 2.

Now, the fault rate for IEEE 118 bus system is evaluated as per Sect. 4 which is shown in Fig. 6. And, this fault rate is used for the assessment of voltage sag. The number of voltage sag due to symmetrical fault is given in Table 4. From the Table 3, it is found that the number of occurrences of voltage sag in the range of 0.9–0.95 pu is more and less for the range 0–0.1 pu.

Table 1 Annual number of voltage sag at bus no.3

Voltage sag range	Voltage sag			
	LLL	LG	LL	LLG
0 to 0.1	0	0	0	0
0.1 to 0.2	0	1	0	0
0.2 to 0.3	0	1	0	0
0.3 to 0.4	0	1	2	2
0.4 to 0.5	2	2	1	5
0.5 to 0.6	1	5	9	2
0.6 to 0.7	1	12	6	8
0.7 to 0.8	2	14	10	9
0.8 to 0.9	2	19	18	10
Total	8 Voltage sag/year	55 Voltage sag/year	46 Voltage sag/year	36 Voltage sag/year

Table 2 Voltage sag per year at bus 5 using analytical method and fault position method with MCS technique

Voltage sag range	Analytical method[6] (Voltage sag/year)	Fault position method with MCS (Voltage sag/year)			
		Number of faults			
		1	5	20	50
$0 < V \leq 0.1$	1	1	1	1	2
$0.1 < V \leq 0.2$	1	1	2	3	3
$0.2 < V \leq 0.3$	1	1	2	3	4
$0.3 < V \leq 0.4$	2	1	4	5	6
$0.4 < V \leq 0.5$	3	2	4	8	10
$0.5 < V \leq 0.6$	3	2	6	12	14
$0.6 < V \leq 0.7$	4	2	10	22	30
$0.7 < V \leq 0.8$	3	5	8	29	41
$0.8 < V \leq 0.9$	5	4	19	64	99
Simulation Time in sec	0.2167	0.31605	1.334367	15.3415	97.84634
Total number of Voltage sag/year	23	16	56	147	209

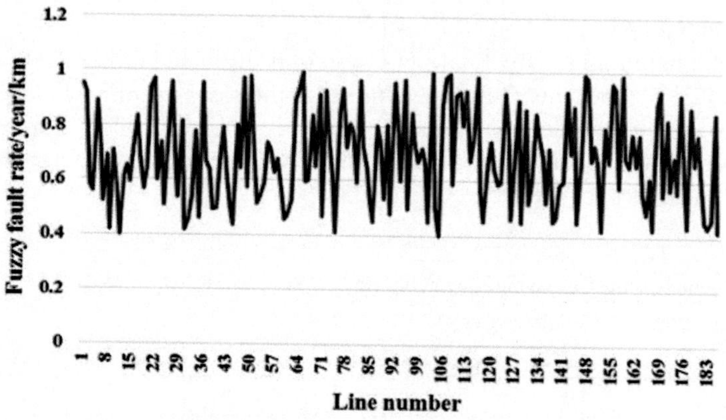

Fig. 6 The fault rate for IEEE 118 bus system

Table 3 Total number of voltage sag for IEEE 118 bus system due to the occurrence of symmetrical fault

Voltage sag range (pu)									
0–0.1	0.1–0.2	0.2–0.3	0.3–0.4	0.4–0.5	0.5–0.6	0.6–0.7	0.7–0.8	0.8–0.9	0.9–0.95
138	212	328	360	430	530	728	732	740	764

Table 4 Total number of voltage sag for IEEE 118 bus system due to the occurrence of unsymmetrical fault

Voltage sag range (pu)									
0–0.1	0.1–0.2	0.2–0.3	0.3–0.4	0.4–0.5	0.5–0.6	0.6–0.7	0.7–0.8	0.8–0.9	0.9–0.95
216	383	583	920	967	1510	2181	2337	3526	9364

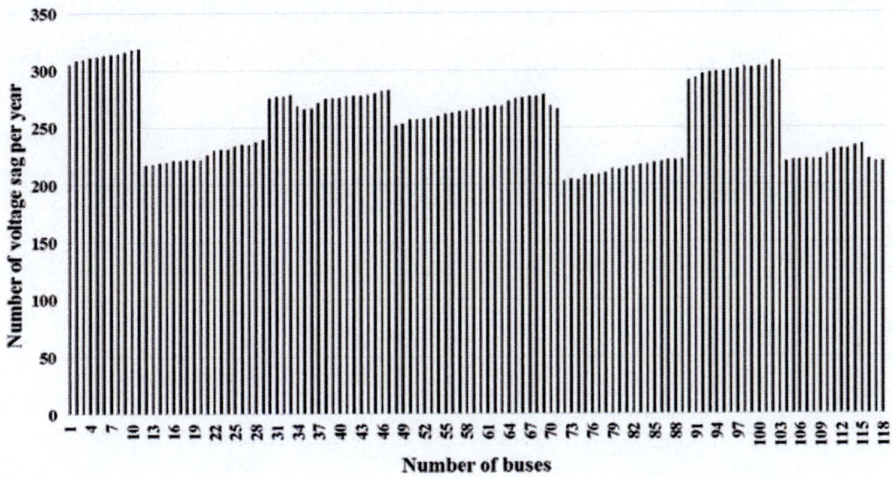

Fig. 7 The voltage sag at each bus when unsymmetrical faults are simulated randomly

Now, with the fault rate shown in Fig. 6, the unsymmetrical faults are simulated, and the number of voltage sag at each bus is shown in Fig. 7. The total number of voltage sag for different ranges is given in Table. 4. It is found that the total number of voltage sag due to unsymmetrical fault is more compared to symmetrical fault.

In Fig. 8, the faults are simulated randomly all over the IEEE 118 bus system, and it is found that the occurrence of LLL faults is less compared to other faults and more for LG fault. Therefore, it is a necessary task to detect and mitigate the LG faults by the utility company. Depending upon the simulated voltage sag data, the voltage sag is classified depending upon IEEE 1159 standard. Voltage sag classifier recommended by IEEE 1159 standard is being followed for the IEEE 118 bus system. The voltage sag classification is based upon duration of voltage sag (i.e., instantaneous, momentary, temporary, and sustained sag).

Instantaneous, momentary, temporary, and sustained terms are used when voltage sag magnitude (or residual voltage) sustain for the duration between 0.5 to 30 cycle, 30 cycle to 3 s, 3 s to 1 min and longer than 1 min, respectively, the results are shown in Table 5. Now, each particular type faults are simulated randomly at different location, and depending upon the IEEE 1159 classifier, the number of voltage sag is given in Table 6.

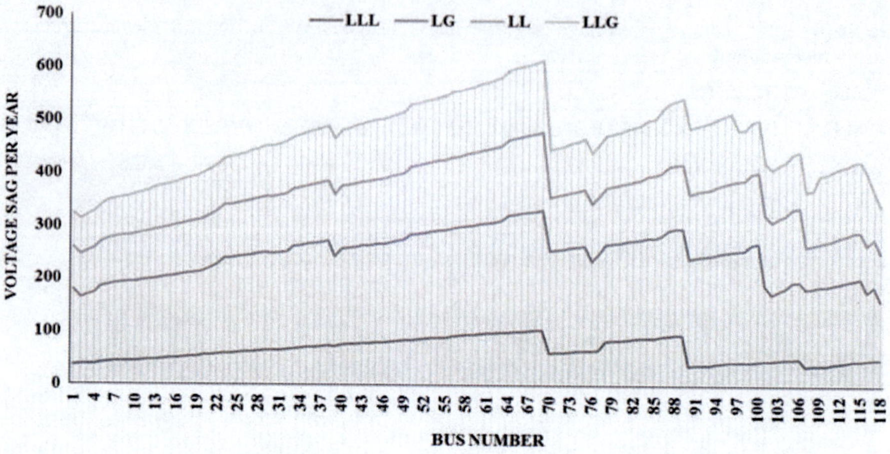

Fig. 8 Occurrence of voltage sag for symmetrical and unsymmetrical fault simulated randomly

Table 5 IEEE 1159 recomoned voltage sag claisfier for IEEE 118 bus sytem

Sl.no	Duration classifier	Number of occurance/year
1	Instatenous sag	14,394
2	Momentary sag	3065
3	Temporary sag	911
4	Sustained sag	428

Table 6 IEEE 1159 recommended voltage sag claisfier for different types of faults randomly simulated

Duration classifier	Type of faults			
	LLLF	LGF	LLGF	LLF
Instantaneous sag/year	23	10,012	4198	1689
Momentary sag/year	121	2193	7219	238
Temporary sag/year	55	743	130	83
Sustained sag/year	2	329	0	0

It is found that the occurrence of instantaneous voltage sag is more for all types of faults and less for sustained sag. The instantaneous voltage sag is not more harmful to the equipments because all the equipments have some fault ride through capability. But, in case of sustained sag, the voltage level fall dip to 0.2–0.1 pu, so if the back-up support is not capable of ride through the faults, then sensitive equipment will definitely trip.

Tables 5 and 6 will provide some important data from the equipment point of view that means which type of voltage sag will be responsible for the malfunction of equipment. The tables also provide relevant information to the insurance company to prepare insurance policy for the industrial machineries.

5 Conclusion

In this paper, the fault positioning method is used to simulate the faults to evaluate the voltage sag for the IEEE 118 bus system. Faults have a different pattern of voltage sag at different systems depending upon their line parameter. To assess the number of voltage sag, fault position method with MCS technique is used to simulate the faults randomly. Using MCS technique, type of faults and its locations can be simulated for both meshed and radial system. Along with MCS technique, voltage sag evaluation is also done using an analytical method. The only drawback of MCS technique is it is time consuming because the simulation time increases as the number of fault increases. The fault rate at each bus is also presented in this paper, and IEEE 1159 is used to classify different types of voltage sag occurring at the IEEE 118 bus system. The outcome of this paper helps the utility companies and industrial customers to come up with a mitigation strategy for the reduction of voltage sag so that future financial loss can be avoided.

References

1. Transmission–Distribution power quality report (TPQ-DPQ III). Report no: 000000003002003905, 2014 Program 1 Power Quality
2. IEEE recommended practice for monitoring electric power quality, IEEE standard 1159–2009 (Revision of IEEE standard 1159–1995), 2009
3. Conrad L, Little K, Grigg C (1991) Predicting and preventing problems associated with remote fault-clearing voltage dips. IEEE Trans Ind Appl 27(1):167–171
4. Lim YS, Strbac G (2002) Analytical approach to probabilistic prediction of voltage sags on transmission networks. IEE Proc Gener Transm Distrib 149(1):7–14
5. Juarez EE, Hernandez A (2006) An analytical approach for stochastic assessment of balanced and unbalanced voltage sags in large systems. IEEE Trans Power Delivery 21(3):1493–1500
6. Behera C et al (2020) A probabilistic approach for assessment of financial loss due to equipment outage caused by voltage sag using cost matrix. Int Trans Electr Energy Syst 30(3):e12202.
7. Martinez JA, Martin-Arnedo J (2006) Voltage sag studies in distribution networks-part I: system modeling. IEEE Trans Power Deliv 21(3):1670–1678
8. Martinez JA, Martin-Arnedo J (2006) Voltage sag studies in distribution networks-part II: voltage sag assessment. IEEE Trans Power Deliv 21(3):1679–1688
9. Martinez JA, Martin-Arnedo J (2006) Voltage sag studies in distribution networks-part III: voltage sag index calculation. IEEE Trans Power Deliv 21(3):1689–1697
10. Qader MR, Bollen MHJ, Allan RN (1999) Stochastic prediction of voltage sags in a large transmission system. IEEE Trans Ind Appl 35(1):152–162
11. Park CH, Jang G (2007) Stochastic estimation of voltage sags in a large meshed network. IEEE Trans Power Deliv 22(3):1655–1664
12. Park CH, Jang G, Thomas RJ (2008) The influence of generator scheduling and time-varying fault rates on voltage sag prediction. IEEE Trans Power Deliv 23(2):1243–1250

13. Liao H, Abdelrahman S, Guo Y, Milanović JV (2015) Identification of weak areas of power network based on exposure to voltage sags—part I: development of sag severity index for single-event characterization. IEEE Trans Power Deliv 30(6):2392–2400
14. Liao H, Abdelrahman S, Guo Y, Milanović JV (2015) Identification of weak areas of network based on exposure to voltage sags—part II: assessment of network performance using sag severity index. IEEE Trans Power Deliv 30(6):2401–2409
15. Behera C et al (2018) Assessment of equipment trip probability due to voltage sags based on fuzzy possibility distribution function. IEEE Access 6:76889–76899
16. Behera C et al (2019) Assessment of financial loss due to voltage sag in an industrial distribution system. In: 2019 IEEE 1st international conference on energy, systems and information processing (ICESIP). IEEE
17. Behera C et al (2019) Series compensation technique for the reduction of voltage sag for transmission system. In: International conference on innovation in modern science and technology. Springer, Cham
18. Behera C et al (2018) Voltage sag mitigation using distributed generation for an industrial distribution system. In: 2018 IEEE international conference on power electronics, drives and energy systems (PEDES). IEEE
19. Behera C et al (2020) Diagnosis report on voltage sag for different power distribution networks. In: 2020 international conference on contemporary computing and applications (IC3A). IEEE
20. Mitra R, Goswami AK, Tiwari PK (2017) Voltage sag assessment using type-2 fuzzy system considering uncertainties in distribution system. IET Gener Transm Distrib 11(6):1409–1419. (4 20 2017)
21. Behera C, Banik A, Goswami AK (2020) A novel approach for voltage sag representation in a chemical industry: a case study. Eng Rep 2(7):e12198

Chapter 7
Feasibility Analysis of SEPIC Converter as a PV Balancer for Practical Photovoltaic System

A. Veera Reddy, P. V. R. L. Narasimham, K. Sai Teja, and P. Shiva Kumar

1 Introduction

The idea of PVB is a new concept of solar module integrated converter (MIC) technology and enables the maximum power point tracking (MPPT) on each PV module. It is essentially placed in between the solar module and the dc bus to compensate for the differential voltages rather than differential currents [1]. Many researchers have been adopted to use of PVB in practical applications to overcome the difficulties of conventional MIC, such as high voltage transformation under mismatching conditions, the power rating of the converter, and unable to provide required output voltage under mismatching conditions [2].

The authors have proposed two different architectures of PVB, which are integrated into each module through a common DC bus configuration. The authors have claimed to use the second architecture due to higher efficiency and share the frontend converter as a flyback converter. This flyback converter feeds low voltage levels as an input to PVB. Moreover, this PVB is essentially working as a buck converter to further reduce the voltage level to interface the dc bus and SPV module. Thus, the PVB has been analyzed and verified through simulation as well as experimental studies on photovoltaic system modules available in the market. In a similar pattern, the authors in [3] have demonstrated the same concept such that the flyback converter as frontend converter and PVB as a buck converter and successfully tested on solar PV system in the field. Since all these research mechanisms have been carried out with the conventional buck converter as a PVB and flyback converter as a frontend converter. However, the state-of-art shows that there is a need for exploration in the converter topologies to reduce the overall cost of the system and therefore, the present work was motivated towards this direction.

A. V. Reddy (✉) · P. V. R. L. Narasimham · K. S. Teja · P. S. Kumar
Department of Electrical and Electronics Engineering, Velagapudi Ramakrishna Siddhartha Engineering College, Vijayawada 520007, India

S. Dawn et al. (eds.), *Smart and Intelligent Systems*, Algorithms for Intelligent Systems,
https://doi.org/10.1007/978-981-16-2109-3_7

In recent times, various dc-dc converters have been identified to step up/down the voltages as per the requirement of the PV system. In this regard, the dc-dc converter topologies such as buck, boost, buck-boost, cuk, ZETA, single-end primary inductance converter (SEPIC), etc., have been well demonstrated on different solar power applications. The proper selection of the converter configuration plays a vital role in reducing the power rating of PVB as well as frontend converter, which leads to the reduction of the overll cost of the system to demonstrate the PVB application. The buck converter has already been implemented as a PVB in [2, 3] and successfully verified to maintain the required duty ratio and working under various irradiation levels.

The boost converter [4] may not be required for this PVB application because the converter has to step down voltage levels rather than step up for the reduction of power rating. Similarly, the buck-boost converter, ZETA converter, CUK converter are not suitable for this specific PVB application on the practical solar photovoltaic system. SEPIC converter is capable of stepping up/down the voltages with non-inverter output voltage and also, low ripple content due to the presence of inductors and capacitors [5, 6]. Among all these, SEPIC converter is preferable in order to maintain the required duty ratio and balancing the differential voltages under different irradiations. However, the application of SEPIC converter as PVB is not yet reported in the literature.

In the present work, the realization of a SEPIC converter as a PVB instead of buck converter is confirmed on a real-world solar PV system. The performance analysis is carried out through simulation studies in order to show the effectiveness of SEPIC converter in terms of its voltage stress on switching devices, energy loss, equivalent efficiency, and power rating of the overall system. Thus, the simulation studies reveal that the SEPIC converter effectively reduces the power rating of PVB as compared to the conventional buck converter as PVB in a solar PV system.

The research orientation is further expressed as the performance analysis of SEPIC converter presented in Sect. 2. Section 3 contains the simulation studies and discussions followed by the conclusions with future scope of research work in Sect. 4.

2 Performance Analysis of SEPIC Converter

The well-known possible architecture in the literature of PVB is shown in Fig. 1 [2]. Here, photovoltaic balancer is conncected in series with PV module, and such types of combinations are connected in parallel with DC bus. The below sections describe the performance of SEPIC converter, SEPIC as PVB, and the mathematical formation of PVB.

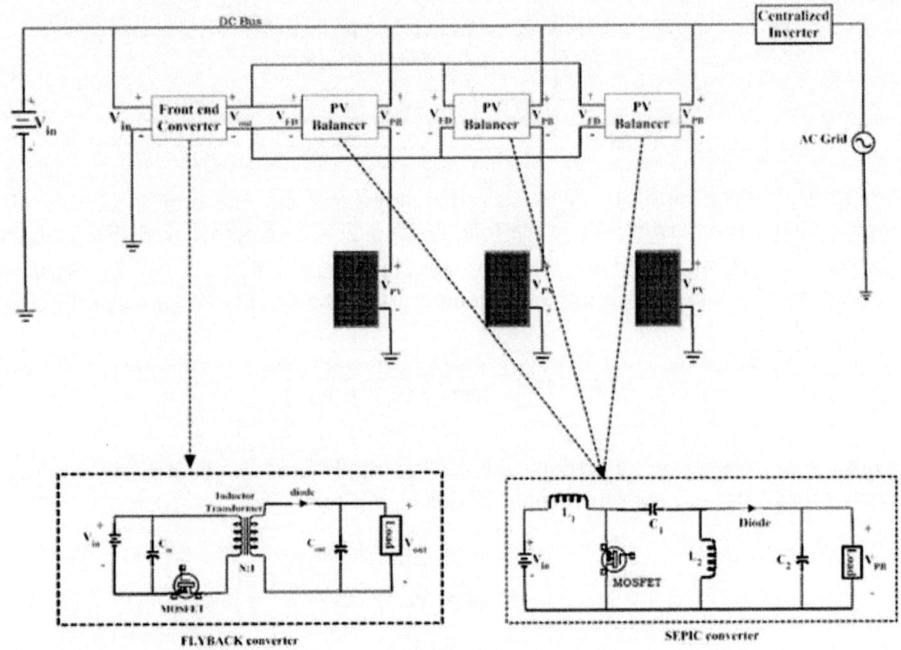

Fig. 1 Architecture of PVB

2.1 *Single-End Primary Inductance Converter*

Single-end primary inductance converter is a DC-DC converter, this type of conversion is handy when the designer uses voltages, i.e., 12 V [7]. It is developed from boost converter with an adverse of non-inverted output. A series capacitor is used to couple energy from input to output. SEPIC converter parameters consist of a coupling capacitor C1, output capacitor C2, inductors L1 and L2, a power MOSFET, and diode as shown in Fig. 1. The SEPIC operates in two modes such as mode-I (MOSFET switch turn ON) and mode-II (MOSFET switch turn OFF). When the switch is ON, the inductors L1 and L2 will linearly charge, the diode in reverse bias and C2 discharges its stored energy through load. When the switch is OFF, both L1 and L2 discharge their stored energy through diode, and load, C1 and C2 both are in charging phase. The percentage of ripple and converter efficiency are influenced by the presence of resistance in the energy storage elements such as inductors and capacitors. Henceforth, it is preferred to choose high-quality inductors and capacitors for minimization of the ripple as well as to prevent the heat build-up. The output voltage and current of SEPIC converter is represented as Eq. (1):

$$\begin{cases} V_o = \left(\frac{D}{1-D}\right)V_{in} \\ I_o = \left(\frac{1-D}{D}\right)I_{in} \end{cases} \quad (1)$$

where D is duty ratio, V_O, I_O is output voltage and current, and V_{in}, I_{in} is the input voltage and current. R_L is the load resistance across output of SEPIC and its impedance matching $R_L^{'}$ with respect to input terminals is $R_L^{'} = \frac{(1-D)^2}{D^2} R_L$. Further, in order to estimate the two inductor elements with the Eq. (2) is given as follows:

$$L_1 = L_2 = \frac{V_{in} V_{out}}{\Delta I_L f (V_{in} + V_{out})} \quad (2)$$

where f is switching frequency and ΔI_L is ripple current. Estimating the two capacitors C_1 and C_2 are represented as Eqs. 3 and 4:

$$C_1 = \frac{V_{in} I_{in}}{(V_{out} + V_{in}) f \Delta V_c} \quad (3)$$

$$C_2 = \frac{V_{in} V_{out}}{8 f^2 L_2 (V_{out} + V_{in}) \Delta V_C} \quad (4)$$

where I_{in} is input current and ΔV_c is ripple voltage.

2.2 *SEPIC Conerter in PVB*

The architecture of PVB with SEPIC converter as shown in Fig. 1

Here, SEPIC converter performs as a PVB, which compensates for the differential voltages between PV module and DC bus. The input of PVB is carried from the output of frontend converter. In this architecture of PVB, the basic circuits of flyback and SEPIC converters are shown in Fig. 1. The dc bus maintaining 28 V and input of PVB is stepped down to 12 V by a flyback converter as frontend converter [8]. The system is disconnected from the DC bus automatically due to fault either in PVB or in PV module.

2.3 *Mathematical Analysis of PVB*

For the feasibility analysis of PVB, the necessary formulations are presented and mathematically expressed below. The output power drawn from the PV module under MPPT condition for any irradiation levels is expressed in Eq. (5).

$$P_{MPP} = V_{MPP} I_{MPP} \tag{5}$$

where $V_{MPP}, I_{MPP}, P_{MPP}$ defines the maximum output voltage of PV modules, maximum output current of PV module, and maximum power output of PV module.

The PVB compensates the differential voltages between DC bus and PV module. Therefore, the PV balancer output voltage is V_{PB} signified in Eq. (6).

$$V_{PB} = V_{DC} - V_{MPP} \tag{6}$$

The same current will flow through PV module and PVB due to their series connection and can be expressed in Eq. (7).

$$I_{PB} = I_{MPP} \tag{7}$$

where the compensating voltage of PVB is V_{PB}, DC bus voltage is V_{DC}, the output current of PVB is I_{PB}, and the power output of PVB is P_{PB}.

The output power of PVB (P_{PB}) is partial power over the full power of a PV module is represented in Eq. (8).

$$P_{PB} = V_{PB} I_{PB} = (V_{DC} - V_{MPP}) I_{MPP} \tag{8}$$

The voltage transformation ratio of PVB will be influenced by the magnitude of DC bus voltage. If the DC bus volttage is too far from PV module voltage, then the voltage transformation ratio is low, while the power rating of PVB will be high, and it is true for vice-versa. The calculation of voltage transformation of PVB R_{PBVTR} is represented in Eq. (9).

$$R_{PBV.TR} = \frac{V_{PB}}{V_{FD}} = \frac{V_{DC} - V_{PV}}{V_{FD}} \tag{9}$$

The ratio of output power of PVB R_{PBR} is represented in Eq. (10).

$$R_{PBR} = \frac{P_{PB}}{P_{MPP}} = \frac{V_{PB}}{V_{MPP}} = \frac{V_{DC} - V_{MPP}}{V_{MPP}} \tag{10}$$

The DC bus voltage is high compared to maximum voltage PV module though the current that flows through the DC bus is less Eq. (11).

$$\text{DCbusloss} = I^2 R \tag{11}$$

where, I is the current and R is the resistance of DC bus. The equivalent efficiency of PVB η_{equ} is expressed in terms of power loss normalized with module output at MPPT as given in Eq. (12).

$$\eta_{equ} = 1 - \frac{P_{loss}}{P_{MPP}} = 1 - R_{PBR}\frac{P_{loss}}{P_{PB}} \tag{12}$$

Since R_{PBR} is less than 20% of the equivalent efficiency of PVB is high.

3 Simulation Results

In order to verify the performance analysis, the simulation work is implemented in MATLAB 2015b with PC configuration, Intel core i3-2350 M CPU, 4 GB RAM, 4-bit operating system. The solar PV module in PV balancer is implemented with practical PV panel, the datasheet of DSP-100 M is illustrated in Table 1 [9].

Table 2 shows the verification process of the PV module operated under different irradiations, i.e., 1000, 500 and 150 W/m^2. While PV module operating under different irradiations obtained 24 V maximum voltage, 4.168 A maximum current, and 100 W maximum power at full irradiation, i.e., 1000 W/m^2. For 500 W/m^2, the obtained maximum voltage, current, and power are 23.35 V, 2.10 A, and 49.16 W. For 150 W/m^2, the maximum voltage is 22.3, maximum current is 0.62, and maximum power is 14.02 W.

To extract as much energy as possible from such a system to make a PV module useful, a PV module is used efficiently only when it is made to operate at its optimal operating point. The amount of power that can be extracted from the module is

Table 1 Practical PV module datasheet

Parameter name	Value
Maximum power (P_{max})	100 W
Maximum voltage (V_{max})	24 V
Maximum current (I_{max})	4.168 A
Open circuit voltage (V_{oc})	27.75 V
Short circuit current (I_{sc})	4.63 A
Operating temperature range	−45 to 80 °C
Temperature co-efficient of P_{max}	0.5%/°C
Short circuit current of the cell at 25^{0}C and 1000 W/m^2 (K_i)	0.0017
Nominal temperature	300 k
Electron charge (q)	1.69e^{-19}
Boltzmann's constant (k)	1.3805e^{-23}
Band gap energy of semiconductor (E_g)	1.1 eV
Number of series cells (N_s)	40
Number of parallel cells (N_P)	1
Series resistance (R_s)	0.0221
Ideal factor (A)	1.2

Table 2 Specification of PV module under different irradiations

S.NO	Irradiation (W/m^2)	V_{OC} (V)	I_{SC} (A)	V_{MPP} (V)	I_{MPP} (A)	P_{MPP} (W)
1	1000	27.75	4.44	24	4.16	100
2	500	27.2	2.23	23.35	2.10	49.16
3	150	25.53	0.66	22.3	0.62	14.02

dependent on the operating voltage of that module. At any moment, the operating point of a PV module depends on varying insolation levels, the load of the system, and temperature. The atmospheric conditions and load variables are changing constantly making it very difficult to extract all of the solar energy available from panels without a controlled system. With the use of maximum power point tracking algorithms along with power electronic converters, maximum power is extracted from the module.

For verification, PV module is implemented in MATLAB with practical module data, and the plotted I-V is shown in Fig. 2 and P–V curves are shown in Fig. 3 under different irradiation. Figure 2 verified the maximum voltage and maximum current under full irradiation. Figure 3 verified the maximum power at full irradiation.

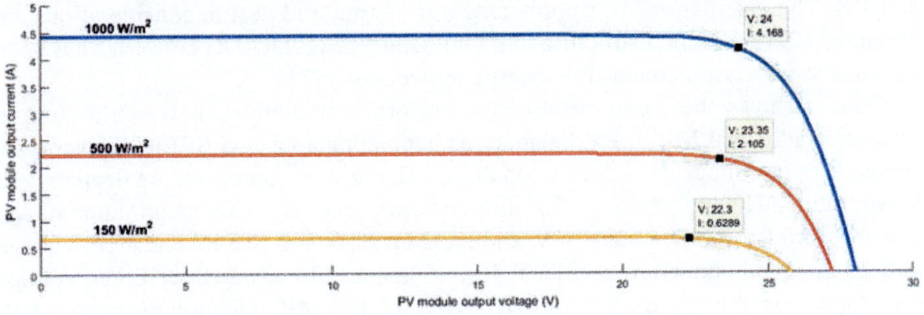

Fig. 2 I-V curves of PV module

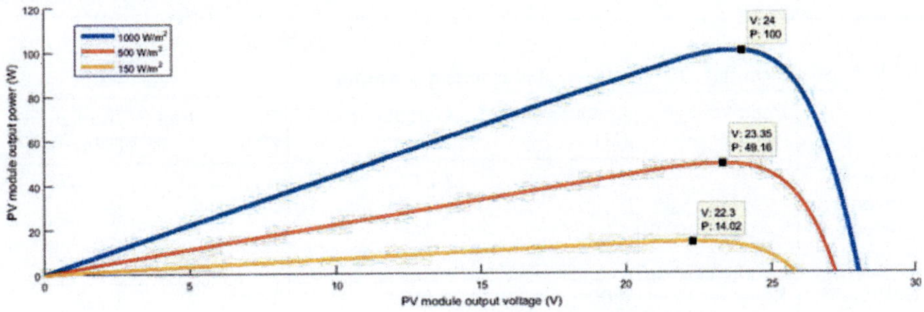

Fig. 3 P–V curves of PV module

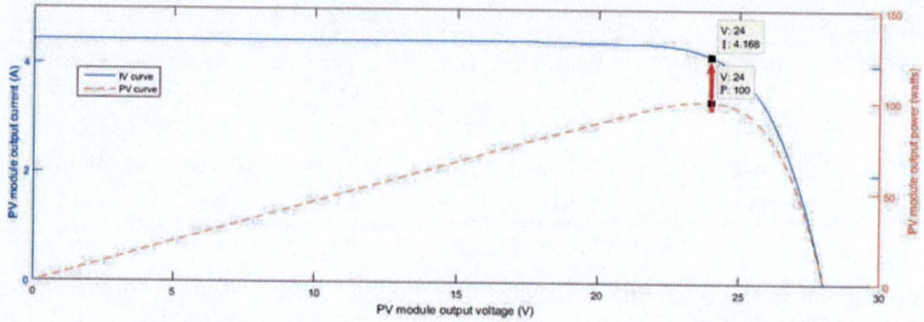

Fig. 4 Examined the MPPT from I–V and P–V curves

The I-V and P–V curves for an irradiation level of 1000 W/m^2 are shown in Fig. 4. The power delivered by a module increases as the current drawn increases. Any additional current drawn from the array will result in the rapid drop off the cell voltages thereby reducing the module output power.

Table 3 shows the specifications of architecture in terms of the input voltage, output voltage, output current, and output power. DC bus specified 28 V dc voltage and flyback converter is specified as frontend converter and SEPIC converter is specified as PVB. The calculations of ripple content in output and design considerations are given in [10] A modified structure of a high gain SEPIC DC-DC converter has been introduced for various renewable energy applications [11].

Table 4 shows the comparison of the performance analysis between the micro inverter, traditional MICs, PV balancer as buck converter, and SEPIC converter in terms of input voltage, transformation ratio, voltage stress on switching devices, and power rating. The input voltage for micro inverter and MIC is the maximum voltage tracked from PV module, i.e., 24 V, and PVB for buck and SEPIC, the input voltage is from frontend converter, i.e., 12 V. Under the partially shaded condition, voltage transformation ratio is high for micro inverters and MIC technologies; for PVB, under partially shaded conditions, the differential voltage is very less and voltage transformation ratio of converter is less. Voltage stress on switching device (12 V) is less on PVB compared to MIC and micro inverter technologies. PVB provides the

Table 3 Specifications of PV balancer and frontend converter

	Input voltage (V)	Output voltage (V)	Output current (A)	Power rating (W)	Irradiation (W/m^2)
Frontend converter (Flyback converter)	28	12	3.61	43.32	
SEPIC converter	12	4.00	4.168	16.672	1000
	12	4.65	2.105	9.788	500
	12	5.71	0.6289	3.591	150

Table 4 Comparison between PV balancer and traditional MICs

	Micro Inverter	MIC	PV balancer (buck)	PV balancer (SEPIC)
Input voltage	V_{MPP}	V_{MPP}	V_{FD}	V_{FD}
Voltage transformation ratio	High	High	Low	Low
Voltage stress on switching devices	High	High	Low	Low
Power rating	High	High	Low	Low

power rating converter is dramatically decrease due to its getting input from external source. In micro inverter and MIC technologies, the converter gets the input from PV module, so it has to maintain the same power rating [12–15].

The PVB compensates the differential voltages between PV module and DC bus as shown in Fig. 5 for different irradiations, here, upper waveform represents the PV module output voltage and lower waveform represents the PVB output voltage for different irradiation. Figure 5a represents the output voltage waveforms of PV module and PV balancer at 1000 W/m^2. Figure 5b represents the output voltage waveforms of PV module and PV balancer at 500 W/m^2. Figure 5c represents the output voltage waveforms of PV module and PV balancer at 150 W/m^2.

The PVB is connected in series with the PV module, so the same current will flow through it. The output currents of PV balancer and the PV modules for different irradiations are shown in Fig. 6.

From the observation of the above curves for different irradiations, the mismatching voltage is supplied by the PV balancer, and due to the series connection of PV balancer and PV module, the output currents are same. The feasibility analysis of SEPIC as PVB is demonstrated and compared qualitatively with the conventional buck converter on same practical system as illustrated in Table 4 in terms of input voltage, output voltage, output current, output power, and operating under different irradiations. From this observation, the power rate of SEPIC converter has been reduced in variance of 2 watts from 18.14 to 16.672 watts by the conventional buck converter at irradiation of 1000 W/m^2. For 500 W/m^2 irradiation, the power rating is extremely reduced in variance of 3.5 watts from 13.31 to 9.788 watts. Whereas, the power rating slightly reduced in variance of 1.2 watts from 4.82 watts as a conventional buck converter to 3.591 watts as a SEPIC converter under the level of 150 W/m^2 irradiation Table 5.

4 Conclusions

In this paper, the SEPIC converter has been successfully realized as a PVB on solar practical PV module. Moreover, the feasibility analysis of SEPIC converter is well demonstrated and compared qualitatively with the conventional buck converter. The power rating of SEPIC converter is reduced to 16.672 from 18.14 watts by the

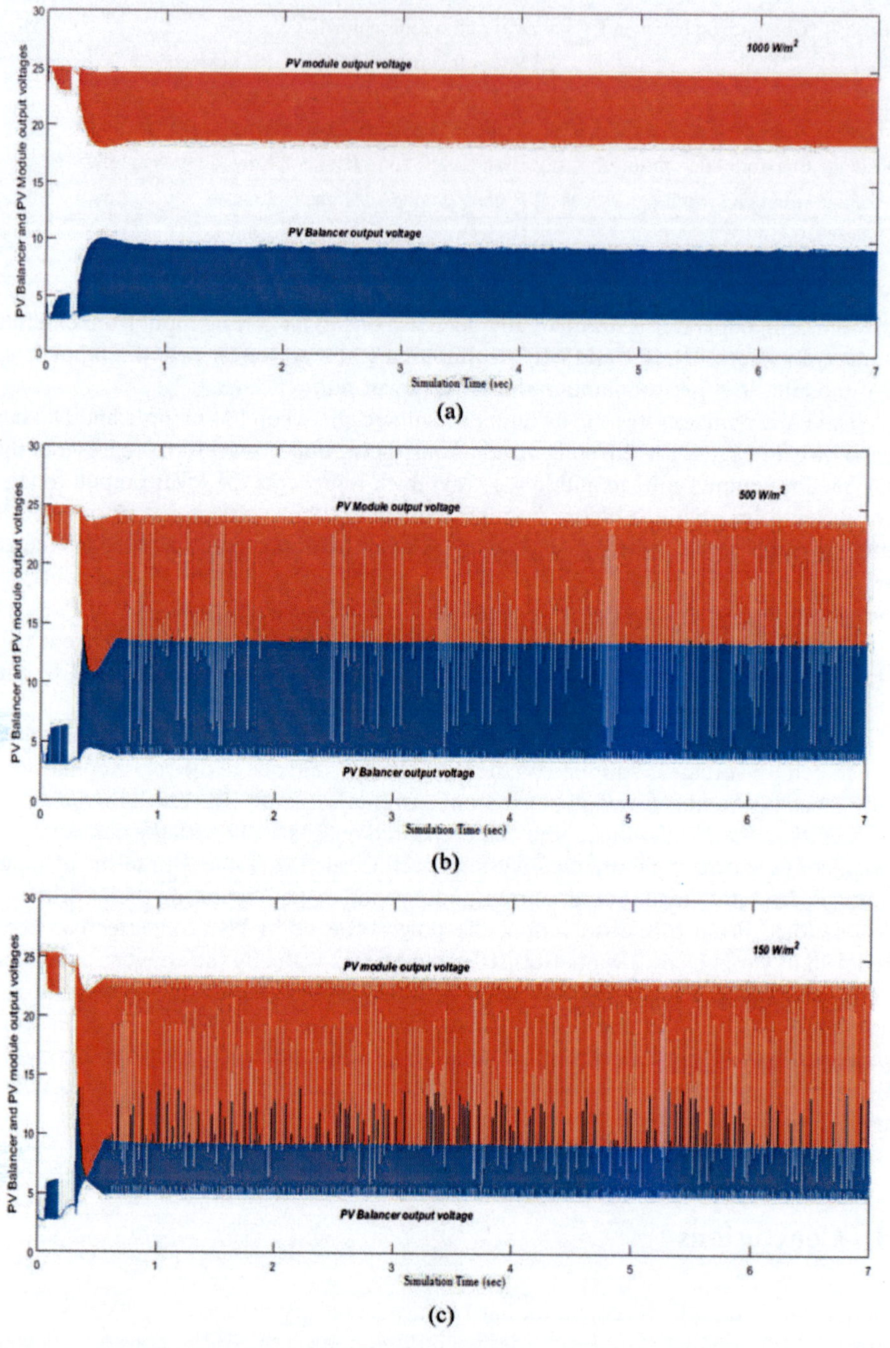

Fig. 5 Differential voltages between PV balancer and PV module **a** irradiation 1000 W/m^2, **b** irradiation 500 W/m^2, and **c** irradiation 150 W/m^2

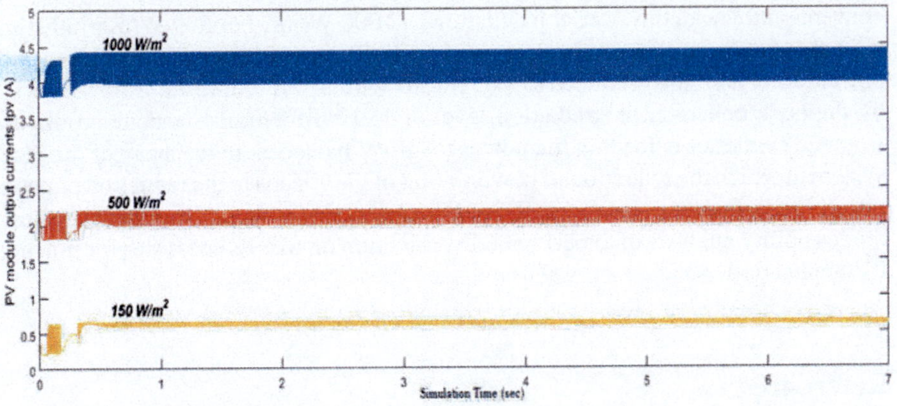

(a)

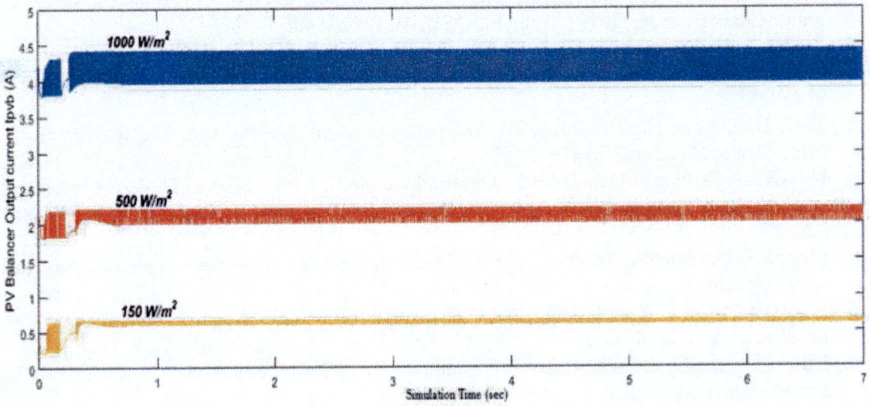

(b)

Fig. 6 Output currents for different irradiations **a** PV module output currents, **b** PV Balancer output currents

Table 5 Comparison between buck converter and SEPIC converter as PVB

PVB	Input voltage (Volts)	Output voltage (Volts)	Output current (Amps)	Power rating (Watts)	Irradiation (W/m^2)
Buck converter	12	4.29	4.23	18.14	1000
		5.39	2.47	13.31	500
		6.79	0.71	4.82	150
SEPIC converter	12	4.00	4.168	16.672	1000
		4.65	2.105	9.788	500
		5.71	0.6289	3.591	150

conventional buck converter at irradiation of 1000 W/m^2. For 500 W/m irradiation, the power rating is drastically decreased from 13.31 to 9.788 watts. W^{2}hereas, the power rating is slightly reduced to 3.591 watts with SEPIC converter from 4.82 watts by the buck converter at irradiation level of 150 W/m^2. It also concluded that the frontend converter is feeding the power to all PV balancers in the array of the given system in which the reduction of power rating of PVB leads to the reduction of power rating in the frontend converter. Thus, the obtained simulation results have shown the feasibility analysis of experimental verification on SEPIC converter for practical PV application.

References

1. Kjaer S, Pedersen J, Blaabjerg F (2005) A review of single phase grid connected inverters for photovoltaic modules. IEEE Trans Ind Appl 41:1292–1306
2. Linares L, Erickson R, Macalpine S, Brandemuehl M (2009) Improved energy capture in series string photovoltaic via smart distributed power electronics. Proc Appl Power Electron Conf 904–905
3. Zhou H, Zhao J, Han Y (2015) PV balancers: concept, architectures and realization. IEEE Trans Power Electron 30:3479–3487
4. UdayKiran D, Narasimham PVRL, Gouthamkumar N, Sudheerkumar D (2016) Investigation of PV balancer architectures on practical solar photo voltaic system. IEEE Conf ICSITech 226–231
5. Mohan N, Undland M, Robbins P (2003) Power electronics: converters, application and design willy 3rd edn
6. Kasper M, Bortis D, Kolar JW (2014) Classification and comparative evaluation of the PV panel-integrated dc-dc converter concept. IEEE Trans Power Electron 29:2511–2526
7. Falin J Designing DC-DC converters based on SEPIC topology. Power management. Texas instruments incorporated
8. Dinwoodie L (1999) Design review: isolated 50 watt flyback converter using the UCC3809 primary side controller and the UC3965 precision reference and error amplifier. Texas Instruments
9. DSP-100M Panel. www.enfsolar.com
10. Maroti PK, Padmanaban S, Holm-Nielsen JB, Sagar Bhaskar M, Meraj M, Iqbal A (2019) A new structure of high voltage gain SEPIC converter for renewable energy applications IEEE Access 7:89857–89868
11. Babaei E, Seyed Mahmoodieh ME (2014) Calculation of output voltage ripple and design considerations of SEPIC converter. IEEE Trans Ind Electr 61:1213–1222
12. SEPIC equations and component ratings. www.maximumintegrated.com
13. Ramki T, Tripathy LN (2015) Comparision of different DC-DC converter for MPPT application of photovoltaic system. IEEE conference 24–25
14. Bellia H, Youcef R, Fatima M (2014) A detail modeling of photovoltaic module using MATLAB. Elsevier, pp 53–61
15. Basic calculation of a buck converter power stage. Texas Instruments, SLVA477B (2015)

Chapter 8
Market Clearing Mechanism by Optimal Scheduling of Electric Power Suppliers

Arup Das, Subhojit Dawn, and Sadhan Gope

1 Introduction

In our daily life, electrical energy has become a fundamental and essential part. We cannot think a single second of our daily life style activity without electricity. Every aspect of our life is related to electricity from house-hold applications to information transmission to industrial applications. For that reason, electrical industry is probably the most complex and largest industry in the world. There are many factors to be taken into account for successful operation of any power system in this domain. Power system industry must have capability to produce and distribute the energy un-interruptedly throughout the world. Electric energy generation and distribution sectors, unlike other business models also want to maximize the profit by ensuring product quality, system efficiency and meet consumer needs [1]. Deregulation is an alteration in the rules and protocols that direct the whole entity of the power system [2]. Before the introduction of deregulation in the overall power market, it is controlled by a single entity, no competition in the market and consumers played almost no role in the system [3]. In deregulated power system, the electrical energy is handled as an item or a product instead of using as a supply or services as in vertically consolidated utilities. The performance of the end user and the distributor are examined using the notion of micro economics. In restructured power system, electricity market is a sophisticated job due to unpredictability in its demand and generation [4]. Lately, significant importance is assigned in increasing or maximizing the profit of the customers; consequently, bidding is of prime importance in independent system operator (ISO) [5]. In this competitive power market, the operational and the physical imperatives of the system have significant danger to the market by the generation

A. Das (✉) · S. Gope
Department of Electrical Engineering, Mizoram University, Aizawl, India

S. Dawn
Department of Electrical and Electronics Engineering, Velagapudi Ramakrishna Siddhartha Engineering College, Vijayawada, Andhra Pradesh, India

S. Dawn et al. (eds.), *Smart and Intelligent Systems*, Algorithms for Intelligent Systems,
https://doi.org/10.1007/978-981-16-2109-3_8

organizations in practicing its market power. The deregulated power market allows the purchase and sells through bidding process. The bidding and selling price depend on overall supply and demand [6]. Electricity price forecasting has become very important for the energy generation companies, depending on which the companies make their decision for rate of production and operational management process [7, 8]. The price of electricity varies over time depending on local conditions, such as availability of source of power, the demand of power and the market design [7]. In this paper, the main objective is to exploit the social welfare or to get the optimum bidding strategy for both the consumers and the producers by optimal scheduling of generators. Here, some of the generators maybe asked to change their generator output depending on the transmission stability. The optimization technique used in this report is Artificial Bee Colony (ABC) and Ant lion optimizer (ALO) algorithm comparably. The optimal scheduling of generators is to provide an economic combination of generators that has to be connected to the load [9].

2 Problem Formulation

The optimal scheduling of generators is to provide an economic combination of generators that has to be connected to the load. Now, for calculating the profit, market clearing price has to be calculated. The optimum settlement arrangement can be acquired where the target work is defined as the benefit of the brokers' profit which is determined as the difference between the purchase cost and sell offer cost.

To optimally clear the market, calculation of market clearing price is very much essential. So, the main objective of this work is to find the market clearing price for maximizing the profit of the producers with optimum generations.

Actually, the objective function of the existing work is by reducing the generation cost of the system, increasing the social welfare. To fulfill all the above situations, the objectives must be diminished. The objective function of the accessible method is given as:

$$E = \sum_{i=1}^{NG} D_i(Q_{Gi}) - \sum_{j=1}^{ND} C_j(Q_{Dj}) \tag{1}$$

where $D_i(Q_{Gi})$ is generation cost curve at the generator bus i, $C_j(Q_{Dj})$ is consumer benefit curve at load bus j, ND is equal to total number of buses, NG is equal to number of generators and ND number of loads.

The equation of Social Welfare (SW) is:

$$SW = \sum_{j=1}^{ND} C_j(Q_{Dj}) - \sum_{i=1}^{NG} D_i(Q_{Gi}) \tag{2}$$

From Eq. (1), it is seen that objective function E consists of social welfare. In this present approach, the main function is to minimize the objective function, E and maximize the SW. For fulfilling this condition, generation cost function $D_i(Q_{Gi})$ has to be minimized, and maximize the consumer benefit curve $C_j(Q_{Dj})$. The objective function 'F' has been minimized subject to the equality and inequality constraints which have been taken from [5].

3 Artificial Bees Colony Algorithm

ABC is an optimization tool that offers a population-based search technique. Bees travel across a multidimensional search field, and some employed and onlooker bees select food sources based on their own experience and change their positions. Some scouts fly and choose food sources at random without using their experience. If the nectar quantity level of the new source higher than the previous one is in their mind, then the new location can be memorized and forget the previous one. Thus, the ABC algorithm combines local search method, performed by employed and onlooker bees, with global search method, managed by onlooker and scouts, aimed at balancing the exploration and exploitation process [10].

4 Ant Lion Optimizer Algorithm

The ALO algorithm replicates the commerce between ant lions and ants in the snare. To display such co-operations, ants are allowed to walk over the inquiry space, and ant lions are permitted to chase them and emerge as fitter the usage of traps. ALO calculation reproduces five primary strides of chases in larvae: arbitrary stroll of ants, constructing traps, entanglement of ants in traps, slipping ants on ant lion, getting preys and re-constructing traps are actualized [11].

5 Result and Discussion

The viability of the proposed system for creating the most efficient bidding approach has been tried on an IEEE 30 bus system (Fig. 1), consisting of 6 sellers and 20 buyers. By using ABC and ALO optimization technique, the presumed bi-degree optimization issue accounting to double-sided offering has been resolved.

The power generating company (GENCOS) is expected to present hourly free-extend bid, while the load entities are expected to offer at minor cost. The decrease and top sure at the bid rate of the GENCOS, whose methodology has been evaluated, were considered as minor expense, also, multiple times of peripheral expense. For every iteration, the bidding techniques of the opponents are fixed by their circulation

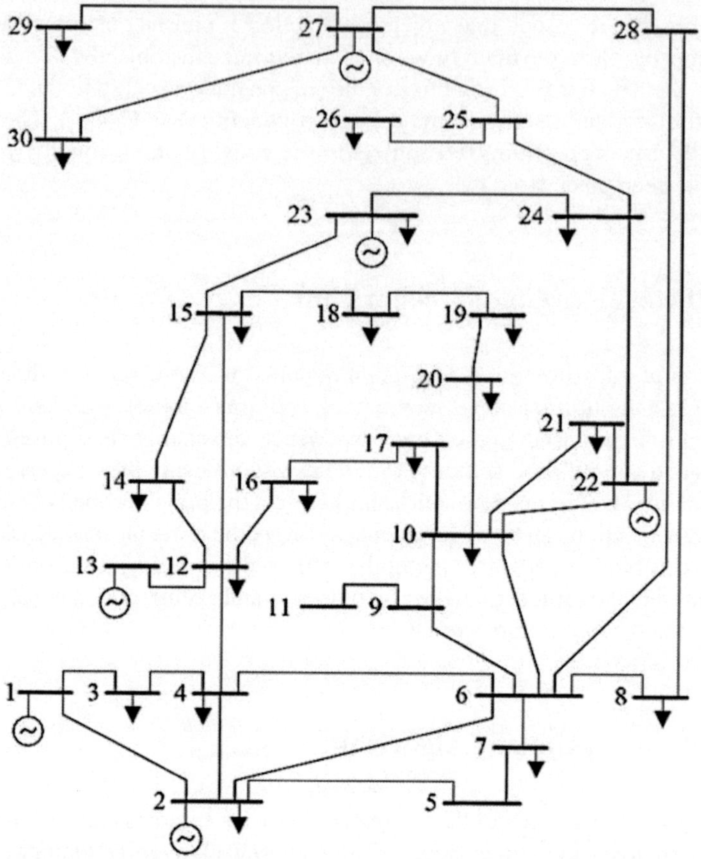

Fig. 1 Single line diagram of IEEE bus test system

capacities, and the offer cost of generator 1 is acquired utilizing ABC and ALO comparably.

Various values of competitor-offered price dissemination parameters and coefficient data of each individual generator are displayed in Table 1.

Subtleties of the load demand and presumed minimum bid cost put together by load substances are shown in Table 2.

The market clearing price and market clearing value for the unconstrained market clearing simulation is shown in Fig. 2, where market clearing price (MCP) and market clearing value (MCV) are acquired from the factor of crossing point of two curves. Where, the two curves are the bidding data of suppliers and bidding data of customer or demand data. The offering information of providers are amassed and arranged in ascending order, and offering information of client are amassed and arranged in descending order.

Table 1 Generator data for IEEE 30 bus system

Gen no	Cost coefficient			Max Generation	Marginal bid price ($/MWh)
	a	b	c		
G_1	0.02	2	0	80	5.2
G_2	0.0175	1.75	0	80	4.55
G_{13}	0.025	3	0	40	5
G_{22}	0.0625	1	0	50	7.25
G_{23}	0.025	3	0	30	3.5
G_{27}	0.00834	3.25	0	55	4.167

Table 2 Load and its marginal bid price of customers of IEEE 30 bus system

Bus no	Bid Price ($/MWh)	Bid quantity (MW)	Bus no	Bid Price ($/MWh)	Bid quantity (MW)
2	9	21.7	17	8.5	9.0
3	3.5	2.4	18	3	3.2
4	20	66.7	19	10	9.5
7	12	22.8	20	3.5	2.2
8	15	30	21	14	17.5
10	9	5.8	23	4.5	3.2
12	12	11.2	24	11	8.7
14	4	6.2	26	5	3.5
15	8	8.2	29	3	2.4
16	3.5	3.5	30	13	10.6

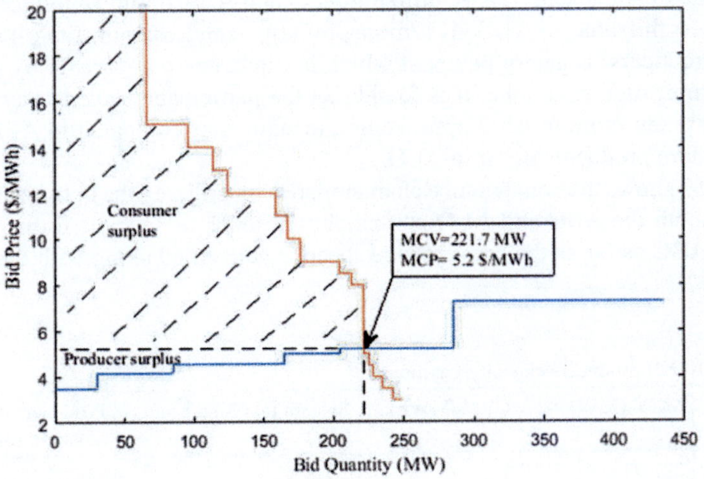

Fig. 2 Supply and demand equilibrium curve

Table 3 Generator generation dispatch for market clearing power simulation

Gen no	Bid price ($/MWh)	Generator generation dispatch (MW)	
		ABC	ALO
G_1	5.2	80.00	80.00
G_2	4.55	60.49	59.02
G_{13}	5	35.56	30.94
G_{22}	7.25	00.00	00.00
G_{23}	3.5	30.00	30.00
G_{27}	4.167	15.48	21.69
SW($/h)	–	441.41	442.57

Here, from the above curve, the market clearing price is found out to be 5.2 Rs/MWh, and the market clearing value is found out to be 221.7 MW. So, in that 221.7 MW range, as a buyer who has bid for more than 5.2 Rs, his bid will get cleared, and whosever the generation company who is ready to sell the power at less than 5.2 Rs, his bid will get cleared. It is visible from the above curve that increment in market clearing price will improve the flexibility of the supply and lessens the demand. Comparably, decline in market clearing price will lessen the supply, furthermore, improve the demand. We can also determine the consumer profit and producer profit by using this figure as denoted above.

'Table 3' presents the optimum generator generation dispatch and social welfare benefit by utilizing ALO and ABC optimization technique. This generator generation dispatches are being utilized according to the demand requirement at the requisition of electricity grid administrators owing to the market demands. It is found that the total profit which is achieved by utilizing ALO and ABC optimization technique is more or less the same, or ALO gives more profit by a small margin. Table 4 shows the overall producers' capacity payment which is calculated comparably by ALO and ABC optimization technique. It is found that the participants have to pay the fixed revenue system more in ALO optimization technique as compared to ABC, hence, gaining more producers profit in ALO.

Figure 3 shows the convergence characteristics by utilizing the optimization technique of both the ABC and ALO comparably. It tends to be spotted that the ALO beat the ABC as far as the convergence speed is concerned along with the solution precision.

Table 4 Overall producers capacity payment

Method	MCV (MW)	MCP ($/MWh)	System marginal price ($/h)	Producers capacity payment ($/h)
ABC	221.7	5.2	1038.95	114.31
ALO	221.7	5.2	1034.62	118.22

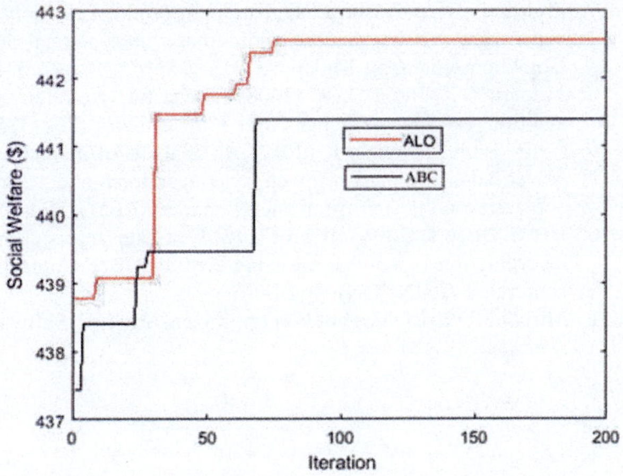

Fig. 3 Comparative convergence profile of social welfare

6 Conclusion

The aim of this work is to benefit the consumer by maximizing the social welfare. For that purpose, the artificial bee colony and ant lion optimization techniques are utilized to solve the market clearing strategy considering double auction competitive market model. This model accepts offers and bid price from the market participants to clear the market. This model displays the market clearing price under double auction market model to maximize the profits of the suppliers. The approach is tested with IEEE 30 bus system, and the results prove the effectiveness of the ALO algorithm implementation in this market environment.

References

1. Grzegorz D (2018) Probabilistic forecasting of electricity prices using kernel regression. In: 15th international conference on the european energy market (EEM), Lodz, Poland, pp 27–29
2. Garg NK, Palwalia DK (2016) Deregulation of transmission pricing:MW-mile method. IEEE 7th power india international conference (PIICON), Bikaner, India, pp 25–27
3. Raikar SB, Jagtap KM (2018) Role of deregulation in power sector and its status in India national power engineering conference (NPEC), Madurai, India, pp 9–10
4. Singh K, Sharma G, Baheti R, Singh AK (2020) Allocation of distributed energy resources in deregulated electric power system. In: 9th power india international conference (PIICON), Sonepat, India
5. Dawn S, Tiwari PK, Goswami AK, Panda R (2018) An Approach for system risk assessment and mitigation by optimal operation of wind farm & FACTS devices in centralized competitive power market. IEEE Trans Sustain Energy 10(3):1054–1065
6. Beigaitė R, Krilavičius T, Man LK (2018) Electricity price forecasting for nord pool data, IN: international conference on platform technology and service (PlatCon), Jeju, South Korea, pp 29–31

7. Prajapati AK, Srivastava SK, Narain A (2020) Electricty pricing forecasting: a bibliographical review. In: international conference on electrical and electronics engineering (ICE3), Gorakhpur, India, pp 14–15
8. Nektaria VK, Bunn D, W (2008) Forecasting electricity prices: the impact of fundamentals and time-varying coefficients Int J Forecast 24(4):764–785
9. Jie B, Tsuji T, Uchida K (2017) An analysis of market mechanism and bidding strategy for power balancing market mixed by conventional and renewable energy. In: 14th international conference on the european energy market (EEM), Dresden, Germany, pp 6–9
10. Deb S, Gope S, Goswami AK (2013) Generator rescheduling for congestion management with incorporation of wind farm using Artificial Bee Colony algorithm. In: Annual IEEE India conference (INDICON), pp 01–06
11. Mirjalili S (2015) The ant lion optimizer. Adv Eng Softw 83:80–98

Chapter 9
Anti Camcorder Piracy Display System

A. V. V. Adithya, G. Sai Kumar, and V. B. K. L. Aruna

1 Introduction

Information is an asset these days; pirating various types of information like text, audio, and video is being done in various ways over the past few decades. From the beginning era of the movie industry in 1901 by EDISON Company until now, theft of video information that is being projected is incurring remarkably huge financial loss to the industry [3]. Pirates initially used to steal the video reels and illegally market them at low prices. Of course, pirating movie reels was not a considerable issue those days because the place where reels are stored was maintained with good security. The issue became very significant after the modernization of filming process. It became even worse after the advent of a digital camera because whatever is projected on the screen can be easily recorded on a handheld camera. Techniques like authentic DVD RERIPPING were also used to extract the film, in which they used to remove read-only properties of DVD and extract video and re-burn it to the number of copies with preserved quality [3]. Of all the methods, camera piracy is the worst black hole in movie industry which drains income to the movie industry. Let us first know some properties of a digital camera and exploit them for good.

The digital camera's key component is a Charge Coupled Device (CCD), a transducer that converts light/photon energy to an electrical charge. An optical lens so adjusted to focus wide angled light over CCD is supported by vertical, horizontal sync sources for capturing frames [1]. This analog charge is encoded to a digital image by an Analog to Digital Converter, and later processed by a Digital Signal Processing circuitry for a better image. The whole operation of the digital camera is done in two intervals. They are.

(1) *Capture time*: It is the time required to capture the frame, digitalizing it and saving it in flash memory.

A. V. V. Adithya (✉) · G. S. Kumar · V. B. K. L. Aruna
Department of Electronics and Communication Engineering, Velagapudi Ramakrishna Siddhartha Engineering College, Vijayawada, India

S. Dawn et al. (eds.), *Smart and Intelligent Systems*, Algorithms for Intelligent Systems,
https://doi.org/10.1007/978-981-16-2109-3_9

(2) *Settling time*: It is the time required to flush the flash memory and clear the charge on CCD and reinitialize the sync sources.

These two intervals repeat one after the other and work in image/video mode. The digital camera makes significant delay in this settling time and may lose some light information. Thus digital cameras work on the principle of discrete frame acquisition whereas the human eye works on continuous light integration [1]. The human eye cannot observe light changes in minute intervals. This is the main flaw in a digital camera one can exploit to tackle the above-stated problem.

1.1 Traditional Methods

The very first and the oldest method proposed to prevent the recording of projected video through a camera was an IR interference technique. This method uses infrared light (invisible to human eyes) source which is focused towards the audience. The human eyes are insensitive to that wavelength whereas a camera picks up the IR which overlaps with actual video resulting in poor video quality [2]. Yet, this technique can be easily tampered with by using a simple IR filter before the camera lens. Later, with the introduction of digital cinemas, watermarking and Disturbance addition techniques were proposed and are being used. These methods were proven to be tamperable by using a straightforward frame low-pass filtering technique [2].

Some complex methods for detecting the cameras using lasers and neutralizing them temporarily were also proposed but are not in practical use because of their complexity and poor efficiency. Some basic ideas also involve creating a temporal aliasing artifact in the recorded video using the mismatch between display frequency and camera sampling frequency. The modulation should be carefully designed that the displayed movie does not contain any noticeable visual artifacts which may degrade the audience experience. However, the disadvantage of the methods of this kind is that these artifacts are very easily removed using simple low-pass filtering. The difference in image formation mechanism between human eyes and imaging sensors can be exploited for good, which is simple and rigid and proven to be effective than above-said techniques. This method was initially proposed in an IEEE 2014 conference and is named as TEMPORAL PSYCHO VISUAL MODULATION (TPVM) [3].

2 Principle and Mathematical Model

A. The RGB Color Model:

An RGB color image is an M * N * 3 array of color pixels, where each color pixel is a triplet corresponding to the red, green, and blue components of an RGB image at a specific spatial location. An RGB image may be viewed as a “stack” of three gray scale images that when fed into the green, red, and blue inputs of a color monitor, produce a color image on the screen [4].

In this model, Red, Green, and Blue are considered basic colors. A pixel can take a different secondary color which is obtained by mixing basic colors in specific proportions. It is graphically represented as a color cube. Referring to Fig. 1, a color cube is a 3-dimensional space having R, G, and B as its axes. All the points inside the cube are corresponding to one of the secondary colors [4].

By convention, the three images forming an RGB color image are referred to as the red, green, blue component images. The corresponding representation is shown in Fig. 2. The data class of the component images determines their range of values. The number of bits used to represent the pixel values of the component images determines the bit depth of an RGB image. For example, if each component is an 8-bit image, the corresponding RGB image is said to be 3 * 8 = 24 bits deep. Generally, the number of bits used to represent basic components R, G, B of a single image is the same [4]. In this case, the number of possible colors a pixel in an RGB image can take is (2

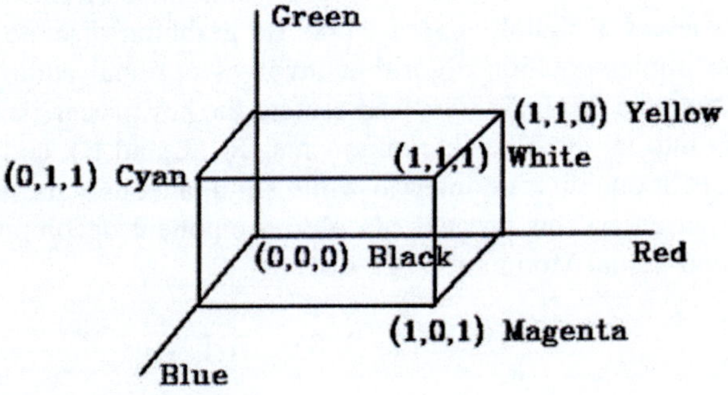

Fig. 1 Schematic of RGB color cube

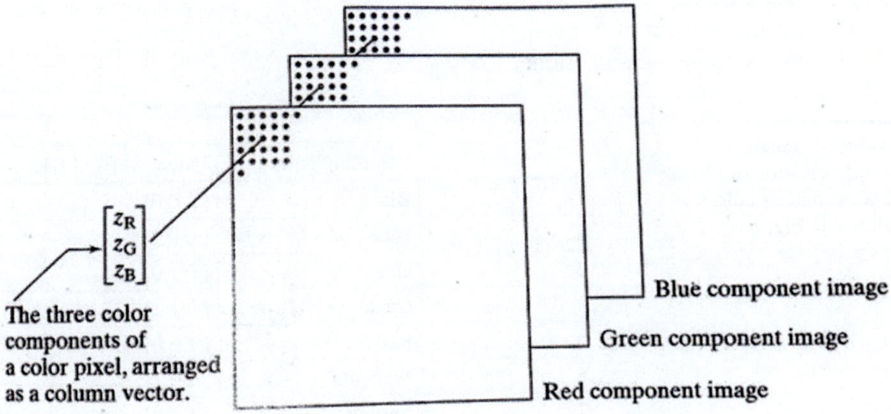

Fig. 2 Pixel of an RGB color image

^ b) ^ 3 where b is the number of bits in each component of the image. For a 24-bit image, each component is of 8-bit, thus the possible colors are 16, 777, 216.

The vertices of the cube are the primary (red, green, and blue) and secondary (cyan, magenta, and yellow) colors of light. The unique combination of the percentage of red, green, blue produces unique colors. As shown in the color cube, equal proportions of three basic colors, i.e., R = G = B = 11,111,111 produces visible white light. Table 1 shows the resultant colors at vertices of the color cube and their respective RGB proportions [4]. All the above data represents digital images whereas analog display systems using VGA (Video Graphic Array) or serial interfaces represents these three basic colors in analog voltage waveforms of specific wavelengths.

B. Implementation:

The basic idea of implementation is to exploit the difference between image acquisition methods of digital cameras and human eye. Speaking in terms of electrical parameters, the physical quantities in the world are analog in nature so as the human eye. The human eye works on the principle of "continuous light integration" over a finite time whereas a digital camera works on grabbing discrete frames at high speed [5]. The implementation algorithm involves optional addition of interface pattern in between the original video frames, and the key process is to split a single frame of video into its three basic components (R, G, and B), and projecting one color component at one discrete interval while simultaneously preserving the audience visual experience. This process of color component decomposition is called Temporal Psycho Visual Modulation (TPVM) [3].

$$O = F(R/G/B) + [D] \quad (1)$$

Where.

F = single frame.
R = red component of frame.
G = green component of frame.

Table 1 Tabular Representation of primary and secondary colors as shown in Fig. 1

S. no	Resulting color	Values of [R G B]
1	Black	[0 0 0]
2	Blue	[0 0 1]
3	Green	[0 1 0]
4	Cyan	[0 1 1]
5	Red	[1 0 0]
6	Magenta	[1 0 1]
7	Yellow	[1 1 0]
8	White	[1 1 1]

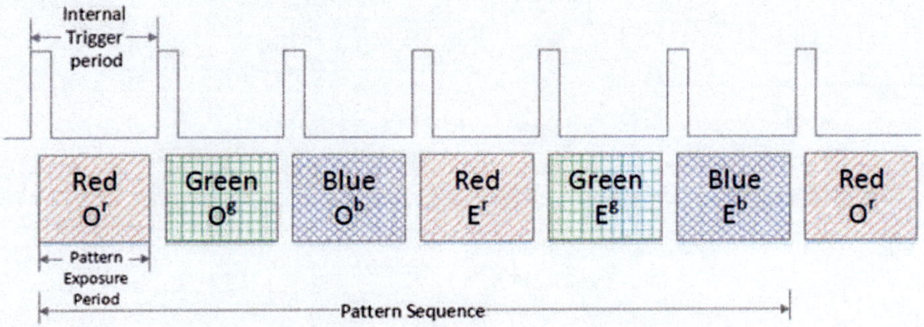

Fig. 3 Timing pattern of decomposed color components of video frames

B = blue component of frame.
D = disturbance component.
O = output video frame in one projection interval.

Thus, if we can use a high frame rate programmable light projector that can perform the above-stated task, due to splitting of color components, digital camera at a single capture instant picks up only one of the color components of frame and some disturbance pattern [3]. Since digital camera has a restoration and buffer flushing time, it misses out on at least more than one color component of a single frame in its capture interval which leads to poor image quality. Since using a Programmable Projector is costly and unreliable, a video pre-processing method is demonstrated using a simulink model. Thus, processed video can be directly projected using traditional projector without any extra cost. So, we can implement anti piracy system in a cheaper way.

Figure 3 shows how basic color components of an individual frame are decomposed and rearranged for projection. Considering the frame to be a 24-bit image, each color component is represented by 8-bit planes.

3 Simulation Using Simulink

As described above, it needs a process flow to modify video contents. So that frame timings and color component timings can be adjusted to achieve TPVM. One such model designed using vision HDL toolbox of Simulink software is shown in Fig. 4. The simulation model is built using Computer Vision Toolbox and Vision HDL Toolbox. The operation of each module and its significance is as follows:

A. Video source:

It is a module that accepts video stored in memory and can output the same for one or more processes. It accepts a black and white video and can output intensity values or

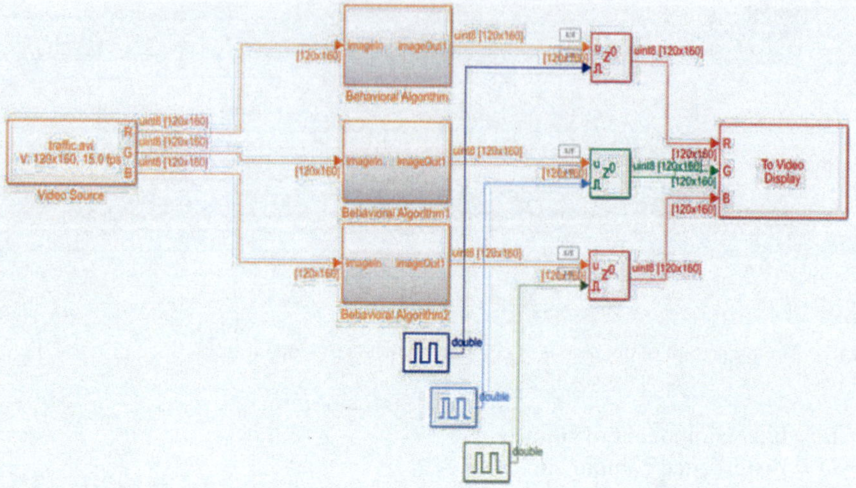

Fig. 4 Simulink process flow diagram of TPVM technique

YCbCr values to drive a display. It accepts a 3-component color video and can output a single multi-dimensional digital signal or three outputs each of which supplies a single color component. In this case, it is used to split color components R, G, B from a full video.

B. Behavioral Algorithm:

It is an HDL-optimized module that can contain any color component correction algorithm. It is an optional module in this case and is just acting as a wire here. This is reserved for future developments like adding a clocked enabler and a temporary component storage buffer.

C. Zero delay output port:

The name specifies the action that it provides no delay for the input signal. The motto behind its use is that it can be enabled using a clock signal which in turn facilitates allowing or blocking input signal with respect to time.

D. Pulse generator:

This is used to produce a clock signal of desired amplitude and time period. It also has a facility to phase shift the signal by given number of seconds. So, it is used as color component selector here.

E. Video display:

It accepts a multi-dimensional signal or each color component separately and displays the video frames as they arrive.

F. Video output:

It can accept multi-dimensional signal or individual color components separately, repackages the video frames as they arrive and stores them in memory.

G. Working:

So far, operation of each module used in the model is described. The following section describes the collective working of all the modules to achieve the stated objective.

4 Results

The main process in the working is the color selection using pulse generator and clock-controlled delay module. As the video frames are split to separate color components at source, they are buffered by behavioral modules to reach clock-controlled module. The pulse generator used here provides a clock signal whose period is inverse of frame rate. Thus, a single clock pulse manages the timing of one frame of video. Now the task is time divided display of three color components of a frame, i.e., only one color component must be displayed at a time, that too without affecting playback speed.

If the "Red" component is being displayed at the moment, the remaining G, B components must be switched off. This way, in a single frame period, three color components are to be displayed one at a time. So, clock pulse timing for each of the three zero delay port is calculated. A hundred percentage of frame period shared among three components results in each component sharing approximately 33.333% of frame period. So, pulse width of each of three clock sources is set to 33.333%. And also, these clocks are phase delayed by fr/3 s, where fr is Frame rate (Figs. 5 and 6).

Fig. 5 Original frame in video before applying TPVM

Fig. 6 Output obtained after applying TPVM

A video display accepts these modified frames and displays video. Or, the frames can be repacked using a video output module to save resulting video data as a file in memory. Now, recorded video quality is expected to be very poor.

5 Conclusion

One can notice severe distortion in the resulting video from the above process. This is because the above method is a pure analogous way of achieving the stated task. The process is tested with a 15 FPS 120 * 160 pixel video in which each of the components in a pixel is represented by 8-bits, which results in poor visual quality after processing. Present-day PC hardware and simulation capabilities are limited, so it is currently not possible to experiment on a HD video of 60 fps or more, which can yield good visual quality and poor recorded quality and i the prime requirement. This is a technical barrier found in this research work. The same task can be achieved with much better efficiency and good audience visual quality by employing digital methods using semiconductor devices. The digital modules which can efficiently process videos are still under development stage and are not HDL-optimized. For the reason which one cannot proceed with making prototype.

References

1. Naimark M (2002) How to ZAP a camera: Using lasers to temporarily neutralize camera sensors
2. Wu X, Zhai G (2013) Temporal psycho visual modulation: a new paradigm of information display [exploratory DSP]. Signal Process Mag IEEE 30(1):136–141
3. Zhai G, Wu X (2014) Defeating camcorder piracy by temporal psycho visual modulation. J Disp Technol 99:1–1

4. Gonzalez RC, Woods RE, Eddins SL (2018) Digital image processing using Matlab 2E, MC Graw Hill Education
5. Gulati RR (2016) Monochrome and color television 5E. New age international publishers

Chapter 10
Performance Evaluation of HAWT-and VAWT-Based WECS with Advanced Hill Climb Search MPPT and Fuzzy Logic Controller for Low Wind Speed Regions

Albert John Varghese, Rejo Roy, and S. R. Awasthi

1 Introduction

Global warming is one of the major issues troubling humanity. Electricity production is one of the major contributors to the emission of greenhouse gases (GHG). Due to this, a change in the way electricity is being generated is underway in which renewable energy plays a major contributing role. Coal-fired electricity generation alone contributes to 30% of the global carbon-dioxide emissions. Hence, a worldwide trend of investing more in renewable energy source can be seen. All the major economies of the world have come together in their efforts to reduce green house gas emissions by signing an international environmental treaty called United Nations Framework Convention on Climate Change (UNFCCC). The major aim of UNFCCC is to reduce the concentrations of GHG in the atmosphere and to prevent their interference with the earth's climate system.

The Ministry of New and Renewable Energy (MNRE) is the nodal ministry in India which deals with development and deployment of new as well as renewable energy so that it can supplement the energy requirements of our country. The world's largest renewable energy expansion programme of 175 GW by 2022 is undertaken by MNRE, this will be achieved by 100 GW from solar, 60 GW from wind, 5 GW from small hydro plants and 10 GW from bio fuels. In view of this, 38.124 GW of wind power capacity has been achieved by September 2020.

Wind is an indirect form of solar energy. Solar energy warms the surface of the earth and water bodies due to which uneven heating occurs, and there is flow of air from cooler area to hotter area which is called as wind. Wind energy is a major source

A. J. Varghese (✉) · R. Roy
Department of Electrical & Electronics Engineering, RNTU, Bhopal, Madhya Pradesh, India

S. R. Awasthi
RNTU, Bhopal, Madhya Pradesh, India

S. Dawn et al. (eds.), *Smart and Intelligent Systems*, Algorithms for Intelligent Systems,
https://doi.org/10.1007/978-981-16-2109-3_10

of green energy. The use of wind energy does not produce any pollution, or it does not require any water for its operation. The wind energy if utilized efficiently has the capacity to generate a large portion of the energy required worldwide.

India has seen a tremendous increase in the development of wind energy. This is made possible because of the following developments:

- Advancements and reduction of cost in Power Electronic Converters
- Development of Variable speed Generators combined with various advanced MPPT algorithms
- Development of high strength and low weight materials
- Improvement in the plant capacity factor due to technical advancements.

Some of the points worth mentioning based on the literature survey are summarized as follows:

The main objective of a WECS is to produce more energy, safe operation and connection to grid [1]. To achieve this, wind turbine dimensions should be increased, and the controllers should respond quickly [1]. The performance comparison of HAWT and VAWT was carried out, and it was found that the parameters that affect wind power output are wind speed, Swept area, Power Coefficient and air density [2]. HAWT tend to produce better efficiency than VAWT [2]. VAWT are used in a variety of applications nowadays, to extract power from lower wind speed, like in Mobile towers [3], multi bladed VAWT are also used to increase efficiency of the WECS [4], etc.

The WECS that work at lower wind speeds can improve power production by using the following techniques:

(a) Flow Augmentation devices like wind boosters are used for VAWT, and shroud and nozzle combined for HAWT [5]
(b) Use of capping and vents in order to reduce thrust and loading effects on VAWT [3]
(c) Super capacitors are used for utilization of wind energy at lower wind speeds [5]
(d) To optimize the wind turbine operation, we use MPPT technique by tracking power, and vary the duty cycle of chopper [5–7]
(e) Parasitic losses account for almost 50% of power produced at low speed [5]
(f) Optimizing blade geometry proves to be effective [5]
(g) Rotors with lesser inertia, lesser weight blades [5].

Generators are widely used for lower wind speed that can work with variable speed PMSG [1, 4, 6–13]. The variable speed generation is widely preferred as energy can be generated over wide wind speed range, and it also removes the use of a gearbox, reducing system complexity and losses [1, 13]. Energy storage systems are used to store excess energy which are produced [6].

MPPT techniques normally used are Perturb and Observe Method [6–8, 11], Optimum Relation Based [6] and Tip Speed Ratio [6].

(a) Supply Maximum power for every variation in wind
(b) Eliminates oscillation problem (current ripples) due to power fluctuation [6, 8, 11, 13]
(c) Increases reliability of the system [6]
(d) Responds to sudden changes in wind speed [9, 11].

Conventionally, a PI controller is used in WECS [14]. For Implementing Intelligent Control, Fuzzy Logic Controller [1] combined with MPPT is used, which supplies maximum power for every condition as well as gives a constant voltage and constant frequency supply [10, 15]. Fuzzy Logic controller combined with ant colony optimization and Perturb and observe-based MPPT is used to generate triggering pulses for the switches of converter [12]. Some advantages of using fuzzy Logic are quick response, insensitivity can be limited and it gives a universal control algorithm, and also, increases efficiency and reliability [15]. ANFIS-based controller was used for Inverter [13].

A standalone low cost wind-based electricity generator can help people in rural and remote areas, and also, improves their living conditions [11]. Almost all Wind Energy Conversion system (WECS) has the parts as listed below [16]. The block diagram for a basic WECS is shown in Fig. 1.

- Wind Turbine—This is used for extracting kinetic energy from wind and converting into mechanical energy; this can be either Horizontal Axis Wind Turbine (HAWT) or Vertical Axis Wind Turbine (VAWT).
- Generator—The turbine is connected to the shaft of the generator, which helps in transfer of energy. This is responsible for converting the mechanical energy into Electrical energy. Generator can be of many types, but for variable speed

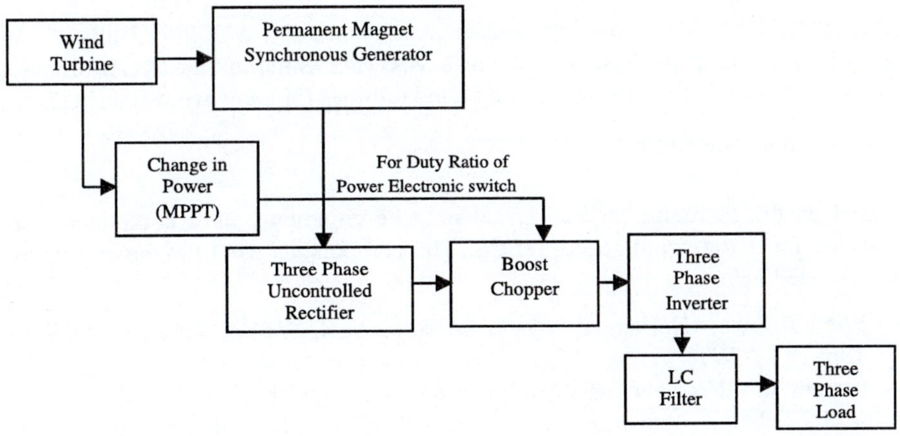

Fig. 1 Basic wind energy conversion system (WECS)

power generation, Doubly Fed Induction Generator (DFIG) or Permanent Magnet Synchronous Generators (PMSG) are normally preferred.
- MPPT Scheme—This is responsible for helping a WECS to decide how to operate so that maximum power can always be extracted from the incoming wind as well all devices are used efficiently and in an optimal manner. Tip Speed Ratio is normally preferred in conventional WECS; other MPPT algorithms are also available.
- Controller—The decisions made by the MPPT are carried out by the Controller. In conventional WECS, Proportional Integral (PI) controller is normally used.
- Power Electronic Converters—This is used so that the power delivered to load will be in a manner that matches its requirements. Load should be supplied with a supply having constant voltage and constant frequency; in case of AC, this is achieved using this converter. Normally, a variable speed WECS will have a Rectifier–Chopper–Inverter in between load and the generator.
- Filter—This is responsible for removing all the ripples and harmonics that can occur in the generated supply.
- Load—The output side of WECS can either have a battery, grid-connected supply or directly connected to an electrical load.

Nowadays, Intelligent controllers are also available for a WECS. One of them is a Fuzzy Logic Controller; this helps the WECS to take decision like a human brain based on a set of predefined rules. Fuzzy logic solves the problem considering all known information and makes decision based on input in an optimal manner. Using Fuzzy Logic, the uncertainty and inaccuracy in any situation can be handled reasonably.

2 Methodology

This study is carried out to determine the performance of a WECS for Bhilai, Chattisgarh [16]. The paper discusses about a MATLAB/Simulink-based performance assessment for a WECS using different wind turbines for a low wind speed regime.

(1) Components Selection:

Based on the literature survey carried out, the commonly used components are selected for a low wind speed WECS. The components used for the simulation are as listed below.

- Wind Turbine—Horizontal Axis Wind Turbine (HAWT) or Vertical Axis Wind Turbine (VAWT)
- Generator—Permanent Magnet Synchronous Generators (PMSG)
- MPPT Scheme—Advanced Hill Climb Search-based Duty Ratio Control
- Controller—Fuzzy Logic Controller

- Power Electronic Converters—Uncontrolled Rectifier–Boost Chopper–Three Phase Inverter
- Filter—LCL Filter
- Load—Three Phase electrical load.

(2) MPPT Scheme:

For extracting Maximum power, Advanced Hill Climb Search MPPT scheme is used. Here, a lookup table is used to decide the duty ratio of the switch used in the DC-DC Boost converter, so that maximum power is extracted. The algorithm for the MPPT scheme is shown in Fig. 2.

The use of Advanced Hill Climb Search simplifies the WECS as it uses electrical measurements and does not require any mechanical measurement devices for monitoring the wind speed.

(3) Fuzzy Logic Controller:

Fuzzy logic cannot be used as a high level artificial intelligence, but can be used for low level machine control operation. The Architecture of a Fuzzy Logic Controller is shown in Fig. 3. Fuzzifier converts the inputs into fuzzy set, the rule base has the if–then rules saved, intelligence simulates human reasoning and with help of rule base makes decisions, and defuzzifier transforms fuzzy sets into outputs.

The steps to implement a Fuzzy Logic Controller are given below.

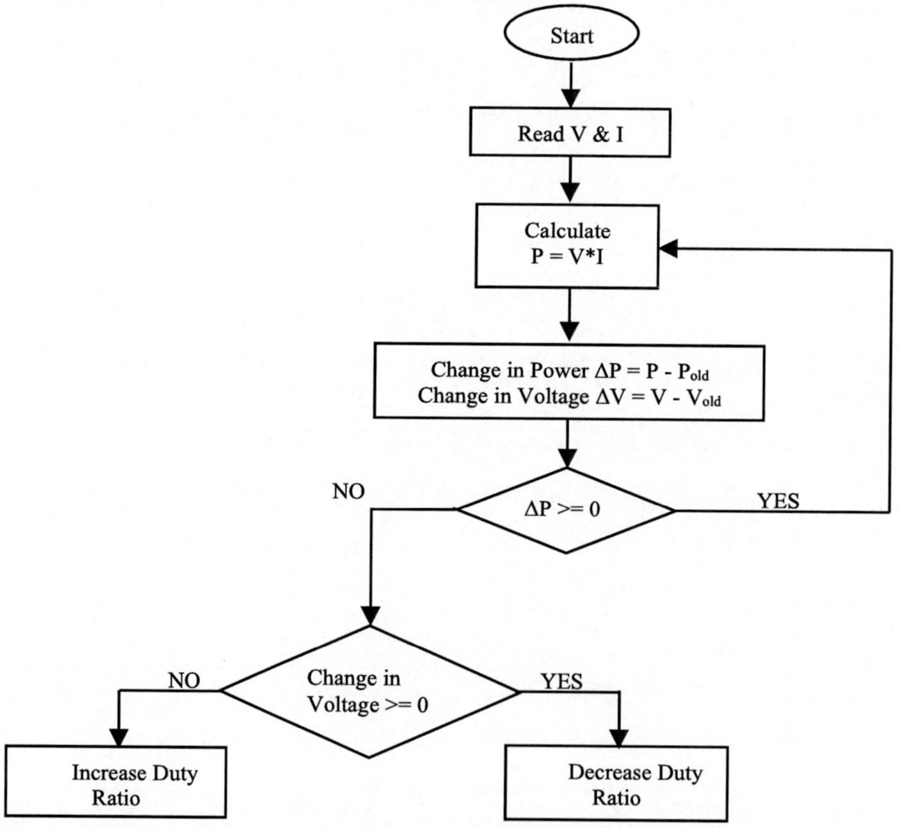

Fig. 2 Flowchart for advanced Hill Climb search MPPT

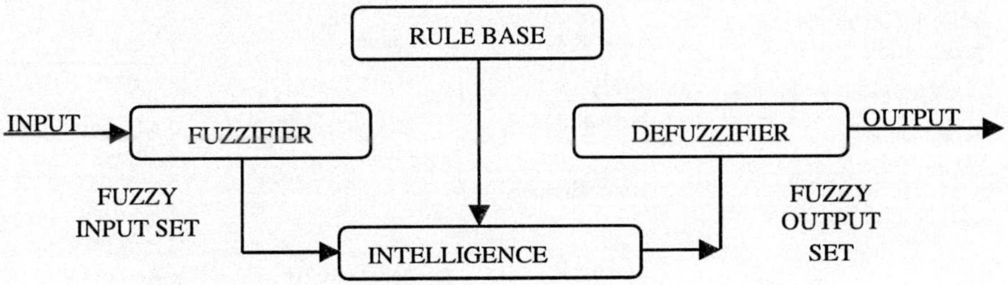

Fig. 3 Architecture of fuzzy logic controller

a. Define the Input Variables based on which decision is to be made
b. Construct a Membership function for them
c. Define a set of rules for the various possible inputs using a logical table
d. Convert the input data into fuzzy data sets
e. Compare the fuzzy data sets with all the values in the rule base
f. Decision has to be made with the help of rule base and inference engine/intelligence
g. Convert the obtained fuzzy data into output

3 Application of Developed Methodology

The study is carried out using MATLAB/Simulink. Two simulation models are used so as to propose a better solution for areas having lower annual wind speeds. Majority of the land area in India fall under this category.

Table 1 Technical specifications

Parameter	Value	Value
Type	HAWT	VAWT
No. of blades	3 (Fibre-Reinforced Plastic)	3 (Aluminium Alloy)
Rotor diameter	4.0 m	3.0 m
		Height–3.6 m
Make	AVATAR-III	Aeolos–V
Cut in wind speed	1.7–2.3 m/s	1.5 m/s
Rated wind speed	11 m/s	10 m/s
Type	Off grid	Off grid
Capacity	3000 Watts	3000 Watts
Generator type	PMSG	PMSG
Rated voltage	230 V	230 V

A 3-bladed HAWT as well as 3-bladed VAWT is considered in order to determine which WECS system has the capacity to extract more power in areas having lower wind speeds. A 3 kW PMSG generator is used so that the system can be a variable speed system, and the use of gearbox is avoided, which will reduce complexity, losses in gearbox and cost. Table 1 lists the technical specifications considered for simulation of turbine and generator.

The power electronic converters used are as follows, Fig. 4 shows their interconnection.

a. A three phase uncontrolled rectifier is connected to the output of the PMSG so that the three phase AC can be converted to a DC output.

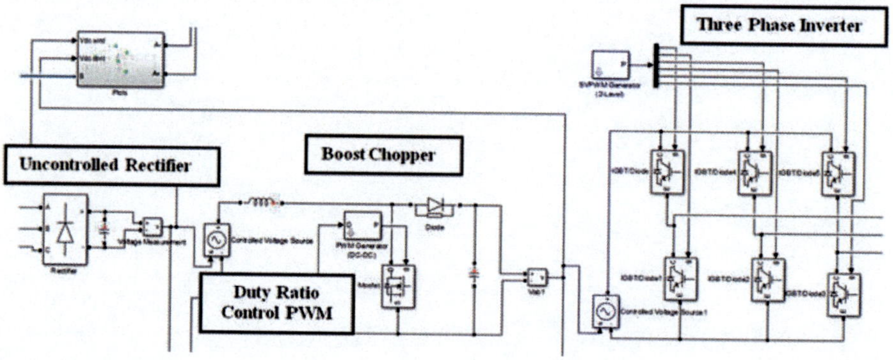

Fig. 4 Power electronic converter (Uncontrolled Rectifier–Boost Chopper–Three Phase Inverter)

b. The DC output is fed to an IGBT-based boost chopper, the duty ratio of which is modified by using the MPPT scheme; for the purpose of switching here, pulse width modulation (PWM) is used.

A three phase inverter is used after the boost chopper which gives an AC output at constant voltage and constant frequency. So that quality power can be fed to the three phase load connected. Here, for the switching of the thyristors (IGBT), space vector pulse width modulation (SVPWM) technique is used.

The controller used is a Fuzzy Logic Controller (FLC) which is connected with the Advanced Hill Climb Search MPPT and is used for changing the duty cycle of the boost converter. Using Fuzzy Logic Controller, it helps in implementing an intelligent control scheme so that the system can take decisions on its own with regard to extracting maximum power always. The implementation of the FLC-based MPPT scheme is shown in Fig. 5.

The membership functions used for defining the logic used for Fuzzy Logic Controller are shown in Figs. 6a–c. Here, change in voltage and change in power are the inputs based on which the decision regarding duty ratio for the boost converter is taken. The condition used is 'If' Change in Power 'and' Change in Voltage 'Then' Duty Ratio.

The output after the power electronic converters is fed to the LC filter so that any disturbances can be smoothed out, and quality power can be finally fed to the

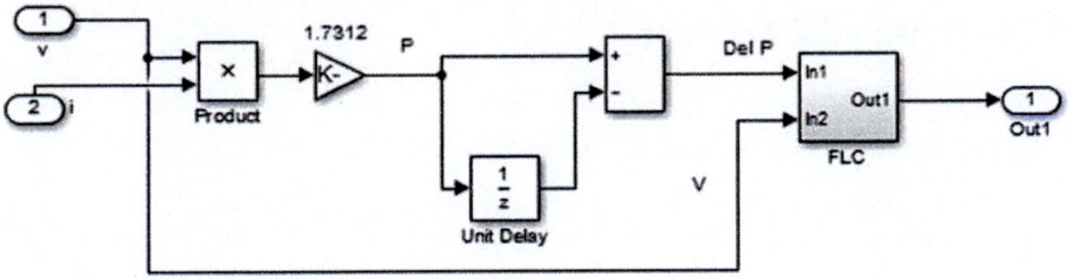

Fig. 5 FLC-based MPPT

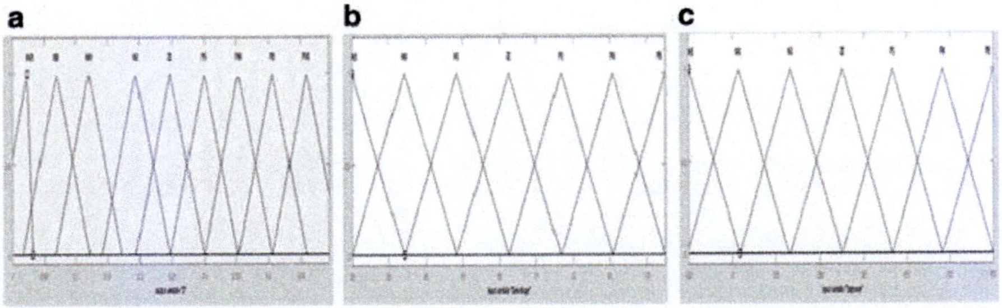

Fig. 6 **a** Membership function for duty ratio. **b.** Membership function for change in voltage. **c** Membership function for change in power

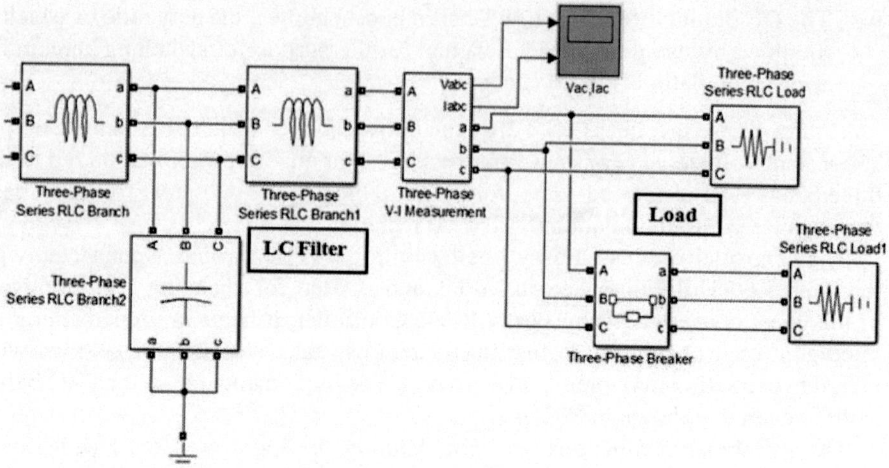

Fig. 7 LC filter and three phase load

load. In this system, as we have considered a standalone system, a three phase load is connected, the connections are shown in Fig. 7.

4 Results and Discussion

Bhilai has a wind speed in the range of 2–6 m/s based on past meteorological data; hence, input wind speed range is selected to be 2–6 m/s. These are selected keeping in mind the areas with lower annual wind speed and to determine which turbine will be feasible with regard to overall power generation and better settling time. The starting speed for the HAWT considered is around 1.7 m/s and the VAWT is around 1.5 m/s. The input speeds are considered the same for simulation of both the WECS models having FLC-based MPPT control scheme. The components and converters used are also similar to provide better comparisons. Figure 8 shows the input wind speed with respect to time.

The various outputs got during the simulation of both the models in MATLAB/Simulink are shown below. Figure 9a shows the generator output voltage V_{ab} and current I_{line} values for a VAWT-based system, and Fig. 9b shows generator output voltage V_{ab} and current I_{line} value for HAWT-based system. Figure 10a shows the generated power for a VAWT-based system while Fig. 10b shows generated power for a HAWT-based system. The output power is almost similar, but at lower wind speeds, VAWT generates more power, and also, when settling time is considered, the VAWT-based system settles much faster.

The outputs at the intermediate stages are shown. Figure 11a shows the DC link voltage output after the boost converter stage for a VAWT-based system, and Fig. 11b shows the DC link voltage output after the boost converter stage for a HAWT-based system. Figure 12a shows the three Phase voltage and current values fed to the load after inverter and LC filter for a VAWT-based system, and Fig. 12b shows the three Phase voltage and current values fed to the load after inverter and LCL filter for a HAWT-based system. Here also, all the values are almost similar with an exception of settling time, the VAWT-based system settles faster comparatively.

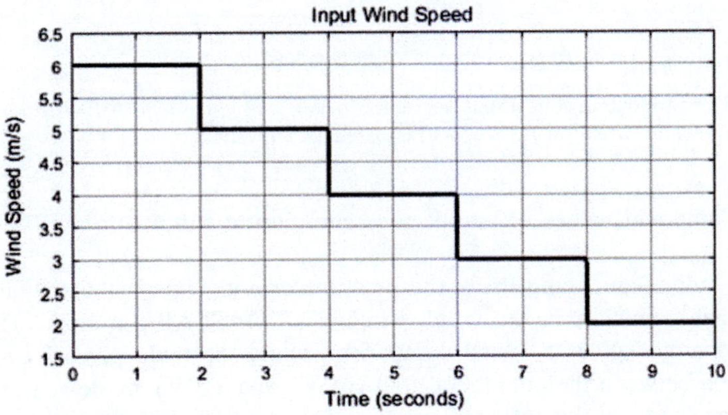

Fig. 8 Input wind speed versus time

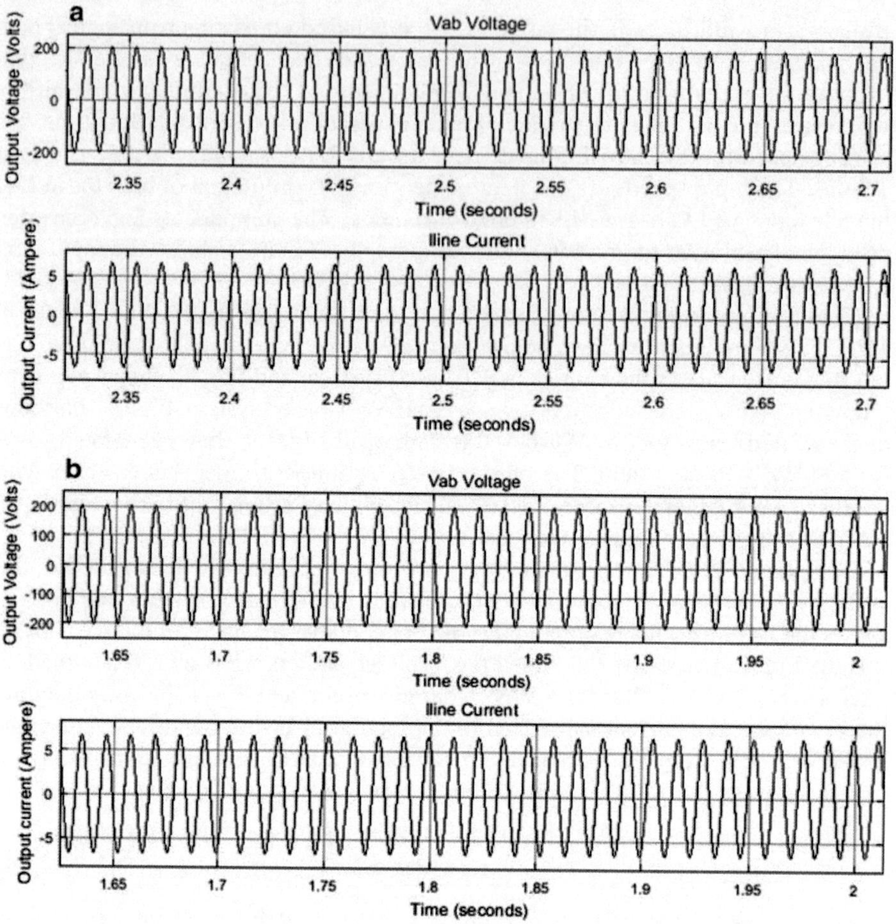

Fig. 9 **a** Generator output voltage and current for VAWT + FLC-based MPPT WECS. **b** Generator output voltage and current for HAWT + FLC-based MPPT WECS

The Numerical values of output generated during simulation are illustrated in Table 2.

These values are found using the peak finder and Signal statistics and cursor measurement function in the graph window of MATLAB/Simulink. The output graphs generated gives a visual comparison while the table gives a quantitative comparison between the outputs of both HAWT and VAWT models. These values and graphs state that the performance of a VAWT-based WECS is better for lower wind speeds.

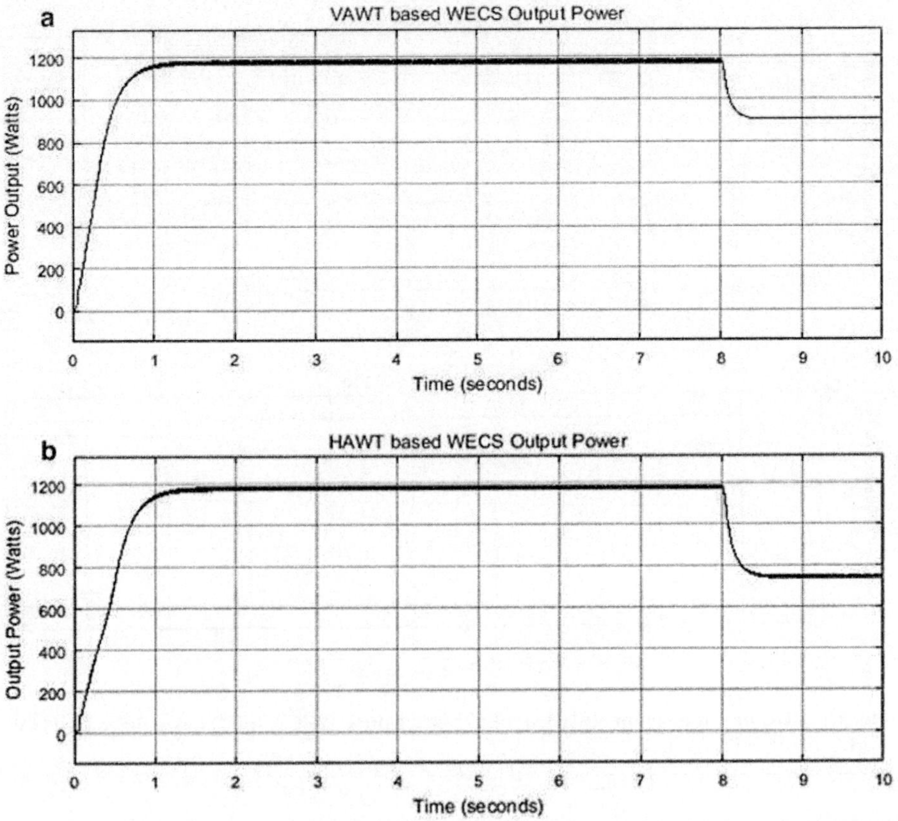

Fig. 10 **a** Generator output power for VAWT + FLC-based MPPT WECS. **b** Generator output power for HAWT + FLC-based MPPT WECS

5 Conclusion

In this paper, HAWT- and VAWT-based WECS have been compared for performance in areas where the turbines cannot reach rated speed for most of the time. Fuzzy Logic Controller combined with Advanced Hill Climb Search MPPT scheme is used, and it helps in controlling the DC link voltage of the Boost Chopper. The Fuzzy rules are based on monitoring the Change in Voltage and Change in Power values. The performance of both the VAWT-based WECS and HAWT-based WECS is done using the same scheme of Fuzzy Logic controller and Advanced HCS-based MPPT. The capacity Utilization factor of VAWT for the simulation is 36.2%, while for HAWT is 34.8% for wind speeds of 2–6 m/s. The simulation results clarify that the VAWT-based system can perform better in regions having lower wind speeds. If the VAWT-based system proves to be feasible, it can help in energy production using wind energy, and

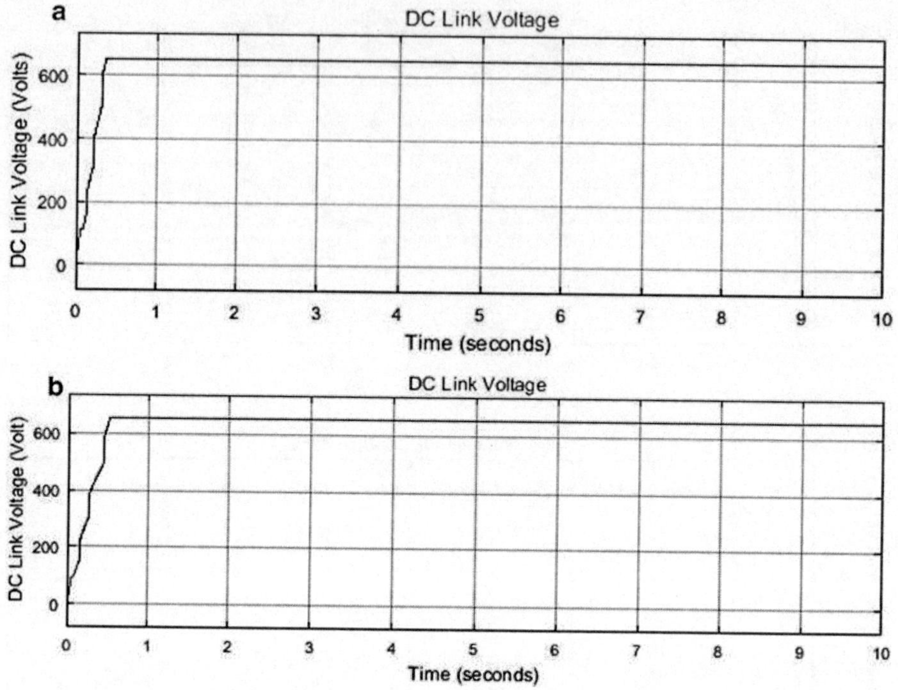

Fig. 11 **a** DC link voltage for VAWT + FLC-based MPPT WECS. **b** DC link voltage for HAWT + FLC-based MPPT WECS

will have tremendous impact like the solar pump scheme currently implemented by the government of India.

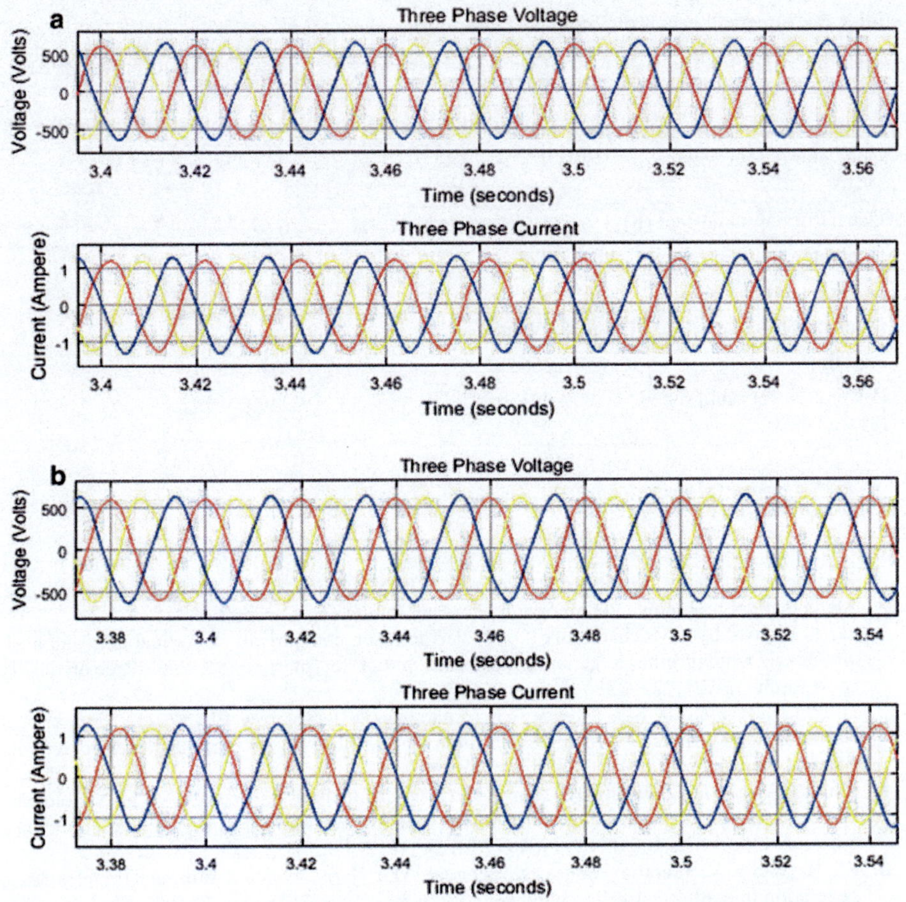

Fig. 12 **a** Three phase voltage and current given to the load for VAWT + FLC-Based MPPT WECS. **b** Three phase voltage and current given to the load for HAWT + FLC-Based MPPT WECS

Table 2 Numerical values of output generated during simulation

Stages of WECS	VAWT + FLC based MPPT WECS	HAWT + FLC based MPPT WECS
	Maximum value	Maximum value
Generator output voltage (V_{ab})	203.4 V @ 2.542 s	203.4 V @ 1.987 s
Generator output current (I_l)	6.673 @ 2.401 s	6.672 @ 1.987 s
Generator power output	1187 W @ 3.768 s	1187 W @ 5.155 s
DC link voltage	654.3 V @ 7.905 s	655.7 V @ 7.848 s
Mean power value	1087 Watts	1046 Watts
Rise time	542.266 ms	676.088 ms
Power at lower wind speed (around 3m/s)	902.6 Watts	749.6 Watts

References

1. Soriano LA, Yu W, de Jesus Rubio J (2013) Modelling and control of wind turbine. Mathematical problems in engineering. Hindawi Publishing Corporation
2. Jazuli F, Soedibyo, Mochamad A (2017) Performance comparison of vertical axis and horizontal axis wind turbines to get optimum power output. In: International conference on quality in research (IEEE), 429–233
3. Abraham JP, Plourde BD, Mowry GS, Minkowycz WJ, Sparrow EM (2012) Summary of Savonius wind turbine development and future applications for small scale power generation. J Renew Sustain Energy 4:1–21
4. Yanto HA, Lin C-T, Hwang J-C, Lin S-C (2009) Modelling and control of household-size vertical axis wind turbine and electric power generation system. In: International conference on power electronics and drive systems (IEEE), pp 1301–1307
5. Sri Ragunath V, Jitendra Pandey K, Mondal AK, Karn A, Wind turbines for electricity generation operating in the low wind velocity regime. SSRN Electron J 1–23
6. Kumari S, Kushwaha V, Gupta TN (2018) A maximum power point tracking for a PMSG based variable speed wind energy conversion system. In: International conference on power energy, environment and intelligent control (IEEE)
7. Sahin P, Resmi R, Vanitha V (2016) PMSG based standalone wind electric conversion system with MPPT. In: International conference on emerging technological trends
8. Ramadoni S, Indah S (2019) Performance improvement for small-scale wind turbine system based on maximum power point tracking control. Energies 12:2–18
9. Sl-Subhi A, Alsumiri M, Alalwani S (2017) Novel MPPT algorithm for low cost wind energy conversion systems. In: International conference on advanced control circuit systems (IEEE)
10. Pamuji FA, Miyauchi H (2016) A new control design of Maximum Power Point Tracking by Fuzzy Logic Controller for wind Turbine connected to low voltage Grid. Int Sem Intell Technol Appl (IEEE)
11. Gómez M, Ribeiro E, Estima J, Boccaletti C, Antonio Marques Cardoso J (2016) Development of an effective MPPT method suitable to a Stand-Alone, low-cost wind turbine system. In: International conference on emerging technological trends
12. Santhoshini P, Maniraj P (2016) SEPIC converter based dynamic power tracking from wind energy conversion system. J Chem Pharm Sci
13. Luz YO, Güney E, ÇalJk H (2013) Power quality control and design of power converter for variable-speed wind energy conversion system with permanent-magnet synchronous generator. Sci World J (Hindawi Publishing Corporation)

14. Hima Bindu G, Nagaraju Mandadi P (2014) Design and modelling of induction generator wind power systems by using MATLAB/SIMULINK. Int J Adv Res Electr Electron Inst Eng 3:11472–11478
15. Srivastava BB, Sudhanshu Tripathi E (2014) Tracking of maximum power from wind using fuzzy logic controller based on PMSG, Int J Mod Eng Res
16. Varghese AJ, Roy R, Awasthi SR (2020) Emerging trend in horizontal axis wind turbines to exploit low wind resource, Anusandhan—RNTU Journal X

Chapter 11
A Novel Asymmetric Multilevel Inverter with Low THD

M. Revathi, K. Aravinda Shilpa, and K. Rama Sudha

1 Introduction

An inverter converts DC voltage to an AC voltage. DC to AC power conversion is a key technology in the modern setup of production, transmission, giving out and use of electric power. To obtain an eminence output voltage or a current waveform with less amount of ripple content, they need high switching frequency along with various PWM strategies [3].

A multilevel inverter is a power electronic device which is able to provide preferred alternating voltage level at the output using several lower level DC voltages as input. Multilevel inverters generate output near to sinusoidal voltages, as similar to a stepped voltage waveform using many direct current (DC) voltage sources. Fundamental switching scheme can be used to power semiconductor switches in the inverter to create a nearly sinusoidal output.

So as to improve the proficiency and to change over low voltage DC source into usable AC source. Thus, multilevel inverter acts as a link between renewable power generation on one side and high-power load on the other side.

The three major types of multilevel inverters applied in industrial applications are flying capacitor, diode clamped and H-bridge multilevel inverter [1, 2]. Cascaded H-bridge multilevel inverter is simpler than other inverter topologies because of reduced number of switches, low harmonic distortion for higher switching frequencies, circuit complexity is reduced, low cost and size [4]. The proposed topology uses H-connect and produces 15-level yield voltage with a smaller number power components and fewer gate drive circuits, notwithstanding less circuit design multifaceted nature. As the quantity of yield levels build to diminish harmonics content.

M. Revathi (✉) · K. A. Shilpa · K. R. Sudha
Department of Electrical Engineering, Andhra University College of Engineering (A), Visakhapatnam, Andhra Pradesh, India

S. Dawn et al. (eds.), *Smart and Intelligent Systems*, Algorithms for Intelligent Systems,
https://doi.org/10.1007/978-981-16-2109-3_11

2 Proposed Topology

The configuration of anticipated multilevel inverter topology for 15-level (15L) operation is depicted in Fig. 1. It comprises of three voltage sources and ten switches. The voltage magnitudes are indicated from V1 to V3.

The unidirectional switches (IGBT) with anti-parallel diodes are used. The switching patterns of the anticipated topology are provided in Table 1.

The anticipated topology is the modification of symmetric structure improved H-connect (SSEHB) inverter which uses bidirectional switches and symmetric number of voltage sources on either side.

The pre-eminent drawback of SSEHB is that it does not generate complementary voltage levels, and the levels are lapsed. The intended configuration eliminate the drawback of SSEHB and enhance the number of voltage levels in output compared to SSEHB.

2.1 Operation

The operation of the proposed topology is explained in fifteen modes as shown in the figures below. Each mode explains about the circulation of current through the different switches and voltage sources in order to obtain different voltage levels. By

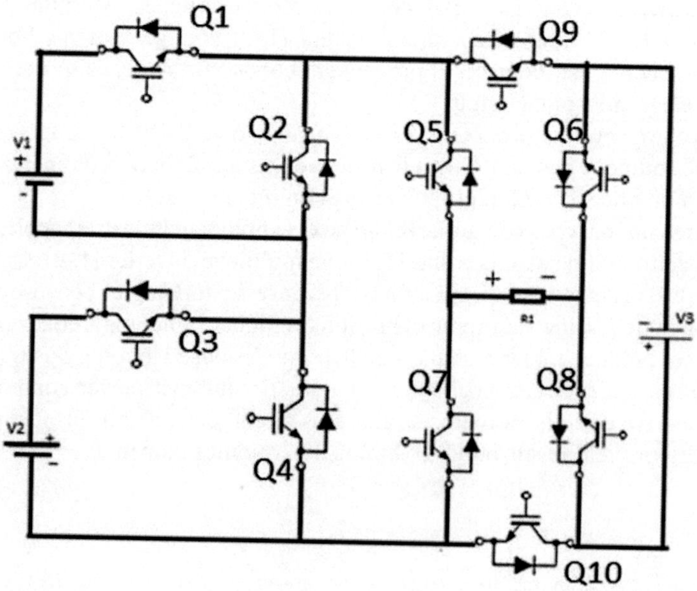

Fig. 1 Circuit diagram of 15-level inverter

Table 1 Switching sequence of 15-level inverter

S1	S2	S3	S4	S5	S6	S7	S8	S9	S10	Vo (V)
1	0	1	0	1	1	0	0	0	1	7
0	1	1	0	1	1	0	0	0	1	6
1	0	0	1	1	1	0	0	0	1	5
0	1	0	1	1	1	0	0	0	1	4
1	0	1	0	1	0	0	1	0	1	3
0	1	1	0	1	0	0	1	0	1	2
1	0	0	1	1	0	0	1	0	1	1
0	0	0	0	0	0	1	1	0	1	0
1	0	0	1	0	1	1	0	1	0	−1
0	1	1	0	0	1	1	0	1	0	−2
1	0	1	0	0	1	1	0	1	0	−3
0	1	0	1	0	0	1	1	1	0	−4
1	0	0	1	0	0	1	1	1	0	−5
0	1	1	0	0	0	1	1	1	0	−6
1	0	1	0	0	0	1	1	1	0	−7

changing the magnitude of voltage sources, different voltage levels can be obtained. For the proposed topology, the voltage magnitudes chosen are V1 = 1 v, V2 = 2 v, V3 = 4 v.

2.1.1 Mode 1: (V0 = +1 V)

In this mode, the switches Q1, Q5, Q8, Q10, Q4 are turned ON, and the remaining switches are turned OFF. These switches enable voltage source V1, and the obtained load voltage is given by V0 = + V1 = +1 V. Figure 2 shows the current paths that are active in this mode.

2.1.2 Mode 2: (V0 = +2 V)

In this mode, the switches Q3, Q2, Q5, Q8, Q10 are turned ON, and the remaining switches are turned OFF. These

switches enable voltage source V2, and the resulting load voltage is given by V0 = + V2 = +2 V. Figure 3 shows the current paths that are active in this mode.

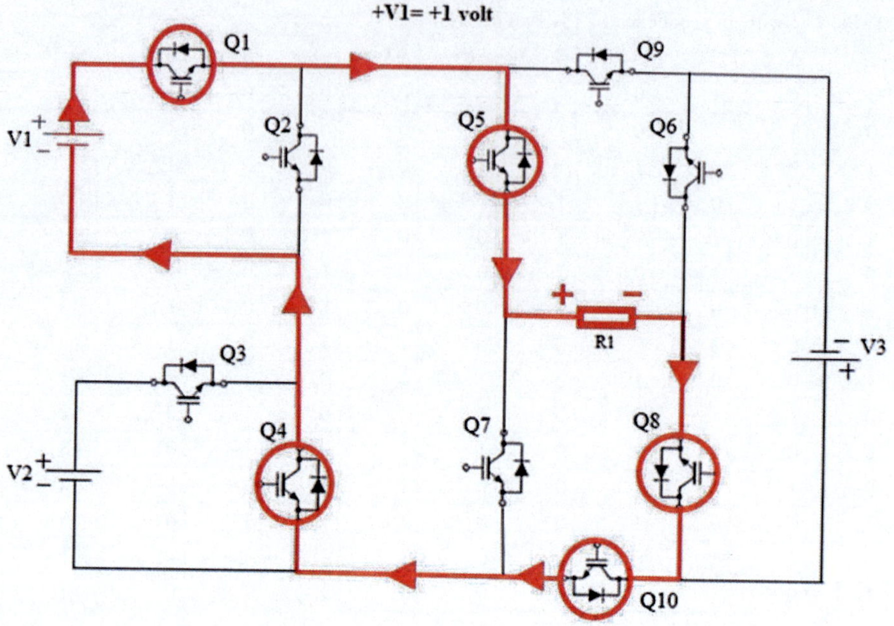

Fig. 2 Switching arrangement for +Vdc

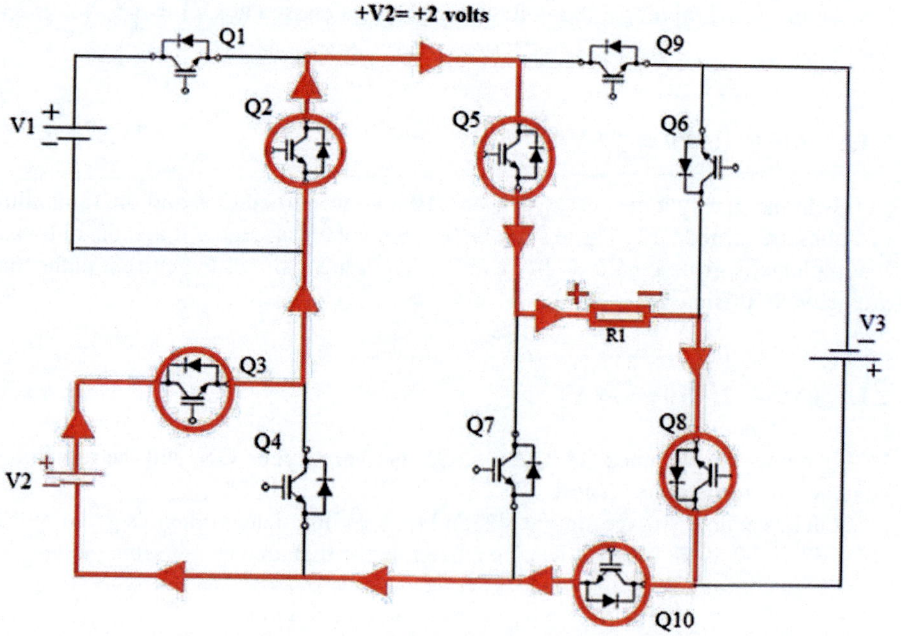

Fig. 3 Switching arrangement for +2Vdc

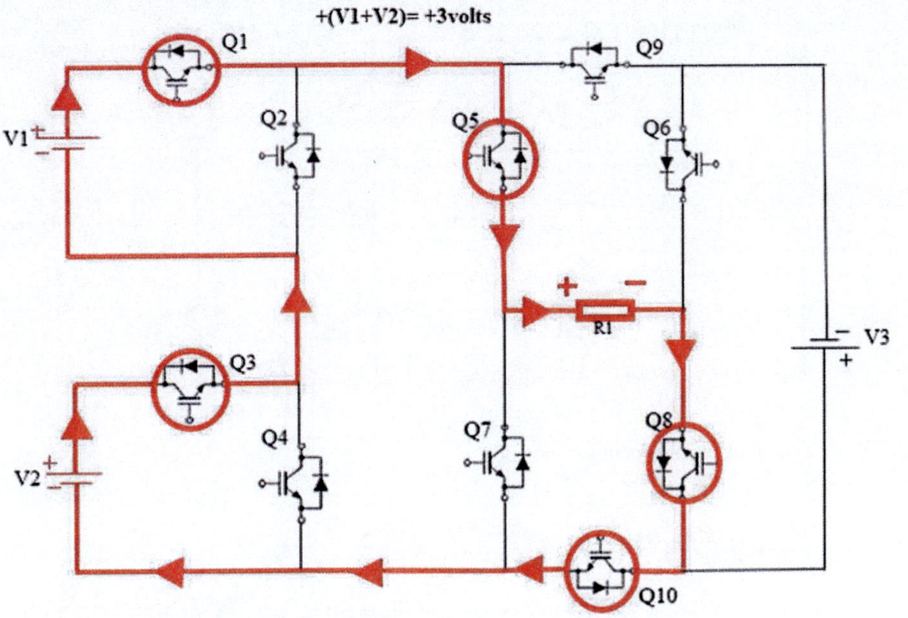

Fig. 4 Switching arrangement for +3Vdc

2.1.3 Mode 3: (V0 = +3 V)

In this mode, the switches Q1, Q5, Q8, Q10, Q3 are turned ON, and the remaining are turned OFF. These switches enable voltage sources V1, V2, and the resulting load voltage is given by V0 = +(V1 + V2) = +3 V. Figure 4 shows the current paths that are active in this mode.

2.1.4 Mode 4: (V0 = +4 V)

In this mode, the switches Q10, Q4, Q2, Q5, Q6 are turned ON, and the remaining switches are turned OFF. These switches enable voltage source V3, and the resulting load voltage is given by V0 = +V3 = +4 V. Figure 5 shows the current paths that are active in this mode.

2.1.5 Mode 5: (V0 = +5 V)

In this mode, the switches Q1, Q5, Q6, Q10, Q4 are turned ON, and the remaining switches are turned OFF. These switches enable voltage sources V1, V3, and the resulting load voltage is given by V0 = +(V1 + V3) = +5 V. Figure 6 shows the current paths that are active in this mode.

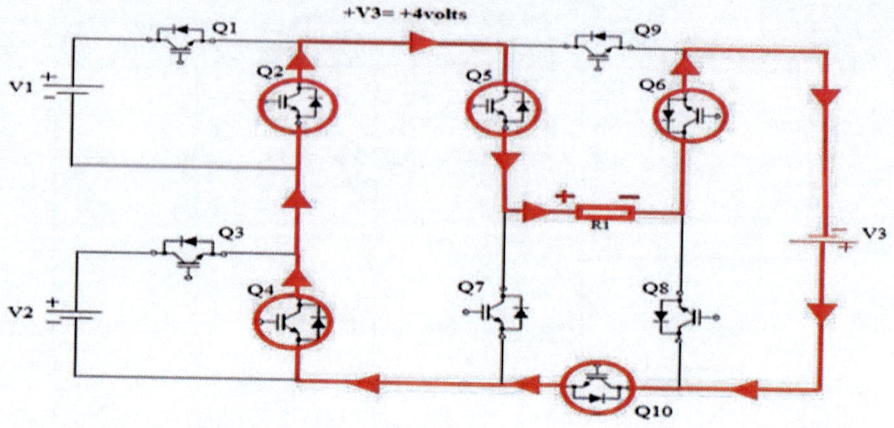

Fig. 5 Switching arrangement for +4Vdc

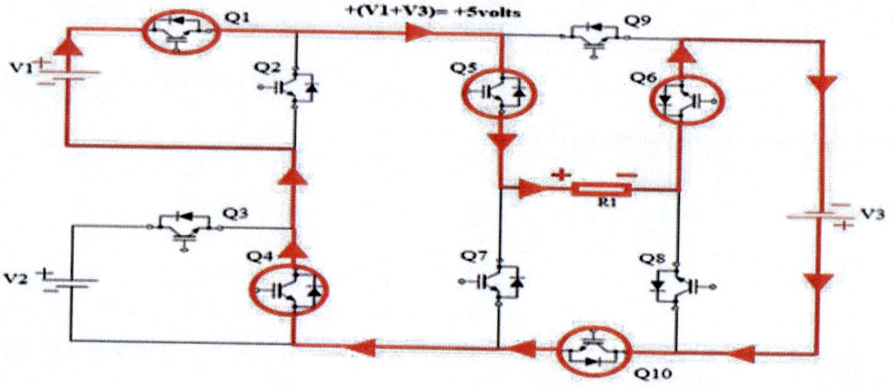

Fig. 6 Switching arrangement for +5Vdc

2.1.6 Mode 6: (V0 = +6 V)

In this mode, the switches Q3, Q2, Q5, Q6, Q10 are turned ON, and the remaining switches are OFF. These switches enable voltage source V2, and the resulting load voltage is given by V0 = +(V2 + V3) = +6 V. Figure 7 shows the current paths that are active in this mode.

2.1.7 Mode 7: (V0 = +7 V)

In this mode, the switches Q1, Q5, Q6, Q10, Q3 are turned ON and the remaining are turned OFF. These switches enable voltage sources V1, V2, V3, and the load voltage is given by V0 = +(V1 + V2 + V3) = +7 V. Figure 8 shows the current

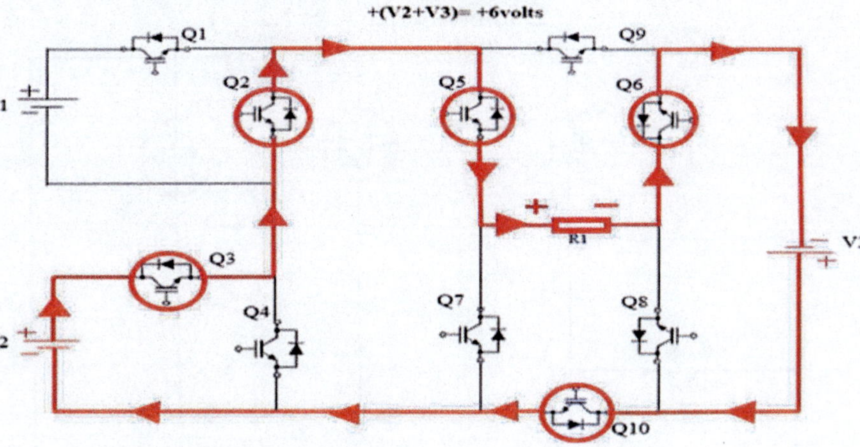

Fig. 7 Switching arrangement for +6Vdc

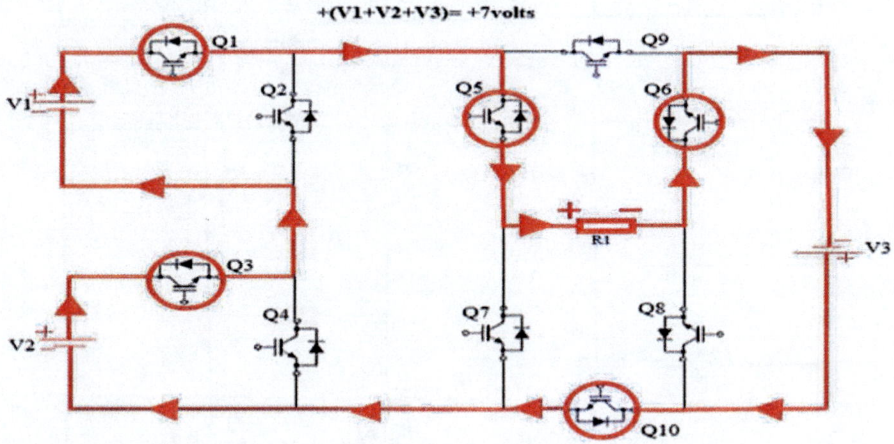

Fig. 8 Switching arrangement for +7Vdc

paths that are active in this mode.

2.1.8 Mode 8: (V0 = 0 V)

In this mode, either Q5, Q6, Q9 are ON or Q7, Q8, Q10 are on with the remaining switches at OFF state. The load terminals are short circuited, and the resulting load voltage is zero, i.e. V0 = 0 V. Figures 9a, b show the current paths that are active in this mode.

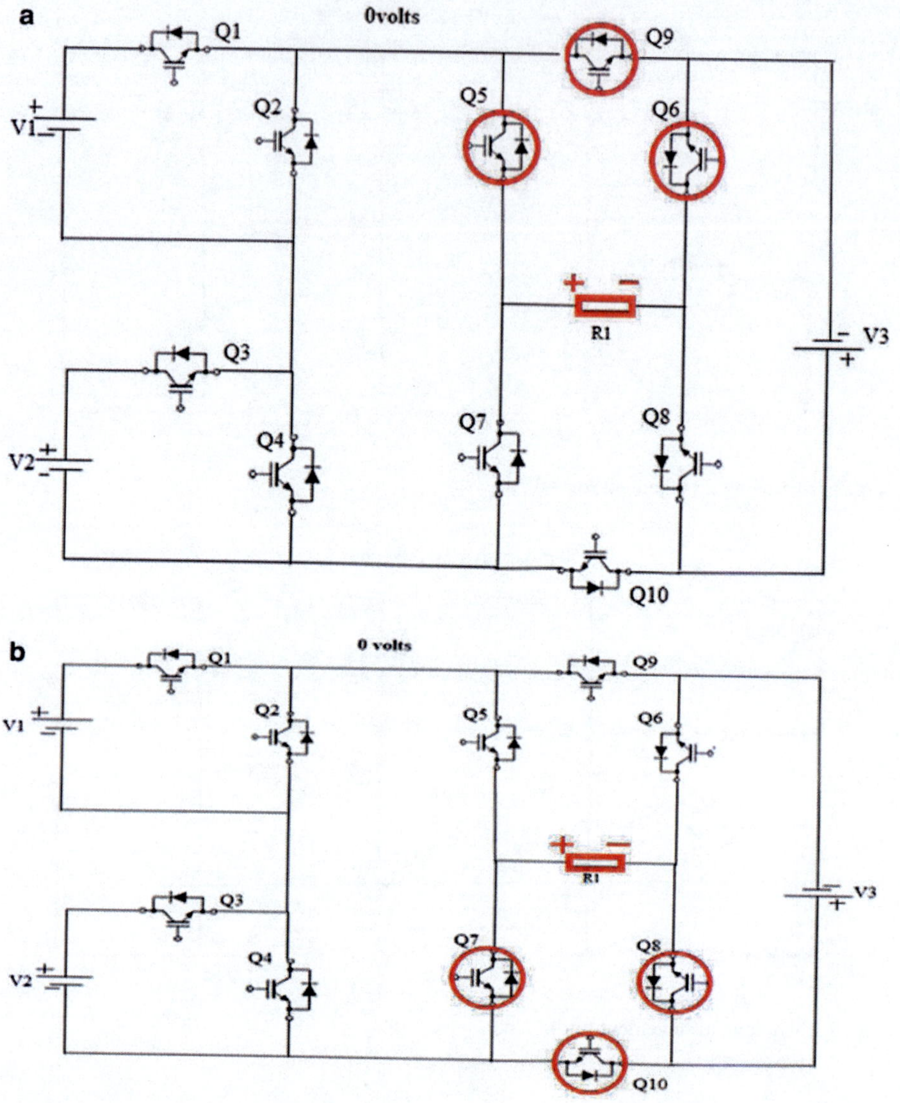

Fig. 9 **a, b** represents switching arrangement for 0Vdc

2.1.9 Mode 9: (V0 = −1 V)

In this mode, the switches Q1, Q9, Q6, Q7, Q4 are turned ON, and the remaining switches are turned OFF. These switches enable voltage source V1, and the resulting load voltage is given by V0 = −V1 = −1 V. Figure 10 shows the current paths that are active in this mode.

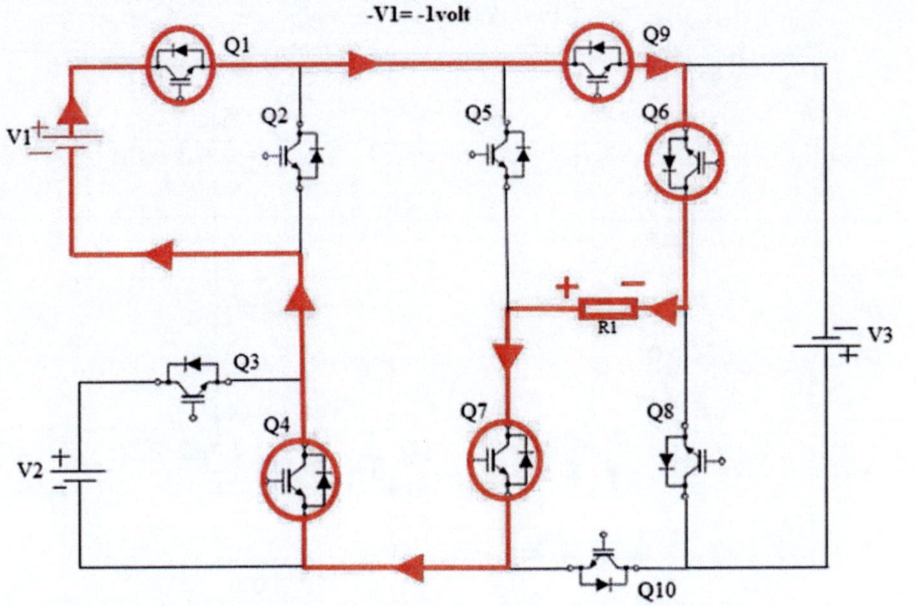

Fig. 10 Switching arrangement for −Vdc

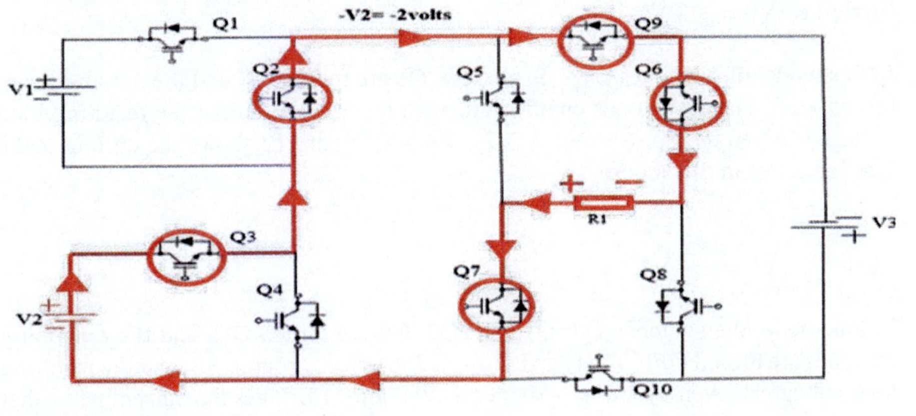

Fig. 11 Switching arrangement for −2Vdc

Mode 10: (V0 = −2 V)

In this mode, the switches Q3, Q2, Q9, Q6, Q7 are turned ON, and the remaining switches are turned OFF. These switches enable voltage source V2, and the resulting load voltage is given by V0 = −V2 = −2 V. Figure 11 shows the current paths that are active in this mode.

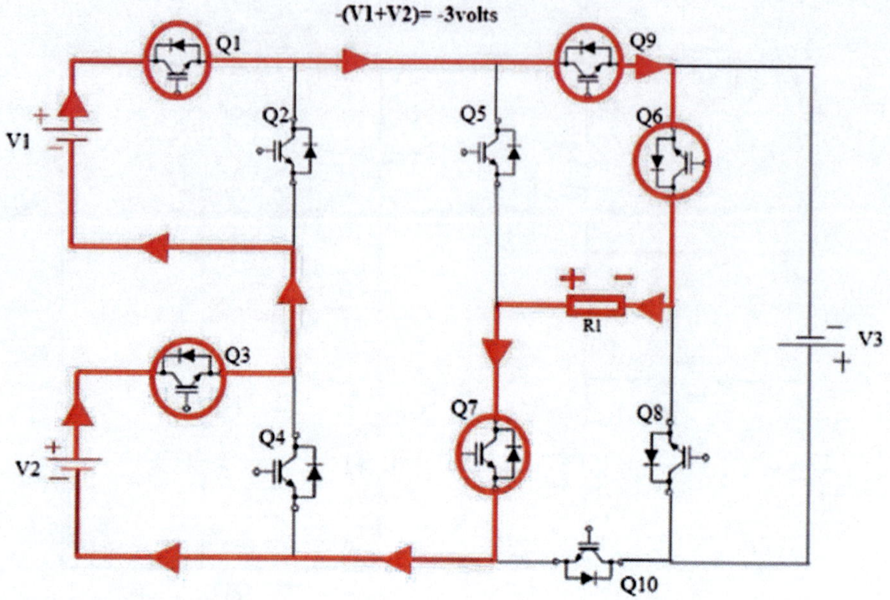

Fig. 12 Switching arrangement for −3Vdc

Mode 11: (V0 = −3 V)

In this mode, the switches Q1, Q9, Q6, Q7, Q3 are turned ON and the remaining are turned OFF. These switches enable voltage sources V1, V2, and the resulting load voltage is given by V0 = −(V1 + V2) = −3 V. Figure 12 shows the current paths that are active in this mode.

Mode 12: (V0 = −4 V)

In this mode, the switches Q8, Q7, Q4, Q2, Q9 are turned ON, and the remaining switches are turned OFF. These switches enable voltage source V3, and the resulting load voltage is given by V0 = −V3 = −4 V. Figure 13 shows the current paths that are active in this mode.

Mode 13: (V0 = −5 V)

In this mode, the switches Q8, Q7, Q4, Q1, Q9 are turned ON and the remaining are turned OFF. These switches enable voltage sources V3, V1, and the resulting load voltage is given by V0 = −(V3 + V1) = −5 V. Figure 14 shows the current paths that are active in this mode.

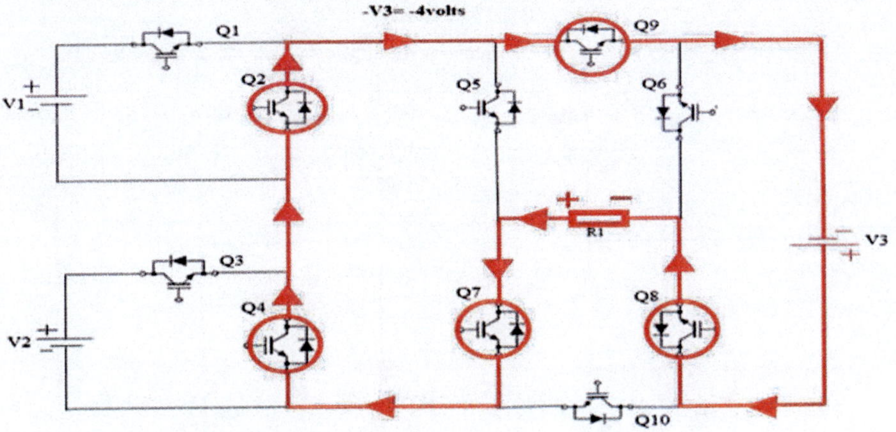

Fig. 13 Switching arrangement for −4Vdc

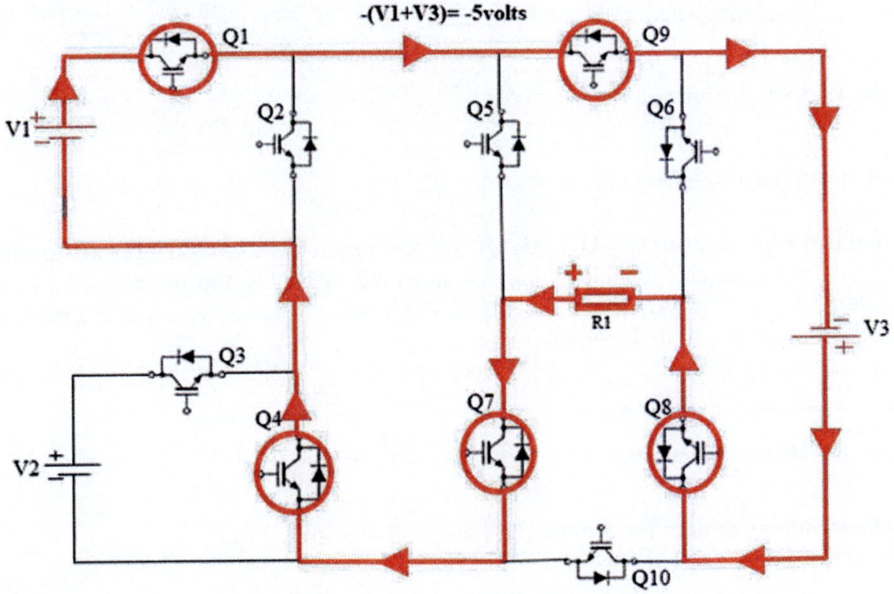

Fig. 14 Switching arrangement for −5Vdc

Mode 14: (V0 = −6 V)

In this mode, the switches Q8, Q7, Q3, Q2, Q9 are turned ON and the remaining are turned OFF. These switches enable voltage sources V3, V2, and the resulting load voltage is given by V0 = −(V3 + V2) = −6 V. Figure 15 shows the current paths that are active in this mode.

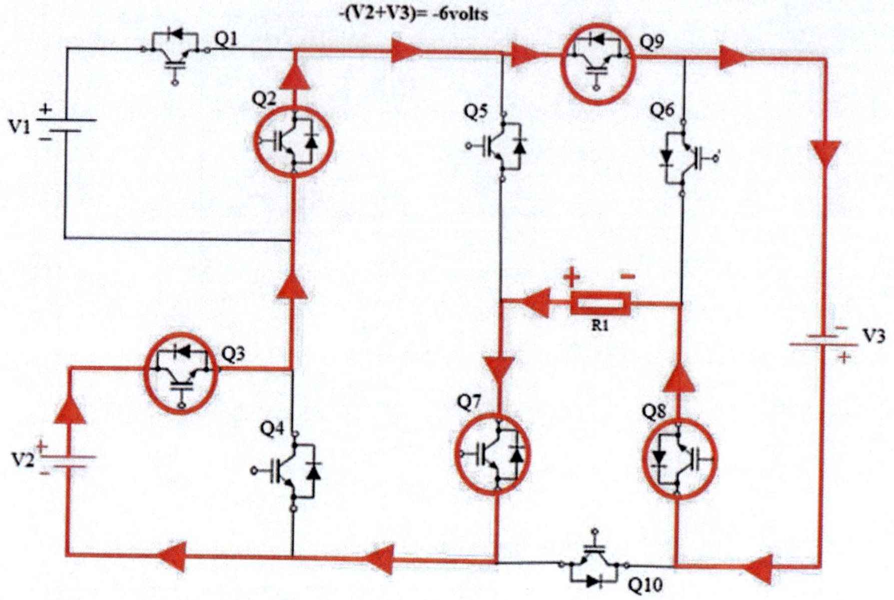

Fig. 15 Switching arrangement for −6Vdc

Mode 15: (V0 = −7 V)

In this mode, switches Q8, Q7, Q3, Q1, Q9 are turned ON and remaining are turned OFF. These switches enable voltage sources V3, V2, V1, and the obtained load voltage is given by V0 = −(V3 + V2 + V1) = −7 V. Figure 16 shows the current paths that are active in this mode.

3 Switching Table

The switching pattern for proposed topology is depicted as follows:
V1 = 1 v, V2 = 2 v, V3 = 4 v

4 Triangular Carrier Waveform

The gate pulses for the control of each switch to obtain the desired output wave form, modulation wave (sinusoidal wave) is related with quantity of carrier waves (triangular waves) equal to the quantity of required levels except zeroth level. Thus, the obtained waveform, by comparing carrier and modulation waves, is used as a

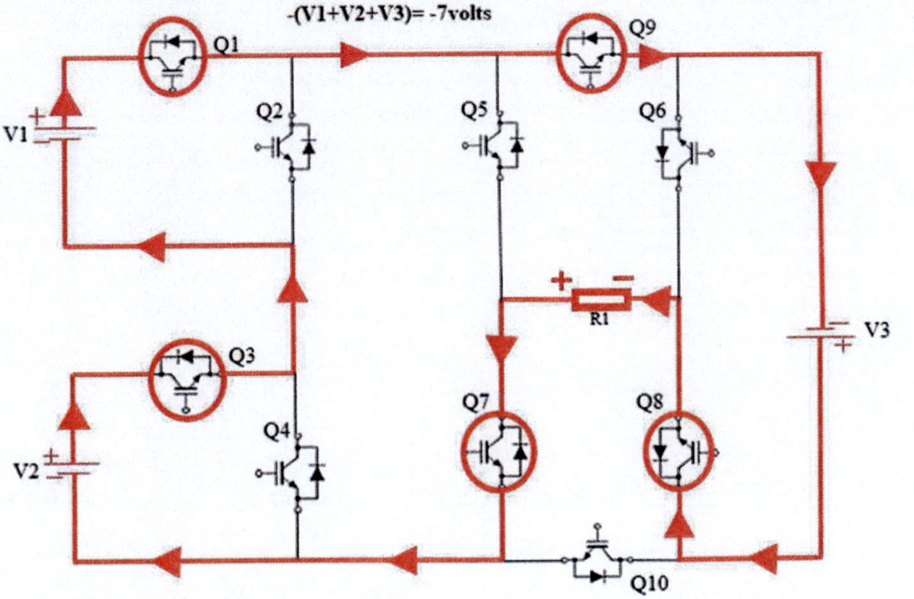

Fig. 16 Switching arrangement for −7Vdc

reference wave for the output, and the clock pulses are generated according to it (Fig. 17).

5 Mathematical Calculations

Levels that can be produced using the topology are

$$\text{No. of levels } (L) = (2(n + 1)) - 1 \tag{1}$$

where n = number of voltage sources

Switches used are

$$\text{No. of switches (S)} = 2(n + 2) \tag{2}$$

where n = number of voltage sources

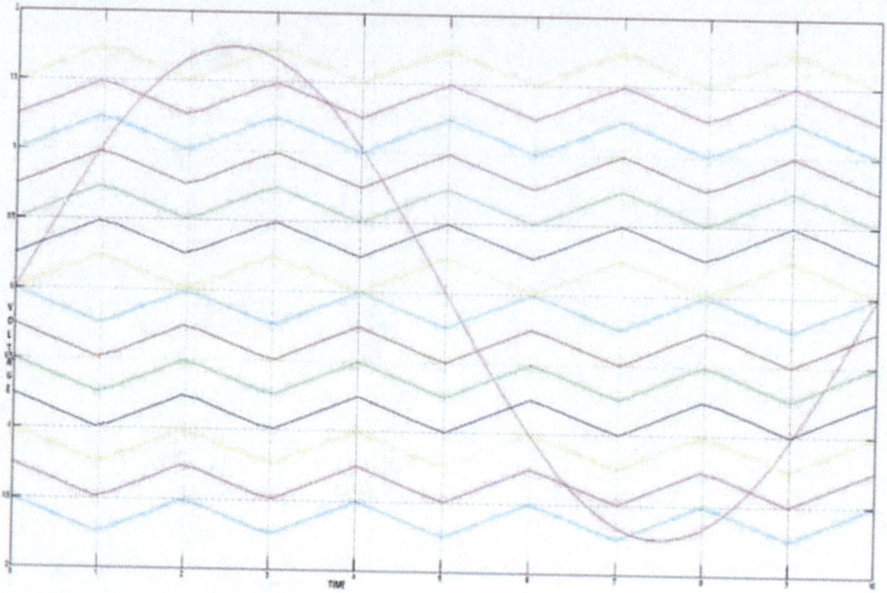

Fig. 17 Carrier sine wave modulation

5.1 Voltage Stress

The most extreme voltage that every switch encounters at its OFF state, without getting harmed, is called voltage stress. The voltage stress is determined without interfacing the heap. The voltage worry over each switch is surrendered in Table 2.

V1 or V2 in Table 2, implies switch 2, switch 4 experiences maximum of V1 or V2 depending on the magnitude of V1, V2.

Table 2 Voltage stress across each switch

Switch	Relating voltage stress
S1	V1 + V2 + V3
S2	V1orV2
S3	V1 + V2 + V3
S4	V2 or V1
S5	V1 + V2
S6	V3
S7	V1 + V2
S8	V3
S9	V1 + V2 + V3
S10	V1 + V2 + V3

5.2 Conduction and Switching Loss

The losses in the IGBT consist of Conduction loss and Switching loss.

(1) Conduction loss:

Conduction power loss

$$\mathrm{Ploss} = \mathrm{Vce(sat)} * \mathrm{D} * \mathrm{lc} \tag{3}$$

where,

Vce(sat) = collectorto emitter saturation voltage
D = dutycycleIc = collector current

(2) Switching loss:

It is the product of switching energies and the switching frequency.

The energy loss during turn-on process is given by

$$\mathrm{Won} = (\mathrm{Vce} * \mathrm{lc} * \mathrm{ton})/6 \tag{4}$$

The loss of power during turn-on time is given by

$$\mathrm{Pon} = ((\mathrm{Vce} * \mathrm{lc} * \mathrm{ton})\mathrm{fs})/6 \tag{5}$$

The loss of power during turn-off time is given by

$$\mathrm{Poff} = ((\mathrm{Vce} * mathrmlc * \mathrm{toff})\mathrm{fs})/6 \tag{6}$$

The losses in the IGBT consist of Conduction and Switching losses.

The total switching power loss can be calculated as:

$$\mathrm{Psw(total)} = ((\mathrm{Vce} * \mathrm{lc} * (\mathrm{ton} + \mathrm{toff}))\mathrm{fs})/6 \tag{7}$$

where,

Vce = collector to emitter voltage
Ic = collector current ton = turn-on time and toff = turn-off time
fs = switching frequency.

5.3 Simulation Result

Figure 18 shows the simulated output voltage waveform of a proposed 15-level

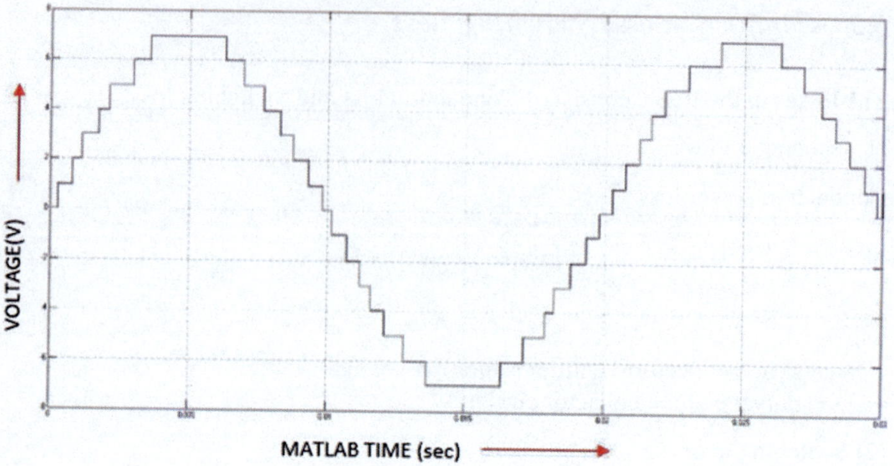

Fig. 18 Simulation output of 15-level multilevel inverter

multilevel inverter. The results show that the performance of multilevel inverter increases with increase of levels with reduced total harmonic distortion.

6 Total Harmonic Distortion

See Figs. 19 and 20.

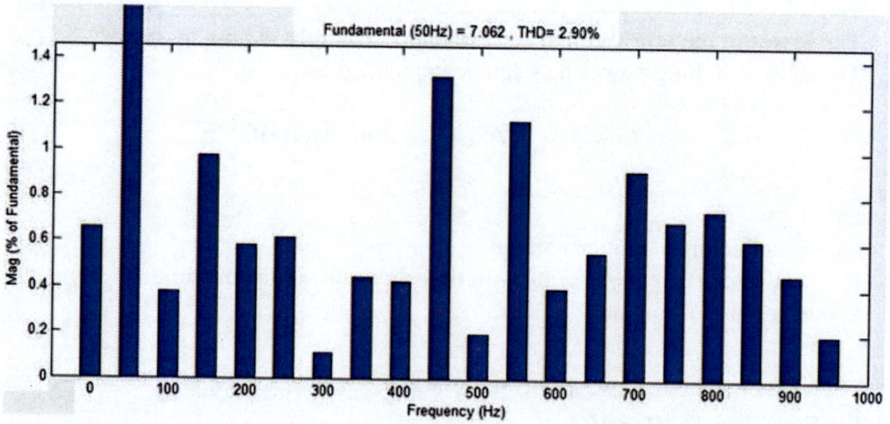

Fig. 19 FFT analysis

```
Sampling time      = 0.0001875 s
Samples per cycle  = 107
DC component       = 0.0465
Fundamental        = 7.062 peak (4.993 rms)
THD                = 2.90%

   0 Hz   (DC):          0.66%     270.0°
  50 Hz  (Fnd):        100.00%       0.5°
 100 Hz   (h2):          0.38%     209.3°
 150 Hz   (h3):          0.97%     214.5°
 200 Hz   (h4):          0.58%     123.0°
 250 Hz   (h5):          0.62%     238.4°
 300 Hz   (h6):          0.11%      10.4°
 350 Hz   (h7):          0.45%     266.1°
 400 Hz   (h8):          0.43%      29.8°
 450 Hz   (h9):          1.33%     199.2°
 500 Hz  (h10):          0.19%      -7.7°
 550 Hz  (h11):          1.13%      12.1°
 600 Hz  (h12):          0.40%     196.1°
 650 Hz  (h13):          0.56%     139.5°
 700 Hz  (h14):          0.91%     166.6°
 750 Hz  (h15):          0.69%     119.0°
 800 Hz  (h16):          0.74%     169.7°
 850 Hz  (h17):          0.61%      83.8°
```

Fig. 20 Parameters list

7 Conclusion

In the present paper, a novel hybrid 15-level multilevel inverter is designed and verified with the performance of the carrier-based LSPWM technique. The model outcome across load with fifteen levels is obtained, and the total harmonic distortion (THD) is obtained as 2.90%. The proposed topology is extremely suitable for active filters, var compensators and grid connected systems. Additionally, the switching technique plays a significant role to improve the FFT spectrum of the output waveform with less switching frequency. Moreover, it can apply to all hybrid and asymmetrical topologies.

References

1. Akagi H (2017) Multilevel converters: fundamental circuits and systems. Proc IEEE 105(11):2048–2065
2. Leon JI, Vazquez S, Franquelo LG (2017) Multilevel converters: control and modulation techniques for their operation and industrial applications. Proc IEEE 105(11):2066–2081

3. Prabaharan N, Palanisamy K (2017) A comprehensive review on reduced switch multi-level inverter topologies, modulation techniques and applications. Renew Sustain Energy Rev 76:1248–1282
4. Venkataramanaiah J, Suresh Y, Panda AK (2017) A review on symmetric, asymmetric, hybrid and single DC sources based multilevel inverter topologies. Renew Sustain Energy Rev 76:788–812

Chapter 12
Video-Based Facial Expression Recognition: A Deep Learning Approach

Jeena Jacob and J. Jeba Sonia

1 Introduction

Face expressions and emotions are formed by the motions of facial muscles. These motions can deliver the emotional state of a person to the observer. The facial expression conveys nonverbal information. It is a nonverbal mode of communication. Verbal information can be complemented by these facial expressions. During the last few years, a large number of active researchers came forward to analyze human emotions and facial expressions in various trending domains because video-based facial expression recognition has its applications in healthcare, human emotion perception, and robotics [1–5].

Video-based face expression recognition (FER) techniques are classified according to their feature representations: static and dynamic. The static method is image-based, and it aims to extract spatial information while dynamic is video-based; it extracts temporal features from adjacent frames in a video sequence. In the past decades, facial expressions are identified from spatial features obtained from still images [1–4]. Image-based methods in the previous years can attain spatial information from motion-less images, but they are not capable of capturing temporal information in successive frames in video sequences. We know that video sequences carry considerably more information for facial expression recognition than still images. Therefore categorizing facial expressions from successive frames in a video is more characteristic. One challenging issue with video-based FER methods is the efficient encoding of video sequences as input into a suitable feature representation. At present, the conventional methods apply Gabor motion energy [6], Local Binary Patterns from Three Orthogonal Planes (LBP-TOP) [7], Local Phase Quantization from Three Orthogonal Planes (LPQ-TOP) [8], etc. In order to distinguish temporal

J. Jacob (✉) · J. Jeba Sonia
Amal Jyothi College of Engineering, Kanjirappally, Kottayam, India

J. Jeba Sonia
e-mail: jjebasonia@amaljyothi.ac.in

S. Dawn et al. (eds.), *Smart and Intelligent Systems*, Algorithms for Intelligent Systems,
https://doi.org/10.1007/978-981-16-2109-3_12

facial expressions, these prevailing feature representations are inadequate. In the past few years, deep neural networks based feature learning came forward which may provide considerably superior performance without having need of domain expertise [9–15].

Deep neural networks are famous for their ability to learn strong features from data. Taking inspiration from this, we bring forward a new method: video-based facial expression recognition method which makes use of a deep neural network. In this paper, we put forward a hybrid deep learning model which is a combination of three deep models. Because of the ability of Convolutional Neural Networks (CNNs) [16] to learn important features automatically with no human supervision, we use deep convolutional neural networks as our first two deep models. Static facial images are handled by a spatial CNN, and optical flow images are handled by a temporal CNN. Video-based spatial and temporal features are individually learned by these two CNNs. The third and last deep model is a fusion network which makes use of a Deep DBN [17] network. This network is trained to learn spatiotemporal features. When training is finished by using DBN, average pooling is held on the divided video sequences in order to bring out video feature representations having a fixed length. Then, facial expression in video sequences is classified with the help of a linear Support Vector Machine (SVM).

In [18], actions in videos have been successfully recognized using a two-stream convolutional neural network. But it uses a score level scheme which refers to a shallow fusion method to combine different spatiotemporal features captured by the two-stream CNN. But shallow fusion method is inefficient for several input procedures [19]. To eliminate this, we design deep fusion procedures that hold a deep fusion framework which can administer several feature fusion tasks. A DBN comprises many Restricted Boltzmann Machines (RBMs) [20]. Therefore each RBM is employed to learn feature depictions of various input procedures. So a DBN model can be used as a deep fusion model to incorporate various features obtained by the individual two-stream CNNs. This is the main inspiration that led us to design a hybrid model for learning video-based FER. Experiment outcomes show that our method has better performance on video-based FER tasks on the selected datasets such as BAUM-1 s [21], MMI [22], and RML [23].

The upcoming sessions of this paper are as follows. Section 2 analyzes a review of the literature in short. Section 3 reports the detailed study of the proposed work. Section 4 handles dataset description, Sect. 5 narrates results, and Sect. 6 provides the conclusion.

2 Literature Review

This section holds a review of literature related to the hand-designed method and deep learning based method for video-based face expression feature extraction. Shan et al. [24] analyzed facial expressions based on Local Binary Patterns (LBP). Statistical local features are used to test facial characterization. Tariq et al. [25] conducted classifications based on representations of mid-level features such as feature extraction, encoding, and pooling. In his other work [26], he uses SIFT descriptors to extract features for FER. These are for static expression recognition of face images. Now, for the identification of dynamic expressions, these features are expanded and put in video sequences. Zhao et al. [7] used dynamic texture recognition with the help of local binary patterns. Jiang et al. [8] introduced a method for analyzing temporal features of facial activities using LPQ-TOP. Klaser et al. [27] presented a four folded technique which consists of 3D gradient computation, orientation quantization, histogram computation, and descriptor computation. Scovanner [28] introduced a three dimensional SIFT feature descriptor for action recognition in video sequences. Hayat et al. [29] presented a study that points out the similarities and dissimilarities of execution of different dynamic descriptors such as three dimensional HOG, three dimensional SIFT, and LBP-TOP and realized that LBP-TOP outperforms the others.

In FER, more robust techniques for extracting spatiotemporal features have been added up recently. Liu et al. [30] produced manifold modeling of video sequences using expressionlet. His work is a three-stage process including spatiotemporal manifold (STM), Universal manifold model (UMM), and the fitting of STM to UMM. Using this strategy, he aligned videos both spatially as well as temporally. Fan et al. [31] presented a dynamic feature containing motion history image and entropy, and a spatiotemporal feature based on Zernike moment. Yan [32] presented a discriminative multi-metric learning. For each video sequence, it first calculates multiple feature descriptors, and multiple distance metrics are learned using these feature descriptors. Till now we have discussed hand-designed methods for feature representation. The upcoming section handles various deep learning-based methods.

During the past decades, deep CNNs have made a great influence on a wide variety of computer vision works including image and video classification, object, and action detection. Deep CNNs [16, 33–35], are having several convolution layers and pooling layers. These classical CNN models [36] are extended to get a deep multilayered framework having five convolutional layers and three max-pooling layers.

Traditional CNNs can bring out spatial relations of images and can process static images. But they are not able to design temporal relationships of input in videos. For this problem, Tran et al. [37] presented an effective approach with the help of 3D-CNNs (3D ConvNets). 3D ConvNets are capable of learning spatiotemporal features from video sequences by sliding over both spatial and temporal dimensions concurrently. In order to learn spatiotemporal expressions from consecutive frames in videos, 3D ConvNets have been used widely in the past years. Besides, 3DCNN-Deformable Action Parts [14] are a form of 3D CNNs utilized for dynamic facial expression recognition. However, while extraction happens, these methods are unable to concurrently handle the deep fusion of spatiotemporal features.

As a solution for this problem, Yan et al. [18] proposed video action identification using two-stream CNNs. But it uses a score level scheme to combine different spatiotemporal features captured by the two-stream CNN. This score level scheme refers to a shallow fusion method. But the shallow fusion method is inefficient to frame the complex non-linear joint distribution of several input procedures. Therefore, we propose a deep fusion network set up with a deep belief network model which effectively makes use of the dominance of two-stream CNNs. The DBN model learns the discriminative features which is the output of two-stream CNNs.

3 Proposed Method

The actual framework of our proposed work is given in Fig. 1. Our system is a hybrid deep learning model consisting of two convolutional neural networks and a deep fusion network. There are two separate channels of input streams as depicted in Fig. 1. Static facial frames are processed by a spatial CNN, and optical flow images formed between adjacent frames are processed by a temporal CNN. The output from fully connected (FC) layers of the individual CNNs provides spatio-temporal characteristics. A fusion network which makes use of a DBN network is mapped out to merge the learned spatiotemporal features. The proposed method involves the following key steps.

Steps.

A. Generating CNN inputs.

B. Learning spatial and temporal features with CNNs.

C. Integrating learned features with DBN.

D. Facial expression classification.

The detailed discussion of the above steps is presented in the following.

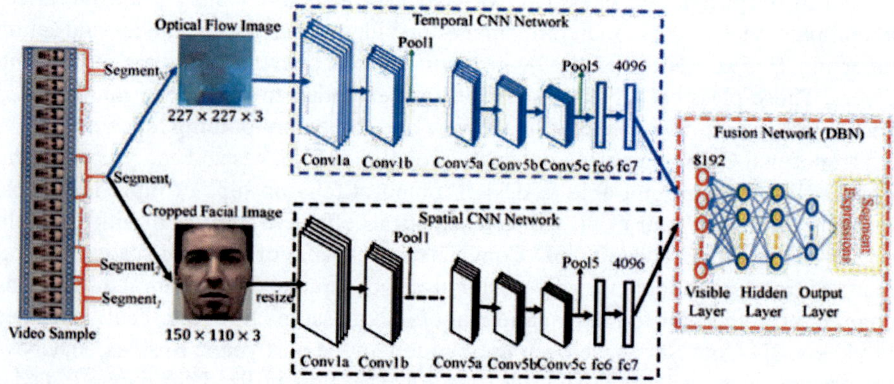

Fig. 1 Depiction of the proposed methodology

3.1 Generating CNN Inputs

Our dataset is a set of video samples of different time durations. Since our CNNs need to have input data of fixed size, each video sample which is having different time durations is then partitioned into fixed-length sequences and feed as inputs of CNN. The size of training data can be extended by this partitioning. The partitioned video segment has a length of 16 (L = 16). For this, we have to consider video samples of sizes less than and greater than length 16. If the length is greater than 16, remove the first and last (video length—16)/2 frames. Likewise, if the length of the video sample is less than 16, then replicate the first and last (16—video length)/2 frames. Like this, we assure that each partitioned segment is having sixteen frames.

3.1.1 Inputs to Temporal CNNs

Firstly, we need to find optical flow images between adjacent frames from each video segment. Optical flow can be calculated from the motion of two frames obtained at consecutive times. As in [38], we configure a three dimensional image by storing the optical flow in the X, Y direction and the magnitude, and a scaling constant of 16 is used. Similarly, we form an optical flow image having size 227 × 227 x 3. Two adjacent frames produce an optical flow image. Therefore, a video segment with a length equal to 16 can produce fifteen optical flow images. This is the input of temporal CNNs.

3.1.2 Inputs to Spatial CNNs

For the inputs of spatial CNNs, we crop a facial image of size 150 × 110 × 3 from each frame in a video segment [24]. Figure 2 shows the original face image and

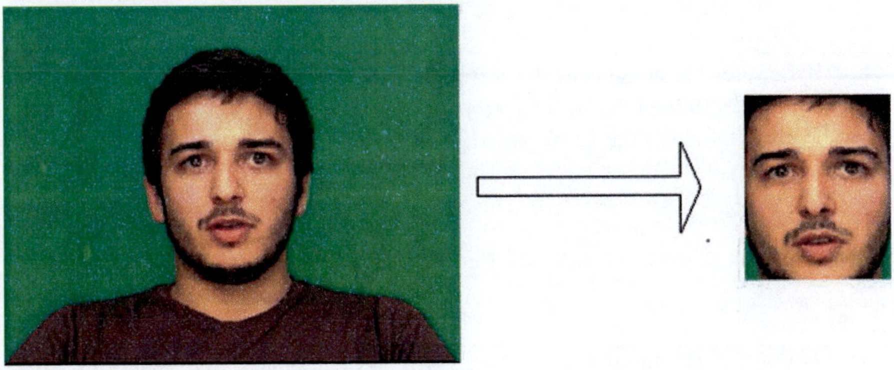

Fig. 2 The original and cropped face

cropped facial image. As in [39], a face detector is employed to find faces from each frame and to crop the facial image. This cropped image is having important facial parts such as head, nose, and mouth. Since the inputs of temporal CNNs have a size of 227 × 227x3, the cropped face image is resized to 227 × 227 x 3 as inputs to spatial CNN. Here we use the number of frames as sixteen; we will have sixteen facial images. But there are only fifteen optical flow images. Therefore we remove the first frame in case of spatial CNNs so as to make the number of frames in temporal CNNs equal to that of spatial CNNs.

3.2 Learning Spatial and Temporal Features with CNNs

The spatial and temporal CNNs depicted in our framework (Fig. 1) have the same structure. This structure is adopted from the VGG16 network [16] which is having five convolution layers, five max pooling layers, and three fully connected layers. In our framework, the fully connected layers fc6 and fc7 have 4096 units. The third fully connected layer fc8 represents the six data categories: anger, happiness, disgust, sadness, fear, and surprise. But in the actual VGG16 network, the fc8 layer represents 1000 image categories.

To learn spatiotemporal features with CNNs, the VGG16 network is pre-trained and then fine-tuned on the target data, which is video-based face expression data. To make spatial and temporal CNN network initialized, the prevailing parameters of the VGG16 network, which is pre-trained on ImageNet data, are copied. The fc8 layer in the VGG16 network is put back with our six face expression categories.

Eventually, we train again the spatial and temporal CNNs separately. For that, we use a standard backpropagation technique so that it reduces error rates and thereby makes the model more reliable.

$$w, \theta^{\min} \sum_{i=1}^{N} H(\text{softmax}(w \cdot \Upsilon(a_i, \vartheta)), y_i), \tag{1}$$

where$W$ denotes the weights of the softmax layer, ϑ denotes the network parameters that belong to both the CNNs, $\Upsilon(a_i, \vartheta)$ is the output of fc7, a_i is the input data, and y_i is the class label for the ith segment. The softmax log loss function H *is* defined as

$$H(\vartheta, y) = -\sum_{j=1}^{C} y_j \log(y_j) \tag{2}$$

where C denotes the number of facial expression categories, six categories. When spatial and temporal CNNs are trained, the fc7 layers of them give feature representations that are learned from video segments.

3.3 Integrating Learned Features with DBN

When spatial and temporal CNNs are trained once, fc7 layers of both of them are integrated into as the inputs of the fusion network that is made with a deep belief network model [17] (DBN). We use this DBN model to collect the non-linear relationships over spatiotemporal modalities. DBN model also provide joint characteristic representation for our inputs.

A DBN model is a neural network structure that consists of several distinct layers. It is a framework formed by putting a collection of Restricted Boltzmann Machines [20] (RBMs) together. RBMs are bipartite graphs, and no communication within layers is allowed in RBMs, and hence the name 'restricted'. DBNs make use of these RBMs. In our framework, we used two RBMs composed of one visible layer and two hidden layers (Fig. 1). This composition of RBMs in DBNs helps in fast, layer by layer unsupervised procedure. The set of RBMs always ends with a softmax layer, which helps in an unsupervised scenario to cluster unlabeled data. In Fig. 1, the output layer in our framework corresponds to the softmax layer for clustering. One DBN can accommodate several RBMs for layer-by-layer learning of inputs. Therefore it can find distribution properties as well as feature learning of input data.

To train the DBN model [40], we employ an unsupervised pre-training and a supervised fine-tuning. The unsupervised pre-training is held by making use of a greedy algorithm.

3.4 Facial Expression Classification

After performing the training of the fusion network, the last hidden layer corresponds to the output, which gives the joint spatiotemporal features. According to the learned features, an average pooling is performed to all partitioned videos. This is to form a global feature representation of a constant time slice. A linear SVM is employed to carry out FER classification in the end.

4 Datasets

The studies are held in three public video-based datasets for facial expression, that is, the MMI [22] database, the RML [23] database, and the BAUM-1 s [21] database. About 6000 segments are formed from 521 videos of the BAUM-1 s database. Similarly, the RML database consists of more than 700 video samples, and over 10,000 video segments are formed from it. Likewise, over 3000 video segments are produced from the MMI database which is having 213 videos. Figures 3, 4 and 5 show examples of cropped facial expressions images from the MMI dataset, RML dataset, and BAUM-1 s dataset.

Fig. 3 Cropped face expression images from the MMI database

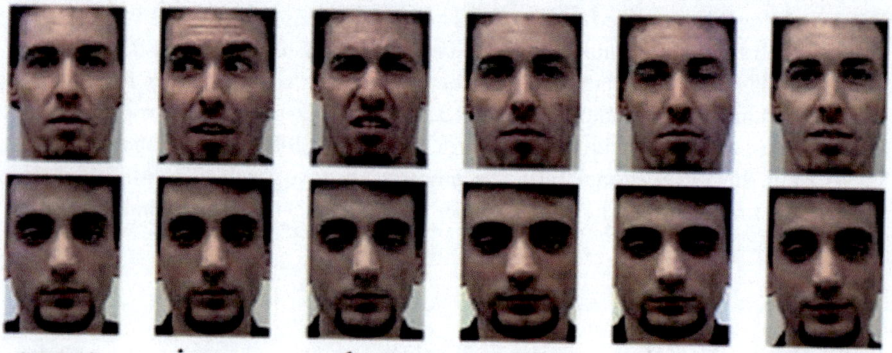

Fig. 4 Cropped face expression images from the RML database

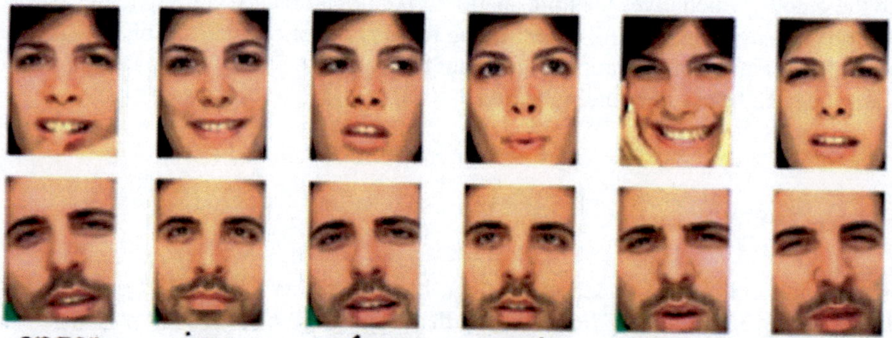

Fig. 5 Cropped face expression images from the BAUM-1 s database

Table 1 Performance evaluation of DBN structures (in percentage)

DBN structure	MMI	RML	BAUM-1 s
DBN-1	62.98	66.45	44.05
DBN-2	67.56	70.28	50.27
DBN-3	70.95	72.70	54.92

Table 2 Accuracy of feature learning (in percentage)

Features	MMI	RML	BAUM-1 s
Spatial CNN	58.55	63.84	49.85
Temporal CNN	47.87	49.28	48.20
DBN fusion	70.82	73.70	54.98

5 Results

At first, we assess the results of structures of fusion networks built with DBN. The performance of DBNs may affect the spatiotemporal features. To check on the performance, we evaluate three different DBNs. They are DBN-1, DBN-2, and DBN-3. The only difference is the number of hidden layers in the DBN model. Table 1 shows the performance evaluation of different DBN structures.

We can clearly understand from Table 1 that DBN-3 outperforms the others by giving better results for the three datasets. DBN-3 provides an accuracy of 70.95% for the MMI dataset, 72.70% for the RML dataset, and 54.92% for the BAUM-1 s dataset, respectively. This shows that the use of multiple RBMs in DBNs provides high performance over feature fusion. Table 2 depicts the accuracy of three methods: features of individual spatial CNNs, individual temporal CNNs, and spatiotemporal fusion with DBNs. We can see that spatiotemporal fusion with DBNs performs better than the individual CNNs. DBN fusion provides an accuracy of 70.82% for the MMI dataset, 73.70% for the RML dataset, and 54.98% for the BAUM-1 s dataset. The ability of DBN in discovering spatiotemporal features and learning feature representations are utilized in our work.

6 Conclusion

We propose a deep learning model involving a spatial CNN, a temporal CNN, and a deep fusion network to identify face expressions from video samples. The execution of the work involves mainly three stages. The first stage is to employ the VGG16 network model which is pre-trained on the large-scale ImageNet data in separately fine-tuning spatial CNNs and temporal CNNs on video-based face expression data. The second stage is to train the DBN network to combine the learned spatiotemporal features. The third stage is to categorize the face expressions using a linear SVM. Experiment on the three datasets shows the advantages of our work.

In future, our plan is to work with micro-expressions on facial expression video datasets.

References

1. Martinez B, Valstar MF, Jiang B, Pantic M (2019) Automatic analysis of facial actions: a survey. IEEE Trans Affective Comput to be published. https://doi.org/10.1109/TAFFC.2017.2731763.
2. Zhao X, Zhang S (2016) A review on facial expression recognition: feature extraction and classification. IETE Tech Rev 33(5): 505–517
3. Corneanu CA, Simon MO, Cohn JF, Guerrero (2016) SE survey on RGB, 3D, thermal, and multimodal approaches for facial expression recognition: history, trends, and affect-related applications. IEEE Trans Pattern Anal Mach Intell 38(8): 1548–1568
4. Sariyanidi E, Gunes H, Cavallaro (2015) A automatic analysis of facial affect: a survey of registration, representation, and recognition. IEEE Trans. Pattern Anal Mach Intell 37(6): 1113–1133
5. Muhammad G, Alsulaiman M, Amin SU, Ghoneim A, Alhamid MF (2017) A facial-expression monitoring system for improved healthcare in smart cities. IEEE Access 5: 10871–10881
6. Wu T, Bartlett MS, Movellan JR (2010) Facial expression recognition using Gabor motion energy _lters. In Proc IEEE Comput Soc Conf Comput Vis Pattern Recognit Workshops. San Francisco, CA, USA, 42–47
7. Zhao G, Pietikainen M (2007) Dynamic texture recognition using local binary patterns with an application to facial expressions. IEEE Trans Pattern Anal Mach Intell 29(6): 915–928
8. Jiang B, Valstar MF, Martinez B, Pantic M (2014) A dynamic appearance descriptor approach to facial actions temporal modeling. IEEE Trans Cybern 44(2): 161–174
9. Zhang T, Zheng W, Cui Z, Zong Y, Yan J, Yan K (2016) A deep neural network-driven feature learning method for multi-view facial expression recognition. IEEE Trans. Multimedia 18(12): 2528–2536
10. Jung H, Lee S, Yim J, Park S, Kim J (2015) Joint fine-tuning in deep neural networks for facial expression recognition. In Proceeding of the IEEE International Conference on Computer Vision (ICCV), 2983–2991
11. Zhang K, Huang Y, Du Y, Wang L (2017) Facial expression recognition based on deep evolutional spatial-temporal networks. IEEE Trans Image Process 26(9): 4193–4203
12. Hasani B, Mahoor MH (2017) Facial expression recognition using enhanced deep 3D convolutional neural networks. In Proceeding of the IEEE Conference on computer vision. Pattern Recognit Workshops. 2278–2288
13. Lopes AT, de Aguiar E, De Souza AF, Oliveira-Santos T (2017) Facial expression recognition with convolutional neural networks coping with few data and the training sample order. Pattern Recognit 61: 610–628
14. Liu M, Li S, Shan S, Wang R, Chen X (2014) Deeply learning deformable facial action parts model for dynamic expression analysis. In Proceeding of the Asian conference on computer vision (ACCV). Singapore, 143–157
15. Zhang S, Zhang S, Huang T, Gao W, Tian Q (2018) Learning affective features with a hybrid deep model for audio-visual emotion recognition. IEEE Trans Circuits Syst Video Technol 28(10): 3030–3043
16. Simonyan K, Zisserman A (2015). Very deep convolutional networks for large-scale image recognition. In Proceeding ICLR. San Diego, CA, USA, 1–14
17. Hinton GE, Salakhutdinov RR (2006) Reducing the dimensionality of data with neural networks. Science vol. 313(5786), 504–507

18. Simonyan K, Zisserman A (2014) Two-stream convolutional networks for action recognition in videos. In Proceedings of the 27th international conference on neural information processing systems, Montreal, QC, Canada, 568–576
19. Ngiam J, Khosla A, Kim M, Nam J, Lee H, Ng AY (2011) Multimodal deep learning. In Proceedings 28th International Conference on Machine Learning (ICML), 689–696
20. Hinton GE (2002) Training products of experts by minimizing contrastive divergence. Neural Comput 14(8): 1771–1800
21. Zhalehpour S, Onder O, Akhtar Z, Erdem CE (2016) BAUM-1: a spontaneous audio-visual face database of affective and mental states. IEEE Trans Affective Comput 8(3): 300–313
22. Pantic M, Valstar M, Rademaker R, Maat L (2005) Web-based database for facial expression analysis. In Proceedings of the IEEE international conference on multimedia expo (ICME). Amsterdam, The Netherlands, 317–321
23. Wang Y, Guan L, Venetsanopoulos AN (2012) Kernel crossmodal factor analysis for information fusion with application to bimodal emotion recognition. IEEE Trans Multimedia. vol. 14(3): 597–607
24. Shan C, Gong S, McOwan PW (2009) Facial expression recognition based on local binary patterns: a comprehensive study. Image Vis Comput vol. 27(6), 803–816
25. Tariq U, Yang J, Huang TS (2012) Multi-view facial expression recognition analysis with generic sparse coding feature. In Proceedings of the European conference computer vision. (ECCV), 578–588
26. U. Tariq et al (2011) Emotion recognition from an ensemble of features. In Proceeding IEEE international conference on automatic face gesture recognition workshops (FG). Santa Barbara, CA, USA, 872–877
27. Klaser A, Marszalek M, Schmid C (2008) A spatio-temporal descriptor based on 3D-gradients. In: Proceedings of the 19th British machine vision conference (BMVC). vol. 275, 1–10
28. Scovanner P, Ali S, Shah M (2007) A 3-dimensional sift descriptor and its application to action recognition. In Proceedings of 15th ACM international. conference multi-media (MM). Augsburg, Germany, 357–360
29. Hayat M, Bennamoun M, El-Sallam (2012) A evaluation of spatiotemporal detectors and descriptors for facial expression recognition. In Proceedings of 5th international conference on human system interaction (HSI). Perth, WA, Australia, 43–47
30. Liu M, Shan S, Wang R, Chen X (2016) Learning expression lets via universal manifold model for dynamic facial expression recognition. IEEE Trans Image Process 25(12): 5920–5932
31. Fan X, Tjahjadi T (2017) A dynamic framework based on local Zernike moment and motion history image for facial expression recognition. Pattern Recognit 64: 399–406
32. Yan H (2018) Collaborative discriminative multi-metric learning for facial expression recognition in video. Pattern Recognit 75: 33–40
33. Krizhevsky A, Sutskever I, Hinton GE (2012) ImageNet classification with deep convolutional neural networks. In: Proceedings advanced neural information process system. 1106–1114
34. Szegedy C et al (2015) Going deeper with convolutions. In Proceedings of the IEEE Conference on the Computer Vision Pattern Recognition (CVPR). Boston, MA, USA, 1–9
35. He K, Zhang X, Ren S, Sun J (2016) Deep residual learning for image recognition. In Proceedings of the IEEE conference on the computer vision pattern recognition (CVPR). Las Vegas, NV, USA, 770–778
36. LeCun Y, Bottou L, Bengio Y, Haffner P (1998) Gradient-based learning applied to document recognition. Proc IEEE 86(11): 2278–2324
37. Tran D, Bourdev L, Fergus R, Torresani L, Paluri M (2015) Learning spatiotemporal features with 3D convolutional networks. In Proceedings of the IEEE international conference on computer vision (ICCV), Santiago, Chile, 4489–4497
38. Gkioxari G, Malik J (2015) Finding action tubes. In Proceedings of the IEEE conference on the computer vision pattern recognition (CVPR). Boston, MA, USA, 759–768
39. Viola P, Jones MJ (2004) Robust real-time face detection. Int J Comput Vis 57(2): 137–154
40. Hinton GE, Osindero S, Teh YW (2006).A fast learning algorithm for deep belief nets. Neural Comput 18(7): 1527–1554

Chapter 13
Bidirectional Buck-Boost Converter in Solar PV System for Supercapacitor Energy Storage System

S. Bhanu Prakash and Gagan Singh

1 Introduction

With the growth in power demand for industries and due to limited sources of fossil fuels, the share of renewable energy resources is growing in distribution systems. Renewable energy resources mostly solar and wind energy play a significant role in power generation. In 2019, India was the second major market in Asia which added an estimated 9.9 GW for a total of 42.8 GW of installed solar power. By the end of 2022, India targets to reach 100 GW (including the solar rooftop) of installed solar PV capacity [1]. The output of the Solar PV system is often fluctuating owing to variation in irradiance and temperature conditions which makes the PV system intermittent in nature and unreliable energy sources for the DC grid. A possible solution to reduce these power fluctuations is made use of an energy storage system like a supercapacitor, which is an efficient storage device for power smoothing applications [2]. Figure 1 represents the block diagram of the Solar PV system with a supercapacitor as an energy buffer. A bidirectional converter allows the power transfer between the supercapacitor and DC grid. Supercapacitors are considered as auxiliary to conventional batteries for energy storage purposes. These are particularly suitable in cases where quick charge and discharge cycles are required for power leveling in a Solar PV system. Supercapacitors have high energy density which quickly absorbs or releases high quantity of energy in very short duration compared to batteries [3]. In a solar PV system, the output power fluctuations are mainly caused due to change in irradiance. The problems associated with an intermittent solar PV system are voltage fluctuation and frequency deviation. These variations can be minimized by means of the supplementary source's like battery, supercapacitor, and MPPT controller [4].

S. B. Prakash (✉) · G. Singh
Department of EE, DIT University, Dehradun, Uttarakhand, India
e-mail: sb.prakash@dituniversity.edu.in

G. Singh
e-mail: gagan.singh@dituniversity.edu.in

S. Dawn et al. (eds.), *Smart and Intelligent Systems*, Algorithms for Intelligent Systems,
https://doi.org/10.1007/978-981-16-2109-3_13

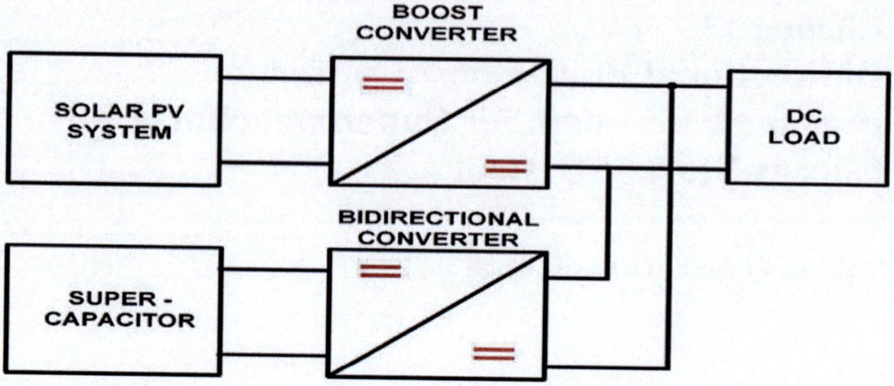

Fig. 1 Block-Diagram of Solar PV system with a supercapacitor-based energy buffer

2 Bidirectional DC-DC Converter

Bidirectional converters have gained more attention because of the increasing need for renewable energy systems with the capability of bidirectional power transfer between the storage system and grid. The outline of the bidirectional DC-DC converter is illustrated in Fig. 2. In general, bidirectional converters are categorized into two types; those are non-isolated type and isolated type. Non-isolated type is most commonly employed DC-DC converters, as there are simple in design, lower cost, compact size, and less weight. Isolated type is mainly preferred in applications like high frequency where isolation is required between the source and load [5].

A summary of different bidirectional DC-DC converter configurations which are associated with energy storage application and control approaches is described in [6]. These bidirectional converters require controllable switches like MOSFET or IGBT with anti-parallel diodes to provide the bidirectional ability. In this paper,

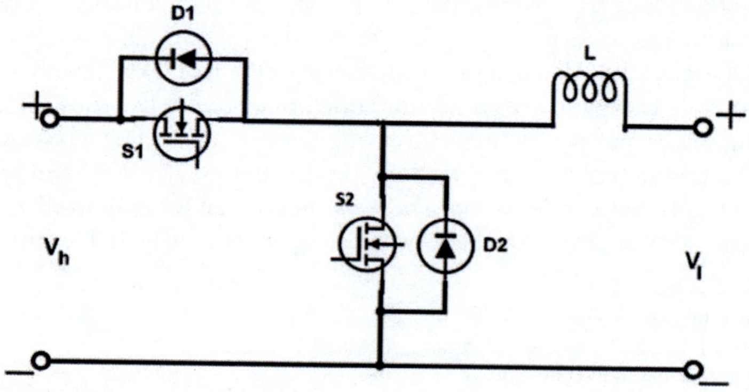

Fig. 2 Outline of Bidirectional DC-DC converter

the bidirectional converter with two-switch topology has been proposed; the detailed analysis and design of parameter values are explained by the same method of analysis as presented in [7].

2.1 Buck Converter Analysis

The elementary circuit of the buck converter circuit is illustrated in Fig. 3. The buck mode always provides the load voltage less than the supply voltage. During this mode of process, switch S_1 is kept On and switch S_2 is kept Off to charge the supercapacitor. The direction of power flow is from the supply voltage side to the load side.

Mode 1. In this mode, switch S_1 is ON and diode D_2 is reverse biased as shown in Fig. 4.

By applying KVL to the current path as shown in Fig. 4, we get

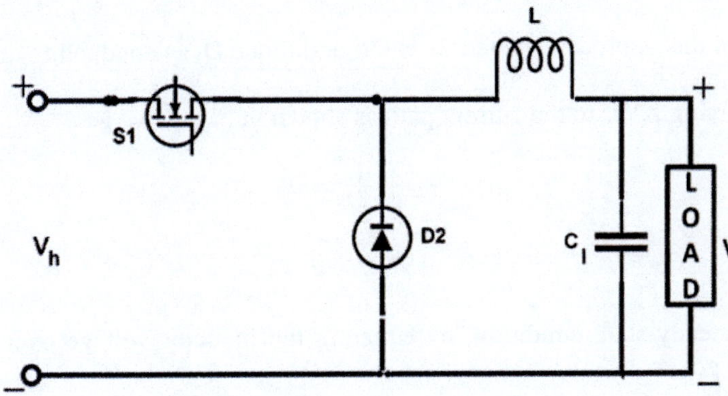

Fig. 3 Buck (step-down) Converter circuit

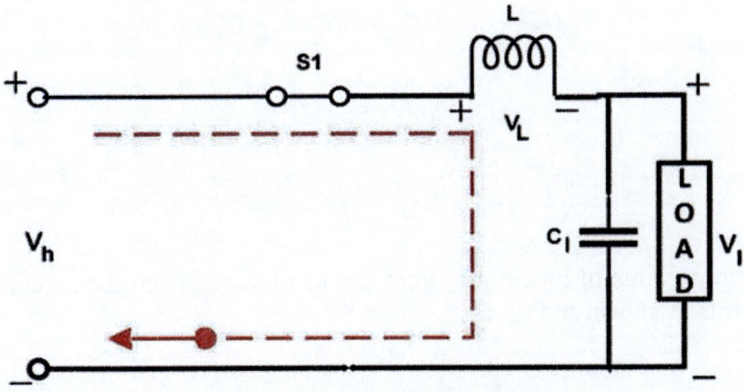

Fig. 4 Equivalent circuit when switch S_1 is conducting (Buck operation)

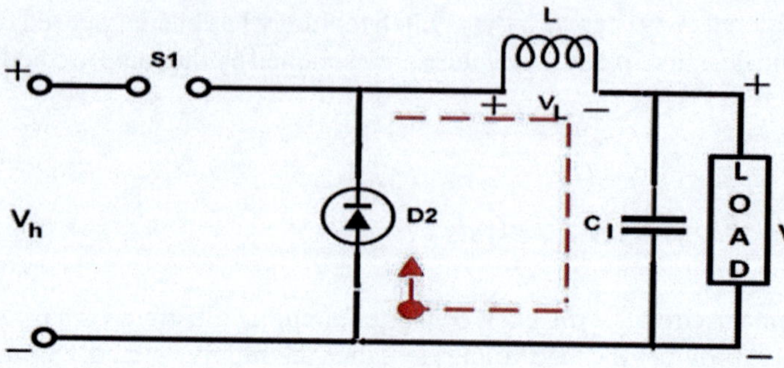

Fig. 5 Equivalent circuit when switch S_1 is OFF (Buck operation)

$$V_h = V_L + V_l$$

$$V_L = V_h - V_l \tag{1}$$

Mode 2. In this approach, switch S_1 is Off and diode D_2 is conducting as shown in Fig. 5.

By applying KVL to the current path as shown in Fig. 5, we get

$$V_L + V_l = 0$$

$$V_L = -V_l \tag{2}$$

Under steady-state condition, by equating the inductor voltage over one time period, we get

$$(V_h - V_l)\, D\, T_s + (-V_l)\,(1 - D)\, T_s = 0$$

$$V_h D T_s - V_l D T_s - V_l T_s + V_l D T_s = 0$$

$$V_h D T_s = V_l T_s$$

$$\frac{V_l}{V_h} = D \tag{3}$$

The average value of inductor current can be obtained from the rectangular area of waveforms as shown in Fig. 6.

$$I_{Lavg} = I_o = \frac{1}{2} i_{L,peak}$$

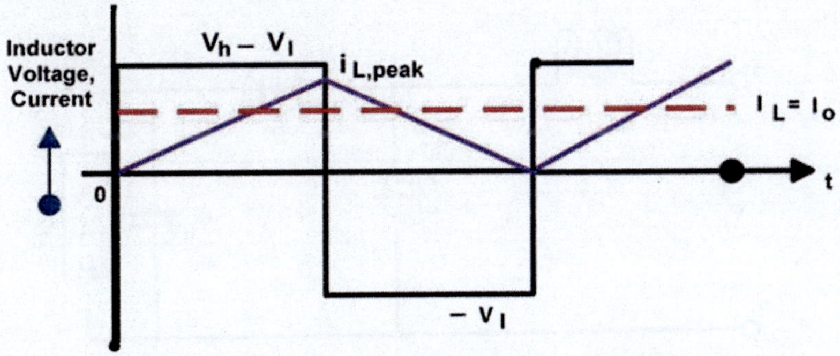

Fig. 6 Graphical representation of Inductor voltage and current (Buck mode)

$$I_o = \frac{DT_s}{2L}(V_h - V_l)$$

$$I_o = \frac{D}{2Lf_s}(V_h - V_l)$$

$$I_o = \frac{V_l}{R} = \frac{D}{2Lf_s}(V_h - V_l)$$

$$L = \frac{D}{2I_o f_s}(V_h - V_l) \tag{4}$$

The capacitor C_l can be determined by the allowed ripple in the load voltage ΔV_l, which is typically 1% of the load voltage and is given by

$$\Delta V_l = \frac{\Delta Q}{C_l}$$

$$\Delta \mathrm{Q} = \frac{1}{2}\frac{1}{2f_s}\frac{V_l(1-D)}{2Lf_s}$$

$$\Delta V_l = \frac{V_l(1-D)}{8C_l\,\mathrm{L}f_s^2}$$

$$C_l = \frac{V_l(1-D)}{8\Delta V_l\,\mathrm{L}f_s^2} \tag{5}$$

2.2 *Boost Converter Analysis*

The elementary circuit of the boost (step-up) converter is illustrated in Fig. 7. The

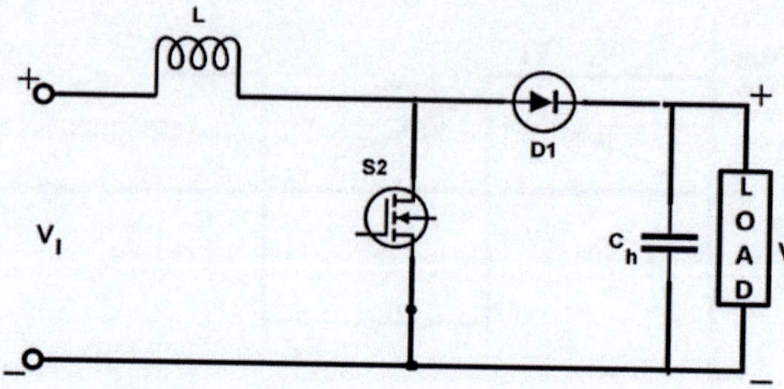

Fig. 7 Boost (Step-up) dc-dc converter circuit

boost mode always provides the load voltage greater than the supply voltage. During this mode of process, switch S_1 is kept Off and switch S_2 is On to discharge the supercapacitor. The direction of power flow is from the supply side to the load side.

Mode 3. In this mode, switch S_2 is ON and diode D_1 is reverse biased. Figure 8 represents the direction of the current path when switch S_2 is conducting (Boost operation).

By applying KVL to the current path as shown in Fig. 8, we get.

$$V_L = V_l \tag{6}$$

Mode 4. During this mode, switch S_2 is Off and diode is conducting. Figure 9 represents the direction of the current path when switch S_2 is Off (Boost operation).

By applying KVL to the current path as shown in Fig. 9, we get

$$V_l = V_L + V_h$$

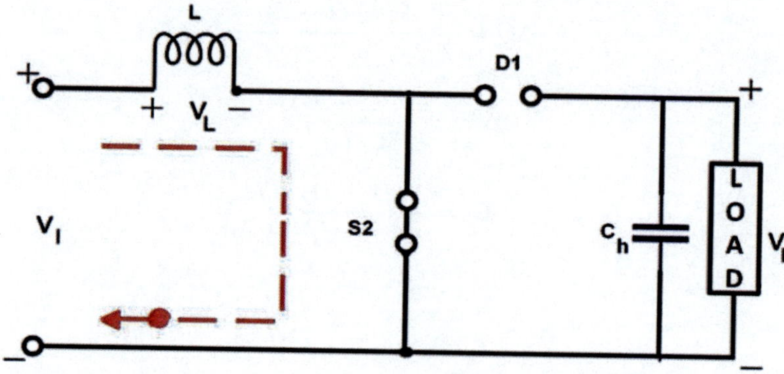

Fig. 8 Represents the direction of current path when switch S_2 is conducting (Boost operation)

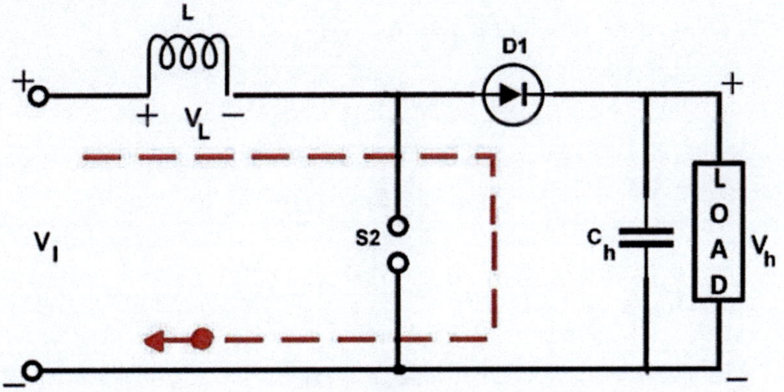

Fig. 9 Represents the direction of current path when switch S_2 is OFF (Boost operation)

$$V_L = V_l - V_h \tag{7}$$

Under the steady-state condition, by equating the inductor voltage over one time period, we get

$$D\ T_s + (V_l - V_h)\ (1 - D)\ T_s = 0$$

$$V_l T_s = V_h\ (1 - D)\ T_s$$

$$\frac{V_l}{V_h} = (1 - D) \tag{8}$$

The average value of inductor current can be obtained from the rectangular area of waveforms as shown in Fig. 10.

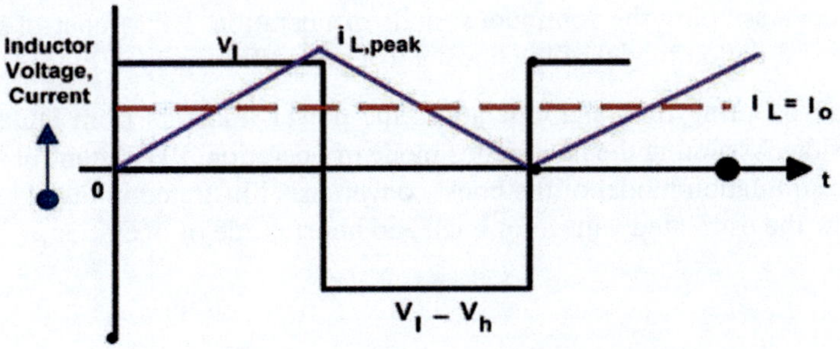

Fig. 10 Graphical representation of Inductor voltage and current (Buck mode)

$$I_{Lavg} = I_o = \frac{1}{2} i_{L,peak}$$

$$I_o = \frac{V_h D T_s}{2L}(1 - D)^2$$

$$I_o = \frac{V_h D}{2L f_s}(1 - D)^2$$

$$I_o = \frac{V_l}{R} = \frac{V_h D}{2L f_s}(1 - D)^2$$

$$\frac{V_l}{V_h} = (1 - D) \tag{9}$$

The capacitor C_h can be determined by the allowed ripple in the load voltage ΔV_h, which is typically 1% of the load voltage and is given by

$$\Delta V_h = \frac{\Delta Q}{C_h}$$

$$\Delta V_h = \frac{I_o D T_s}{C_h}$$

$$C_h = \frac{I_o D}{f_s \Delta V_h} \tag{10}$$

3 Proposed Simulation Models of Buck and Boost Mode of Operation

Boost mode: During this boost operation, the power transfers from DC load to DC supply side. Assuming the continuous mode of operation, PWM control scheme is used. The simulation model of the boost converter is illustrated in Fig. 11.

Buck mode: During this buck operation, the power transfers from DC supply to DC load side. Assuming the continuous mode of operation, PWM control scheme is used. The simulation model of the boost converter is illustrated in Fig. 11. Tables 1 and 2 show the estimated values for buck and boost mode of operation.

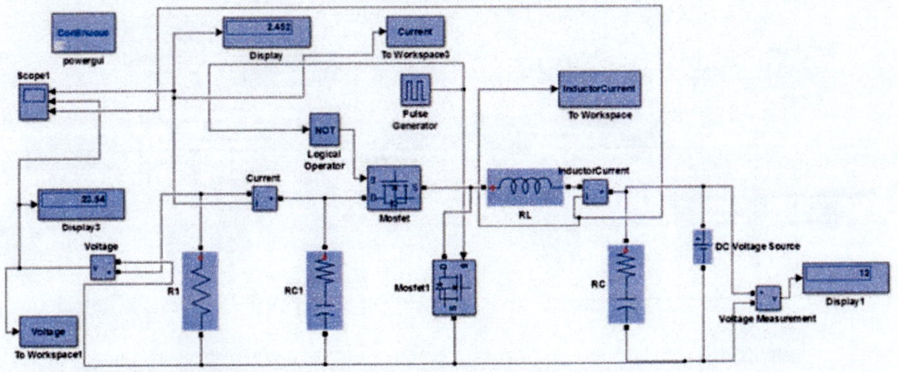

Fig. 11 Simulation model of Boost converter

Table 1 Estimated parameter values of Boost converter

Variable	Parameter	Buck mode
V_l	Supply Voltage	12 V
V_h	Load Voltage	24 V
f_s	Converting Frequency	20 kHz
P	Output Power	120 W
L	Inductor	30 μH
C_l	Capacitors	4.167 mH
Δi	Inductor current ripple	0.1 A
ΔV	Output voltage ripple	3%

Table 2 Estimated parameter values of Buck converter

Variable	Parameter	Boost mode
V_h	Supply Voltage	24 V
V_l	Load Voltage	12 V
f_s	Converting Frequency	20 kHz
P	Output Power	120 W
L	Inductor	30 μH
C_h	Capacitors	4.167 mH
Δi	Inductor current ripple	0.1 A
ΔV	Output voltage ripple	3%

4 Simulation Results and Discussion

Simulink results of inductor current, output voltage, and output current are shown in Figs. 12, 13, 14 and 15. We can conclude that the load voltage is 24 V and load current is 2.5 A, obtained as per the designed values.

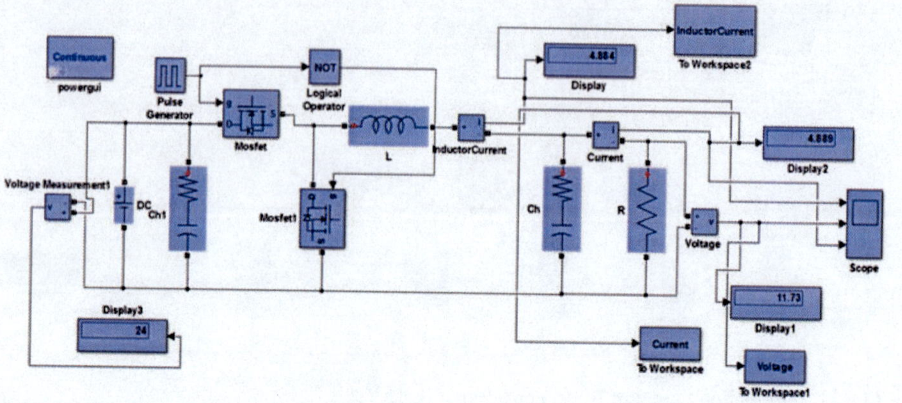

Fig. 12 Simulation model of Buck converter

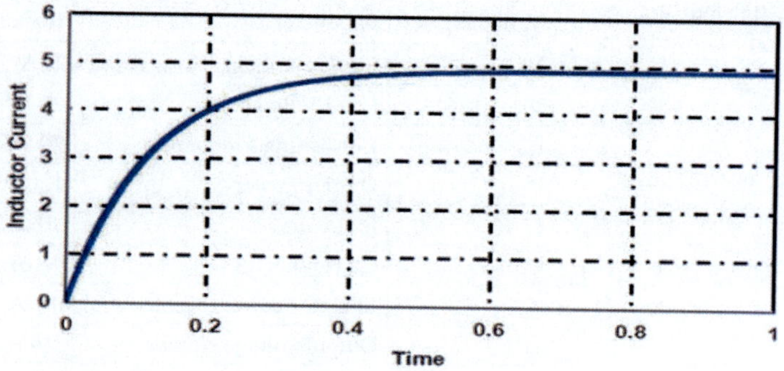

Fig. 13 Simulink result of Inductor current in Boost mode

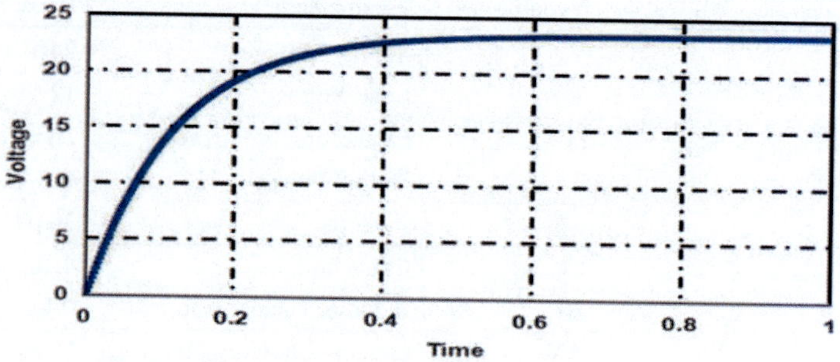

Fig. 14 Simulink result of Output Voltage in Boost mode

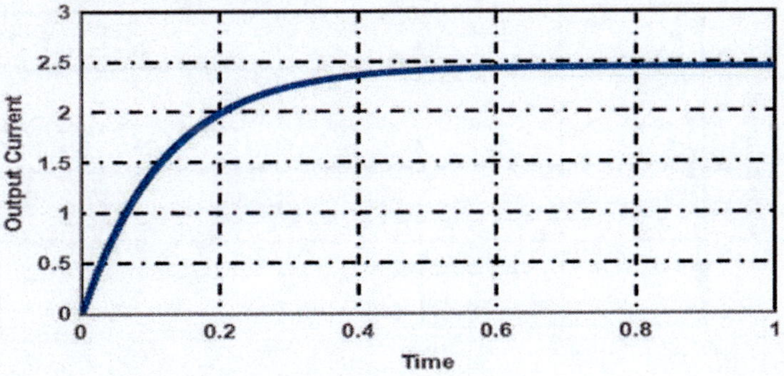

Fig. 15 Simulink result of Output Current in Boost mode

Simulink results of inductor current, output voltage, and output current are shown in Figs. 16, 17 and 18. We can conclude that the load voltage is 12 V and load current is 5.3 A, obtained as per the designed values. The inductor current ripple is 0.1 A.

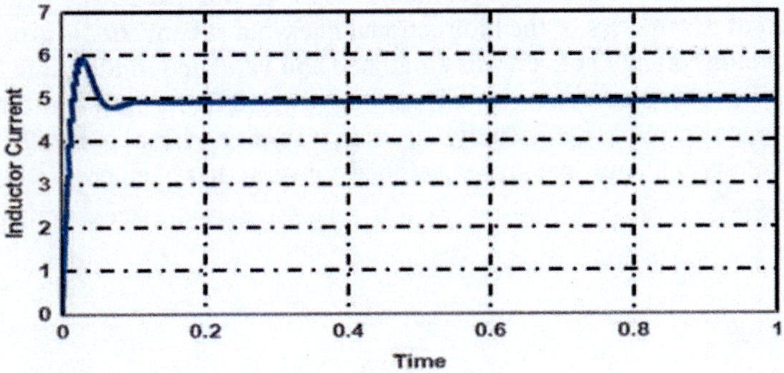

Fig. 16 Simulink result of Inductor current in Buck mode

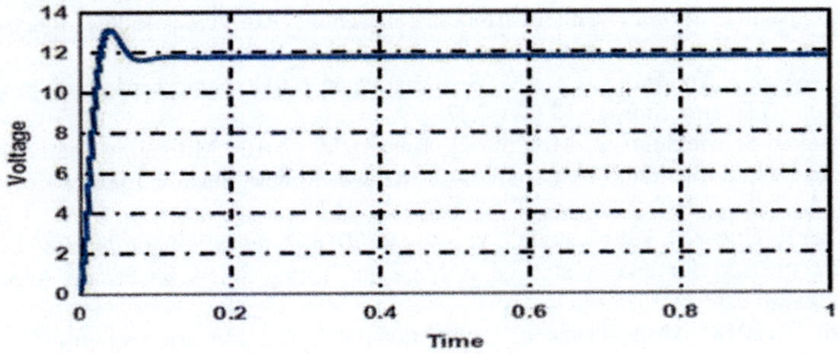

Fig. 17 Simulink result of Output Voltage in Buck mode

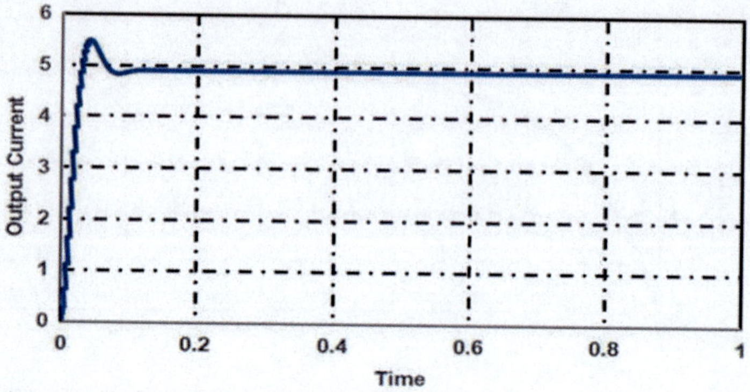

Fig. 18 Simulink result of Output Current in Buck mode

5 Conclusion

Solar PV system with supercapacitor energy storage system can act as an energy buffer for smoothing the PV power fluctuations. In this paper, the detailed study and design of parameters of the bidirectional buck-boost converter is proposed. The developed bidirectional converter is simulated and validated in Simulink MATLAB software. The designed converter allows the power transfer between the grid and energy storage system. In this study, the energy storage system is ideally chosen as constant voltage, and however, the developed converter has been proved for its better performance.

References

1. Renewables 2020 Global Status Report, (Paris: REN21 Secretariat). ISBN 978–3–948393–00–7. REN21 (2020)
2. Pegueroles-Queralt J, Bianchi FD, Gomis-Bellmunt O (2015) A power smoothing system based on supercapacitors for renewable distributed generation. IEEE Trans Industr Electron 62:343–350
3. Libich J, Máca J, Vondrák J, Čech O, Sedlaříková M (2018) Supercapacitors: properties and applications. J Energy Storage 17:224–227
4. Shivashankar S, Mekhilef S, Mokhlis H, Karimi M (2016) Mitigating methods of power fluctuation of photovoltaic (PV) sources—a review. Renew Sustain Energy Revi 59 (2016) 117–1184
5. Tytelmaier K, Husev O, Veligorskyi O, Yershov R (2016) A review of non-isolated Bidirectional DC-DC converters for energy storage systems. Int Young Scientists Forum Appl Phys Eng (YSY), 22–28
6. Kondrath N (2018) An overview of Bidirectional DC-DC converter topologies and control strategies for interfacing energy storage systems in microgrids. J Electri Eng, 11–17
7. Mohan N, Undeland TM, Robbins WP (2007) Power electronics: converters, applications, and design. Wiley, Third edition January

Chapter 14
Speech Separation Using Deep Learning with MATLAB

Chandra Mahesh Saga, V. B K L Aruna, and K. Venkata Ratna Prabha

1 Introduction

Speech separation is separating an input monaural signal into its individual auditory sources. Earlier the attempts are performed out in conduct to suppress the background noise or sound from the audio signal instead of insulating the different speakers individually from the audio signal. Some approaches are discussed which used only hardware models to attain speech signal separation, array signal processing which uses multiple microphones and follows the principle of time–frequency masking and W disjoint Orthogonality. This array signal processing model is one among them. Later, more refined methods are introduced, they include methods which are speaker-dependent, and we separate the different speakers' speech. In other references, we can indicate that such a method relies on the speaker's speech signal characteristics which are distinct for every speaker. We can use those characteristics to compare and separate speech signals from different speakers, and some techniques are implemented on the limited subject matter and a few techniques are grammar-based. Speaker independent models are on the prevailing research tract. They are currently very popular and nevertheless are well-experimented problems in speech processing [6]. In order to attain a consistent result, it demands a preceding expertise and an appropriate microphone configuration has to be designed and constructed in view of settling through signal processing. Another complication that originates with this is the label permutation problem where we cannot determine the corresponding speaker. The cocktail party effect refers to the manner of separating the target speaker source while muting all the other sources in a monaural signal. The task of achieving the separation of the destination source from a monaural signal with merely one microphone is called Monaural Speech Separation. Humans can focus on the target source in the noisy environment because of our natural thinking ability, and this intelligence can be imparted to machines using machine learning.

C. M. Saga (✉) · V. B. K. L. Aruna · K. Venkata Ratna Prabha
Velagapudi Ramakrishna Siddhartha Engineering College, Vijayawada, Andhra Pradesh, India

S. Dawn et al. (eds.), *Smart and Intelligent Systems*, Algorithms for Intelligent Systems,
https://doi.org/10.1007/978-981-16-2109-3_14

The machine can be trained with different instances to bring out the most optimal results. The way that the machines perform this task with the assistance of neural networks is called Cocktail Party Problem. It has been an intractable problem for the machines to perform this task. Earlier the complexity of speech separation is identified as Speech Processing Problem, and several experiments are conducted with many Array signal processing systems [4], but they did not yield the envisioned results. Later it is worked out using machine learning techniques, which made it much complex as it requires the specification of features to figure out the obstacle [5]. Deep Learning became the substantial move for this trouble as we are not expected to show any collection of features for classification, and it identifies its own features as the data passes through the neural network. Deep Learning is referred to as a unique case of Machine Learning. It performs the classification and regression directly from the images, sounds, or text files without specifying the features explicitly by the users. The deep learning is so-called because the term deep refers to the quantity of layers present in the network. The process is carried out with recent advancements in deep learning framework, which is deep clustering where the individual sources are clustered with a trained neural network and can assign an embedding vector to each source to separate auditory sources.

2 Related Work

The Speech separation being the most fundamental problem in audio processing subjected to numerous experiments over the decades. Nozomu Hamada [1] presented an array processing solution to separate multiple speech signals by utilizing a triangular microphone array which is based on the Time–Frequency Masking and W Disjoint Orthogonality. It involves a sophisticated environment and lab setup. Hershey et al. [7] proposed the technique of separating the multiple speakers into individual embedding's which was achieved using a deep clustering framework which is based on K-means clustering and introduced permutation-free function. Deep learning is used instead of Machine Learning because machine learning works efficiently on smaller data, but it does not meet our required expectations when employed on larger data. Deep learning is competent in handling a large dataset which is required to develop a Speaker Independent model by training the model with numerous speakers. In speaker separation, the class-based methods work effectively for speaker-dependent case but they are not relevant in speaker-independent case as it encounters permutation problem so partition-based methods are taken for this task [3]. A Convolutional Neural Network CNN is a type of Deep Neural Network typically utilized in the signal prediction where the input signal is converted into 2D representation. CNNs [2] are used in speech processing often as they are used to converting the signals into the frequency domain and analyze them as a spectrogram image. The network learns the unique features in the samples passed through the network, and classification is performed according to it. It is identified as an image classification problem other than speech classification and identifies the features like color, thickness, and orientation basically in the initial layers and more features in the next layers on the network [8]. A CNN layer performs functions like convolution, ReLU, and pooling. This work demonstrates a speaker-dependent model of speech separation using the binary masking technique and how it can be extended to the three-speaker models by employing variable threshold techniques in masking. Our model is capable of detecting and separating the speech of the speakers that had never been seen before.

3 Data Set

This dataset is created for the estimation of mask and for training purposes with three speakers which are recorded in the real-world environment and saved as.wav files. The audio data is taken from three speakers with each duration of about 100 s. This data is utilized for training as well as the parts of this data are needed for validation purposes. The monaural signal is taken as an input through the microphone with a duration of 20 s.

4 Network Architecture

The training audio clips and the validation audio clips are sampled to an 8 kHz sampling rate, and stereo audio is converted to mono audio by considering solely one channel. Short-time Fourier Transform STFT is implemented utilizing a Hanning Window of window length 128, FFT size 128, and the overlap length 127 to ensure higher overlap on the data which enhances efficiency. This will emerge in the audio input feature. For larger data, the overlap length must be decreased. This will serve as input to the CNN. The layers connected in the network are given in Fig. 1.

The network is implemented in MATLAB, and its incorporated operations are employed for implementing STFT transformations. Sigmoid Activation follows for all the network layers. Batch normalization is performed after all Activation Layers in order to standardize the values based on their mean and standard deviation values. Drop out layer is employed after normalization in order to evade over-fitting of data. Using a minibatch size of 64 samples, the network is trained with Adam Learning Optimizer model.

5 Implementation

The network is originally designed for the separation of two speakers with the monaural signal as input. The model developed is a speaker-dependent model owing

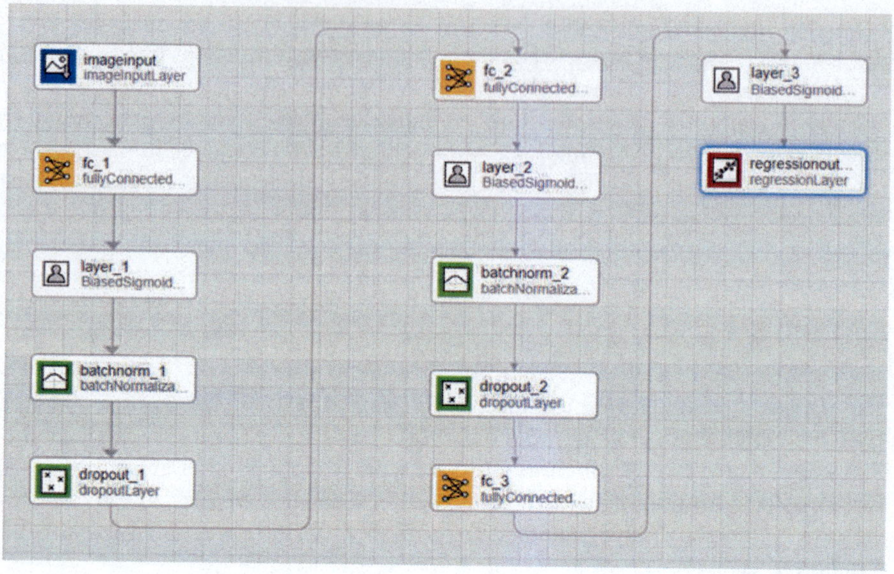

Fig. 1 Neural network based architecture

to hardware limitation as to train a substantial estimate of speakers. The comprehensive process is implemented on a single i5 processor CPU with the NVDIA MX110 graphic card. The training audio clips and validation audio clips are committed to the network which is stored in.wav files and after normalization by applying STFT transformation. The training of the audio clips lasts for about 11 min with an understanding rate of 0.001 along with validation results and 3 min if the validation process is skipped. The separated embedding of spectrogram is modified into an audio file by inverse STFT transformation. The same process is implemented for the three speaker model. A User Interface can be established in order to perform the operations like loading the training data with ease, accepting the input monaural speech, to train the network, single out the desired output speaker, and so on.

6 Experimentation and Results

The network is loaded with the training data. The input monaural signal is applied as a mixture of audio signals of both trained speakers which was a part clipped from the training data. The network implements a state of art separation for validation data. A mask is computed with the training and validation data and when the binary masking is performed, the threshold for separation of two speakers was 0.5 on a scale of 0 to 1. Another input was considered with the two speakers to assess the network. The network separates the two speaker speeches but not as effectively as it does in event of training data with the threshold level of the mask being at 0.5. The

Table 1 Signal to Distortion Ratio analysis between validation and testing data inputs

Two speaker model	Data	Signal strength in dB	Distortion	SDR in dB
Speaker 1	Validation	19.3195	0.756	10.8
Speaker 2		20.2910	1.164	9.48
Speaker 1	Testing	17.4620	1.6168	6.64
Speaker 2		20.5921	3.425	4.949

network is processed again with training data with an increased number of epochs. The separation efficiency increased with reference to the preceding one. This suggests that on increasing the number of epochs in training, the neural network efficiency will be increased. It is likewise observed that the separation threshold for the binary masking need not be at 0.5 always; it differs from speech to speech. For better speech separation, it is relevant to maintain a variable threshold in speech separation. This entails that the productivity of the network depends upon the number of epochs and thresholding values. To investigate the competence of the network for two speaker model, the audio clips are taken independently and the values are gauged and later the signals are mixed as a monaural signal and passed into the network in order to analyze by what amount the signal is distorted and to calculate the signal to distortion ratio for both validation and training clips. The details are presented in Table 1. It is observed that the signal to distortion ratio is extremely high for the validation clips and low for the validation clips. The signal that is trained in the network is recovered easily than the alternative signal of the same speaker.

For better separation, the SDR factor must be higher. The variation in the SDR factor between testing and validation data is because the network did not identify the deep features in it that are expected for the separation. So we have to increase the number of epochs so the network identifies relevant features for separation. The concept of variable thresholding can be extended to multi-level variable thresholding. This makes a basic phenomenon for three speaker speech separation model by exploiting the masking. The input is taken as the monaural signal of three inputs but the network is requested to perform the separation for two speakers alone. When the network clusters the two speakers of the masking with increased epochs, the clusters of different speakers will move apart from the thresholding value 0.5 as all speakers lie at level 0.5 at the start. After training the network, the untrained speaker lies at 0.5 and the trained speakers separate. By applying variable thresholds, the different speaker clusters are taken and the speeches are reconstructed by applying Inverse STFT transformation. The SDR values for three speaker separations are illustrated in Table 2.

Table 2 Signal to Distortion Ratio analysis for three speaker models

Three speaker model	Signal strength in dB	Distortion	SDR in dB
Speaker 1	19.3195	2.12	6.3
Speaker 2	20.2910	4.26	3.8
Speaker 3	18.1275	3.305	3.8

7 Conclusion

Speech separation technique is administered to resolve the ineptitude to comprehend separately two distinct speech signals amid noise cited as the "cocktail party problem". The Segregation of speech signals of specific speakers from a monaural signal is accomplished using Convolutional Neural Networks for two speakers and three speaker models. The SDR improvements can be worked out by training the model with a colossal volume of data and by strengthening the number of epochs. The extension of the neural network model with multi-level variable threshold indicates a significant advancement in settling the speech separation. Deep clustering is employed to isolate speech signals but there are nevertheless some disparities and inconsistencies. When data exceeds 100 s, the software becomes irresponsive and the system hangs. Likewise, there is scant availability of training data for the network which inevitably culminates in warped male output speech and separated female speech output is slightly incorporated with the male signal. Although MATLAB is exceptionally productive, the complication emerges with the computational proficiency. To run this program on MATLAB, we require a minimum of INTEL XEON 22 CORE PROCESSOR and for training circumstances, lots of datasets of speech from Wall Street Journal (WSJ0) are accessible, but their cadence is distinctive when compared to our inflection, accordingly we lack the dataset of our region.

References

1. Hamada N (2008) Separation of multiple speech signals by using triangular microphone array. ECTI Trans
2. Ephrat A (2018) Looking to listen at the cocktail party Google research. ACM Trans
3. Isik Y, Roux JL, Watanabe S (2005) Single channel multi speaker separation using deep clustering. Mitsubishi Res
4. Chowdhury TA, Sehgal A, Kehtarnabaz N, Integrating signal processing modules of hearing aids into a realtime smartphone app
5. Fu T, Yu G, LiliGuo, Wang Y, Liang J (2017) Integrating signal processing modules of hearing aids into a real time smartphone app. 2017 IEEE International conference on FUZZY systems
6. Sehgal A, Kehtarnabaz N (2010) Utilization of two microphones for realtime low latency audio smartphone apps. 2010 IEEE 24th IEEE international conference on advanced information networking and applications
7. Hershey JR, Chen Z, Le Roux J, Watanabe S (2016) Deep clustering discriminative embeddings for separation and segmentation. MERL Res

8. Kim D-S, Lee S-Y, Kil RM (2012) Auditary processing of speech signals for robust speech recognition in real world noisy environments. 2012 IEEE 15th International IEEE conference on intelligent transportation systems

Chapter 15
Maximum Power Extraction from Solar Photovoltaic Strings Using Grey Wolf Optimization Technique Under Partial Shading Condition

T. Nagadurga, P. V. R. L. Narasimham, and V. S. Vakula

1 Introduction

Numerous maximum power extraction techniques were reviewed in the recent articles about the incident of uneven irradiance falling on the photovoltaic panel, ensures the decrease in power output obtained from a solar photovoltaic system and the generated hot-spot amends the solar cell. However, the gesture of the photovoltaic module beneath fractional shading is unstable with respect to time, tracking circuit design for a solar photovoltaic power system must be furnished through aspects, for example, tracking the global optima at the diverse environment, e.g., shading, deterioration of photovoltaic panel, and flexibility toward the adjustment of P–V plot changes in PV module, soft, and stable capture activities. Mohanthy et al. [1] presented the design of a maximum power tracking circuit with grey wolf optimization technique, which abides the limitation of conventional tracking techniques like poor tracking performance, efficiency, and oscillations near and around MPP. The authors developed the experimental prototype of the proposed optimization technique to validate the simulation results. Eltamaly and Farh [2] applied the Gray Wolf technique along with fuzzy Controllers (FC) to mitigate the oscillations near and around GMPP. Crepinsek et al. [3] suggested a novel TLBO algorithm which seems to be a rising star among all

T. Nagadurga
Department of Electrical and Electronics Engineering, Lakireddy Bali Reddy College of Engineering, Mylavaram, AP, India

P. V. R. L. Narasimham (✉)
Department of Electrical and Electronics Engineering, Velagapudi Ramakrishna Siddhartha Engineering College, Kanuru, AP, India

V. S. Vakula
Department of Electrical and Electronics Engineering, JNTUK-University, College Of Engineering, Vizianagram, AP, India

S. Dawn et al. (eds.), *Smart and Intelligent Systems*, Algorithms for Intelligent Systems,
https://doi.org/10.1007/978-981-16-2109-3_15

meta-heuristic approaches. The authors reported that the proposed algorithm outperforms some of the optimization techniques regarding constrained benchmark functions. Finally, the researchers concluded that the TLBO algorithm finding through qualitative and quantitative analysis in their work. Mohammad et al. [4] proposed an adaptive perturb and observe method based on the variation of step size for effective tracking of maximum power in the presence of shading. To validate the performance of the proposed work, the authors considered and pretend in a Matlab environment. The authors compared the execution of the recommended approach with conventional P&O and incremental conductance techniques. The accomplishment rate of the suggested technique and its concert were evidently established under the shading situation. Nagadurga et al. [5] had presented a TLBO algorithm for harvesting a maximum amount of power from a solar PV system under changing solar weather conditions. The authors compared the proposed technique with the PSO technique and expressed that implementation of TLBO is easy due to its simple tuning parameters to the PSO algorithm. Fathy et al. [6] suggested an improved TLBO for incorporating the MPPT tracker in photovoltaic systems. The authors considered the best student as a teacher in the learning stage to improve the TLBO algorithm. An experimental prototype is developed with the help of a PIC microcontroller and Arduino to implement the proposed approach (ITLBO) for tracking GMPP under fractional shading conditions of the PV system. The authors test the algorithm for different ways of shading patterns and concluded that GMPP is achieved with the proposed approach in comparison with TLBO and PSO methods. Reddy et al. [7] presented a particle swarm optimization method for controlling the harmonics in the distributed energy generation system to improve the power quality features in the power grid. The authors developed the proposed controlling technique and implemented it in the Matlab/Simulink background for various types of loads. The authors gave the significant outcome analysis of the projected control technique with the existing techniques. Rao et al. [8] improved the basic teaching–learning optimization design to improve the exploration and exploitation range via offering the idea of a flexible teaching factor. The authors tested the performance of the improved TLBO algorithm on unconstrained benchmark functions and concluded so as to the suggested improved TLBO method gave better results. Rezk et al. [9] reviewed two optimization approaches specifically PSO and Cuckoo search (CS) for obtaining the greatest power output from the solar photovoltaic panel under shaded conditions. The authors compared the proposed algorithms over the INR-based tracker and concluded that a PSO and CS technique promises convergence to the global optima. The tracking time for MPP in case of CS technique is compact correlated to PSO in all the considered cases. Selvamuthukumaran et al. [10] suggested a flexible step size maximum power point tracking by applying the stepped comparison search method. The suggested algorithm is able to efficiently track the optimum peak power point under varying atmospheric conditions. The robustness of the proposed algorithm has been tested for multiple shading cases with the help of an experimental prototype. Nafesh et al. [11] implemented a new maximum power extraction algorithm of the solar PV system by utilizing the photovoltaic module current and irradiation levels at diverse shading situations. Zhang et al. [12] presented sorting particle swarm optimization based on an artificial intelligence algorithm. The authors developed a stand-alone photovoltaic system based on Z-source inverter; experiments are done with the established ZSI. The authors reported that the proposed sorting PSO algorithm reduced the oscillations of PV voltage during the tracking of MPP. Nagadurga et al. [13] had implemented the particle swarm optimization algorithm for tracking maximum power from a solar PV system for different irradiation levels of solar energy. Mao et al. [14] proposed a novel two-stage particle swarm optimization technique with reduced steady-state oscillations. They also used adaptive speed factor into the improved PSO algorithm, and the test results show that the proposed algorithm gave better results far compared to the existing technique.

2 PV System Descriptions

The refined power from the solar PV unit is usually very low; to improve the power rating, there is a need for a merging number of solar cells in series and parallel. The complete requirement of the solar PV system and boost converter used in this proposed work is presented in Table 1. At the time of partial shading conditions, several peaks on the MPP plot were predicted due to the existence of free-wheeling diodes. Fly-back diodes are coupled across each unit at the time of fractional shading condition to minimize the possibility of hotspots. Figure 1 displays the 4S arrangement of PV blocks associated with the series along with P–V characteristics.

Table 1 Specifications of solar PV module (KYOCERA solar KC200GT)

Parameter	Value
Number of cells per module	54
Voc(V)	32.9 V
Isc(A)	8.21 A
VMpp(V)	26.3 V
IMpp(A)	7.61 A
PMpp(W)	200.143 W
Input inductance (L)	10 mH
Input side capacitance(Cin)	10e-6 F
Output side capacitance(Cout)	330 μF
Switching Frequency	25 kHz

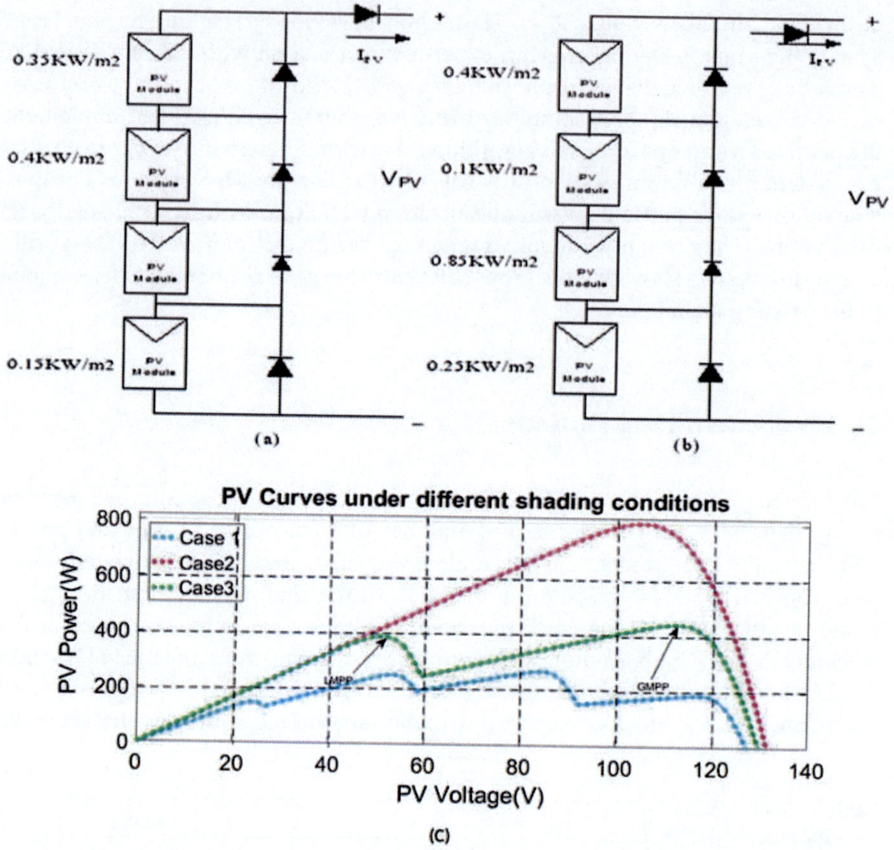

Fig. 1 4S arrangement under various shading patterns. **a** Pattern 1. **b** Pattern 2. **c** P–V characteristics of a solar PV Module under uniform and partial shading condition

3 GWO Technique for Tracking Maximum Power During Partial Shading

GWO technique imitates the hunting procedure and leadership chain of command of wolves offered in a natural world refined by Mirjalili et al. [15]. Wolves are considered to be the major of the food cycle, and they wish to endure in a group. Different kinds of wolves are committed for the simulation of the leadership hierarchy, such as alpha (α), beta (β), delta (δ), and omega (ω). For mathematical analysis of the social hierarchy of wolf, alpha (α) is treated as the fittest solution as a result the next solutions are named as beta (β) and delta (δ), respectively. The residual candidate solutions are simulated to be omega (ω). The major stages of GWO technique, particularly (i) hunting, (ii) chasing, (iii) tracking for prey, (iv) encircling prey, and (V) attacking prey those are carried out to construct grey wolves for achieving global optimum point.

The following equations are for the modeled behavior of wolves during hunting and encircling the prey:

$$\vec{D} = \left|\vec{C} * \vec{X_p}(t) - \vec{X_p}(t)\right| \tag{1}$$

$$\vec{X}(t+1) = \vec{X_p}(t) - \vec{A} * \vec{D} \tag{2}$$

where

T Coefficient vectors
D, A, & C Coefficient Vectors
X_P Position vector of Prey
X Grey wolf Position vector
The vectors A and C are computed using the below equations

$$\vec{A} = 2\vec{a} * \vec{r_1} - \vec{a} \tag{3}$$

$$\vec{C} = 2 * \vec{r_2} \tag{4}$$

where the value of $'a'$ decreases directly from 2 to 0 throughout the emphases and random values r_1, r_2 in the scope of [0, 1]. The hunting procedure is regularly guided by 'α' group wolves called as pioneers succeed via beta (β) and delta which may likewise perform chasing periodically. Delta (δ) and omega wolves are afraid of the enclosed wolves in the group. Hence, 'α' wolf is measured because of the fittest arrangement having better information about the position of the prey. The wolves stop the hunting process assaulting the prey. The flow chart of the Grey wolf swarm optimization is shown in Fig. 2.

3.1 Implementation of GWO for MPP Tracking

The MPPT controllers calculate V_{PV} and I_{PV} of the solar PV module through sensors and compute the power output for the consequent grey wolves', i.e., duty ratios. Figure 2 shows the procedure of tracking the prey in the Grey Wolf Optimization-based MPPT technique. In case of shading, the Power versus voltage characteristic plot is determined through a variety of local and global power points. It is too found that at the time wolves identify the MPP, their resultant coefficient values of D, A, and C turn into nearly equal to zero. The GWO fitness function is given by.

$$\mathrm{P}\left(\mathrm{d}_\mathrm{i}^\mathrm{K}\right) > P\left(\mathrm{d}_\mathrm{i}^{\mathrm{k}-1}\right) \tag{5}$$

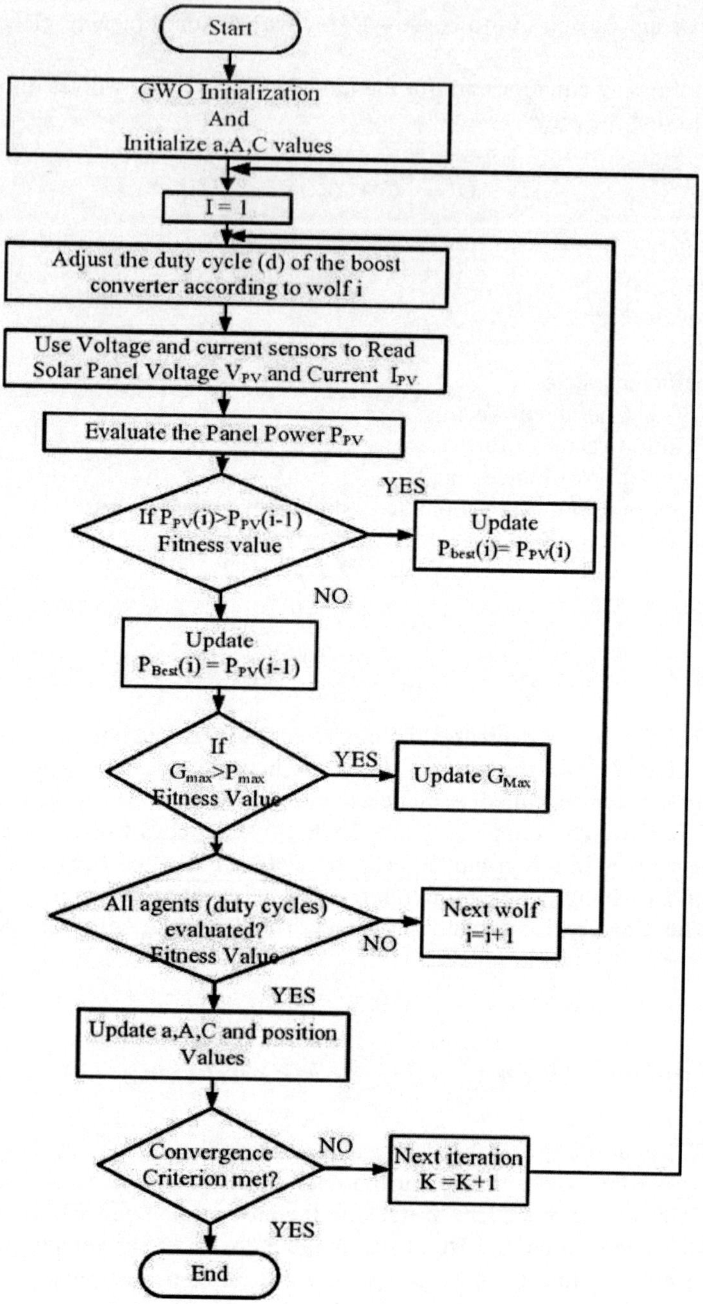

Fig. 2 Flow chart of GWO technique

where

P—Output Power in W; d—duty ratio; i—Total Number of grey wolves; k—Iteration count.

4 Results and Discussion

In this research work, MATLAB/SIMULINK software is used broadly to study the Grey wolf optimization (GWO) technique for getting more power from the solar photovoltaic system during fractional shading circumstance and to estimate the performance of GWO; it was compared with the PSO algorithm. In the GWO algorithm parameter 'a' this is linearly decreased from 2 to 0. Matlab program code is written in M-file for employing these algorithms. The duty cycles are calculated for a boost converter using the GWO algorithm and are linked to the simulation circuit. The simulation circuit of implementation of GWO algorithm for MPP tracking under shading conditions with four series connected PV modules is shown in Fig. 3.

Exhaustive simulation work is carried out for the developed simulation circuit with different partial shading patterns. The P–V curves for the 4S configuration with partial shading conditions applying the GWO algorithm is shown in Fig. 4. For the shading pattern of G1 = 1000 w/m^2, G2 = 1000 W/m^2, G3 = 500 W/m^2, and G4 = 500 W/m^2, the GMPP is 439.5717 W. For second pattern G1 = 800 W/m^2, G2 = 600 W/m^2, G3 = 400 W/m^2, and G4 = 200 W/m^2, the maximum power extraction technique gets resumed and the GWO technique is capable to track the GMPP of 335.63 W. From this simulation graph, it is noticed that the proposed tracking technique GWO produces superior tracking speed and the sustained oscillations dissolve speedily

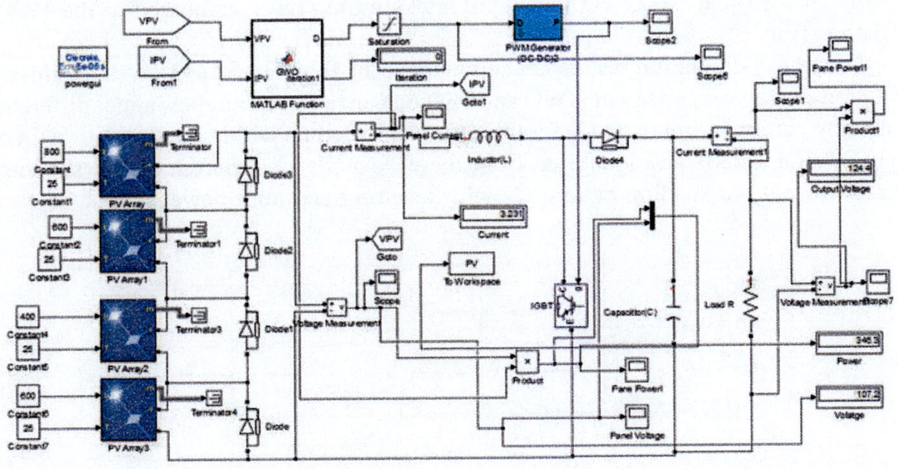

Fig. 3 Simulation circuit of KC200GT series-connected PV module under different shading patterns by implementing GWO algorithm

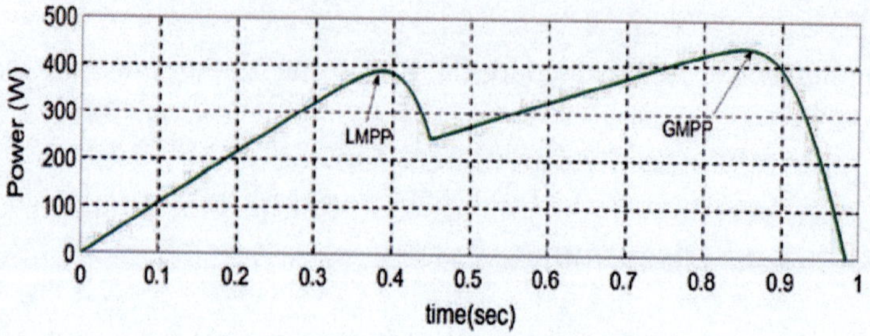

Fig. 4 Power curve under second shading pattern like 1000 W/m^2, 1000 W/m^2, 500 W/m^2, and 500 W/m^2

Table 2 Simulation results of the Solar PV system for shading patterns

Pattern	Global Power Output (W)	MPPT Method	Tracked Power output(W)	Voltage(V)	Current(A)	%η
Pattern-1 [1000,1000,500,500]	439.5717	P&O	298.26	122.12	2.447	67.85
		GWO	399.25	125.6	4.28	94.3
Pattern-2 [800,600,400,200]	335.63	P&O	165.61	90.834	1.816	49.34
		GWO	300.62	84.902	3.95	89.5

when related to P&O technique. Table 2 presents the output results of the GWO and P&O MPP tracking algorithm for different shading patterns. The tracked powers output from solar photovoltaic string by means of grey wolf optimization and perturb and observe (P&O) optimization techniques are shown in Fig. 5, respectively, and it is observed that the GWO technique tracks more power compared to the P&O method (Fig. 6).

Statistical simulation results are summarized in Table 3. Table 3 gives the information on the comparison of GWO and PSO optimization techniques under different shading patterns from G_1 to G_6 for the KC200GT module within the simulation time period of 1 sec. It was identified from the case study of different partial shading patterns that the shading pattern G_1 will develop maximum power output. It was

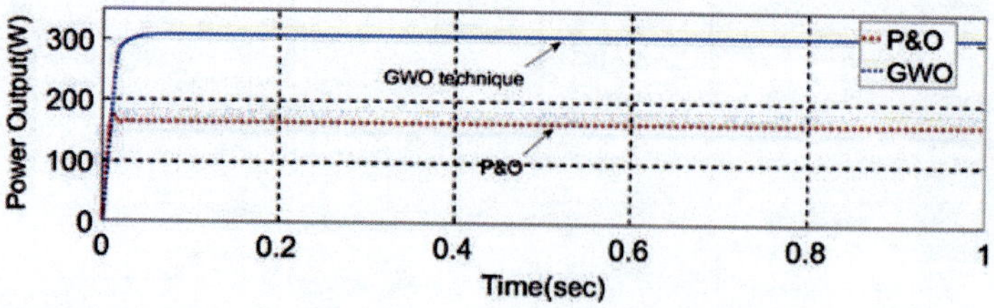

Fig. 5 Simulation result of tracked output power using GWO technique and Perturb and Observe techniques for shading pattern of 800 W/m^2, 600 W/m^2, 400 W/m^2, and 200 W/m^2

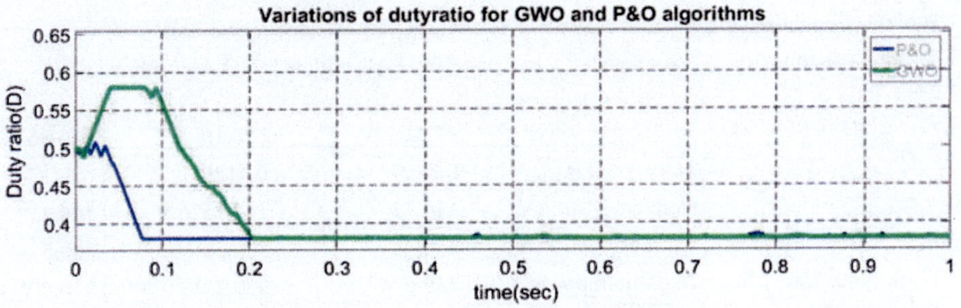

Fig. 6 Simulation result of duty cycle variations using GWO technique and Perturb and Observe techniques for shading pattern of 800 W/m^2, 600 W/m^2, 400 W/m^2, and 200 W/m^2

observed from the simulation results of different shading conditions that GWO will work extensively for partial shading conditions and gave better performance superior to particle swarm optimization in view of speed and certainty. To verify the expected output from the proposed technique both PSO and GWO approaches are simulated in the MATLAB/SIMULINK environment under different partial shading patterns (G_1 to G_6) and related with Particle swarm optimization technique (PSO) from the existing literature. The statistical simulation results show that the GWO technique tracks more voltage and power output from the solar photovoltaic system compared to the PSO and P&O optimization techniques. Summarization of statistical simulation results like power, voltage, and current of the PV module under different shading patterns are presented in Table 3. By the statistical analysis, it is noticed that the switching signal (duty ratio) for boost converter to extract maximum power from the PV module is in the range of 0.2 to 0.6 for different shading patterns (G_1 to G_6).

5 Conclusion

During shading conditions, solar photovoltaic systems establish many peaks on the Power versus Voltage characteristic plot. Conventional maximum power extraction methods are convenient to trace the single peak in uniformly distributed solar irradiation on PV panel, but in case of shading, it may lose the global maximum point (GMPP) and sticks at local maxima (LMPP). To clear up this problem, meta-heuristic optimization techniques are able to apply for extracting more power from solar PV strings. GWO technique has been chosen as one of the various intelligent computing techniques to extract maximum power from solar irradiation beneath various shading patterns. Even though several techniques are excellent in tracking the global optima (GMPP), it undergoes many pitfalls like fatigue to convergence and fails to catch GMPP in case of changing shading pattern, and the probability of getting stuck with one of the local maxima (LMPP) is more. In this work, heuristic optimization technique, namely, Grey wolf Optimization (GWO), is realized for the universal optimization issue of shading of PV module that has GMPP and LMPP on the P–V

Table 3 Simulation results of the Solar PV system for various shading patterns

Different shading patterns	Parameter	GWO Algorithm	PSO Algorithm	P&O
G_1 = [1000, 900, 800, 700]	Maximum power (W)	624.13 W	624.16 W	620.36 W
	Duty @MPP	0.3297	0.4126	0.311
	Voltage @MPP(V)	114.13 V	113.1526 V	109.113
	Current @MPP(A)	5.454 A	5.326 A	3.865
G_2 = [900, 550, 100, 600]	Maximum power @GMPP(W)	336.6 W	330.2 W	316.4 W
	Duty @GMPP	0.3296	0.3021	0.297
	Voltage @GMPP(V)	82.46 V	81.2 V	80.7 V
	Current@GMPP(A)	3.849 A	3.55 A	2.85 A
G_3 = [750, 850, 600, 800]	Maximum power @GMPP (W)	340.0625 W	339.5 W	325.5 W
	Duty @GMPP	0.5127	0.5027	0.407
	Voltage @GMPP(V)	53.6725 V	53.21 V	51.5 V
	Current @GMPP(A)	6.48 A	6.112 A	6.143 A
G_4 = [600, 800, 400, 200]	Maximum power @GMPP (W)	258.29 W	254.2 W	245.2 W
	Duty @GMPP	0.5123	0.512	0.306
	Voltage@GMPP(V)	54.32 V	54.21 V	50.55 V
	Current @GMPP(A)	4.123 A	4.021 A	4.01A
G_5 = [600, 200, 800, 250]	Maximum power @GMPP (W)	171.2 W	168.5 W	164.5 W
	Duty @GMPP	0.412	0.392	0.386
	Voltage @GMPP(V)	86.21 V	85.12 V	84.3 V
	Current @GMPP(A)	2.67A	2.64 A	2.43 A
G_6 = [400, 600, 800, 100]	Maximum power (W) @GMPP	232.52 W	229.32 W	210.5 W
	Duty @GMPP	0.261	0.242	0.22
	Voltage @GMPP(V)	87.443 V	84.26 V	84.14 V
	Current @GMPP(A)	2.82 A	2.807A	2.514 A

characteristic curve. It is an appropriate MPPT technique for harvesting maximum power from a series-connected PV module under different shading patterns due to its better accuracy and tracking speed.

References

1. Mohanthy S, Subudhi B, Ray PK (2016) A new MPPT design using Grey Wolf optimization technique for photovoltaic system under partial shading conditions. IEEE Trans Sustain Energy 7(1):181–188
2. Eltamaly AM, Farh HM (2019) Dynamic global maximum power point tracking of the PV systems under variant partial shading using hybrid GWO-FLC. Sol Energy 177:306–316
3. Matej C, Liu S-H, Mernik l (2012) A note on teaching learning based optimization algorithm. Inf Sci 212(2012):79–93
4. Mohammed ANM, Radzi MAM, Azis N, Shafie S, Mohd Zainuri MAA (2020) An enhanced adaptive perturb and observe technique for efficient maximum power point tracking under partial shading condition. Appl Sci 10:3912
5. Nagadurga T, Narasimham PVRL, Vakula VS (2020) Global maximum power point tracking of solar PV strings using the teaching learning based optimization technique. Int J Ambient Energy. https://doi.org/10.1080/01430750.2020.1721327.
6. Ahmed F, Ziedan I, Amer D (2020) Improved teaching learning based optimization algorithm–based maximum power point trackers for photovoltaic system. Electr Eng 100:1773–1784
7. Reddy NN Chandrashekar O, Srujana A (2019) Power quality enhancement by MPC based multilevel control employed with improved particle swarm optimized selective Harmonic elimination. Energy Sources Part A: Recovery Utilization Environ Effects 22(1):23–34
8. Rao RV, Patel V (2013) An improved teaching –learning – based optimization Algorithm for solving unconstrained optimization problems. Scientia Iranica D 20(3):710–720
9. Rezk H, Fathy A, Abdelaziz AY (2017) A Comparision of different global MPPT techniques based on meta-heuristic algorithms for photovoltaic system subjected to partial Shading conditions. Renew Sustain Energy Rev 74(2017):377–386
10. Selvamuthukumaran R, Kumar YS, Rajesh G (2016) Global Maximum power point tracking of multiple PV modules under partially shaded condition using stepped comparison search. Electric Power Compon Syst. https://doi.org/10.1080/15325008.2016.1157106
11. Nafesh A, El-Shafy A (2010) Novel maximum power tracking algorithm for grid- connected photovoltaic system. Int J Green Energy 7(6):600–614
12. Zhang J, Ding K, Runjie M, Cai Y (2018) Global maximum power point tracking method based on sorting particle swarm optimizer. https://doi.org/10.1080/15435075.2018.1529579
13. Nagadurga T , Narasimham PVRL , Vakula VS (2019) Harness of maximum solar energy from solar PV strings using particle swarm optimization technique. Int J Ambient Energy https://doi.org/10.1080/01430750.2019.1611643
14. Mao M, Zhang L, Duan Q, Oghorada OJK, Duan P, Hu B (2017) A two-stage particle awarm optimization algorithm for MPPT of partially shaded PV arrays. Int J Green Energy 14(8):694–702. https://doi.org/10.1080/15435075.2017.1324792
15. Mirjalili S, Mirjalili SM, Lewis A (2014) Grey wolf optimizer. Adv Eng Softw 69:46–64. https://doi.org/10.1016/j.advengsoft.2013.12.007

Chapter 16
Impedance Source Inverter Based Asynchronous Motor Drive Using Different Modulating Signals

M. Ranjit, R. Giridhar Balakrishna, V. Ramesh Babu, and Jalluri Srinivasa Rao

1 Introduction

The ever-increasing global power demand urges the renewable energy sources to be extensively used along with conventional fossil fuel sources. The primary implication of low output voltage with the renewable energy sources puts a glitch on their use in high voltage applications. Generally, a boost converter which involves the complex circuitry or a bulky step-up transformer is used to enhance the voltage to a certain suitable level before feeding it to inverters. This makes the system more complicated and uneconomical. The Impedance Source Inverter (ISI) can be an alternative to surpass the above-specified limitations. An ISI is made with two passive elements (X-Shaped) and interfaced between the power source and inverter shown in Fig. 1. An ISI provides buck-boost conversion in a single stage, and hence, it is more reliable and economical [1–3].

The boosted dc voltage obtained from ISI is powered to an inverter which drives the AC load. The voltage source inverter (VSI) has six active states and two zero states. The power is transferred from source to load in the active states. On the other hand in a zero state, there is no power transfer from source to load. But in ISI, it has six active states and two zero states, and the shoot-through state is also present. An additional Shoot Through state is also obtained by shorting the load terminals.

M. Ranjit (✉) · V. Ramesh Babu · J. Srinivasa Rao
Department of EEE, VNRVJIET, Hyderabad, Telangana, India
e-mail: ranjit_m@vnrvjiet.in

V. Ramesh Babu
e-mail: rameshbabu_v@vnrvjiet.in

J. Srinivasa Rao
e-mail: srinivasarao_j@vnrvjiet.in

R. Giridhar Balakrishna
Department of EEE, VRSEC, Vijayawada, Andhra Pradesh, India
e-mail: balu.eee214@vrsiddhartha.ac.in

S. Dawn et al. (eds.), *Smart and Intelligent Systems*, Algorithms for Intelligent Systems,
https://doi.org/10.1007/978-981-16-2109-3_16

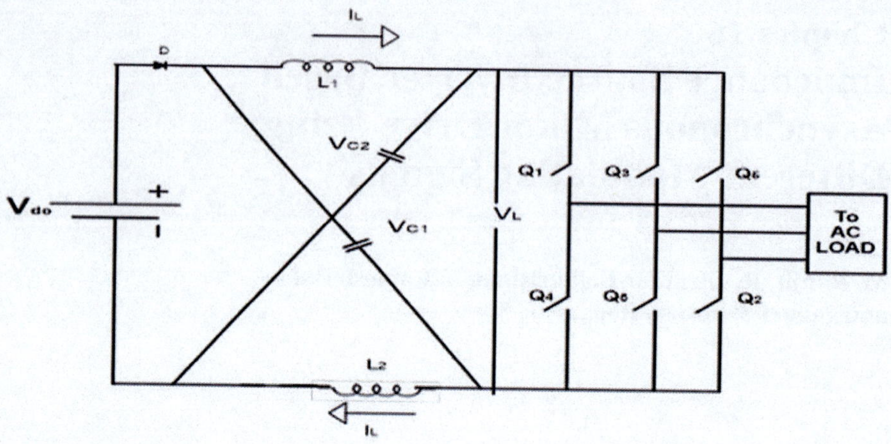

Fig. 1 Impedance. *Source* Inverter[1]

This condition occurs whenever both the switches in any one or more leg(s) conduct simultaneously.

In the past, different shoot-through control algorithms have been reported in [3, 4]. Those are as follows: Simple boost control [4], constant boost control [4], and maximum boost control [4].

Conventional asynchronous motors suffer from the problems of zero sequence current and electromagnetic interference. It results from the failure of bearings and hence reduces the life of the motor. To overcome these drawbacks, open-ended asynchronous motors have been used [5–7]. It could be obtained by connecting two conventional inverters on either side of OEWAMD. To generate the control signals to the inverters, various PWM techniques have been used [7–13].

In the proposed work, different modulating signals are generated using different SVPWM techniques. These modulating signals are compared with a carrier signal to generate the control pulses to the inverters. Using ISI, shoot-through operation is also analysed using simple boost control. An attempt has been made to simulate the ISI-based OEWAMD and also verify the boost functionality of the ISI.

2 Methodology

The proposed configuration schematic is shown in Fig. 2. Two ISIs are fed along with the conventional DC-AC converters on either side of the open-ended asynchronous motor drive. The two ISIs are used to boost the input voltage fed to conventional DC-AC converters. The control pulses for the DC-AC converters are generated from the various SVPWM techniques.

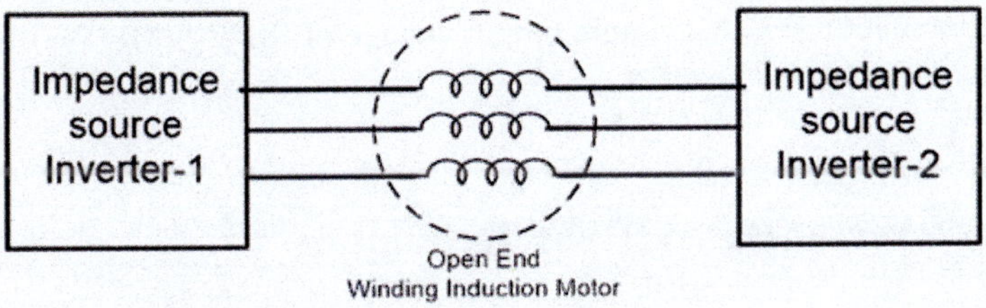

Fig. 2 Schematic of proposed configuration

(i) Simple Boost Control (SBC): In this control scheme, two straight lines are used as upper and lower shoot through (ST) lines as shown in Fig. 3. Whenever the triangular waveform is greater than the upper peak value, all the upper switching devices of the inverter are switched on, whereas the lower devices are already on causing ST. Whenever the lower peak value is greater than the triangular waveform, all the lower switches are on and result in the shoot through [3, 4]. In this control, the ST duty ratio decreases with an increase in modulation index (MI).

Calculation of design parameters from SBC:

Maximum output voltage = MI*V_{dc} / 2.

Maximum output voltage of ISI = B* MI*V_{dc} / 2.

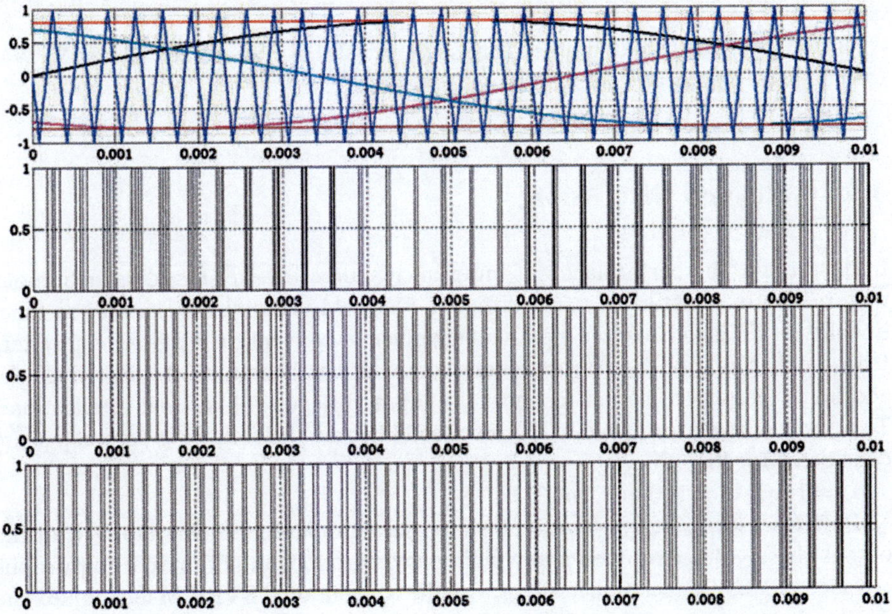

Fig. 3 Control pulses using SBC

Boosting Factor B = Maximum output voltage of ISI*2 / (MI*V_{dc}).
$V_L = V_C = ((B + 1) / 2)* V_{dc}$.

3 Space Vector Based PWM Techniques

SVPWM techniques are used to generate the control pulses to the conventional inverters. To generate control pulses using various SVPWM techniques requires modified modulated waves. The new set of modulated waves are generated by append the zero sequence phasor to the reference voltage sinusoids.

$$V_{In}^{*} = V_{In} + V_{Zs} \tag{1}$$

$$V_{In} = V_{Ref}\cos(\theta - \frac{2(q-1)\Pi}{3}) \tag{2}$$

where i = a, b, c and q = 1, 2, 3.

$$V_{Zs} = \frac{V_{DC}}{2(2a-1)} - a * V_{Max} + (a-1) * V_{Min} \tag{3}$$

Here V_{Max} and V_{Min} are the maximum and minimum values of V_{In}.

Different modulating signals are generated by varying the constant "a" in the above equations [7].

Based on the distribution of the zero sequence signal, various advanced SVPWM techniques are generated as shown in Fig. 4.

4 Results and Discussion

In this article, the performance of impedance source inverter-based asynchronous motor drive is investigated using different SVPWM techniques. The unique functionality of ISI is to boost the input voltage given to conventional DC-AC converters fed on either side of the Asynchronous motor drive. Therefore, it will reduce the additional step-up converter to boost the input voltage, unlike solar applications. To estimate the performance of the proposed inverter, the following parameters are considered for ISI.

L = 1mH, C = 1000μF, and B = 0.8

Figures 5, 6, 7, 8, 9, 10 and 11 show the modulating signals along with the output phase voltage of open-end winding asynchronous motor drive using various advanced SVPWM techniques along with the conventional SVPWM technique. The input voltage given to ISI is 100 V. It will boost the input voltage to 130 V as shown

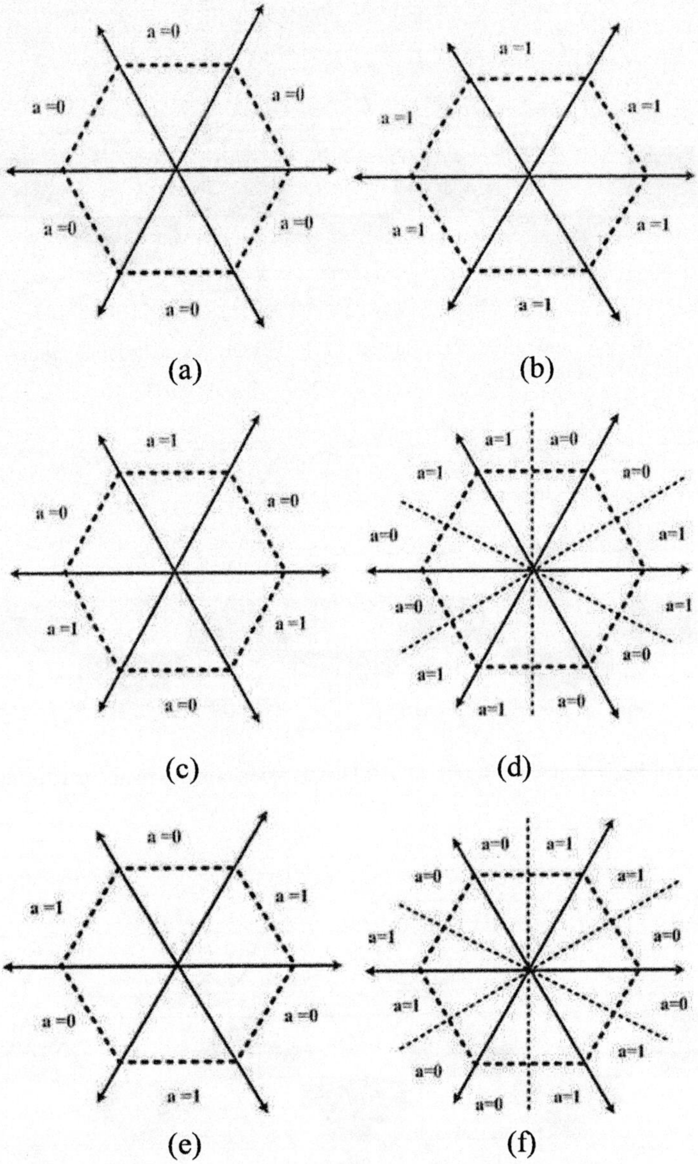

Fig. 4 Variation of constant "a" to generate various SVPWM techniques. **a** SVPWM1, **b** SVPWM2, **c** SVPWM3, **d** SVPWM4, **e** SVPWM5, **f** SVPWM6

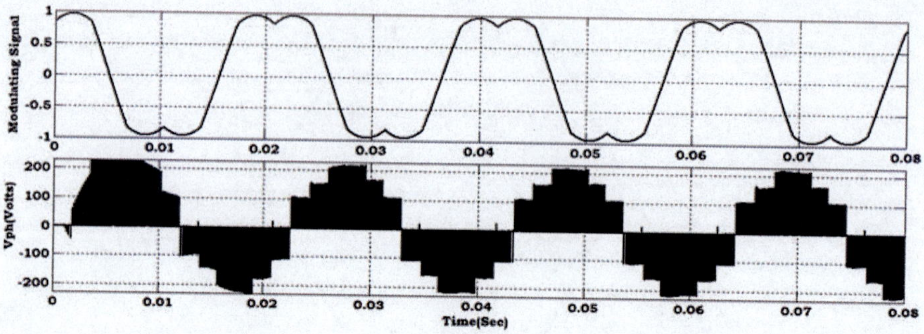

Fig. 5 Trace of V_{ph} and modulating signal of ISI based Asynchronous motor drive using Conventional SVPWM technique

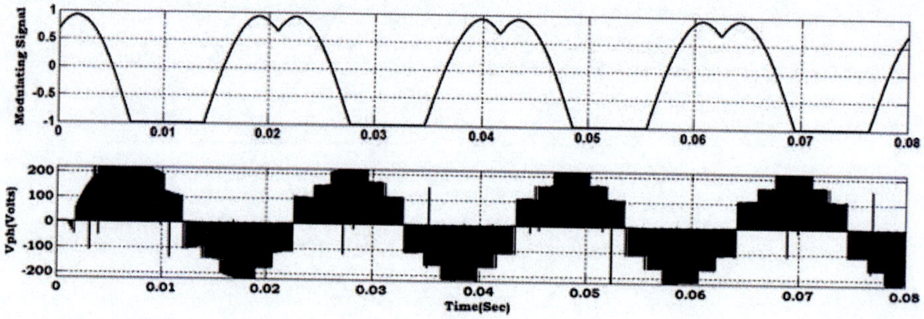

Fig. 6 Trace of V_{ph} and modulating signal of ISI based Asynchronous motor drive using SVPWM1 technique

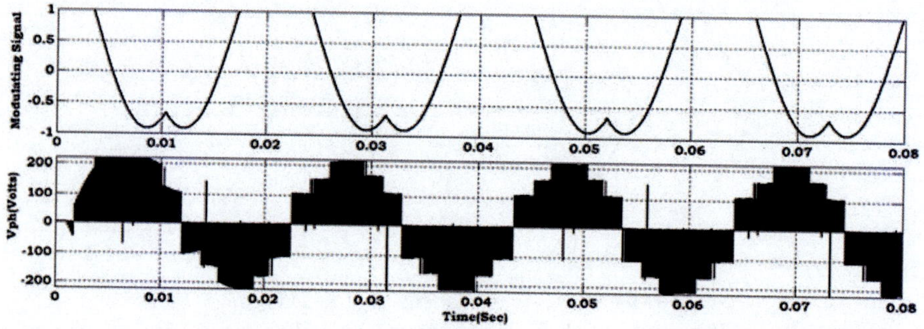

Fig.7 Trace of Vph and modulating signal of ISI based Asynchronous motor drive using SVPWM2 technique

in Fig. 12. The main advantage of ISI is to boost the input voltage in single-stage conversion. The boosted input voltage is given to the two conventional inverters fed on either side of the asynchronous motor drive. The control signals to the two inverters are generated using different modulating signals along with a simple boost

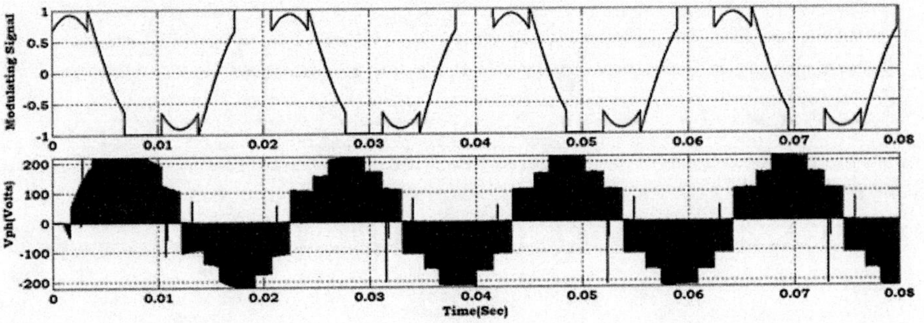

Fig. 8 Trace of Vph and modulating signal of ISI based Asynchronous motor drive using SVPWM3 technique

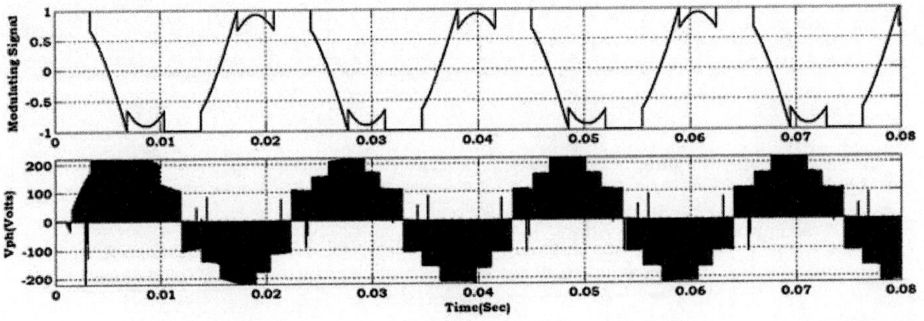

Fig. 9 Trace of Vph and modulating signal of ISI based Asynchronous motor drive using SVPWM4 technique

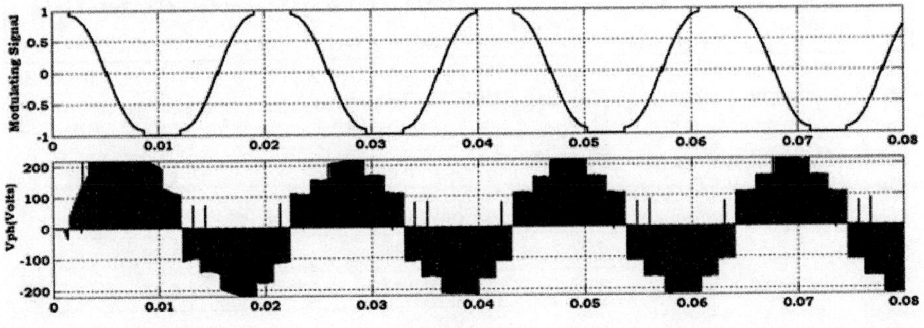

Fig. 10 Trace of Vph and modulating signal of ISI based Asynchronous motor drive using SVPWM5 technique

control technique. Figure 13 shows the harmonic samples of output voltage using the SVPWM4 technique. It has been noticed from Table 1, SVPWM-4 technique gives less THD in output voltage compared to other techniques. An ISI-based asynchronous

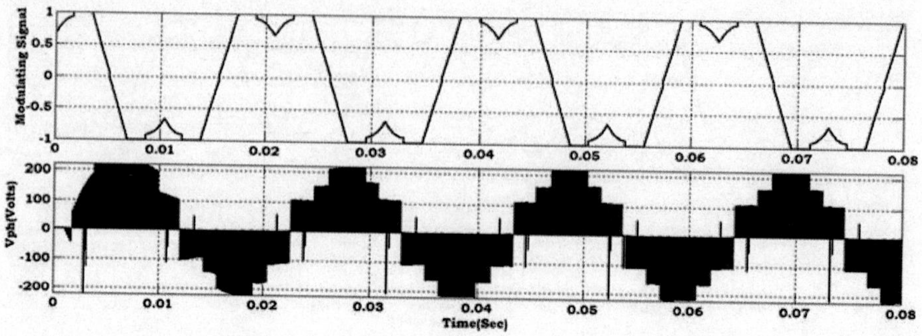

Fig. 11 Trace of Vph and modulating signal of ISI based Asynchronous motor drive using SVPWM6 technique

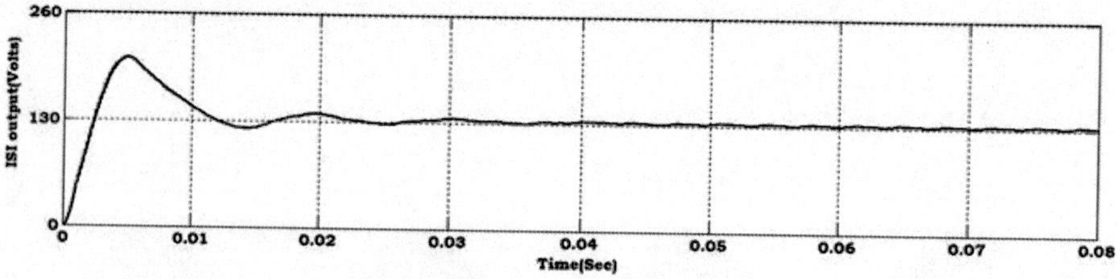

Fig. 12 Trace of boosted input voltage to the inverter obtained from ISI

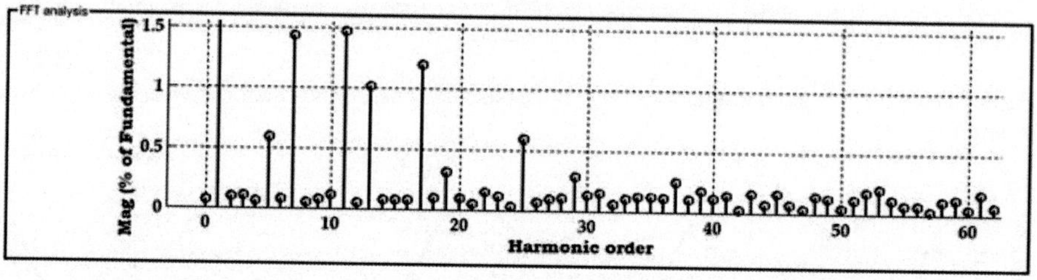

Fig. 13 Harmonic Samples of Vph Using SVPWM4 Technique

Table 1 Harmonic values of Vph

S.No	Type of SVPWM	Vph(%Thd)
1	Conventional SVPWM	5.72
2	SVPWM1	6.42
3	SVPWM2	6.46
4	SVPWM3	4.97
5	SVPWM4	2.87
6	SVPWM5	4.90
7	SVPWM6	8.57

motor drive generates three-level output voltage with increased magnitude compared to conventional three-level output. This is the main advantage of this proposed work.

5 Conclusion

In this work, the performance of ISI-based open-ended asynchronous motor drive has been analyzed using numerous SVPWM techniques. To generate the control pulses for two inverters fed on either side of open-end asynchronous motor drive, a simple boost control method has been used along with the modulating signals. The performance analysis of the proposed configuration has been carried out with 100V input voltage to ISI and 0.8 boost factor. The superior performance with the use of SVPWM techniques has been witnessed from the results. The THD values got reduced remarkably in output voltages with SVPWM4 over other PWM techniques.

References

1. Peng FZ (2003) Z-source inverter. IEEE Trans Ind Appl 39(2):504–510
2. Loh PC, Vilathgamuwa DM, Lai YS, Chua GT, Li YW (2005) Pulsewidth modulation of Z-source inverters. IEEE Trans Power Electron 20(6):1346–1355
3. Shen MS, Peng FZ (2006) Control of the Z-source inverter for fuel cell-battery hybrid vehicles to eliminate undesirable operation modes. In: Proceedings IEEE IAS, pp 1667–1673
4. Husodo BY, Anwari M, Ayod SM, (2010) Taufik "Analysis and simulation of ZSI control methods". IPEC 2010 @2010 IEEE
5. Stemmler H, Guggenbach P (1993) Configurations of high power voltage source inverter drives. EPE 1993, Brighton, U.K., pp 7–12
6. Chen S, Lipo TA, Fitzgerald D (1996) Modeling of motor bearing currents in PWM inverter drives. IEEE Trans Ind Applicat 32:1365–1370
7. Somasekhar VT, Gopakumar K, Shiva Kumar EG, Sinha SK (2002) A space vector modulation scheme for a dual two level inverter fed open-end winding induction motor drive for the elimination of zero sequence currents. EPE J 12(2):26–36
8. Baiju MR, Mohapatra KK, Kanchan RS, Gopakumar K (2004) A dual two-level inverter scheme with common mode voltage elimination for an induction motor drive. IEEE Trans Power Electron 19(3):794–805
9. Narayanan G, Krishnamurthy HK, Zhao D, Ayyanar R (2006) Advanced bus-clamping PWM techniques based on space vector approach. IEEE Trans Power Electron 21(4):974–984
10. Das S, Narayanan G (2012) Novel switching sequences for a space-vector modulated three-level inverter. IEEE Trans Ind Electron 59(3):1477–1487
11. Shiny G, Baiju MR (2009) Space vector PWM scheme without sector identification for an open-end winding induction motor based 3-level inverter. In: Industrial electronics, 2009. IECON '09. 35th annual conference of IEEE, pp 1310–1315
12. Chung DW, Kim JS, Sul SK (1998) Unified modulation technique for real-time three phase power conversion. IEEE Trans Ind Applicat 34:374–380
13. George DS, Baiju MR, Space vector based random pulse width modulation scheme for a 3-level inverter. In: Open-end winding induction motor configuration "international symposium on industrial electronics" (ISIE-2012) IEEE, pp 742–747

Chapter 17
Z-Source Inverter for RES-EVS with Flexible Energy Control Functions

Shaik Yalavarthi Hussain and K. Radha Rani

1 Introduction

There is an expanding requirement for sustainable Renewable Energy Sources (RES) with auxiliary highlights, especially in low voltage circulation frameworks. Subordinate highlights incorporate consonant pay, receptive Power remuneration, low voltage ride through capacity, and so forth [1]. This is because of the reality that there is expanded infiltration of nonlinear force gadgets-based burdens. These heaps infuse consonant flows into the network which can cause contortion at Power coupling, especially in powerless lattice frameworks.

In addition, because of the irregular idea of the perfect vitality sources, for example, wind and sun powered vitality, their expanded infiltration lead network voltage variances rely on power age and request. These voltage changes can influence delicate force electronic loads, for example, flexible speed drives, lighting frameworks, and so forth which can prompt continuous stumbling, mal activity, and consequently prompting expanded upkeep costs. Sustainable power source coordination with power quality upgrading frameworks, for example, dynamic voltage restorer (DVR), bound together force quality conditioner, and dissemination static compensator, gives a perfect arrangement by joining advantages of clean.

DSTATCOM is a shunt VSC which for load power quality issues, for example, current sounds, load responsive force, unbalance, and so forth. DVR is an arrangement VSC which ensures touchy burdens against network voltage unsettling influences, for example, droops/swells, shunt interference, and so forth. UPQC is a flexible device as it makes up for both power side and lattice side force quality issues. A point-by-point survey of different UPQC setups and control has been yielded. The arrangement VSC of UPQC comes into activity under lattice voltage hangs/swells,

S. Y. Hussain · K. R. Rani (✉)
Department of Electrical and Electronics Engineering, R.V.R & J.C. College of Engineering, Chowdavaram, Guntur, Andhra Pradesh, India

S. Dawn et al. (eds.), *Smart and Intelligent Systems*, Algorithms for Intelligent Systems,
https://doi.org/10.1007/978-981-16-2109-3_17

flash, and unbalance which are brief span varieties. Contrasted and shunt VSC which compensator, the arrangement VSC use is a lot lesser.

The significant segments of an electric vehicle framework are the engine, controller, power supply, charger, and drive Train [2]. Great execution in factor speed dc drives relies vigorously upon the control system and controller structure. These exhibitions incorporate various angles, for example, quick ascent time, least overshoot, least consistent state mistake, high productivity, unwavering quality, and economy. The ordinary straight controllers, for example, Proportional Integral and Proportional Integral Derivative, have been utilized in numerous applications [3].

The Integral Proportional controller has been applied with dc drives. Be that as it may, these controllers are touchy to framework parameter varieties and burden aggravation. The exhibition shifts with working conditions, and it is likewise hard to 15 tune controller increase both online and disconnected. The expanded profitability and improved item quality request quick reaction and parameter–obtuse vigorous drive frameworks [4].

With the extension in propels, mechanical Electric vehicles and cross-breed electric vehicles are progressively concerned these days because of their capable activity. In this broaden, the force can be produced by the Solar and put away in batteries. When the vehicle is under the running condition, the force is traded on the motor and draws the current from the battery. In spite of the fact that they are just at a moderately undeveloped stage as far as the market entrance, electric vehicles speak to the most earth agreeable vehicle fuel, as they have definitely no outflows The vitality created to control the Electric vehicles and the vitality to move the vehicle is 97 percent cleaner as far as toxic contaminations [5].

The upside of electric engines is their capacity to give power at practically any motor speed. One of the huge contentions made via vehicle organizations against electric vehicles is that Electric vehicles are fueled by power plants, which are controlled essentially by coal or Hydra. In any event, expecting that the power to control the Electric vehicles is not delivered from housetop sun oriented or flammable gas, it is still a lot cleaner than fuel created from oil.

The electric vehicle is driven by the battery. On exchanging the vehicle, the engine takes current from the battery which is gathered from the sun-powered and put away in a battery. The engine changes over the electrical vitality put away in the battery into mechanical vitality, and subsequently, the vehicle pushes ahead. At the point when the vehicle is turned on, the engine additionally turns over pivoting which is thus associated with the generator which starts creating the force.

The generation of the synchronous will begin as the electric vehicle begins driving. Here synchronous generator has been utilized in light of the fact that it can work at low force. The yield of the generator is an Alternating sort; the equivalent is put away in the battery. Consequently, it tends to be changed over into DC with the assistance of a rectifier circuit. The rectifier circuit changes over this AC into DC. The DC segment is gone through the channel circuit which expels music. Then the DC is put away in the ultracapacitor. Subsequently, the force can be produced with no outside powers and this procedure is called self age.

2 Literature Survey

Power electronics and motor drives in electric and plug-in electric vehicles. A. Emadi, L. Young-Joo, and K. Rajashekara. With the growing call for environmentally friendlier and better fuel economy cars, automotive corporations are specialized in electric automobiles, electric-powered motors, plug-in electric automobiles, and fuel-mobile automobiles. These cars could also enable meeting the demands for electrical electricity due to the growing use of digital capabilities to enhance vehicle overall performance, gasoline economic system, emissions, passenger consolation, and protection. In electric motors, HEVs, PHEVs, and gas-cellular automobiles, the demanding situations are to acquire excessive-efficiency, ruggedness, small sizes, and low prices in electricity converters and electric-powered machines, in addition to in associated electronics.

In unique, in fuel-cellular vehicles, an energy-conditioning unit which includes a dc–dc converter for matching the gas-mobile voltage with the battery % will also be vital. In steer-through-wire and brake-by means of-cord applications, a fast-reaction motor, inverter, and control gadget are important and need to be capable of function in unfavorable environmental situations. Furthermore, the integration of actuators with power electronics now not best improves the general machine reliability however also reduces the cost, length, and so forth. In addition to power electronics, the era of the electric motor plays a prime function inside the vehicle's dynamics and the form of a power converter for controlling the vehicle running traits.

3 Structure of Electric Vehicle

All electric vehicles have four principle building blocks. They are as per the following: A. Battery to create a DC voltage, B. A DC to AC converter to change over the DC voltage to a high-recurrence AC voltage, C. An AC engine coupled to the drive train, and D. The battery charger circuit to charge the batteries. Now and then, an extra DC to DC converter is likewise needed to venture up the low voltage from the batteries (Fig. 1).

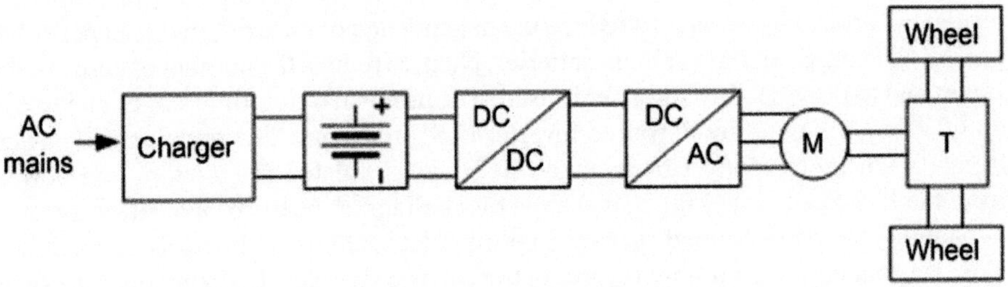

Fig. 1 Block diagram of an electric vehicleBattery

The battery details for the electric vehicles contrast for various kinds of electric vehicles. The majority of the vehicles use lithium-particle batteries with 370 V as ostensible DC voltage. The battery limit ranges from 20 to 100 kWh. Higher is the battery limit, more is the driving scope of the vehicle. The driving extent for the flow of electric vehicles ranges from 60 miles for every charge to 380 miles for each charge.

4 Z-Source Inverter for Renewable Energy Sources and EV Charging

An inverter can be associated with a battery to change over the PV-produced DC power into AC power. It empowers the utilization of household apparatuses without mains power. The particular segments required may incorporate significant segments, for example, battery controller, assistant vitality sources, and once in a while the specific electrical burden.

Practically all mass electric force is created, transmitted, and devoured in a rotating flow (AC) organize. Components of AC frameworks create and devour two sorts of intensity: genuine force (estimated in watts) and responsive force (estimated in volt-amperes receptive or changes). Genuine force achieves helpful work (e.g., running engines and lighting lights). Receptive force underpins the voltages that must be controlled for framework Unwavering quality.

The voltage profile is improved by controlling the creation, retention, and stream of receptive force all through the system. Receptive force streams are limited to diminish framework issues. Transmission misfortunes can be determined dependent on the normal properties of parts in the influence framework: opposition, reactance, capacitance, voltage, current, and influence, which are routinely determined by service organizations as an approach to indicate what segments will be added to the frameworks, so as to diminish misfortunes and improve the voltage levels.

The concentrated voltage receptive control is one such control which can help not exclusively to keep the framework voltages inside indicated confines yet, in addition, to safeguard the responsive force adjusts for improved security and to diminish the transmission misfortunes for the productive framework activity. Basically, solar energy is the source of renewable energy which is mainly used in the applications of residential and commercial. To reduce the dependence of the grid, in this paper solar energy will charge electric vehicle batteries. Next, to reduce the number of conversion stages and to provide, the converter is used. The removal of multiple stages and boost up of voltage is done by Z-source inverter (ZSI) topology. For tracking the power from PV, an Incremental Conductance (IC) was simulated for tracking the power from the PV. The below Fig. 2 shows the block diagram of the proposed system.

Stability of grid and load management of electrical are the board issues that are likewise concentrated broadly regarding the electric vehicles. Utilizing the battery in

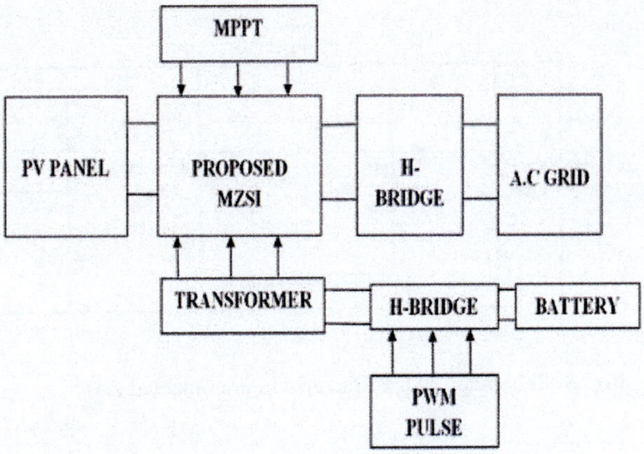

Fig. 2 Block Diagram of Z-Source Inverter for renewable energy sources and Ev charging

electric vehicles, network vitality from the sustainable can be put away, and furthermore, a similar battery can be utilized by the framework administrator to enable the matrix to recuperate from momentary voltage hangs and plunges brought about by load changes. Regardless of this scholastic level research on different angles, the whole development in the capacity gadget-driven electric vehicle industry in the business fragment is centered around a solitary issue. This issue is to broaden its driving separation with longer charge lengths.

The activities of solar oriented PV framework clarified here with two phases. Sunlight-based vitality is produced, and it will be changed over to power in the main stage. In the subsequent stage, a support converter is utilized to help up the voltage. A lift converter is a DC-to-DC power converter. This converter ventures up the voltage, while the current is ventured down.

The reproduction outline made sun-powered, diesel, inverter circuit, and burden. From the sun-powered, PV produces dc flexibly that will be converter to AC by utilizing an inverter. Diesel generator yield voltage is AC flexibly; connect rectifier changes the voltage to DC. Each source exclusively has a support converter. The voltage level sun-based and DG are increments to the ideal voltage. At long last inverter changes over the DC flexibly from help converter to AC and it will be given to the heap (Figs. 3 and 4).

5 Results

The below Fig. 5 shows the comparison of grid current for energy sources of EV and Z-Source inverter of RES-EVs.

The below Fig. 6 shows the comparison of grid power for energy sources of EV and Z-Source inverter of RES-EVs.

The below Fig. 7 shows the comparison of PV power for energy sources of EV and Z-Source inverter of RES-EVs.

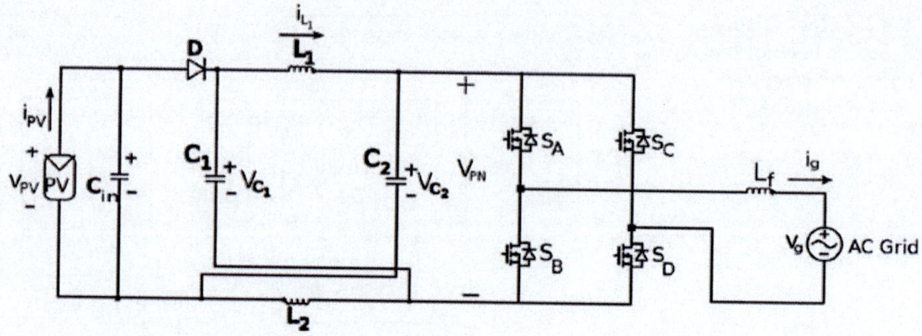

Fig. 3 Schematic of a PV/ac grid interconnected ZSI

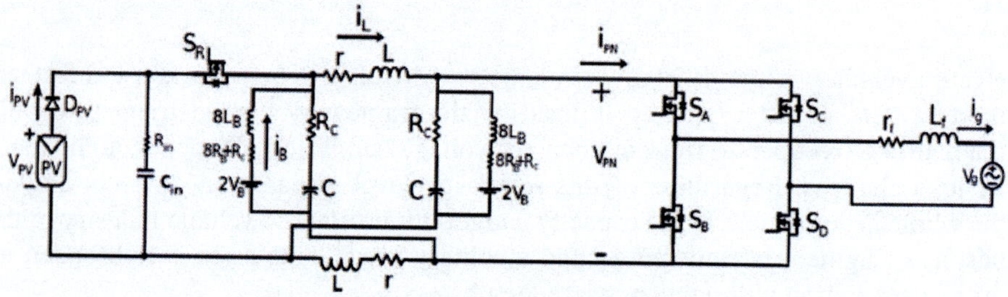

Fig. 4 Equivalent model of the proposed MZSI with a battery

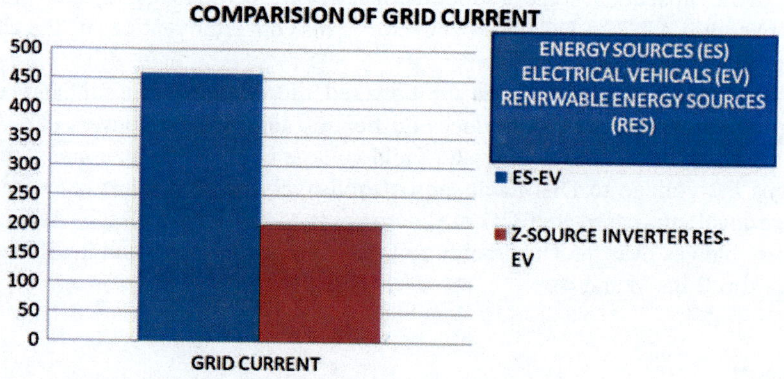

Fig. 5 Comparision of grid current

6 Conclusion

Hence in this paper, the high performance of Z-source Inverter for renewable energy sources and EV charging was implemented. Basically, solar energy is the source of renewable energy which is mainly used in the applications of residential and

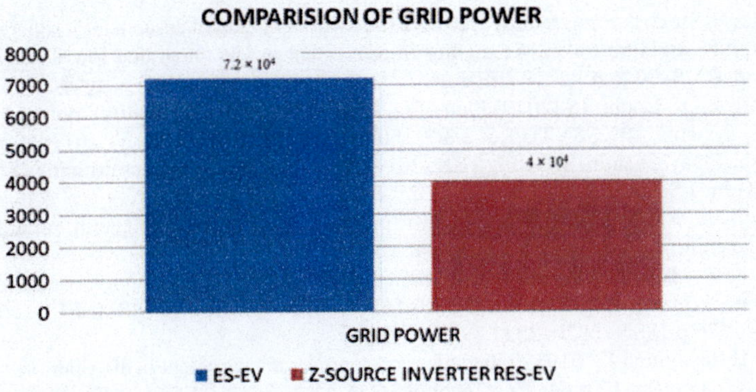

Fig. 6 Comparision of grid power

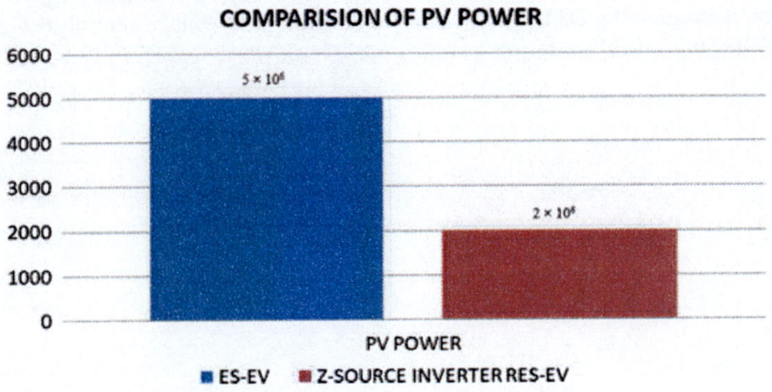

Fig. 7 Comparision Of PV power

commercial. To reduce the dependence of the grid, in this paper solar energy will charge electric vehicle batteries. Next, to reduce the number of conversion stages and to provide, the converter is used. The removal of multiple stages and boost up of voltage is done by Z-source inverter (ZSI) topology. For tracking the power from PV, an Incremental Conductance (IC) was simulated for tracking the power from the PV.

References

1. Zhang J, Yuan R, Yan D, A non-helpful game based charging power dispatch in electric vehicle charging station and charging effect analysis. 978–1–5386 - 8549 - 5/18/$31.00 ©2018 IEEE
2. Shuanglong S, Zhe Y (2018) Study on group control charging system and cluster control technology of electric vehicle. 978–1–5386 - 8549 - 5/18/$31.00 ©2018 IEEE

3. Kumar N, Singh B, Panigrahi BK (2018) Framework of gradient descent least squares regression based NN structure for power quality improvement in PV integrated low-voltage weak grid system. 0278–0046 (c) 2018 IEEE
4. Badoliya KK, Lodhi RS (2019) Plan of D-STATCOM for power quality improvement in wind power systems. 107 CSVTU Res J 8(2). https://doi.org/10.30732/RJET.20190802002
5. Anusha P, Jayachandra S, Reddy TSK (2019) Alleviation of harmonics utilizing D-STATCOM. IJSRD Int J Scientific Res Dev 7(6):2321–0613. ISSN (on the web)
6. Awasthi A, Rai V (2020) Force quality improvement using UPQC with variable DC Link voltage control. Int J Innov Res Technol Manag 4(1)
7. Pattathurani L, Dwibedi RK, Sivachidambaranathan P (2015) A voltage controlled dstatcom for power quality improvement. IOSR J Electr Electron Eng (IOSR-JEEE) 10(6). e-ISSN: 2278–1676,p-ISSN: 2320–3331
8. Baby H, Jayakuma J (2019) Update of reactive power management in distribution system using D-STATCOM. Int J Trends Eng Technol 36(1), MAY 2019 - ISSN: 2349 – 9303.
9. Singh B (2012) Prelude to FACTS controllers in wind power farms: a technological review. Overall diary of sensible power source research bindeshwarsingh 2(2)
10. Prasad KK, Myneni H, Ganjikunta Siva Kumar S (2018) Power quality improvement and PV power injection by DSTATCOM with variable DC link voltage control from RSC-MLC, 1949–3029 (c) 2018 IEEE

Chapter 18
Assessment of Single and Two-Stage Optimization Processes on Optimal Capacitor Placement in Power Distribution Systems

Soumyabrata Das and Tanmoy Malakar

1 Introduction

In a Power System, the interconnected systems primarily comprise generation, transmission, and distribution arrangements. Altogether the foremost loads like commercial, industrial, and domestic are linked to the power system via distribution networks. The Distribution System (DS) feeds all the inductive loads to the consumer end; therefore, more power losses and poor voltage regulation occur in the DS. In DS, around 13% of entire power generations are lost as ohmic loss [1]. The reasons behind this ohmic loss are high R/X ratio, low voltage, and lengthy radial structures. Hence, it is necessary to place suitable procedures to overcome these problems and elevate the quality of the supply at the customer end.

It has been noted that insufficient amounts of reactive power supply in DS is the main reason behind the high power loss and poor voltage regulation. Proper installations of the shunt capacitors (SC) [2] at desired places can overcome these noted problems. Henceforth, in the recent decade, researchers have put more emphasis on optimal capacitor placement in RDS. Thereafter, various optimization techniques are introduced to identify the Optimal Capacitor Location and Sizes (OCLS). As noted in the literature earlier, researchers used classical techniques [3, 4] to solve the OCLS problem. But those techniques encountered difficulties in handling discrete control variables.

Furthermore, due to large search space requirements, such techniques' computational complexity was much more eminent. To overcome that, researchers started proposing numerical methods [1, 5] to solve power system problems. Apart from numerical methods, researchers also have proposed various heuristic optimization techniques to solve the OCLS problems like Particle Swarm Optimization (PSO) [6], Plant Growth Simulation Algorithm (PGSA) [7], Flower Pollination Algorithm (FPA) [8], Direct Search Algorithm (DSA) [9], Teaching Learning Based

S. Das (✉) · T. Malakar
National Institute of Technology Silchar, Silchar, Assam, India

S. Dawn et al. (eds.), *Smart and Intelligent Systems*, Algorithms for Intelligent Systems,
https://doi.org/10.1007/978-981-16-2109-3_18

Optimization (TLBO) [10], Differential Evolution-Pattern Search (DE-PS) [11], Heuristic [12], and Competitive Swarm Optimizer (CSO) [13]. In all applications of population-based search approaches, it is observed that they give near-optimal solutions. Therefore, huge test runs are necessary to get an optimum solution, increasing the computing time. To overcome this, the researchers have projected several sensitivities-based approaches to identify potential candidate nodes for capacitors at first and then used the meta-heuristic algorithms to determine the optimum sizes in the second stage. The literature survey reveals that there are distinct advantages by finding probable capacitor locations in the first stage and then optimizing the sizes and locations in the second stage. Authors in [9, 10] solved the OCLS problems as a single-stage optimization problem. Whereas the work in [6–8, 11–14] mainly comprises two-stage optimization. The author in [6–8] used the Loss Sensitive Factor (LSF) to identify the high potential candidate buses. Similarly, work based on Voltage Stability Index (VSI) [8], Power Loss Index [11], Node sensitivity Index [12], and Relative Emission Index [13, 14] are other methods for selecting the potential candidate buses. Then the authors are used PSO [6], PGSA [7], FPA [8], DE-PS [11], Heuristic [12], and CSO [13] to solve the capacitor placement problem.

In this study, the authors wish to investigate the impact of single-stage and two-stage optimization methods in capacitor placement problems. Initially, the OCLS problem is solved as a single-stage optimization method using the OCSO algorithm. Thereafter, the same problem is solved as a two-stage problem. In the first stage, probable candidate buses are selected through the REI parameter [13, 14]. In the second stage, the optimal locations and sizes of SCs are identified using the newly developed algorithm named 'Opposition based Competitive Swarm Optimizer (OCSO)' [14]. The OCSO is a population-based optimization algorithm, and it is derived from CSO. The OCSO algorithm has superior search capability and faster convergence than any other contemporary method. Moreover, the OCSO algorithm has not been experienced considerably to solve the distribution network problems. The proposed OCLS has been verified on the standard IEEE 10 bus and 69 bus Radial Distribution System (RDS). The results of the OCSO are compared with other methods stated in the literature.

2 Problem Formulation

2.1 *Objective Function*

The objective function for the capacitor placement problem in the RDS network is to minimize the annual cost, which can be determined by the following Eq. (1) [8].

Minimize,

$$\text{Annual Cost} = K_{p*} P_{Loss} * T + D(K_I * CB + \sum_{i=1}^{CB} Q_{Ci}) + K_o CB, \quad (1)$$

Table 1 The used parameters

Items	Description	Value
Kp	Cost / kW-hour	$0.06/kW-hours
Kc	Cost / kVAr	$25/kVAr
Ki	Cost / SC installation	$1600/location
Ko	Operating cost	$300/location
D	Depreciation factor	0.2
T	Total time	8760 hours

where P_{LOSS} is the active power loss in the RDS networks. K_P is the power loss component. Whereas K_I and K_O are the installation cost and operating cost components, respectively. T is the total time. CB is the total number of active shunt capacitors, and D is the deprecation factor. The value of all cost components is given in Table 1.

2.2 *Constraints*

Equality Constraints (ECs): ECs are shown in (2) and (3).

$$P_{ss} = \sum_{i=1}^{nb} P_{di} + \sum_{j=1}^{nl} P_{lossj}, \tag{2}$$

$$Q_{ss} = \sum_{i=1}^{nb} Q_{di} + \sum_{j=1}^{nl} Q_{lossj} - \sum_{i=1}^{CB} Q_{ci}, \tag{3}$$

where P_{ss} and Q_{ss} are the real and reactive power of the substation and nb and nl are the total number of load buses and lines. P_{di} and Q_{di} are the demand for real and reactive power at bus '*i*', and P_{lossj} and Q_{lossj} are the real and reactive power line loss at branch *j*.

Inequality Constraints (ICs): The purpose of the OCLS problem is to find the stable operating state for power system operation by modifying the various power controllers within their allowable range. This fact is modeled as ICs in this OCLS problem.

The number of Capacitor constraint:

$$\sum_{n=1}^{N_{CB}} L_n \leq N_{CB}. \tag{4}$$

Capacitor size constraint:

$$Q_{cn}^{\min} \leq Q_{cn} \leq Q_{cn}^{\max}. \tag{5}$$

Total reactive power constraint:

$$\sum_{n=1}^{N_{CB}} L_n Q_{cn} < Q_d^{Total}. \tag{6}$$

Bus voltage constraint:

$$V_i^{\min} \leq |V_i| \leq V_i^{\max}. \tag{7}$$

Line capacity constraint:

$$S_{li} \leq S_{li}^{\max}, \tag{8}$$

where S_{li} and S_{li}^{max} are the absolute value of complex power flow and its maximum limit through line 'i'.

In this paper, bus voltage and thermal limit (line capacity) constraints are managed by imposing the penalty factor methods. Therefore, the final objective function with the penalty factor is presented in (9).

$$\begin{aligned} Annual\ Cost = K_{p*}P_{Loss} * T + D(K_I * CB + \sum_{i=1}^{CB} Q_{Ci}) + K_o CB \\ +\lambda_{vi}|V_i - V_i^{\lim}| + \lambda_{sli}|S_{li} - S_{li}^{\lim}| \end{aligned} \tag{9}$$

3 Overview of OCSO Algorithm

The OCSO algorithm is developed from the CSO algorithm. The primary purpose of the OCSO algorithm is to improve the search capability of the CSO algorithm and the faster convergence [14]. In the OCSO algorithm, three-fourth of the total population is updated in each iteration. At each iteration, a random couple is engaged, and competition is completed between each couple's particles. The particle with better fitness is declared as a winner and divided into two subcategories. The first category of the winner is directly transferred to the next generation swarm. Whereas the second category of the winner is passed through the opposition-based learning process [14, 15]. On the other hand, the losers update their position and velocity from the winner's experience using (10) and (11).

$$U_{lj}(g+1) = R_1(j,g)U_{lj}(g) + R_2(j,g)(X_{wj}(g) - X_{lj}(g)) + \varphi R_3(j,g)(\overline{X_j}(g) - X_{lj}(g)). \quad (10)$$

$$X_{lj}(g+1) = X_{lj}(g) + U_{lj}(g+1). \quad (11)$$

The updated velocity and positions are expressed in (10) and (11), respectively. $X_{wj}(g)$, $X_{lj}(g)$, $U_{wj}(g)$, and $U_{lj}(g)$ are the positions and velocities of the winners and losers in the jth round, respectively. $R_1(j,g)$, $R_2(j,g)$, and $R_3(j,g)$ are random quantities.

4 Results

The proposed capacitor placement problems are solved in two standard RDS networks with a single-stage and two-stage optimization approach in this work. For the single-stage case, the search space is spread over to all the load buses of the RDS. However, the potential candidate buses are selected through REI based approach in the two-stage case. The minimization of annual cost is taken as the objective function in both the cases, and the OCSO algorithm is used for optimization. The parameters used in the annual cost calculation are provided in Table 1. The optimal results of the OCSO algorithm are compared with other literature.

4.1 *Case Study 1: 10 Bus Test System*

The data for the 10 bus system is taken from [16]. The total real (reactive) power load is 12368 kW (4186 kVAr). In this work, the possible candidate buses are selected according to the REI-based method. After that, the OCSO is utilized in the second stage to minimize the annual cost by determining SC's optimal location and sizes. The results of 10 bus RDS with single-stage and two-stage optimization are discussed here. The OCSO is used to obtain both stage optimization algorithms' results, as shown in Table 2. Table 2 confirms that the average computational time in two-stage optimization is 12.61% less than the single-stage optimization, which is obvious as the search space dimension is more in single-stage optimization than two-stage optimization. Therefore the convergence speed is more in two-stage optimization. The convergence plot for annual cost using OCSO for both stage optimization is shown in Fig. 1. Figure 1 shows that the OCSO algorithm converges at the 71st iteration and the 54th iteration for single-stage and two-stage optimization, respectively. Therefore it can be concluded that the two-stage optimization converges faster as compared single-stage optimization approach. But the optimum results obtained with both stage optimization algorithms are almost the same. Moreover, the candidate bus selection based on LSF and VSI is also reported in Table 2. From Table 2, it is clear that the REI-based method provides the best result in terms of candidate

Table 2 Comparison between single-stage and two-stage 10 bus RDS

Items	Single-stage optimization	Two-stage optimization		
		LSF [8]	VSI [8]	REI
Optimal location and sizes(kVAr)	5(1350)	5(1500)	7(500)	5(1350)
	6(1400)	6(600)	8(600)	6(1400)
	9(450)	9(600)	9(1050)	9(450)
		10(1100)	10(1050)	
Total kVAr	3200	3800	3200	3200
Annual cost ($/year)	381,411.90	382,821.50	396,741.10	381,411.90
Net saving ($/year)	30,537.61	29,128.39	15,208.44	30,537.61
% saving	7.41	7.07	3.69	7.41
Av. CPU time (Sec)	15.39	NR	NR	13.45

NR = Not Reported

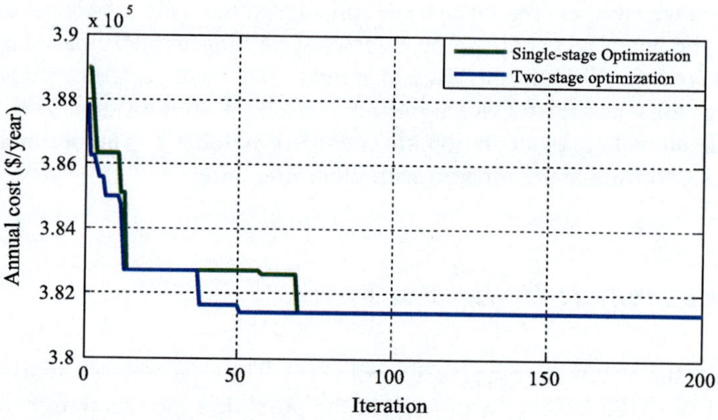

Fig. 1 Convergence plot of annual cost using OCSO for 10 bus RDS

bus selection and annual cost minimization. The usefulness of the OCSO to minimalize the annual cost by determining SC's locations and sizes is demonstrated in Table 3. Thereafter the OCSO algorithm's results are compared with Fuzzy [17], PSO [6], PGSA [7], and FPA algorithm [8]. With such SC placement, the annual cost is $381,411.90 using OCSO, the least among all other methods. The percentage reduction in yearly cost with OCSO is 7.41% w.r.t uncompensated case. Similarly, power loss also decreased significantly after the proper placement of SC compared to the uncompensated method. The system's voltage profile after capacitor placement is compared with its uncompensated case shown in Fig. 2.

Table 3 Results of 10 bus RDS for different algorithm

Items	Uncompensated	Compensated				
		Fuzzy [17]	PSO [6]	PGSA [7]	FPA [8]	OCSO
Total losses(kW)	783.77s	704.883	696.21	694.93	688.28	691.69
% loss reduction	–	10.065	11.17	11.33	12.18	11.74
Optimal location and sizes(kVAr)	–	4(1050)	5(1182)	5(1200)	5(1500)	5(1350)
		5(1050)	6(1174)	6(1200)	7(300)	6(1400)
		6(1950)	9(264)	9(200)	9(600)	9(450)
		10(900)	10(566)	10(407)	10(1100)	
Total kVAr		4950	3186	3007	3500	3200
Annual cost ($/year)	411949.5	397716.5	384338	382770.2	381740	381411.9
Net saving ($/year)	–	14233	27711.5	29179.3	30209.54	30537.61
% saving	–	3.46	6.7	7.08	7.33	7.41

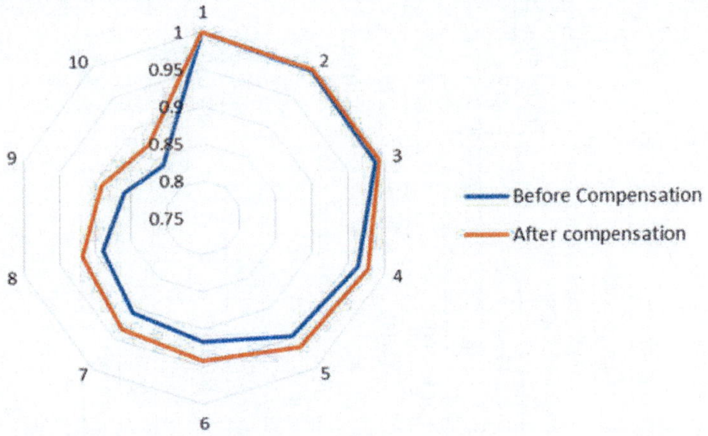

Fig. 2 Effect of SC on voltage profile for 10 bus RDS

4.2 *Case Study 2: 69 Bus Test System*

To compare the usefulness of the projected method on a bigger dimensional problem, the OCLS problem is further tested on the 69 bus RDS that comprises the main feeder and seven branches. The data for 69 bus RDS are taken from [4]. The total real (reactive) power demand is 3802.19 kW (2694.6 kVAr). Like earlier, here also the OCSO algorithm is used to obtain the results for both single-stage and two-stage optimization algorithms, as shown in Table 4. Table 4 shows that the average

Table 4 Results of 69 bus RDS for different algorithm

Items	Uncompensated	Compensated					
		Single-stage optimization with OCSO	Two-Stage Optimization				
			Heuristic [12]	DSA [9]	TLBO [10]	DE-PS [11]	OCSO
Total losses (kW)	224.8949	145.52	148.48	147	146.35	151.3763	145.52
% loss reduction	–	35.29	34	33.19	–	32.7	35.29
Optimal location and sizes(kVAr)	–	12(500)	8(600)	61(900	12(600)	57(150)	12(500)
		50(1200)	58(150)	15(450)	61(1050)	58(50)	50(1200)
			60(1050)	60(450)	64(150)	59(100)	
						60(150)	
						61(1000)	
Total kVAr		1700	1800	1800	1800	1450	1700
Annual cost ($/year)	118204.8	86228.91	88901.1	88123.2	87781.56	89913.4	86228.91
Net saving ($/year)	–	31975.9	29303.67	30081.56	30423.2	28291.4	31975.9
% saving	–	27.05	24.79	25.45	25.74	23.93	27.05
Av. CPU time (Sec)	–	23.99	NR	NR	NR	NR	20.46

NR = Not Reported

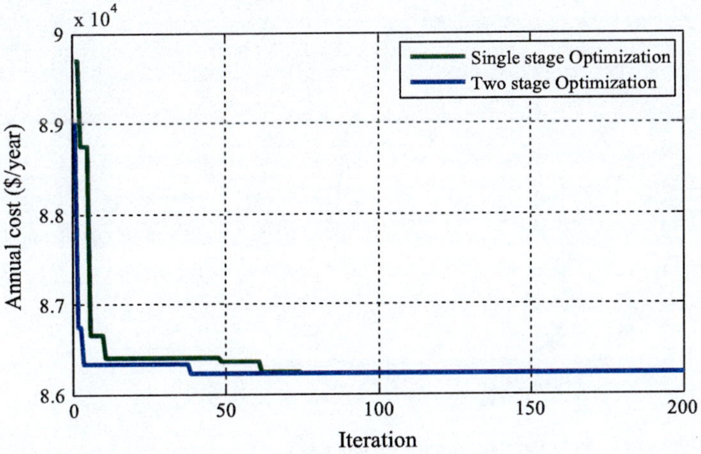

Fig. 3 Convergence plot of annual cost using OCSO for 69 bus RDS

computational time in two-stage optimization is 20.46 s, which is 14.71% less than the single-stage optimization. The convergence plot for annual cost with both-stage optimizations is shown in Fig. 3. It is observed from Fig. 3 that the single-stage optimization and two-stage optimization are converged at 76th and 40th iteration, respectively. The results analysis shows that two-stage optimization is converging faster than single-stage optimization for an optimum solution. The best result attained through the OCSO algorithm as mentioned and compared with other state-of-art literature is given in Table 4. The optimal values of reactive power injections by the capacitor using the OCSO algorithm are 500 and 1200 kVAr at buses 12 and 50. With such SC placement, the annual cost is $86,228.91 using the OCSO, the least among all other methods. The system's voltage profile after capacitor placement is significantly increased and compared with its uncompensated case shown in Fig. 4. It has been observed from Fig. 4 that the lowest voltage founded on bus 54 is increased to 0.9317 p.u. from 0.9092 p.u.

5 Conclusion

This paper has successfully executed the OCSO algorithm with single-stage and two-stage optimization to solve the capacitor placement problem in two different RDS networks. In single-stage optimization, all load buses are considered as potential candidate buses. Whereas, in two-stage optimization, SC's possible candidate buses are determined by the REI parameter, and the optimum capacitors' locations and sizes are obtained with the OCSO algorithm. From the analysis of the results, it has been observed that the two-stage optimization provides faster convergences as compared to single-stage optimization with the same accuracy. The simulation results reveal

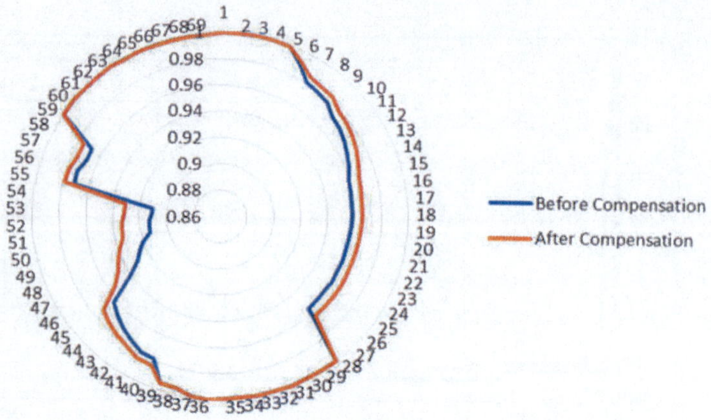

Fig. 4 Effect of SC on voltage profile for 69 bus RDS

that the computational time required for a two-stage optimization algorithm reduces up to 14.71% compared to single-stage optimization. Moreover, the effectiveness of REI based candidate bus selection method is also compared with other methods. It is observed that the identification of possible candidate buses based on REI may lead to the discovery of healthier results as compared to other methods. Then, the accurateness of the OCSO algorithm is compared with other recognized techniques. It is found that the OCSO algorithm beats other methods in the reduction of the annual cost for solving the OCLS problem.

References

1. Nojavan S, Jalali M, Zare K (2014) Optimal allocation of capacitors in radial/mesh distribution systems using mixed integer nonlinear programming approach. Electr Power Syst Res 107:119–124
2. Araujo D, Ramos L, Penido DRR, Carneiro S Jr, Pereira JLR (2018) Optimal unbalanced capacitor placement in distribution systems for voltage control and energy losses minimization. Electric Power Syst Res 154 (2018):110–121
3. Grainger JJ, Lee SH (1981) Optimum size and location of shunt capacitors for reduction of losses on distribution feeders. IEEE Trans Power Apparatus Syst 3:1105–1118
4. Baran ME, Wu FF (1989) Optimal capacitor placement on radial distribution systems. IEEE Trans Power Delivery 4(1):725–734
5. Franco JF, Rider MJ, Lavorato M, Romero R (2013) A mixed-integer LP model for the optimal allocation of voltage regulators and capacitors in radial distribution systems. Int J Electr Power Energy Syst 48:123–130
6. Prakash K, Sydulu M (2007) Particle swarm optimization based capacitor placement on radial distribution systems. In: 2007 IEEE power engineering society general meeting, pp 1–5. IEEE
7. Rao, Srinivasas R, Narasimham SVL, Ramalingaraju M (2011) Optimal capacitor placement in a radial distribution system using plant growth simulation algorithm. Int J Electr Power Energy Syst 33(5):1133–1139

8. Abdelaziz AY, Ali ES, Abd Elazim SM (2016) Flower pollination algorithm for optimal capacitor placement and sizing in distribution systems. Electric Power Compon Syst 44(5):544–555
9. Raju, Ramalinga M, Ramachandra Murthy KVS, Ravindra K (2012) Direct search algorithm for capacitive compensation in radial distribution systems. Int J Electr Power Energy Syst 42(1):24–30
10. Sultana S, Roy PK (2014) Optimal capacitor placement in radial distribution systems using teaching learning based optimization. Int J Electr Power Energy Syst 54:387–398
11. El-Fergany AA (2013) Optimal capacitor allocations using evolutionary algorithms. IET Gener Transm Distrib 7(6):593–601
12. Hamouda A, Lakehal N, Zehar K (2010) Heuristic method for reactive energy management in distribution feeders. Energy Convers Manage 51(3):518–523
13. Das S, Malakar T, Optimal capacitor placement and sizing in distribution system using competitive swarm optimiser algorithm. Int J Adv Intell Paradigms. https://doi.org/10.1504/IJAIP.2021.10034544) [Article in press]
14. Das S, Malakar T (2019) An emission constraint capacitor placement and sizing problem in radial distribution systems using modified competitive swarm optimiser approach. Int J Ambient Energy, 1–24
15. Das S, Malakar T (2020) Estimating the impact of uncertainty on optimum capacitor placement in wind-integrated radial distribution system. Int Trans Electr Energy Syst 30(8):e12451
16. Das D, Nagi HS, Kothari DP (1994) Novel method for solving radial distribution networks. IEE Proc Gener Transmiss Distrib 141(4):291–298
17. Su C-T, Tsai C-C (1996) A new fuzzy-reasoning approach to optimum capacitor allocation for primary distribution systems. In: Proceedings of the IEEE international conference on industrial technology (ICIT'96), pp 237–241. IEEE

Chapter 19
Harmonic Distortions Mitigating in an ELCr with Hybrid Hydro Electric Network Based on Fuzzy Controller

M. Divya and R. Vijaya Santhi

1 Introduction

Over the past few decades, the massive utilization of conventional fuels has resulted in a persistent depletion of fossil resources and has also adversely affected the climate. It results in concentrating on certain sustainable energy sources to fulfill energy needs. In remote or faraway areas, it is difficult and expensive to build grid systems; therefore, stand-alone systems are considered a promising choice [1]. Also, these areas have ample resources in nature like a waterfall, wind, biomass, rivers, etc. Efforts and attempts should be made for optimal usage of these sources to produce electrical power.

Transmitting power to long-distance owes so much of cost especially for faraway places, so by installing these devices, it acquires durable and much efficiency. For hydropower energy, microturbines or turbine pumps were used as prime movers. These are costly and complex for speed controls when compared to hydrosystems which are used to regulate the flow of water entering the turbines, as the demand for load varied, and also ELCr eliminates the governor for the turbines [2].

An'ELCr designed to control and maintain the desired output for a small power system. These are the simple systems that operate satisfactorily if the variable load is small when compared to fixed load. This is to allow the generator and turbine to run at their maximum power or partial power which is set manually and hold the electrical load at the correct speed. The principle of the ELCr is that it regulates the generated voltage and frequency. It acts like an electronic governor that functions as a frequency and voltage regulator on a generator. If the consumer demand is decreased for some purpose, then the dump load is allowed to handle the extra power that the consumer is not using; this maintains the total power constant. And the main component of ELCr is a chopper circuit [3]. The chopper duty cycle is balanced such

M. Divya (✉) · R. Vijaya Santhi
Electrical Engineering Department, Electrical Engineering Department, Andhra University College of Engineering, Visakhapatnam, Andhra Pradesh, India

S. Dawn et al. (eds.), *Smart and Intelligent Systems*, Algorithms for Intelligent Systems,
https://doi.org/10.1007/978-981-16-2109-3_19

that the generator's output power remains constant [4, 5]. The strength of utilizing ELCr in modest hydroelectric plants is that they evacuate the governor's prerequisite for turbines and maintains the correct speed. A steady frequency and voltage are produced by maintaining load almost constant, sensed, and regulated by ELCr on the turbine.

In this paper peak overshoots get reduced by using the FUZZY controller when compared to the PI controller which will improve the power quality of output voltage.

2 Hybrid System of Renewable Energy Sources

Figure 1 shows a generator model connected to ELCr. The hybrid systems of renewable energy are being implemented more in remote or faraway areas due to recent development in renewable energy technology. From wind and solar the output power is not continuously supplying to load because the available time is different for wind and solar. Also, the security and stability of the power grid are weakened. To defeat these issues, hydro energy is used. In the hybrid network at least two sources are joined so that if there is inaccessibility of one source, the other source fills the gap by this the system becomes more reliable. Thus the entire power system can realize a stable output.

2.1 Wind Power Generation

The turbine of wind technology is still among the most promising renewable energy technologies. Wind energy defines the mechanism by which wind is used for electricity generation. Wind power is made up of moving air molecules that have mass. Any object moving with mass carries a certain sum of kinetic energy. The turbine of the wind essentially a mechanical energy that changes over kinetic energy present in the wind through mechanical energy. It is then changed over by the generator into

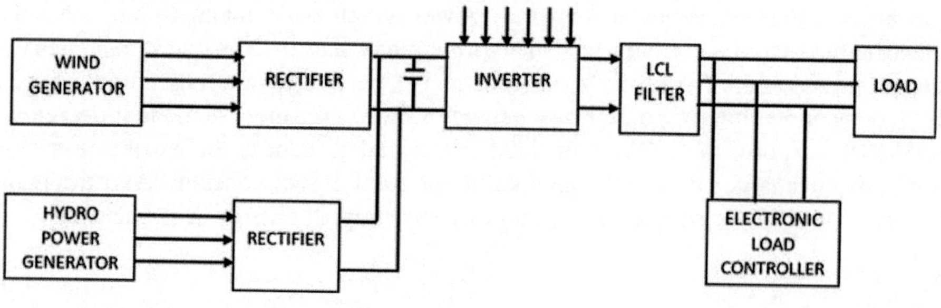

Fig. 1 Model of ELCr connected to the generator

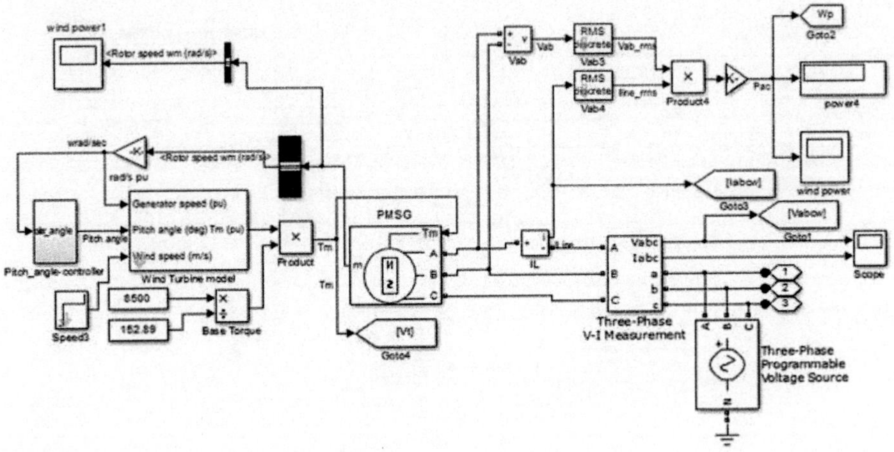

Fig. 2 Wind system connected to PMSG

electrical energy. The mechanical output power got by a turbine relies on the velocity of the wind, thickness of the air, swept area of the rotor, and power coefficient of a wind turbine.

$$P_t = \rho A C_p(\lambda, \beta) v^3. \tag{1}$$

Here in Fig. 2 simulation model of the Permanent Magnet Synchronous Generator (PMSG) is used to produce electricity from wind power. For the most part, the wind is at a speed of 10–12 m/s from PMSG.

2.2 *Hydropower Generation*

A self-excited induction generator is used to be driven by a constant hydropower turbine feeding local loads. The power available is defined by (2).

$$P = \rho A C_p(\lambda, \beta) v^3. \tag{2}$$

Self Excited Induction Generator(SEIG) is used for generating electricity with mini-hydro as shown in Fig. 3 which is concentrated on a stand-alone operation. These are almost always absorbed reactive VARs so to compensate for this issue capacitor banks connected across terminals of the stator. And the torque for the generation of hydropower which is expressed as (3)

$$T = K_1 \omega + K_2, \tag{3}$$

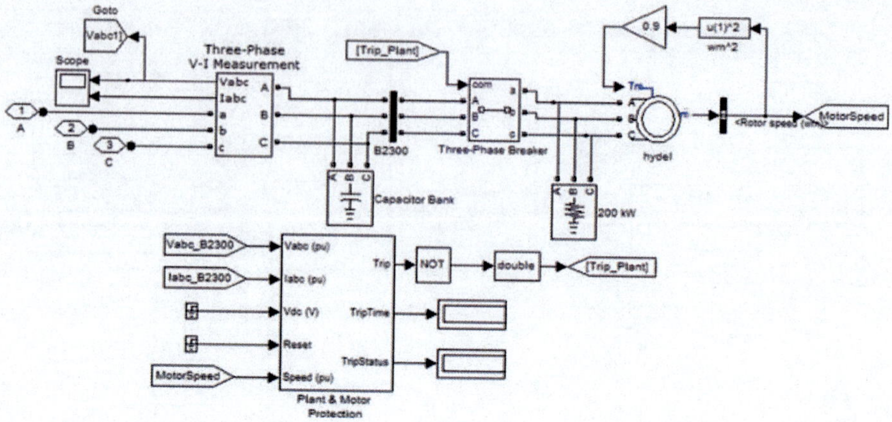

Fig. 3 Hydro system connected to SEIG

where K_1 and K_2 are values to obtain frequency and desired magnitude of the voltage.

3 Method of Control to Maintain the Desired Output

The fundamental point of the ELCr control plan in Figs. 4 and 5 is to keep up the desired frequency and voltage which relies on the variation of wind speed and the speed at which the water streams. To main constant voltage and frequency, the development of a voltage source converter is used to regulate the output and maintains active and reactive power flow to the load in the area of renewable energy sources. The three-phase currents are converted into a two-phase reference frame here two decoupled currents are obtained; one is the direct axis component current, and another is the quadrature axis component current from the machine. These two currents are controlled independently. To obtain direct axis component current, the difference between reference voltage DC capacitor link and actual voltage are compared and error gets processed with fuzzy controller and the output is given to PLL based inverse park transform. To obtain quadrature axis component, the current is the peak value difference between reference line-line voltage and actual voltage is compared and error get processed with fuzzy controller and the output is given to PLL based inverse park transform [6].

The unit template voltages are calculated by division of line to line voltages by their amplitude ‘Vt’ [6]. The unit template voltages are derived as

$$u_a = \frac{Vab}{Vactual(peak)}, \quad (5)$$

$$u_b = \frac{Vbc}{Vactual(peak)}, \quad (6)$$

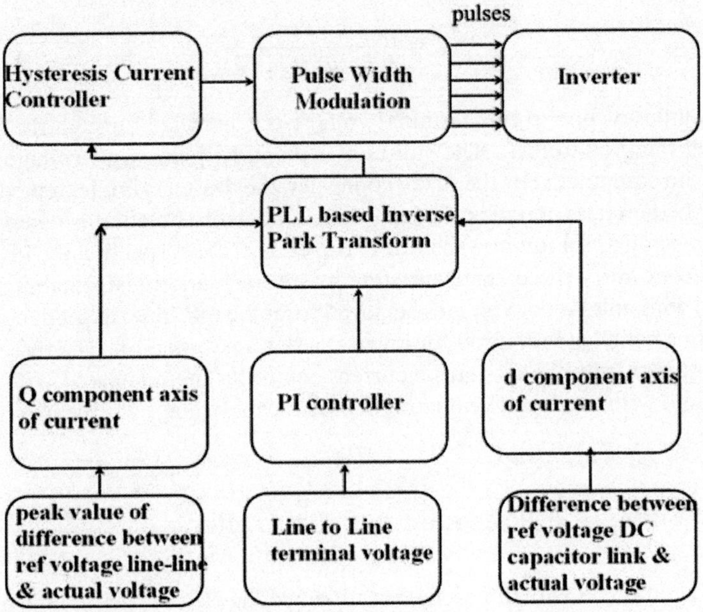

Fig. 4 Schematic control block diagram with PI

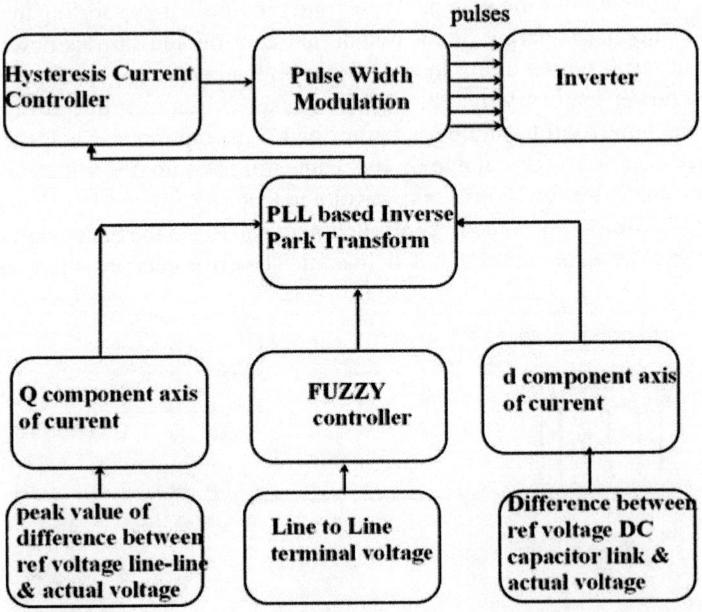

Fig. 5 Schematic control block diagram with FUZZY

$$u_c = \frac{Vca}{Vactual(peak)}. \quad (7)$$

The output of line to line terminal voltage is controlled through the PI/FUZZY controller then fed to PLL. The values of Kp and Ki gains with voltage open-loop systems are computed. For the fuzzy controller, the design is implemented in an easy way; for better characteristics, a triangular membership function is taken. Mamdani scheme is applied for tunning the rules. The current axes, specifically, ids*, iqs*,are changed over into a three-stage structure by inverse park transformation[7]. PLL is provided with unit layout voltages acquired from the machine, to produce cosωt and sinωt terms which are required for inverse park transformation. The three reference flows acquired from the hysteresis current controller are compared with the actual load current with produce terminating pulses for the voltage source inverter.

4 Modelling of Electronic Load Controller

The ELCr comprises with three-phase rectifier, voltage DC capacitor link, dump load, and IGBT chopper is shown in Fig. 6. A three-phase rectifier is an IGBT current-controlled voltage source converter, across a capacitor link that is connected which acts as a filter to remove the voltage ripples and provide self-supporting to dc system. The IGBT chopper is acting like a switch that controls the surplus power and this switch is in series with a dump load. This dump load absorbs generated power and consumed power load varying. Here ELCr acts as a circuit for power diverters and redirects the power left to dump load after load consumption and helps to maintain the balance of power. The voltage dc-link capacitor is compared with the voltage of a reference capacitor for control output voltage load [8].

Along these lines, Fig. 4 the PI controller input and Fig. 5 the fuzzy logic controller input is given the error signal that is delivered. The error gets processed with the PI

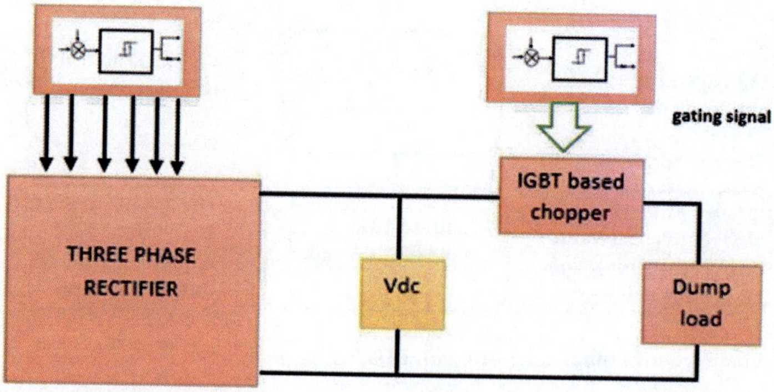

Fig. 6 Electronic Load Controller

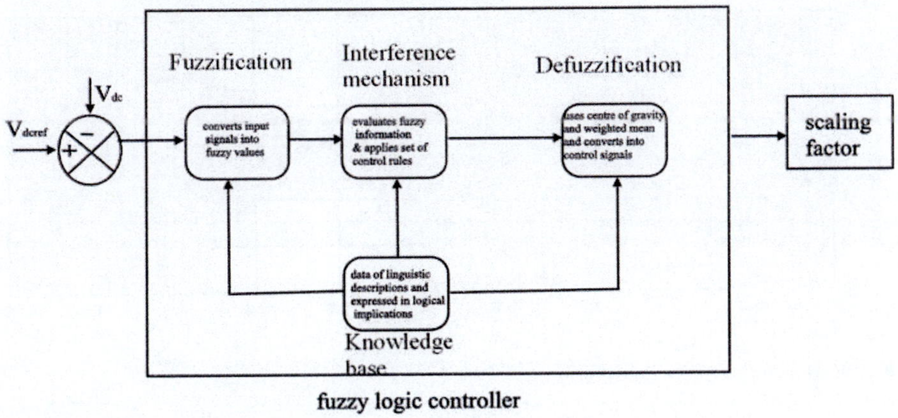

Fig. 7 Fuzzy Logic Detailed Structure

controller in Fig. 4 and the fuzzy controller in Fig. 5 is then compared with triangular pulses in the hysteresis control loop. That output is given to PWM (pulse width modulation) to generate the pulses. These pulses are given to the controlling chopper circuit to monitor the surplus power move to the dump load. The ELCr observes the variations in load at the consumer-side. Whenever unexpected load reduction occurs, an abrupt voltage swell happens and ELCr senses this voltage swell, and the dump load is connected through an IGBT chopper. So the excess power is stored in dump load. Figure 7 shows the detailed structure of fuzzy-based on fuzzification, defuzzification, interference mechanism, and knowledge base components.

5 Results and Observation

Figures 8 and 9 show voltage and current waveforms that read across the loads on the consumer side. Hereby observing the current waveform shows inequitable because

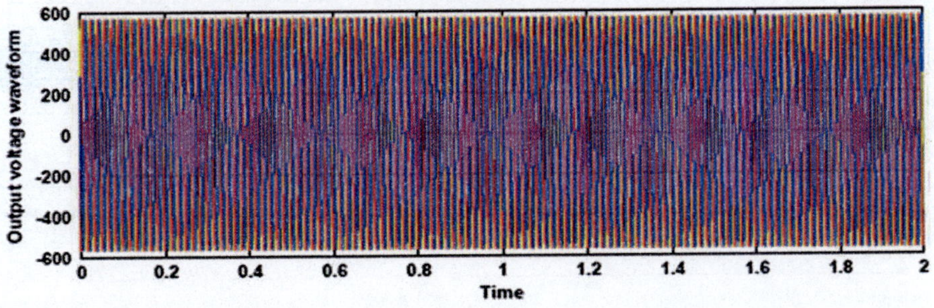

Fig. 8 The output voltage across consumer loads

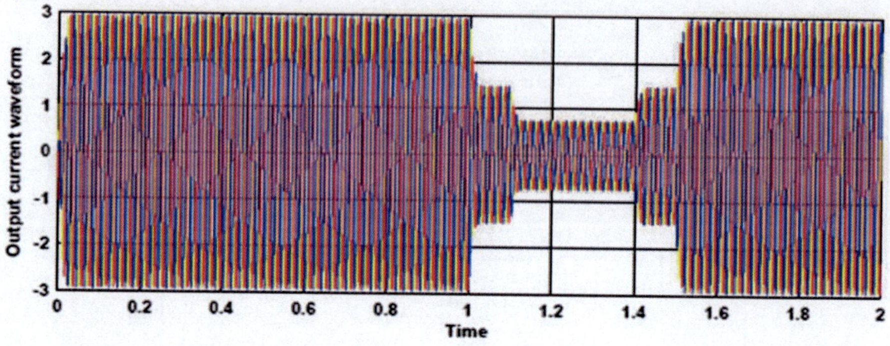

Fig. 9 The output current waveform

at 1 s some amount of power is reduced by using the action of switching, but still, the voltage remains constant in the duration of 1–1.1 s. After 1.1 s again 200w is isolated and it was prolonged in the duration from 1.1–1.5 and slowly it started to increase from 1.5 s. The inequitable situation caused from 1–1.45 s in this case also the voltage remains constant which shown in the figure.

By observing in Fig. 10 whenever the power of the load at the consumer side is unequitable, that is, power drops at time 1 s from 1100 W to 800 W and at time 1.1 s again drops from 800 W to 600 W during this period the load variation is sensed by ELCr and the dump load is connected through IGBT chopper; there some amount of power is building in the dump side. In Fig. 11, at dump load, the power starts absorbing and raising from 210 W to 220 W and 1.1 s is 220 W to 300 W.

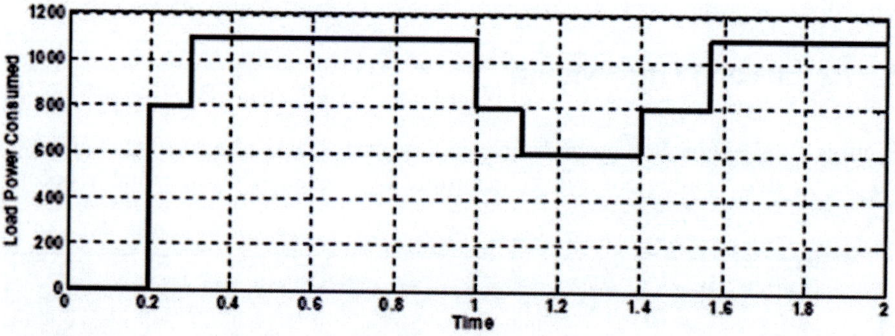

Fig. 10 Load power consumed

Fig. 11 Power consumed by dump resistor

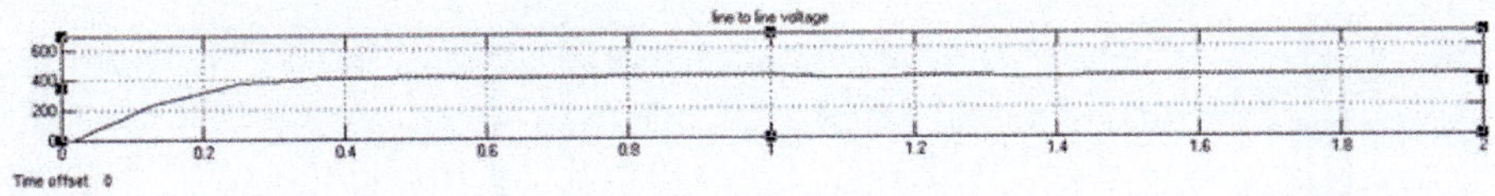

Fig. 12 Line-line voltage

At time 1.4-sec power raised to 600 W to 800 W and at 1.57 s 800 W to 1100 W during these periods ELCr controls the surplus power and maintains the voltage of the line side in Figs. 12 and 13 the line-line voltage and the frequency maintains constant. Next, in Fig. 14 the percentage of THD is reported by using the filter LCL of the preceding and succeeding at an unchangeable frequency which improves the performance and harmonics reduced from 30.11% to 5.53% in ELCr with PI controller, and in Fig. 15, the harmonics reduced from 28.05% to 2.03% in ELCr with the FUZZY controller. Hence the control method is compared with PI and FUZZY controllers which underline the dominance of the FUZZY logic controller.

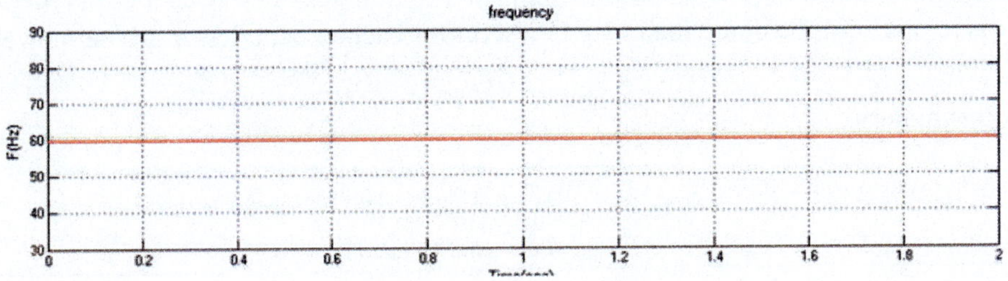

Fig. 13 Frequency

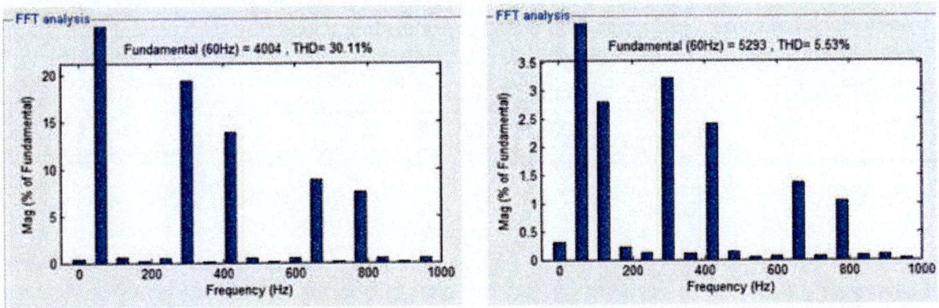

Fig. 14 Analysis of before and after filtering LCL with PI controller

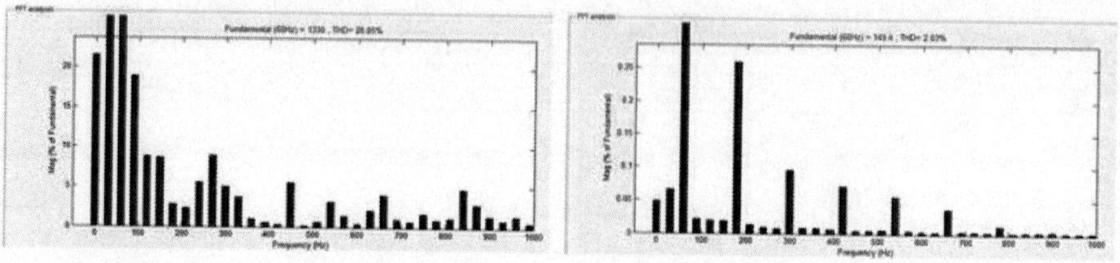

Fig. 15 Analysis of before and after filtering LCL with FUZZY controller

6 Conclusion

The key intention used for this study is to reveal an effective and rapid power transferring both the discharge load, mitigating the harmonics and the working load to uphold a steady voltage of terminals across the load. This ELCr performs the same as anticipated. Converter circuit in ELCr functions to monitor the exchange of reactive and real power to preserve the steady voltage on the consumer side. This device is being positioned to provide power supply at the rural locations as well as the dissipated power in the dump load may be useful for backup batteries or driving power tools.

Parameters

Wind system with PMSG	1.5 MW, 415 V, 60 Hz, 4 poles. 3 Ω,4.9 mH,0.001 N-m-s Speed of the wind 12 m/s
Hydro system with SEIG	7.5KW, 415 V, 60 Hz, 4 pole 30micro farad excitation capacitor, Stator parameters = 0.9Ω, 4.6mH Rotor parameters = 0.6Ω, 5.7mH
back to back converter	2000 μF dc capacitor link,d axis PI-0.04,0.0004,q axis PI-0.012,0.002
LCL filter	45mH, 25μF
ELCr	3000μF,50 Ω

References

1. Rajagopal V, Singh B (2010) Electronic load controller using IcosΦ algorithm for standalone induction generator. In: Power electronics, drives and energy systems (PEDES) & power India, Joint International Conference on IEEE, pp 1–6, Dec 2010
2. Singh B, Murthy SS, Madhusudan, Goel M, Tandon AK (2006) A steady-state analysis on voltage and frequency control of self-excited induction generator in micro-hydro system. Power Electronics, drives and energy systems, 2006. PEDES'06. International conference on. IEEE, pp 1–6, Dec 2006

3. Singh PK, Chauhan YK (2013) Performance analysis of multi-pulse electronic load controllers for self-excited induction generator. Energy efficient technologies for sustainability (ICEETS), international conference on. IEEE, 2013, pp 1299–1307, Apr 2013
4. Ramirez JM, Torres M E(2007) An electronic load controller for the self-excited induction generator. IEEE Trans Energy Convers 22(2):546–548
5. Singh B, Murthy SS, Gupta S (2005) Transient analysis of isolated asynchronous generator with electronic load controller supplying static and dynamic loads. IEEE Trans Ind Appl 41(5):1194–1204
6. Pudur R, Gao S (2016) Performance analysis of Savonius rotor based hydropower generation scheme with electronic load controller. J Renew Energy 2016:7
7. Gao S, Pudur R (2013) Harnessing hydroelectric power using Savonius rotor coupled with an asynchronous generator connected to the grid. In: Proceedings of the IEEE PES asia-pacific power and energy engineering conference (APPEEC '13), pp 1–4, Kowloon, Hong Kong, December 2013
8. Singh B, Kasal GK, Chandra A, Kamal-Al-Haddad (2011) Electronic load controller for a parallel operated isolated asynchronous generator feeding various loads. J Electromag Anal Appl 3:101–114

Chapter 20
Comparative Study of Wireless Power Transfer and Its Future Prospective

Amit Kumar Baghel, Chinmaya Behera, Shankar Amalraj, Ajit Singh, and Sisir Kumar Nayak

1 Introduction

Wireless power transfer (WPT) is the method to transfer power from the source to load without the use of wires or cables. There are different ways of WPT which can be classified broadly into radiative and non-radiative WPT, as shown in Fig. 1. Radiative WPT is in which the antenna radiates electromagnetic wave where electric and magnetic fields are orthogonal to each other and related by free-space wave impedance, i.e.,

$$H = \frac{E}{\eta} \tag{1}$$

where
H is the time and space varying magnetic field,
E is the time and space varying electric field, and
η is the free space impedance $\approx 377\ \Omega$.

The broad category of radiative WPT is microwave whose working frequency is in GHz and light waves or laser having a working frequency in THz. In non-radiative WPT, the transfer of power is basically through magnetic coupling of the transmitter and receiver coils where the maximum power transfer happens at a particular resonant frequency. Inductive and resonant inductive coupling are the two broad categories of

A. K. Baghel (✉) · C. Behera · S. Amalraj
Electronics and Electrical Engineering Department, IIT Guwahati, Guwahati, India

A. Singh
Electronics Engineering, Deen Dayal Upadhyaya College, New Delhi, India

S. K. Nayak
EEE Department, IIT Guwahati, Guwahati, India
e-mail: sknayak@iitg.ac.in

S. Dawn et al. (eds.), *Smart and Intelligent Systems*, Algorithms for Intelligent Systems,
https://doi.org/10.1007/978-981-16-2109-3_20

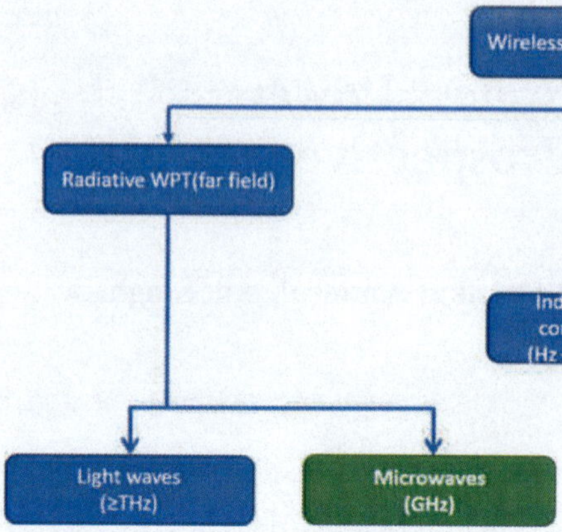

Fig. 1 Different methods of WPT

non-radiative WPT having a working frequency ranging in Hz-MHz and MHz-GHz, respectively.

In this paper, various technologies used in WPT are discussed in Sect. 2. The concept and various technologies available for non-radiative and radiative WPT have been discussed in Sects. 3 and 4, respectively. A comparative study of different WPT technologies based on power transfer, distance of power transfer, and mobility of the receiver system is discussed in Sect. 5, and the concluding remark is given in Sect. 6.

2 Work Done till Now in WPT Technology

2.1 WPT Using MM

WPT is soon becoming the need of the present society due to its advantage of transferring power without the use of wires. This can help in getting rid of the large grid of wires and charging single or many devices at a time. Far-field microwave and mid-range WPT [1–6] have been demonstrated and reported in the literature. It is observed that there is an increase in WPT efficiency from 3.36% to 13.58% with the spiral super conducting metamaterial (MM). Improvement in magnetic resonance WPT using superconducting ferromagnetic MM is shown in [7]. In the lower frequency of operation, a solenoid which is loaded by ferrite is designed as a unit cell [8]. An improvement of 10% is seen at a distance of 4.5 cm in WPT at the operating frequency of 5.57 MHz. It is observed that all the studies are related to non-radiative or near-field radiative WPT using MM, thus opening the research direction field for far-field radiative WPT using MM. The MM has been used to increase the gain and aperture efficiency of the horn antenna [9–18].

According to the properties the of MM lens, various MM lenses are classified as gradient refractive index(GRIN) lens MM [14–18], negative index MM lens [9–11], and zero index MM lens [12, 13]. GRIN MM as the name suggests whose refractive index varies spatially and perpendicular to the direction of propagation [19]. The spatial variation of the refractive index is in such a way in which the maximum refractive index is at the center and minimum at the corner of the lens. MM has the refractive index variation $n(r)$ given by $n(r) = n0 - \sqrt{r^2 + F^2 - F}/T$, where $n0$ is the refractive index at the center, F is the focal length of the antenna, T is the thickness of the lens, and r is the position of the MM unit cell from the center of the lens. To get the broadband property of MM, the zero index MM is designed which has shown the improvement in the directivity of the antenna [17]. Having a directional antenna and then using MM to make it more directional will improve the far-field WPT many times.

2.2 *Improvement in the Received Power at the Rx Antenna Side*

Having discussed various ways of increasing and improving the gain of the Tx antenna, the Rx antenna gain improvement is also the area of discussion for increasing the overall efficiency of the WPT system. The various micro-strip patch antenna is designed and reported. For example, a cross dielectric resonator on the microstrip patch antenna is designed to give better gain and efficiency 9 dBi [20–23].

A circularly polarized monopole patch antenna with parasitic shorted patches is designed which give the conical beam. The vertically polarized electric field $E_{\theta 1}$ is provided by monopole patch, and the horizontally polarized electric field $E_{\varphi 1}$ is provided by parasitic shorted patches and annular ring [24]. In [25], the sequential phase feeding for the truncated patch antenna which is sandwiched between meta-surface and ground plane helps in increasing the 10-dB impedance bandwidth of 58.06% and the axial 3 dB ratio is 41.67% with a peak antenna gain of 12.08 dBi. However, for the far-field WPT applications, the radiation intensity on the aperture of the micro-strip patch antenna should be higher so to convert it into useful power.

To extract more power from the receiver antenna, the radiation density at the point of application (in this case, the location of the sensors, drone, robot) should be more so to capture more radiated power at the Rx antenna surface. This can only happen by changing the plane wave into the spherical wave and placing the low-profile patch antenna at the point of higher radiation density. Parabolic dish (PD) antenna helps to achieve the same by focusing the plane wave at the focus. Research to make the PD antenna which is lighter in weight, low profile, and flexible enough is carried so to be used with the Rx antenna of sensors and drone, i.e., using Polydimethylsiloxane (PDMS) as a substrate with the aluminum coating acting as the perfect metal reflector.

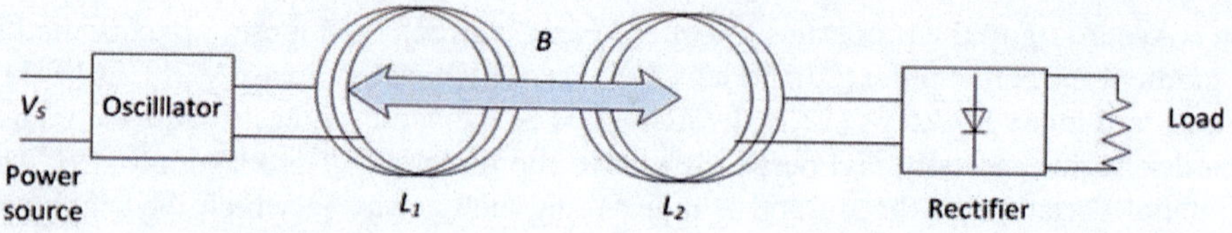

Fig. 2 Inductive coupling WPT

3 Non-Radiative WPT

3.1 *Inductive Coupling WPT*

In inductive WPT, the power is transferred between coils of conductive wire by the magnetic field. Various applications of inductive WPT are proposed and shown like battery charging of electric vehicles, an application dealing with avionics, etc [26–31]. The advantage of inductive WPT is higher efficiency. However, the distance of power transfer is in a few cms and depends on the position of the transmitter (Tx) and receiver (Rx) coil. The power transfer reduces or becomes zero if the Tx and Rx coils are not in the line of sight or there is a misalignment between the coils. Figure 2 shows the schematic of inductive WPT.

3.2 *Magnetic Resonant Coupling WPT*

The difference between inductive and magnetic resonant coupling WPT is only the addition of resonant circuit at Tx and Rx side having a particular frequency of resonance which is used for mid-range WPT applications [32–35]. The advantage of this method of WPT is that simultaneously many devices can be charged and the distance of power transfer is more than inductive power transfer, but the issue of coil misalignment is still there in magnetic resonant coupling WPT. Figure 3 shows the schematic of magnetic resonant coupling WPT where C is the capacitance.

4 Radiative

4.1 *Laser or Light Wave WPT*

Laser or light wave WPT is one of the radiative WPT in which the light wave emitted by the laser is used to transfer power to a larger distance. The advantage of using laser WPT is that the kWs of power is transferred to a larger distance to the receiver having a smaller aperture efficiently. The property to give a highly coherent beam

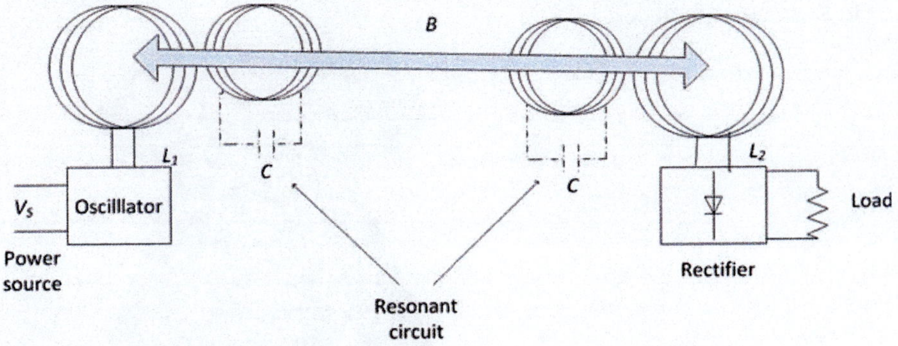

Fig. 3 Magnetic resonant coupling WPT

makes laser WPT a good candidate for charging unmanned aerial vehicles(UAVs) [36], robots [37], and orbiting satellites [38]. However, the attenuation through the atmosphere changes with the weather conditions, and the efficiency of the laser diode and photovoltaic array is less. The intensity of the laser is also a concern for the researchers because of which if anything comes under its line of sight, for example, birds and animals, will be heated up and destroyed. Also, the line of sight is a problem. Various developments to increase the efficiency of power transfer using laser are being carried out by researchers [39]. Figure 4 shows the schematic of laser WPT.

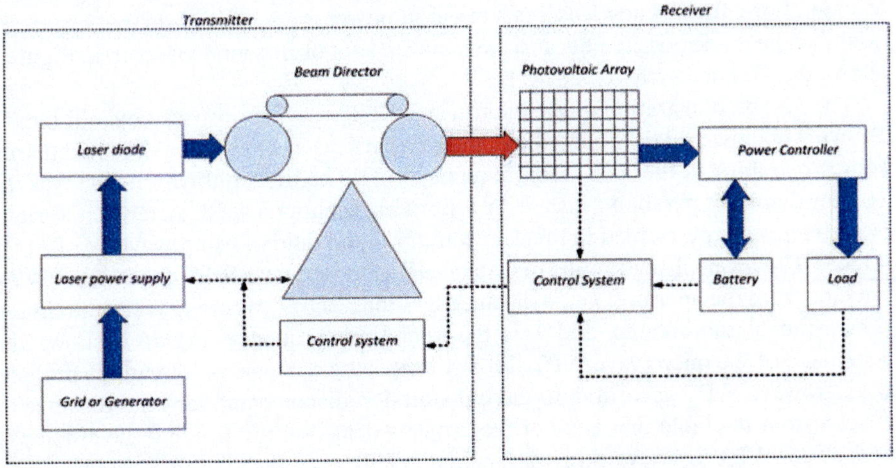

Fig. 4 Laser WPT

Fig. 5 Wanderclyffe Tower

4.2 *Microwave WPT*

The idea and the practical demonstration of WPT were demonstrated by great engineer and innovator Nicolas Tesla in the year 1897 [40]. He built the 187 feet long Wanderclyffe Tower known as Tesla tower to power the whole world, but his dream could not be made possible because he incurred a lot of financial losses in it. Figure 5 shows the Wanderclyffe tower [41].

The first demonstration of the microwave WPT was demonstrated by William C. Brown who powered the helicopter at the height of 50 feet with the 5-kW magnetron source operating at the 2.45 GHz frequency [42]. The transmitting antenna was the 10 feet diameter parabolic dish with a horn antenna as a feed. After that various experiments are performed in the 21st century in the field of Microwave WPT [1–6, 43–47]. However, all are limited to distance less than the near field of the transmitting antenna from the aperture, i.e., a distance less than $2D^2/\lambda$, where D is the maximum dimension of the antenna and λ is the operating frequency. Figure 6 shows the schematic of the microwave WPT. Table 1 shows the comparison of various projects in microwave WPT according to the transmission distance and far-field distance. It is seen from the table that none of the projects deals with improving the efficiency of far-field WPT or are working in far-field WPT.

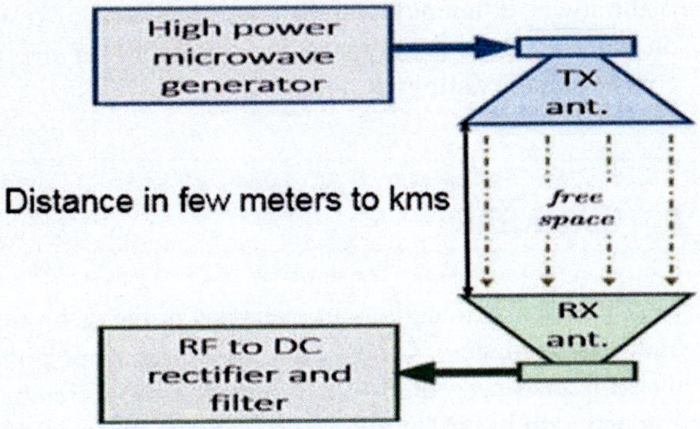

Fig. 6 Microwave WPT

Table 1 Comparison of various projects in microwave WPT Project

	Japan	USA	South Korea	China	China
	2015 [43]	2016 [44]	2018 [45]	2108 [46]	2019 [47]
Frequency (GHz)	5.8	5.8	2.45	10	5.8
Size of Tx antenna (m^2)	1.44	0.047	1	0.0625	1
Transmission distance (m)	54	0.4	1	4	10
Far-field distance (m)	111	3.6	16.3	8.3	77.33

5 Comparative Analysis of Various WPT Technology

In this section, a comparative analysis of different WPT technologies is discussed. Table 2 shows the comparison of different WPT technologies based on the efficiency of power transfer, the distance of power transfer, and the mobility of the receiver system. From the table, it can be referred that the inductive WPT gives higher WPT efficiency, but for very smaller distance, the mobility of the receiver system is not there, i.e., if the receiver system is not in the line of sight of the transmitter system, than the WPT efficiency is zero. However, in case of microwave WPT, the efficiency is lower but the distance of power transfer is very long in meters or sometimes in km. It also has no issue of receiver mobility as the beam can be steered according to the position of the receiver. Comparing all the available technologies, microwave WPT is and will be the best candidate for far-field WPT to charge sensors present in aerostat which are used for various purposes like weather monitoring, surveillance, robots, and drone charging for defense applications. The drawback of this technology

Table 2 Comparison of various WPT technology Type of WPT technology Efficiency Distance Mobility

Inductive coupling	Very high	Very short	No
Magnetic resonant coupling	High	Short	Difficult
Microwave	Low	Very long	Yes
Laser or light waves	Low	Long	Difficult

is the lower efficiency which is an open area of research. This work tries to focus on various ways to increase the efficiency of the WPT system compared to the conventional existing systems.

6 Conclusion

This research article gives an overview of the different WPT techniques, advantages, and disadvantages of the same. Also, the work going on in different parts of the world is also presented. For improving the efficiency of microwave WPT, the Tx antenna gain has to be increased by using metamaterial. Also, the flexible parabolic dish reflector antenna has to be fabricated to change the plane wave to a spherical wave to focus it at the receiving antenna aperture. By improving the design of the transmission and receiving antenna, the overall efficiency of microwave WPT is improved. The future research goals in this direction are presented.

References

1. Shinohara N, Matsumoto H (1998) Experimental study of large rectenna array for microwave energy transmission. IEEE Trans Microw Theory Tech 46(3):261–268
2. Valenta CR, Durgin GD (2014) Harvesting wireless power: Survey of energy- harvester conversion efficiency in far-field, wireless power transfer systems. IEEE Microwave Mag 15(4):108–120
3. Dickinson RM (1976) Performance of a high-power, 2.388-ghz receiving array in wireless power transmission over 1.54 km. In: 1976 IEEE-MTT-S international microwave symposium, pp 139–141
4. Chen J, Hu Z, Wang S, Cheng Y, Liu M (2016) Investigation of wireless power transfer for smart grid on-line monitor- ing devices under hv condition. Procedia Comput Sci 83:1307–1312. The 7th international conference on ambient systems, networks and technologies.
5. Cheng YZ, Jin J, Li WL, Chen JF, Wang B, Gong RZ (2016) Indefinite-permeability metamaterial lens with finite size for miniaturized wireless power transfer system. AEU Int J Electron Commun 70(9):1282–1287
6. Cho Y, Kim JJ, Kim D, Lee S, Kim H, Song C, Kong S, Kim H, Seo C, Ahn S, Kim J (2016) Thin pcb-type metamaterials for improved efficiency and reduced emf leakage in wireless power transfer systems. IEEE Trans Microw Theory Tech 64(2):353–364
7. Wang X, Wang Y, Fan G, Hu Y, Nie X, Yan Z (2018) Experimental and numerical study of a magnetic resonance wireless power transfer system using superconductor and ferromagnetic metamaterials. IEEE Trans Appl Super Cond 28(5):1–6
8. Gz Rodrez ES, RamRakhyani AK, Schurig D, Lazzi G (2016) Compact low-frequency metamaterial design for wireless power transfer efficiency enhancement. IEEE Trans Microw Theory Tech 64(5):1644–1654
9. Iyer AK, Eleftheriades GV (2007) A multilayer negative-refractive-index transmission-line (nri-tl) metamaterial free-space lens at x-band. IEEE Trans Antennas Propag 55(10):2746–2753
10. Navarro-Cia M, Beruete M, Campillo I, Ayza MS (2011) Beamforming by left- handed extraordinary transmission metamaterial bi- and plano-concave lens at millimeter-waves. IEEE Trans Antennas Propag 59(6):2141–2151

11. Das S, Nguyen H, Babu GN, Iyer AK (Nov 2015) Free-space focusing at c-band using a flat fully printed multilayer metamaterial lens. IEEE Trans Antennas Propag 63(11):4702–4714
12. Turpin JP, Wu Q, Werner DH, Martin B, Bray M, Lier E, Near-zero-index metamaterial lens combined with amc metasurface for high-directivity low-profile antennas. IEEE Trans Antennas Propag 62(4):1928–1936
13. Yuan LH, Tang WX, Li H, Cheng Q, Cui TJ (2014) Three-dimensional anisotropic zero-index lenses. IEEE Trans Antennas Propag 62(8):4135–4142
14. Lin Q, Wong H (2018) A low-profile and wideband lens antenna based on high- refractive-index metasurface. IEEE Trans Antennas Propag 66(11):5764–5772
15. Mei ZL, Bai J, Niu TM, Cui TJ (2012) A half maxwell fish-eye lens an- tenna based on gradient-index metamaterials. IEEE Trans Antennas Propag 60(1):398–401
16. Xu H, Wang G, Tao Z, Cai T (2014) An octave-bandwidth half maxwell fish- eye lens antenna using three-dimensional gradient-index fractal metamaterials. IEEE Trans Antennas Propag 62(9):4823–4828
17. Ma HF, Cai BG, Zhang TX, Yang Y, Jiang WX, Cui TJ (2013) Three- dimensional gradient-index materials and their applications in microwave lens antennas. IEEE Trans Antennas Propag 61(5):2561–2569
18. Erfani E, Niroo-Jazi M, Tatu S (May 2016) A high-gain broadband gradient re- fractive index metasurface lens antenna. IEEE Trans Antennas Propag 64(5):1968–1973
19. Smith DR, Mock JJ, Starr AF, Schurig D (2005) Gradient index metama- terials. Phys Rev E 71:036609
20. Nasimuddin, Esselle KP (2007) A low-profile compact microwave antenna with high gain and wide bandwidth. IEEE Trans Antennas Propag 55(6):1880–1883
21. Zhang H, Abdallah Y, Chantalat R, Thevenot M, Monediere T, Jecko B (2012) Low-profile and high-gain yagi wire-patch antenna for wimax ap- plications. IEEE Antennas Wirel Propag Lett 11:659–662
22. Cao Y, Cai Y, Cao W, Xi B, Qian Z, Wu T, Zhu L (2019) Broadband and high-gain microstrip patch antenna loaded with parasitic mushroom-type structure. IEEE Antennas Wirel Propag Lett 18(7):1405–1409
23. Yang Z, Liang F, Yi Y, Zhao D, Wang B (2019) Metasurface-based wideband, low-profile, and high-gain antenna. IET Microwaves Antennas Propagation 13(4):436–441
24. Xu H, Zhou J, Zhou K, Yu Z (2018) Low-profile circularly polarised patch antenna with high gain and conical beam. IET Microwaves Antennas Propa- gation 12(7):1191–1195
25. Ta SX, Park I (2017) Compact wideband circularly polarized patch antenna array using metasurface. IEEE Antennas Wirel Propag Lett 16:1932–1936
26. Dai J, Ludois DC (Nov 2015) A survey of wireless power transfer and a critical comparison of inductive and capacitive coupling for small gap applications. IEEE Trans Power Electron 30(11):6017–6029
27. Wang C-S, Stielau OH, Covic GA (2005) Design considerations for a contactless electric vehicle battery charger. IEEE Trans Ind Electron 52(5):1308–1314
28. Ou C, Liang H, Zhuang W (2015) Investigating wireless charging and mobility of electric vehicles on electricity market. IEEE Trans Ind Electron 62(5):3123–3133
29. Xu Q, Hu D, Duan B, He J (July 2015) A fully implantable stimulator with wireless power and data transmission for experimental investigation of epidural spinal cord stimulation. IEEE Trans Neural Syst Rehabil Eng 23(4):683–692
30. Athalye P, Maksimovic D, Erickson R (2003) High-performance front-end con- verter for avionics applications [aircraft power systems]. IEEE Trans Aerosp Electron Syst 39(2):462–470
31. Huang S, Lee T, Huang T (Dec 2014) Inductive power transfer systems for pt-based ozone-driven circuit with flexible capacity operation and frequency-tracking mechanism. IEEE Trans Ind Electron 61(12):6691–6699
32. Sohn YH, Choi BH, Lee ES, Lim GC, Cho G, Rim CT (Nov 2015) Gen- eral unified analyses of two-capacitor inductive power transfer systems: Equiva- lence of current-source ss and sp compensations. IEEE Trans Power Electron 30(11):6030–6045

33. Chwei-Sen Wang, Covic GA, Stielau OH (2004) Power transfer capability and bifurcation phenomena of loosely coupled inductive power transfer systems. IEEE Trans Ind Electron 51(1):148–157
34. Chwei-Sen Wang, Stielau OH, Covic GA (2005) Design considerations for a contactless electric vehicle battery charger. IEEE Trans Ind Electron 52(5):1308–1314
35. Jegadeesan R, Guo Y (2012) Topology selection and efficiency improvement of inductive power links. IEEE Trans Antennas Propag 60(10):4846–4854
36. Raible DE (2008) High intensity laser power beaming for wireless power transmis- sion, Master's thesis. Cleveland State University, Dept. Elect. Comput. Eng.
37. Kawashima N, Takeda K (2005) Laser energy transmission for a wireless energy supply to robots. In: Proceedings Symposium on Automation and Robotics Construction, p 373–380
38. Shi D, Zhang L, Ma H, Wang Z, Wang Y, Cui Z (2016) Research on wireless power transmission system between satel- lites. In: 2016 IEEE wireless power transfer conference (WPTC), pp 1–4
39. Jin K, Zhou W (2019) Wireless laser power transmission: a review of recent progress. IEEE Trans Power Electron 34(4):3842–3859
40. Tesla N (1900) System of transmission of electrical energy
41. Tesla N (1902) A new tesla laboratory on long island.Electr World Eng 16:98–99
42. Brown WC (1969) Experiments involving a microwave beam to power and position a helicopter. IEEE Trans Aerosp Electron Syst AES-5(5):692–702
43. Takahashi T, Sasaki T, Homma Y, Mihara S, Sasaki K, Nakamura S, Makino K, Joudoi D, Ohashi K (2016) Phased array system for high efficiency and high accuracy microwave power transmission. In: 2016 IEEEinternational symposium on phased array systems and technology (PAST), pp 1–7
44. Gowda VR, Yurduseven O, Lipworth G, Zupan T, Reynolds MS, Smith DR (2016) Wireless power transfer in the radiative near field. IEEE Antennas Wireless Propag Lett 15:1865–1868
45. Khang S, Lee D, Hwang I, Yeo T, Yu J (Jan 2018) Microwave power transfer with optimal number of rectenna arrays for midrange applications. IEEE Antennas Wirel Propag Lett 17(1):155–159
46. Chen Q, Chen X, Duan X (2018) Investigation on beam collection efficiency in microwave wireless power transmission. J Electromag Waves Appl 32(9):1136–1151
47. Yi X, Chen X, Zhou L, Hao S, Zhang B, Duan X (2019) A microwave power transmission experiment based on the near-field focused transmitter. IEEE Antennas Wirel Propag Lett 18(6):1105–1108

Chapter 21
Monopole Antenna for UWB Applications with DGS

K. V. Prasad, M. V. S. Prasad, and Padarti Vijaya Kumar

1 Introduction

Ultra-wideband (UWB) wireless communication technology has received significant attention from both research and industry since the Federal Communication Commission (FCC) released the frequency band from 3.1 to 10.6 GHz for commercial communication applications in February 2002. In a UWB system, the UWB antenna is one of the most critical components and has lured significant research in recent years. The wideband antenna is simpler than dual-band or triple-band designs, including narrowband elements, which have the tendency to get complicated and may be prone to detuning in some circumstances. Ultra-wideband (UWB) antenna is a proper alternative in these cases. Moreover, UWB is an emerging new technology for broadband internet access and public safety applications, employing the spectrum region of 1.9 GHz to 10.6 GHz at significantly low power levels. UWB technology has many benefits: high data rate (>100Mb/s), low power consumption, compactness, low cost, excellent immunity to multipath interference, and reduced hardware complexity. The microstrip patch antenna is preferred due to its compactness, relative ease of fabrication, and considerable low volume. With simple geometry, the patch antenna offers many advantages not commonly exhibited in other antenna configurations in the UWB range because of its compatibility with microwave frequencies. However, despite the many benefits, they do have some considerable drawbacks that restrict their applications. They are inherently narrow bandwidth due to resonance nature (typically 1–5%), low gain, spurious feed radiation, insufficient polarization purity, and manufacturing tolerance problems.

K. V. Prasad (✉) · P. Vijaya Kumar
V R Siddhartha Engineering College, Vijayawada 520007, India
e-mail: prasadkv@vrsiddhartha.ac.in

M. V. S. Prasad
R V R & JC College of Engineering, Guntur 520007, India

S. Dawn et al. (eds.), *Smart and Intelligent Systems*, Algorithms for Intelligent Systems,
https://doi.org/10.1007/978-981-16-2109-3_21

1.1 Literature Survey

De and Sarkar [1], deal with an ultra-wideband monopole antenna. The ground plane is optimized, considering fractal concepts and showing possible effects of the ground plane on antenna performance. Sonar and Mishra [2], discussed the basic design of a Circular Patch Monopole antenna for wideband applications. Effects of radius variations of a circular patch are thoroughly discussed, and the basic equation for patch radius calculation for the first resonant frequency of a UWB antenna is specified. Sharma [3], studied various parameters affecting the performance of a Circular Ring Monopole antenna and Defective Ground Structures, along with those different surface wave effects are discussed. Haraz and Sebak [4], presented a UWB antenna's essential operation. Circular patch monopole operation for wideband performance is analyzed and implemented. Tran and Haider [5], dealt with implementing an SWB antenna, an extension of the current UWB antennas. The fractal concept to establish an SWB antenna is discussed, and limiting factors of antenna performance are studied. It presents the properties of the FR-4 substrate used to fabricate the antenna and discusses the limitations of different substrates. Liu and Cheung [6], discussed the minor effects of impedance mismatch on the UWB antenna implementation. Antenna losses from the feed line and feed line radiation effects are elaborated. Ashtanker and Dethe [7], worked on a circular monopole antenna that gave a glimpse at the possibility of the evolution of a wideband antenna from a basic circular monopole. Mohamed and Shafai [8], performed a study on modern ultra-wideband antennas. Various current design implementations are discussed, and their differences are explained. Parametric effects of Ring Patch implementation from Circular patch monopole are discussed. It provides the evolution of ring patch antenna from circular monopole antenna. Kwaha and Inyang [9], provided basic equations for calculating the dimensions of a circular patch and provided a solution model and theoretical analysis of circular patch monopole implementation. It gives formulae and fundamental concepts for a circular patch's resonant frequency and the relation between equivalent radius and actual radius. Kumar and Dhubkarya [10–12], gave the analysis and approach toward a ring patch antenna. The ring patch's width equations are given based on the characteristic impedance. Equations for the microstrip patch's outer radius and inner radius are given, and the ring patch antenna is analyzed theoretically. Lu and Huang [13], published works relating to the effects of rectangular slots in the ground plane on an antenna's performance. DGS through rectangular slot implementation is discussed, and the impact of increasing the dimensions of the ground plane is presented. The current flow and distribution in the ground plane due to DGS implementation are explained, and possibly affected frequency ranges are given. Liu et al. [14], reported that on reducing ground plane effects on monopoles through DGS elucidates the impact of various slot shapes on the antenna parameters. Slot implementation results and resonance shifts due to implemented slots near the patch are explained. The antenna results with slot implementation and without are presented as part of the parametric analysis. FCC report [13–21] gives us the guidelines of spectrum utilization. Frequency parameters to be considered while modeling antenna and radiation limits are elaborated. Safety guidelines to be considered are devised.

2 Antenna Design and Specifications

The possibility of improving antenna through employing different techniques in necessary steps is presented through recent literature [1, 7]. The UWB antenna is evolved from a Circular Patch monopole in four steps. In the first iteration, a circular monopole antenna is designed for a low resonant frequency of 3 GHz, which is the lowest frequency of the UWB range, and dimensions are optimized. In the second iteration, an annular ring is developed for better impedance matching and resonance characteristics. In the third iteration, the ground plane is optimized using the fractal concept for better resonance characteristics and improved impedance bandwidth (Fig. 1).

2.1 *Evolution of Antenna Design*

The antenna is designed after considering several parameters. The antenna depends on the microstrip antenna's monopole operation to ensure wideband performance. Circular patch monopole antennas are widely used for UWB applications since they have a simple structure, easy fabrication, wideband characteristics, and Omni-directional radiation patterns. Monopole is generally obtained by replacing the infinite ground plane with a finite ground plane [3, 4, 8].

2.2 *Microstrip Feed Line*

Microstrip feed is the feeding technique employed. A conducting strip is connected to the microstrip patch's edge in this technique. The microstrip is smaller in width than the patch, and this technique has the advantage that the feed can be etched on the same substrate to provide the planar structure. Hence this is an easy feeding scheme since it allows for ease of fabrication and simplicity in modeling.

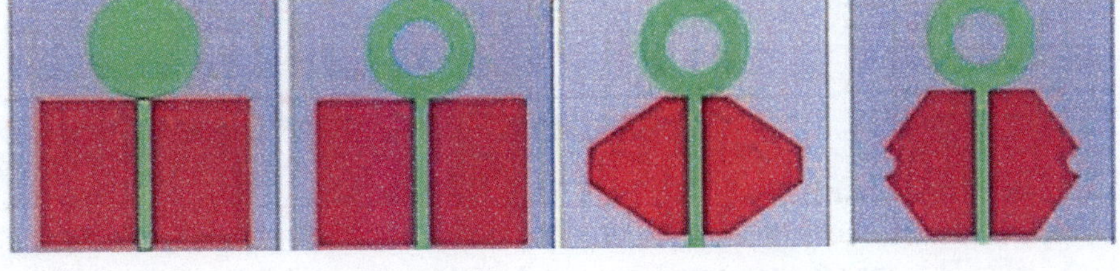

Fig. 1 Proposed monopole circular Patch

2.3 Defected Ground Structures

Considering the fractal concept, the antenna ground plane is modeled as an octagon to improve the current distribution. Defected Ground Structures are employed to enhance the antenna parameters further. DGS implementation through the slots ensures reduced surface waves caused by the radiation patch on the ground plane [3]. The improvement is mainly observable in low frequencies. Rectangular slots are employed to improve the current distribution toward the ground plane's edges. This causes a significant shift of impedance bandwidth toward the lower frequency range. The rectangular slots implemented did not exhibit substantial low-frequency effects in the UWB range, resulting in reduced impedance bandwidth. Due to the undesirable effects of rectangular slots, circular slots are considered and employed. Variation of circular slots' position is also considered, which showed significant improvement of antenna return loss over the impedance bandwidth. The slot centers are kept constant at a distance of 12 mm from the center of the feed line, being in line with the radiating patch's circumference and the ground plane's horizontal axis. It is observed that with an increase in the radius of slots, improvement in bandwidth and maximum return loss at a lower frequency than the UWB range is observed. Hence optimum slot radius is considered as 3 mm. The slot positions are varied, expecting a change in the lower frequencies. Slots are moved away from the feed line in small steps.

Distance from the center of the feed line to the center of the circle is considered as d. After numerous simulations and observations, the final design is made using slots and fractal techniques. The proposed antenna is fabricated using low-cost glass epoxy (FR4) substrate material having a thickness (h) of 1.6 mm, dimensions of 60 mm × 60 mm, and permittivity $\varepsilon r = 4.4$. The material is chosen due to its low cost and local availability.

The optimal parameters of the antenna above are bestowed in Table 1.

3 Results and Discussion

3.1 Return Loss

S11 parameter indicates return loss, and it is defined as a maximum reflection of power from the given antenna. Return Loss was well below −10 dB in the proposed application frequencies. The return loss characteristics confirm an impedance bandwidth of 14.4 GHz (1.96 GHz–16.4 GHz) (Fig. 2).

Return loss characteristics show that the antenna can work around 2.8 GHz to 17.6 GHz. This makes the impedance bandwidth to be 14.8 GHz. It can also be observed that at low frequency, the return loss of antenna is significantly improved while compared to simulation results showing that the FR-4 substrate is best suitable for low frequencies.

Table 1 Design specifications

Specification	Measure
Radius of an inner circle of Patch	7 mm
Radius of the outer circle of Patch	12 mm
Microstrip feed dimension of Patch	35 mm × 3 mm
Vertical side length of Octagon	2 mm (cannot be shown due to slot)
Horizontal side length of Octagon	24 mm
Slant sides lengths of Octagon	20 mm
Feed of Ground Plane	1 mm × 3 mm
Radii of Circular slots	3 mm
Horizontal Position of Centre of circular slots	At a distance of 22.5 mm from the center of the microstrip feed line
Vertical Position of Centre of circular slots	The midpoint of the vertical side of a ground plane

3.2 *Fabricated Antenna Design*

See Fig. 3.

4 Conclusion

A Monopole antenna with DGS is designed and fabricated successfully to work in the UWB range. The simulation results were well in agreement with the fabricated antenna measured parameters. The antenna finds its applications in WLAN, WiMAX, WPAN Networks, and other short-range commercial networks and sensor systems. This work has opened numerous areas for future work that could be done to optimize the working of an antenna using defective ground structures. The work done on circular slots and their varying position effects can be extended to the design of other monopole antennas for significant improvement of results. Substrate effects and design concepts studied can help develop better working antennas by implementing better substrate materials. While the antenna has improved characteristics, work can be further continued to design filters and avoid harmonics radiation. The transmitter and receiver system based on current work can be developed to form the core of commercially feasible UWB systems.

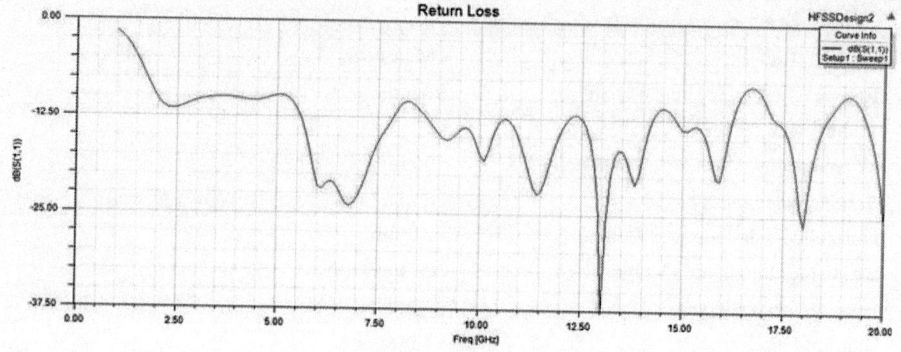

Frequency/GHz

(a)

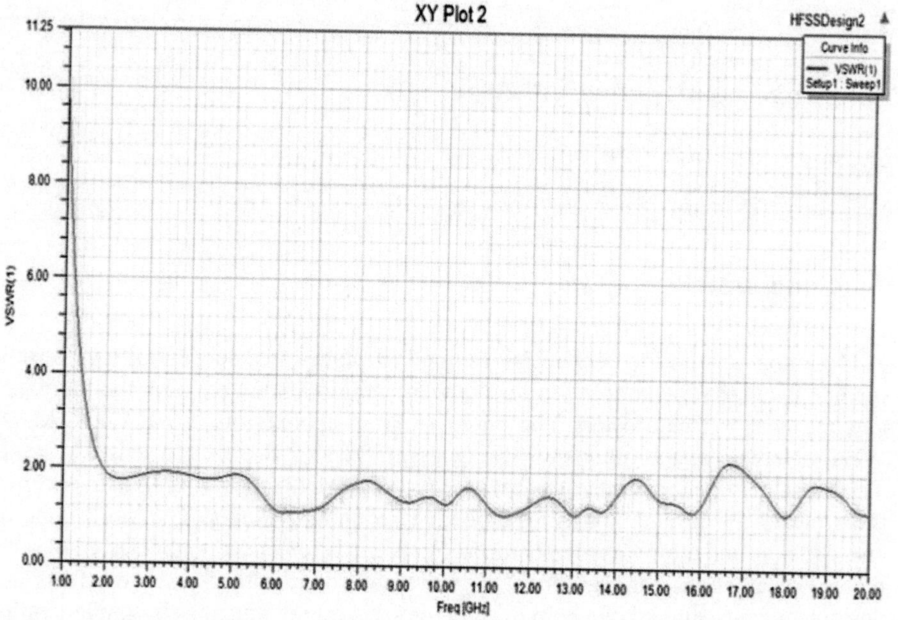

(b)

Fig. 2 Simulated plots of monopole antenna for UWB applications. **a** Return loss and **b** VSWR

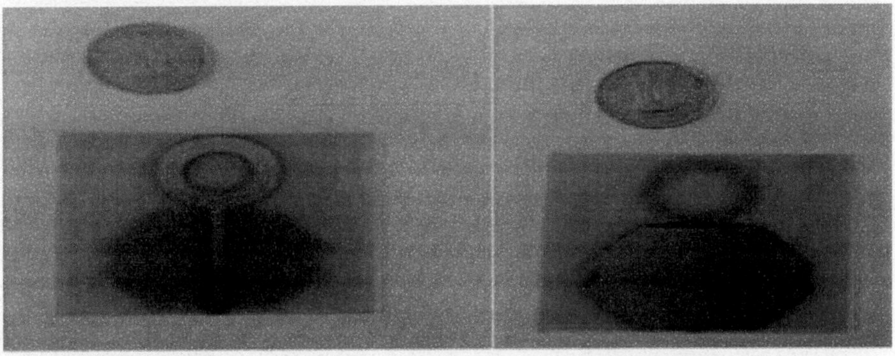

(a)

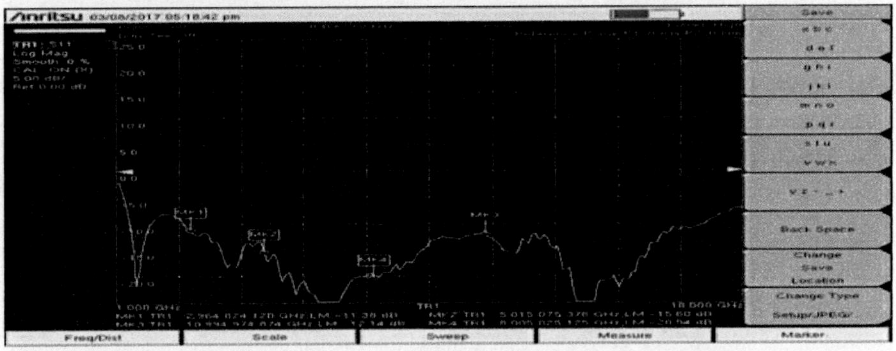

(b)

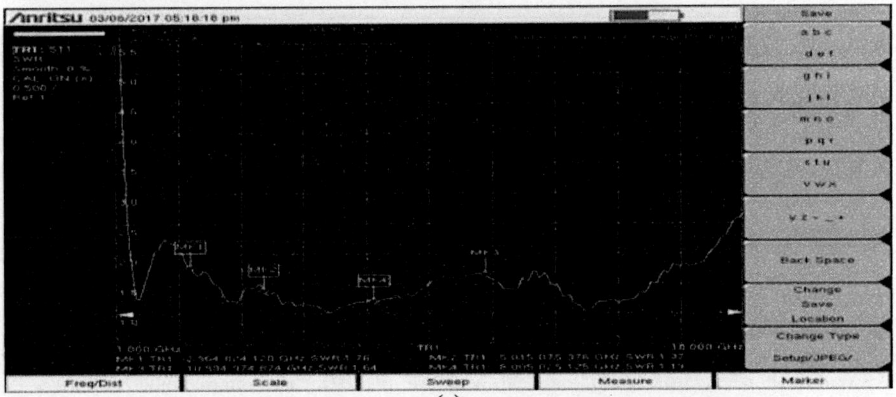

(c)

Fig. 3 Fabricated model and plots of monopole antenna for UWB applications. **a** Return loss and **b** VSWR

References

1. De S, Sarkar PP (2015) A high gain ultra-wideband monopole antenna. Int J Electron Commun
2. Sonar AF, Mishra NS (2013) UWB circular monopole antenna 1(1) ISSN:23208945
3. Sharma R (2013) Study of circular ring microstrip antennas for wideband applications using DGS. INFLIBNET
4. Haraz O, Sebak A-R (2103) UWB antennas for wireless applications. Licensee Intech
5. Tran D, Haider N (2012) Architecture and design procedure of a Generic SWB antenna with superb performances for tactical commands and ubiquitous communications. Licensee Intech
6. Liu L, Cheung SW (2012) Cable effects on measuring small planer UWB monopole antennas. Licensee InTech
7. Ashtankar PS, Dethe CG (2012) Design and modification of circular monopole UWB antenna for WPAN application. IISTE
8. Mohamed A, Shafai L (2011) Performance study on modern ultra-wideband monopole antennas. Ultra-wideband communications: novel trends—antennas and propagation, Dr. Mohammad Matin (Ed.), InTech, https://doi.org/10.5772/17795
9. Kwaha BJ, Inyang ON (2011) The circular microstrip patch antenna—design and implementation. IJRRAS 8(1)
10. Prasad KV, Prasad MVS, Kumar MS, Alekhya B (2018) Surface wave suppression in patch arrays using EBG structures. In: 2018 Conference on signal processing and communication engineering systems, SPACES, 2018, January, pp 99–104
11. Kudumu VP, Mokkapati VSP (2020) A slot-shaped ebg structure for improving the isolation between patch arrays. Int J Microwave Opt Technol 153):269–278. ISSN No. 1553–0396
12. Kumar R, Dhubkarya DC (2011) Design and analysis of circular ring microstrip antenna. Global J Res Eng
13. Lu Y, Huang Y (2011) Reducing ground-plane effects on UWB monopole antenna. IEEE Antennas Wirel Propagation Lett 10
14. Liu L, Cheung SW, Yuk TI (2011) "Bandwidths improvements using ground slots for compact UWB microstrip-fed antennas. PIERS Proceedings, Suzhou, China
15. Padarti VK, Rao NV (2020) Adaptive SOICAF algorithm for PAPR mitigation in OFDM systems. Wirel Pers Commun 113:927–943. https://doi.org/10.1007/s11277-020-07260-y
16. Kumar PV, Nandanavanam VR (2018) A novel method for joint- PAPR mitigation in OFDM based massive MIMO downlink systems. Int J Eng Technol 7(3):1185–1188
17. Padarti VK, Nandanavanam VR (2021) An improved ASOICF algorithm for PAPR reduction in OFDM systems. Int J Intell Eng Syst 14(2):352–360. ISSN No: 2185–3118
18. Padarti VK, Nandanavanam VR (2019) Performance evaluation of coexistence of Wi- Fi and LTE licensed–assisted access to unlicensed spectrum using markov chain analytical model. Int J Eng Adv Technol 8(6):2032–2037. ISSN No: 2249–8958
19. Kumar V, Rao NV (2018) A novel method for ICI cancellation in OFDM systems. In: 2018 Third international conference on electrical, electronics, communication, computer technologies and optimization techniques (ICEECCOT)
20. Kumar PV, Nandanavanam VR (2019) Performance analysis of OFDM-based massive MIMO downlink system. Int J Recent Technol Eng 8(3):7161–7165. ISSN No: 2277-3878
21. Prasad KV, Prasad MVS, Raja Sekhar A (2018) Surface wave reduction in micro strip array With E-shape electromagnetic bandgap structure. In: 2018 Third international conference on electrical, electronics, communication, computer technologies and optimization techniques (ICEECCOT) 14–15, December 2018

Chapter 22
High Resolution Spatial Data Analysis and Haze Removal for Remote Sensing Images

M. Padmaja, B. Yogichandar, and M. Karishma

1 Introduction

Spatial information investigation is the cycle that incorporates the conventional methods to contemplate objects utilizing the geographies, mathematical and geological properties [1]. The spatial examination utilizes an assortment of procedures, numerous still in their initial turn of events, utilizing distinctive insightful methodologies as demonstrated in Fig. 1. In a more limited sense, the spatial examination is the strategy applied to structures at the human scale, most eminently in the investigation of topographical information.

In spite of the fact that spatial information investigation can be utilized for a wide scope of utilization, it additionally has not many computational mistakes, and there are different assemblages in the examination. Lately, an expanding number of tasks and applications are done depending on moderate or high-goal satellite pictures [6]. Albeit high-goal pictures are accessible, the accessibility top-notch picture frequently relies upon luck. Cloud and haze are the two barometrical impacts that cause picture tainting. Given the limitations of satellite orbital qualities and barometrical conditions, exhaustive satellite information with cloudiness influenced scenes are typically acquired.

Because of the presence of cloudiness, the picture is debased by the dispersing of the environment pretty much, bringing about a decrease of differentiation and trouble in recognizing objects. Cloudiness expulsion from the satellite pictures would be treated as a pre-preparing venture for ground data extraction. Hypothetically, it is doable to eliminate dimness from the murky pictures through environmental methods, of which the attractive qualities ought to include strength, ease-to-utilize. In early examination, some differentiation improvement calculations were utilized to upgrade the difference and highlights of the debased pictures [2]. The differentiation

M. Padmaja (✉) · B. Yogichandar · M. Karishma
ECE Department, Siddhartha Engineering College, Vijayawada, India
e-mail: drpadmaja@vrsiddhartha.ac.in

S. Dawn et al. (eds.), *Smart and Intelligent Systems*, Algorithms for Intelligent Systems,
https://doi.org/10.1007/978-981-16-2109-3_22

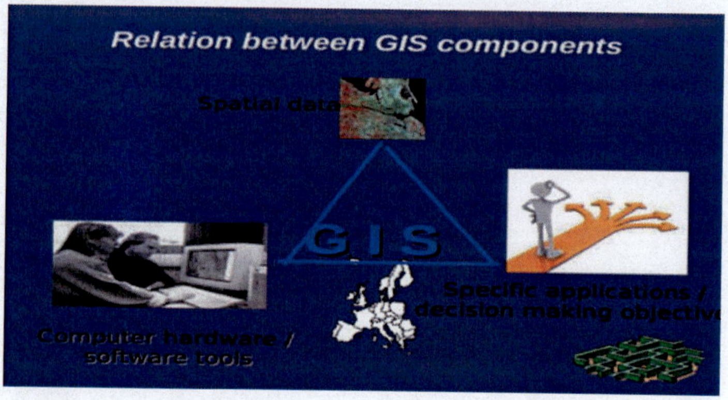

Fig. 1 Relation between GIS components

upgrade techniques, in any case, may cause loss of picture data and furthermore, picture twisting. In any case, these strategies cannot manage the circumstance that the densities of the cloudiness and the profundity of the scene objects are vibrational in the first picture [3]. A portion of the techniques proposed before are:

1. Cluster matching technique for Landsat TM data based on an assumption that each land cover cluster has the same visible reflectance in cloudy and hazy regions [4].
2. Haze removal method that calculates the haze thickness map based on the local non-overlapping search of dark objects. Assuming an additive model of the haze influence, the haze-free signal at the sensor is restored by subtracting the HTM from the hazy images [5].
3. Efficient haze removal and depth map estimation method for outdoor coloured RGB images based on dark channel prior calculation. Dark channel prior was applied for haze removal of remote sensing images. In order to eliminate the colour distortion and oversaturated areas in the restored images, the transmission is recomputed, which can lead to good results and sufficient speed. Few of the methods along with the Dark channel prior method are studied based on the contrast to noise metric. Dark channel prior method is one of the best methods for removal of haze.

1.1 Object Identification and Algorithms

In antiquated days, the land overviews were led physically to digitize the land cover and land use data. Manual land reviews were unsafe, costly and required more exertion and time taking. Later, airborne pictures were utilized to extricate the land cover and land use data extensively; yet, these pictures incorporate uproarious information and are troublesome in planning with the geographic directions [6]. Numerous specialists endeavoured to computerize the digitization cycle that will raise the speed and decrease the assets required. This was conceivable with the assistance of far off detecting [7].

Remote sensing is obtaining information about an object without making direct physical contact with it. Remote detecting development has made it conceivable to obtain the land use and land cover data decisively and capably over an enormous scope [7]. This was exceptionally valuable in recognizing the assets that are accessible on the world's surface [8]. For the most part, land cover and land use were not the equivalent. Land cover was the common assets on the world's surface like mountains, backwoods, water bodies, deserts, soils, etc. Land use was man-made assets like structures, ranches, streets and so on [9].

In the beginning, the remotely sensed images consisted of very poor resolution, so it was very difficult for processing. However, because of the headways in far off sensor innovation today, we can get pictures with Very High Resolution (VHR) [10]. The impact of people on the climate can be concentrated with the assistance of these pictures. The primary point of the model is to consider a couple of the pictures and eliminate disturbances, noise in the image and characterize it into various items [11–13].

2 Methodology

The implementation there involves two major steps, namely, the dark channel prior method for removal of noise and object identification.

2.1 Dark Channel Prior Method

The removal of external disturbances is being performed by the dark channel prior method. It is one of the best methods to analyse the hazy image. Dark channel means a minimum of low-intensity value. Here, initially average intensity is assumed to be zero; the intensity of the image needs to be increased to a significant value. The detailed explanation of haze removal by Dark channel prior method and its implementation is explained as follows which is shown in Fig. 2. The image taken contains disturbances, noise, which is treated as haze. After accepting the image to analyse it properly, the RGB components need to be separated by using functions that involve two steps.

1. Converting RGB into HSV (Hue Saturation and Value)
2. Representing these values with a random number.

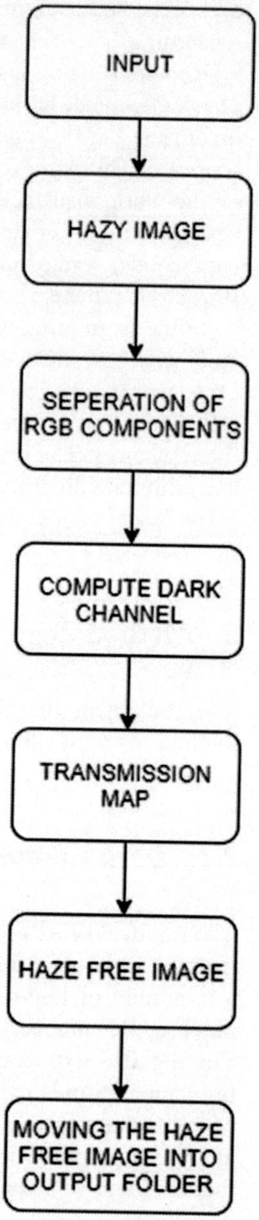

Fig. 2 Implementation of dark channel method

2.2 Computation of Dark Channel

Parting the RGB picture in Red Green Blue channels and choosing the base power pixels from each of the three diverts will bring about a type of dull picture. Again, dull channels can be alluded to, choosing the most minimal power pixels from three channels of a RGB picture. The way towards choosing the base serious pixels from a RGB picture can be alluded to as term earlier, and we call it dull channel earlier. This picture acquired is dissected by utilizing the crate and guided channel.

The box channels are utilized to figure the relationship between the pictures. Here, in the execution of the aggregate total, contrasts are being figured and utilized as required. The guided channel can perform smoothing the picture, and furthermore, safeguards the sharp edges. Also, the guided channel is a lot quicker, and the outcomes are exceptionally exact. So, guided channels are primarily used to eliminate fog rather than bilinear channels, laplacian channels and so forth.

2.3 Transmission Map

A haze image can be mathematically modelled as

$$\text{A.}\quad I(x) = J(x)\,T(x) + (1 - T(x))a.$$

where $T(x)$ is represented as transmission map $T(x) = e - \beta d(x)$. In clear weather conditions $\beta = 0$, but becomes negligible for hazy images. Here, the atmospheric light A needs to be estimated in order to obtain the transmission map $T(x)$. For high-intensity patches in the image, A should be estimated, and it is done in the dark channel prior method. The pixel value of the dark channel $J^{\text{dark}(x)}$ is likely to be zero, and the transmission map is obtained. If the $J^{\text{dark(x)}}$ is not equal to zero, then the transmission may not be obtained. After completion of the Dark channel method, we get the cloudiness free picture, which is saved in a different folder that is being utilized for object investigation to show the vegetation in the picture taken. The accompanying advances are engaged with examining the picture.

2.4 Segmentation

For the input image taken for classification, it needs to be converted into a greyscale image and then segmented into pixels for identification. Segmentation is the procedure of partitioning an image into several numbers of fragments. The segmentation techniques used in this approach were multiresolution segmentation and chessboard segmentation, each has its own role. Multiresolution segmentation is partitioning the image based on a specific threshold value. It is performed by using the gaussian filter.

In the multiresolution segmentation, a canny edge detector is used in the analysis, which is used to extract the image edges more accurately.

The chessboard segmentation is the simpler one in which the image is divided into equal portions and is analysed by using the NDVI and WVI indexes.

2.5 Band Ratios

The Normalized Difference Vegetation Index (NDVI) is an index of plants that can be used to evaluate the remote sensing measurements and judge whether the object contains live green vegetation or not. They are evaluated as shown in Formulas 1–8 as shown in Table 1.

The sub-objects classification was performed by selecting parameters from the output of band ratios. The NDVI and NDVI-Green usually ranged from −1 to 1, for all other ratios it ranges from 0 to ∞. The output of NDVI shows how vegetation and water can be detached from other features. The Water Ratio Index was not useful in classifying any feature. The Blue Red Ratio gives refinement between neighbouring objects. The results of Green NDVI are almost similar to NDVI apart from that water gets discriminated well from other classes when using Green NDVI. Blue NIR ratio shows clear discrimination for water. Green Red Ratio output is similar to Red Green Ratio and gives good refinement for soil. The Water Vegetation Index is used to separate vegetation, water very well.

Table 1 Band ratios

S. no	Band ratio parameter	Formula
1	Normalized difference vegetation index (NDVI)	$\text{NDVI} = \frac{DNir-DNr}{DNir+DNr}$
2	Water ratio index (WRI)	$\text{WRI} = \frac{DNnir+DNr+DNg}{DNb}$
3	Blue red ratio (BRratio)	$\text{BRratio} = \frac{DNb}{DNr}$
4	NDVI green ratio	$\text{NDVI green} = \frac{DNnir-DNg}{DNnir+DNg}$
5	Blue near infrared ratio	$\text{BNirratio} = \frac{DNb}{DNnir}$
6	Green red ratio	$\text{GRratio} = \frac{DNg}{DNr}$
7	Water vegetation index	$\text{WVI} = \frac{DNr+DNg}{DNnir}$
8	Red green ratio	$\text{GRratio} = \frac{DNR}{DNg}$

Table 2 Classification based on WVI

Classes	Membership functions
Water	WVI > 3
Trees	WVI < 1.5 and green entropy > 2.4
Grass	WVI <= 1.5 and greater entropy <= 34
Soil	WVI > 1.5 and <= 1.8 and red green ratio > 0.91
Shadow	Intensity 2 <= 70 & WVI > 1.6

2.6 *Membership Functions*

The membership functions are very useful in identifying the feature objects from the image. The Water Vegetation Index (WVI) threshold values for identifying water vegetation are classified based on Table 2.

2.7 *Classification*

The primary part of image classification is to identify, detect, diagnose and classify features of an object in an image depending on the type of class. The classification techniques used here are Support Vector Machines (SVM).

2.8 *Support Vector Machines (SVM) and Multi-class Support Vector Machine (MVSM)*

The SVM is basically a supervised algorithm, so it requires some training data for classifying the feature objects. It is a binary classifier, so it can classify only two classes at a time. The objects in the SVM were classified by framing a hyper plane in the feature space so that all the positive vectors will reside on one side and all the negative vectors will reside on the other side of the hyper plane.

In Fig. 3, H1, H2 and H3 are the three linear hyper planes separating the training samples in the feature space.

The SVM was a binary classifier, so the classification is limited for the classes. For the images which require more number of classes, MSVM is used with the help of the following methods as mentioned:

1. *One-Against-One (OAO)*: In this, the SVM classifiers are created for all class pairs. The class that appeared mostly among the two classes in pairs is the outcome of classification. If both the classes in pairs appeared equally, then one class is selected randomly as the output label. The main drawback is that the number of classifiers increases with the number of classes.

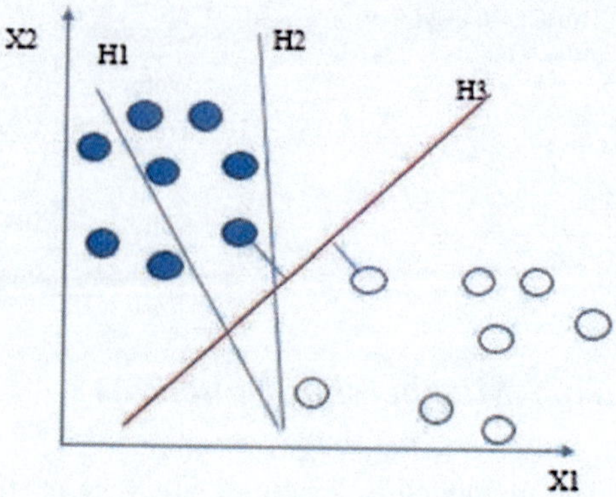

Fig. 3 Hyper planes separate correctly the training samples

2. *One-Against-All (OAA)*: In this, the number of binary SVM classifiers created will be equal to the number of classes. Each classifier is trained to differentiate one class from the remaining classes. The final output is the class that has the largest margin with the SVM classifier. If two or more classes are close, then they are unclassified or rejected.

The object identification and classification are done by training the SVM classifier; once the SVM gets trained, it classifies the objects based on feature calculation, and it computes WVI and classifies the red area, blue area and green area from the image.

2.9 Accuracy

Accuracy assessment of object detection was done by utilizing a pixel-based assessment strategy. In this strategy, every single distinguished object was checked for convergence with outwardly translated vector layers by performing a mapping of a total number of unmatched classes by a total number of classes. Accuracy determines the quality of information. The accuracy can be calculated as

$$\text{Accuracy} = 1 - \frac{\text{Total no. of unmatched classes}}{\text{Total no. of classes}}$$

3 Results

To evaluate the implemented code, first open Matlab and set the path (depends on where code file is placed in the pc) and select the 'newmain.m' file. The steps for evaluating the file is as mentioned below:

1. Evaluate the file by pressing F5
2. A pop-up opens for the selection of input image
3. Select .png formatted image
4. SVM evaluates the image and classifies it into the red area, blue area and green area
5. The obtained results are shown in the Matlab interface

The code is tested for a few on various images, and the results are shown in Fig. 4. The red area, blue area and green area in the interface denote the average pixel count in the image. The error in the images is computed after the successful classification of the image.

The accuracy and error for the above images are in Table 3.

By looking at the classified image, it can be found that the water, grass and shadow classes were classified well. The intensities of trees and grass are near, so the trees are treated as grass. When the image with high resolution is taken for the analysis, the percentage of error is very minimal, and the image has been retrieved successfully. In Fig. 6.6, the cloud area was lighter due to which it is identified as shadow, and percentage error has been increased.

From the literature survey, it has been observed that the Ml (Maximum likelihood) classifiers have an accuracy of 88%. From the results observed by using the SVM classifiers, the work has an accuracy of up to 95%. The retrieval of the image was successful along with the concept of haze removal and has extended the accuracy up to 96% by removing the external disturbances.

4 Conclusion and Future Scope

The principle objective of the paper is accomplished by eliminating the haze from the input picture taken from the dark channel prior method, and an efficient image classification technique was proposed and implemented for the classification of multispectral remotely sensed satellite images. The feature objects in the image were acquired by utilizing multiresolution segmentation and chessboard segmentation. Multiresolution segmentation was utilized for distinguishing the edges and following the limits. Chessboard segmentation was utilized to get ready preparing tests by sectioning the picture into pixels of size individually. The classes for required feature objects are made utilizing NDVI band proportions and WVI record. The groupings of the element classes are finished utilizing the SVM classifier. Precision up to 95% was achieved by utilizing the SVM classifier. The acquired outcomes are near the genuine accessible information.

The work can be utilized progressively for applications like disaster management. At the point when characteristic disasters happen, people cannot get to the spot, and by then, the pictures taken from the satellites can fill in as the best asset to investigate the circumstance. This model can be utilized in different applications, for example, checking of crop's wellbeing and development purposes.

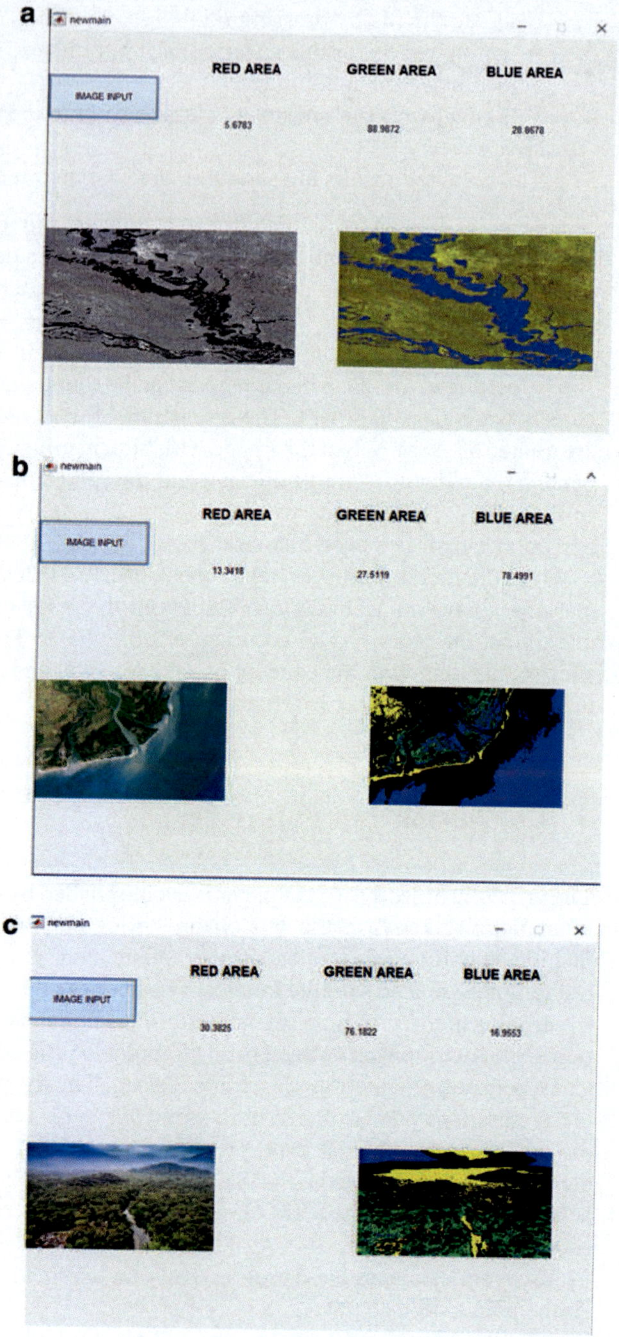

Fig. 4 Classified output image

Table 3 Results of classified images

Figure number	Image dimensions	Green retrieved portion	Blue retrieved portion	Error in the green area retrieved	Error in the blue area retrieved
4a	4967 * 5936	95.1	92.6	0.049	0.074
4b	399 * 250	81.5	92.2	0.185	0.078
4c	304 * 364	83.37	54.6	0.163	0.454

The future work of the paper can be carried out in the direction of online-based analysis, which can be done by the object-based image analysis (OBIA), and if good data sets are available, the work can be further extended by providing a handy application for the user to access the data from other different places.

References

1. Karnatak HC, Singh H, Garg RD (2017) Online spatial data analysis and algorithm development for geoscientific applications using remote sensing data. Proc Natl Acad Sci India, Sect A 87(4):701–712. https://doi.org/10.1007/s40010-017-0442-x
2. Moro GD, Halounova L (2014) Haze removal for high-resolution satellite data: a case study. Int J Remote Sens 28(10):2187–2205. https://doi.org/10.1080/01431160600928559
3. Praveen P, Babu CJ, Rama B (2016) Big data environment for geospatial data analysis. Int Conf Commun Electr Syst (ICCES). https://doi.org/10.1109/CESYS.2016.7889816
4. Liu C, Hu J (2015) Haze detection, perfection, and removal for high spatial resolution satellite imagery. Int J of Remote Sens 32(23):8685–8697. https://doi.org/10.1080/01431161.2010.547884
5. Wu D, Zhu Q, Wang J, Xie Y, Wang L (2014) Image haze removal: status, challenges and prospects. In: 2014 4th IEEE international conference on information science and technology. https://doi.org/10.1109/icist.2014.6920524
6. Ablin R, HelenSulochana C (2013) A survey of hyperspectral image classification in remote sensing. Int J Adv Res Comput Commun Eng 2(8):August 2013
7. Li Z, Hayward R, Zhang J, Jin H, Walker R (2010) Evaluation of spectral and texture features for object-based vegetation species classification using support vector machines, Austria, July 5–7, 2010, IAPRS, vol XXXVIII, Part 7A
8. Nishii R, Tanaka S (1999) Accuracy and inaccuracy assessments in land-cover classification. IEEE Trans Geosci Remote Sens 37(1):January 1999
9. Bruzzone L, Prieto DF (2001) Unsupervised retraining of a maximum likelihood classifier for the analysis of multi-temporal remote sensing images. IEEE Trans Geosci Remote Sens 39(2):February 2001
10. Koc San D, Turker M (2010) Building extraction from high-resolution satellite images using Hough transform. In: International archives of the photogrammetry, remote sensing and spatial information science, vol XXXVIII, Part 8, Kyoto Japan
11. Vijjapu P (2013) Classification and building detection using fuzzy-sub object class densities and shadow influences, March 2013
12. Karnatak H, Pandey K, Oberai K et al (2014) Geospatial data sharing, online spatial analysis and processing of Indian biodiversity data in internet GIS domain—a case study for raster-based online geo-processing. Int Arch Photogramm Remote Sens Spat Inf Sci ISPRS Arch XL-8:1133–1137. https://doi.org/10.5194/isprsarchives-XL-8-1133-2014

13. Aswathy S, Binu VP (2016) Review on Haze removal methods. http://www.ijsrp.org/research-paper-0716/ijsrp-p5522.pdf

Chapter 23
Energy Monitoring Framework Utilizing Internet of Things

Battula Veera Vasantha Rao and K. Padmavathi

1 Introduction

Each Electrical Circuit needs to comply with the law of preservation of energy standards. Brilliant structure is an arrangement of interconnected frameworks which has significant level control and sensors alongside robotization, correspondence, and examination capacities [1]. It gives instruments to anticipate and reacts to solace, upkeep, and energy execution issues which prompt better hardware upkeep, higher inhabitant fulfillment, decreased energy utilization, and cost. In [2], a system is characterized to recognize the most helpful techniques for checking and information translation in keen conditions. A few measurements such as goals, testing rate, fleeting utilization, division, excess, and so forth are investigated, with regards to shrewd structures.

The checking framework for building energy utilization comprises of electrical force framework checking, climate observing, and warming ventilation cooling (Central air) framework observing. Energy checking includes different advances, for example, information assortment, correspondence, and information bases. These ought to be progressed enough to get a handle on the status of energy utilization progressively [3].

Sub metering gives the neighborhood electrical utilization design. Capable energy utilization can be refined by analyzing the recorded data [4]. It could help in scheduling power cost-based administration of the activity of burdens and in getting data with respect to the nature of intensity. Sub metering can help in diminishing the energy costs in the colossal structures by performing energy assessing.

B. Veera Vasantha Rao (✉)
Ph.D Scholar, EEE Department, BMS College of Engineering, Karnataka Bengaluru, India

K. Padmavathi
EEE Department, BMS College of Engineering, Bengaluru, India
e-mail: kp.eee@bmsce.ac.in

S. Dawn et al. (eds.), *Smart and Intelligent Systems*, Algorithms for Intelligent Systems,
https://doi.org/10.1007/978-981-16-2109-3_23

The majority of the energy meters and multi work meters introduced for estimating the energy utilization in a family unit or some other business structures convey to the unified SCADA frameworks utilizing RS485 norm of sequential correspondence. Information Analytics is done on the information got from these meters. In this manner, the initial step is to separate the information from these meters to do further investigation. Truth be told, the significance of information and investigation on information has been perceived very early where non-nosy recognizable proof of electrical burdens was made just by breaking down the information [5].

Whereas advanced savvy meters are accompanying inbuilt IP network, the customary multifunction meters have RS485 correspondence abilities. However, since IP-based meters are modern developments, as of now, a greater part of existing systems have multifunction meters, and it is therefore important to look at how information can be separated from them. Furthermore, RS485 over Ethernet has a bit of flexibility. Ochai et al. note that [6] RS485 is quite possibly the most ideal correspondence media for office organizing, instead of Ethernet where numerous UTP links and exchanging centers should be sent. A multifunction meter introduced in panel board measures frequency, line to line, phase voltages, currents in all phases, power factor, power usage, energy usage of all the three phases.

In the current scenario, in order to step towards the concept of more genius frameworks, several existing structures point to the implementation of an energy observation system. For the purpose of data extraction from the meters, developers face difficulties. Furthermore, in a few situations, the meters themselves are not transferable and should be replaced. In any case, information on the best way to extract information from them is not common when transferable meters are available, resulting in dependency on providers that charge vigorously. The purpose of this paper is to discuss the key factor in data extraction from multifunction meters, assuming that they are communicable and support the RS485 protocol.

The paper is structured as follows: Sect. 2 presents the architecture of basic framework of energy monitoring. The process used for data extraction is described in Sect. 3. Section 4 presents responses. Observations are provided in Parts 5 and 6 future work follows.

2 Architecture of Energy Monitoring Frame Work

Energy Monitoring frame work is outfitted with a Multifunction Meter with RS485 port empowered, CT coils, RS485 to USB Connector, and Raspberry Pi. Figure 1 speaks to the association of multifunction meter associations with the auxiliary transformer. Utilizing Modbus Protocol in RTU mode RS485 is utilized to permit electrical information transmission. Modbus Protocol is a sequential-based expert and slave design. It is utilized in the Building Management System (BMS) and Industrial Automation framework (IAS). In Modbus, RS485 network correspondence starts when the expert gadget sends a question to slave gadgets. Slave gadgets invest their energy, observing the organization for inquiries explicitly routed to it. At the point

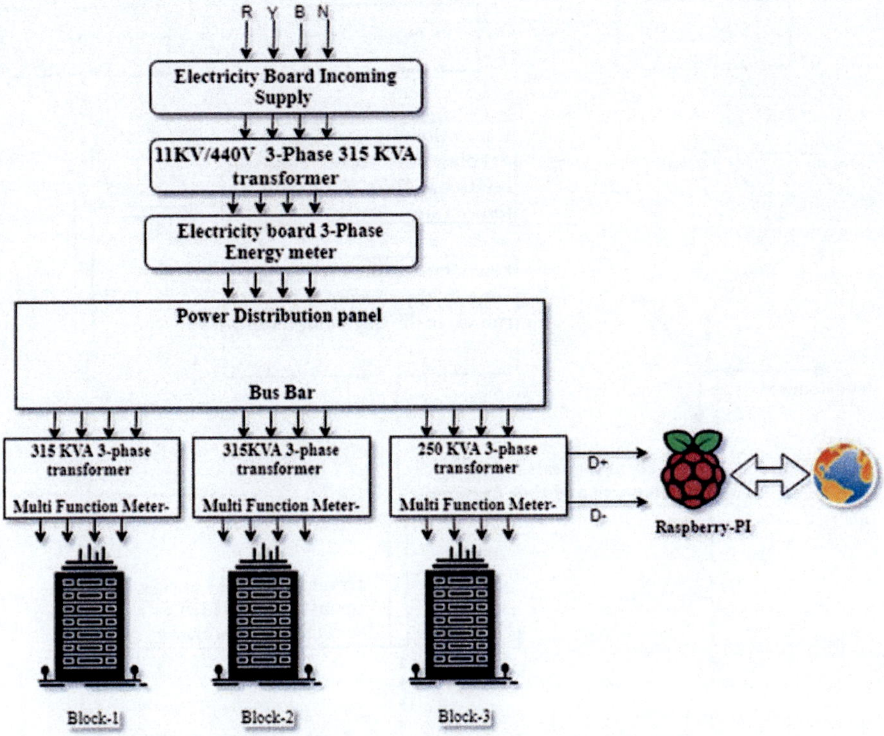

Fig. 1 Architecture of energy monitoring frame work

when an inquiry is gotten, it will either play out an activity or answer with a reaction to the expert. Questions are started by the expert gadget; Rs485 can have up to 32 slaves without a repeater. All slaves and experts are expected to have similar Baud rate signals per sec. All the information removed from put away in the information base for future examination. Python code for extraction utilizing a protected document move convention and secure distant login and record move convention.

3 Methodology for Data Extraction

Figure 2 represents stream diagram to extract data from Multifunction Meter to Raspberry PI utilizing Modbus convention. Python insignificant Modbus module is utilized to peruse the registers data from the port (ttyUSB) of Raspberry pi. Raspberry pi read information from the registers (100, 118, and 3) by writing program using python script language. Unlock the meter information for the necessary meters. Link all the read meter data and set up a string. Rehash a similar information recovery consistently. Compose the information to the record with the timestamp prefixed to

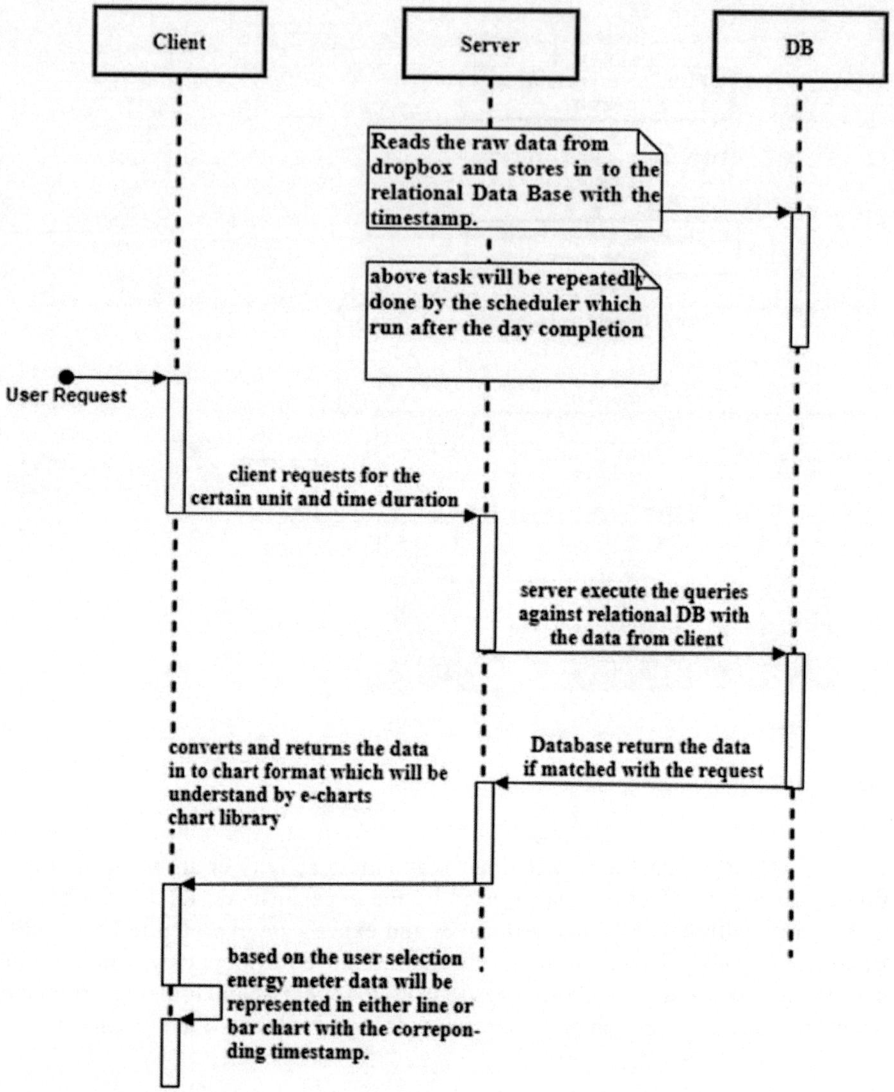

Fig. 2 Sequence diagram

the units. Figure 3 speaks to Flow Diagram to Push Data in Raspberry Pi to Dropbox utilizing secure document move convention and secure far off login and record move convention. Crude information is changed over into social information utilizing a spring boot micro service and put away in the h2 data set. The scheduler will transfer each day's meter information to Dropbox Per day 1440 records persisted in the database.

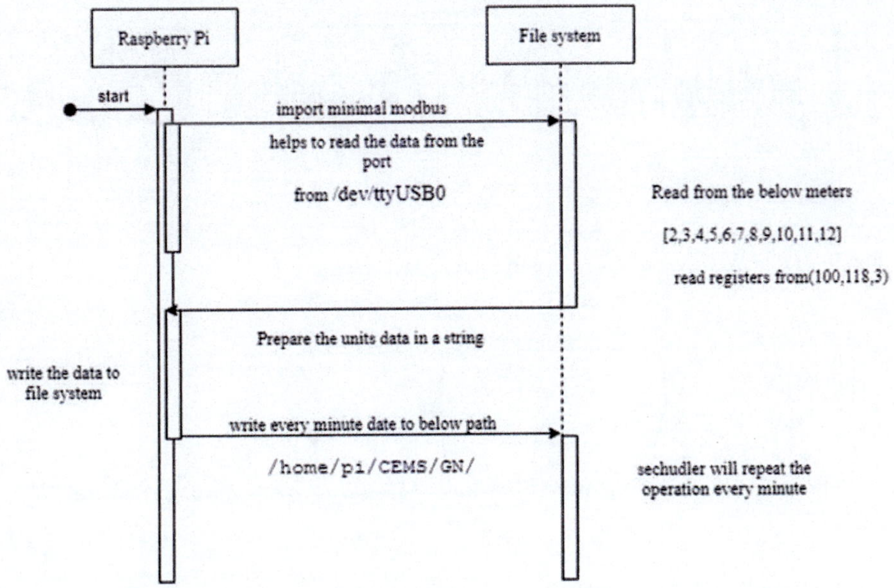

Fig. 3 Flow diagram to extract data from MFM to raspberry PI

4 Results

Observed exactly at 19.58 h, fault happened on Y_ phase due to rain fall in the campus, identified exact time, day, month, year, type of fault happened as shown in Fig. 4. Due to Covid19, the weekly load duration curve for the 315KVA transformer varies from 14 to 18 KW during 04 May 2020 to 11 May 2020 as shown in Fig. 5. recognized the base load such as UPS, Servers and LAN switches. Due to partial lockdown, weekly load duration curve for 315KVA transformer during 04 January 2021 to 11 January 2021 is varying between 20 to 50 KW as shown in Fig. 6. Due to Covid19, currents in I_R I_Y, I_B phase currents are uniform during 04 May 2020 to 11 May 2020 ($I_R = 77$A mps, $I_Y = 68$ Amps, $I_B = 60$ Amps) as shown in Fig. 7. Due

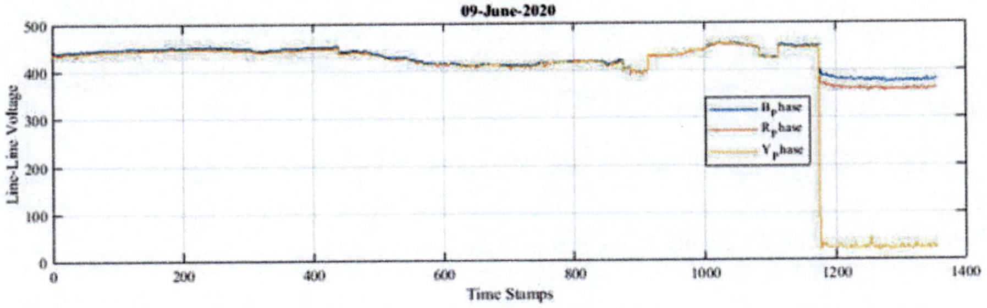

Fig. 4 Plot of line to line voltages

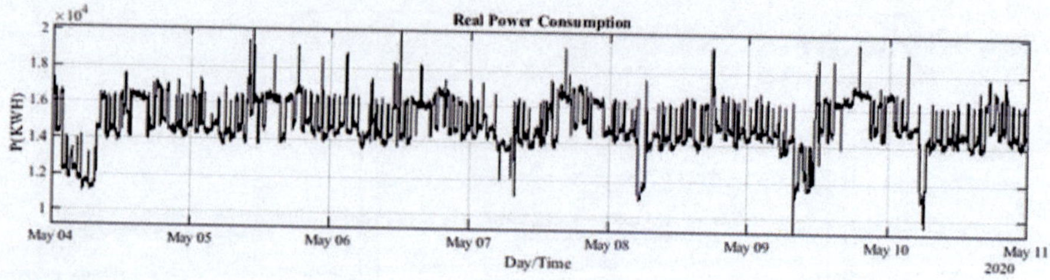

Fig. 5 Plot of Weekly load duration curve during lockdown

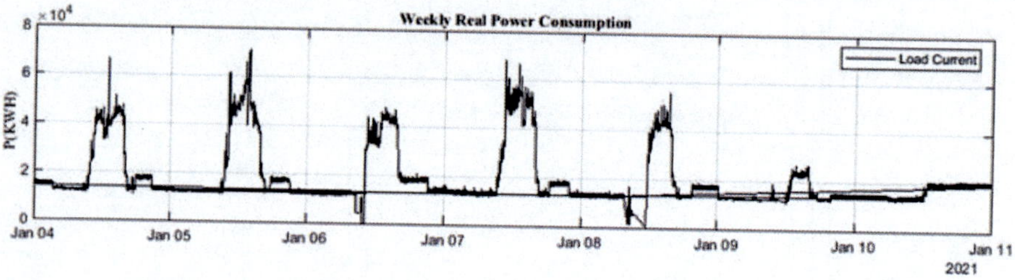

Fig. 6 Plot of Weekly load duration curve during partial lock down

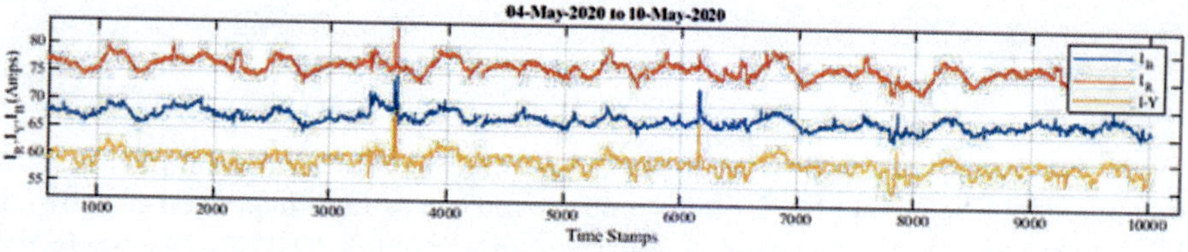

Fig. 7 Plot of currents in RYB phases during lockdown

to partial lock down currents in I_R I_Y, I_B phase are not uniform during 04 January 2021 to 11 January 2021 as shown in Fig. 8.

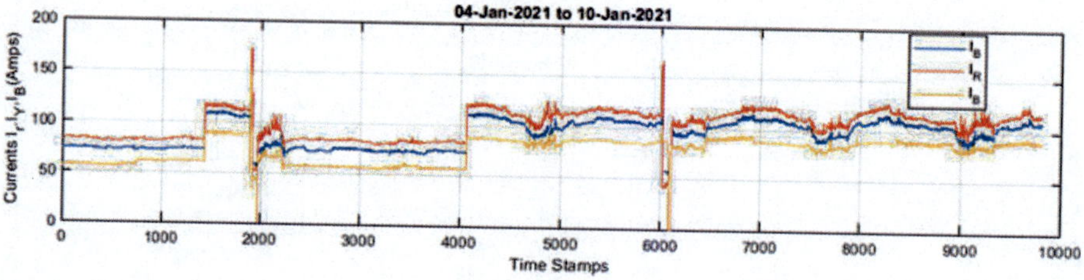

Fig. 8 Currents in RYB phases during partial lockdown

5 Conclusions

This paper presents an energy observing structure accomplished for the 315 KVA transformer. Energy monitoring frame work is outfitted with MFM with RS485 port empowered, CT coils, RS485 to USB Connector, and Raspberry Pi. By composing Python program, information is extricated from energy meter utilizing Modbus convention and put away into raspberry pi which is a low structure factor PC. By composing a python program, the information from the information base is moved to utilize secure record move convention and secure far off login and document move convention. Crude information is changed over into social information utilizing a spring boot micro service and put away in an h2 database. Plotting is done between time stamps to extracted parameters.

6 Future Scope

Using energy monitoring frame work discerns real time consumption using web-based SCADA, Based on Time of Use Tariffs, Critical Peak Pricing and Demand response systems, peak load control is achieved. It is becoming increasingly difficult to sustain the quality of supply with the proliferation of distributed and variable generation tools, such as solar PV and wind turbines, which operate intermittently. It has capacity to proficiently distinguish and resolve blackouts and report important authentic data.

Acknowledgments The creators here by recognize the monetary help given by Department of Electrical and Electronics Engineering, RVR&JC College of Engineering, Guntur, Andhra Pradesh, India" research project no: RVR&JC/EEE/IRP/FY/2019-20/02.

References

1. King J, Perry C (2017) Smart buildings: using smart technology to save energy in existing buildings. American council for an energy-efficient economy
2. Bolchini C, Geronazzo A, Quintarelli E (2017) Smart buildings: a monitoring and data analysis methodological framework. Build Environ 121:93–105
3. Zhao L, Zhang J-L, Liang R-B (2013) Development of an energy monitoring system for large public buildings. Energy and Build 66:41–48
4. Krishnanand K, Chinh HD, Gupta M, Panda S (2016) Electrical sub metering with repurposable applications for the built environment. In: Industrial electronics society, IECON 2016–42nd annual conference of the IEEE. IEEE, pp 3948–3953
5. Cole I, Albicki A (1998) Data extraction for effective non-intrusive identification of residential power loads. In: Instrumentation and measurement technology conference. IMTC/98. Conference Proceedings. IEEE, vol 2. IEEE, pp 812–815

6. Ochiai H, Nakagami H, Teranishi Y, Esaki H (2014) Facility networking with ip over rs485: Packet control for master-slave cascaded networks. In: 2014 IEEE international conference on smart grid communications (SmartGridComm). IEEE, pp 410–415

Chapter 24
Enhancement of Performance Parameters in Wireless Mobile Adhoc Networks Using DSR and Cache-Modified DSR Routing Protocols

P. Satyanarayana, M. Vani Pujitha, G. Venkata Subbaiah, and Mugada. Srivani

1 Introduction

Mobile Ad Hoc Network (MANET) is group of nodes configured randomly with no permanent infrastructure [1–4]. Those networks are usable when infrastructure is damaged or costly or not available. These networks show regular topology modify besides regular host movement as shown in Fig. 1. Cellular infrastructure is not needed by mobile Adhoc networks. On the other hand, it employs multi-hop wireless links where information is routed using intermediate nodes. Every node behaves as receiver and transmitter. In most of the real world scenarios, where backbone infrastructure is not required, at that place MANETs are preferred, which are simple to install, and the most excellently used where the infrastructure is missing or impractical. In many applications, mobile Adhoc networks are used. Some of the applications are in ear phone, personal area network, laptop, wrist watch, and cell phone that can contribute to network.

MANET facility can be availed in tanks, military environment planes, and by soldiers. MANETs can be used in boats, meeting rooms, small air crafts, sport stadiums, and taxi cab networks. MANETs can also be used in various crisis operations such as policing, firefighting, search, and rescue operations [5–9]. MANETs

P. Satyanarayana (✉) · G. V. Subbaiah
ECE Department, V. R. Siddhartha Engineering College, Vijayawada, AP 520007, India

M. V. Pujitha
CSE Department, V. R. Siddhartha Engineering College, Vijayawada, AP 520007, India
e-mail: pujitha@vrsiddhartha.ac.in

Mugada. Srivani
IT Department, Vijaya Institute of Technology for Women, Vijayawada, AP 520007, India

S. Dawn et al. (eds.), *Smart and Intelligent Systems*, Algorithms for Intelligent Systems,
https://doi.org/10.1007/978-981-16-2109-3_24

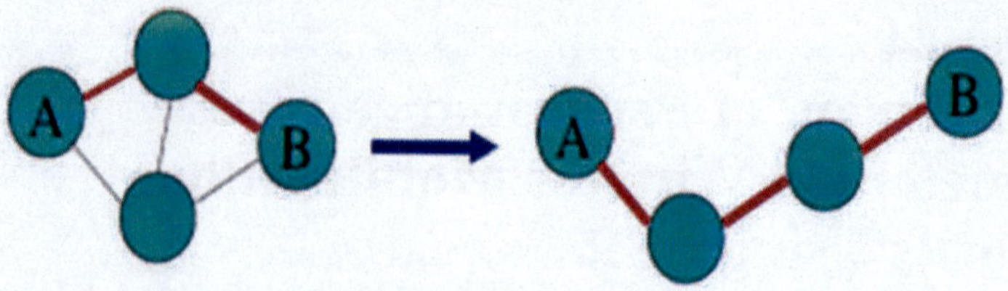

Fig. 1 Illustrates mobility and topology change in MANET

are facing so many challenges, some of which are limitations imposed by mobility, limited resources, and limitations of wireless networks.

Generally, there are so many limitations for wireless mobile Adhoc networks; these include transmit environment, packet failure, regular partitions, limited bandwidth, and changeable capacity links. The major limitations of the mobile Adhoc networks include small battery life time and restricted capabilities. Particularly in MANET, the mobility environment of nodes require fresh adaptations and applications in flow control and congestion in Transport layer, protocol stack in application layer, handoff and media access in Link layer, routing and addressing in Network layer, and broadcast interference and errors in Physical layer.

This article shows importance on MANET caching mechanism that can influence availability of data and enhancement of performance. Several helpful MANET caching mechanisms are identified in literature. For routing purposes, caching is used. Numerous researches concentrate on the routing problems that are rectified with caching. The rest of the manuscript is prepared as follows. Section 2 reviews various caching methods used in MANET. Section 3 reviews the proposed caching approach in MANET. Section 4 provides results of the proposed method. Section 5 concludes the manuscript.

2 Related Work

Du [10] projected and implemented an innovative helpful caching scheme for MANET. The method is called "COOP"; it identifies sources of information that can give small amount of communication expenditure. It gets free of caching duplications as maximum as possible and enhances efficiency and information availability. The experiments are prepared in terms of time efficiency, energy efficiency, and information availability.

Miranda and Leggio [11] stressed on reproduction of information in numerous nodes so as to enhance performance. The algorithm proposed by them was named as PC cache interface latency restrictions with probabilistic method for proficient helpful caching. This method allows a node to have information gathered from different node such as one step neighbor in a decentralized manner.

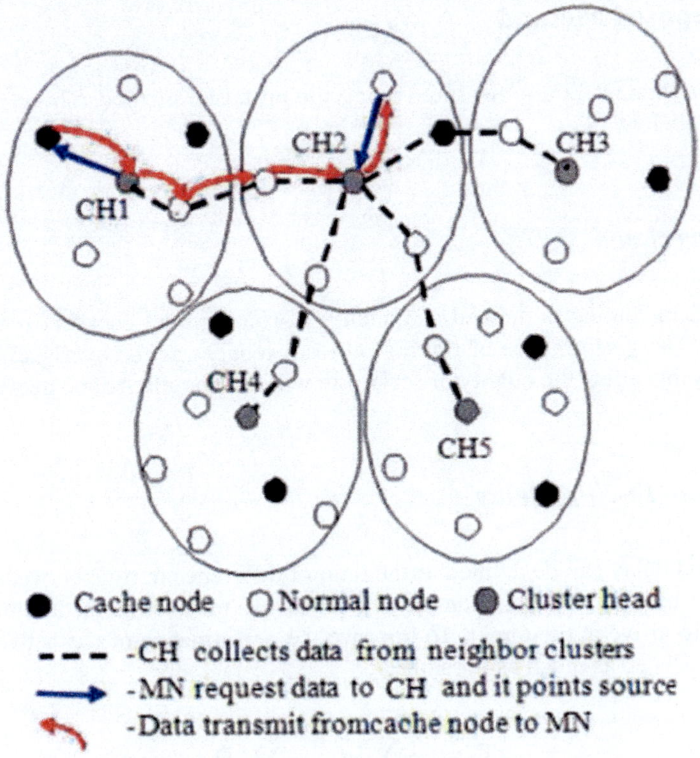

Fig. 2 Cooperative caching in Mobile adhoc networks

Ma et al. [12] implemented a supportive cache-based content distribution structure (CCCDF) among two nodes to contain supportive caching for proficient content delivery. Two strategies are provided such as minimum, maximum, and optimal. The earlier takes care of contented deliverance efficiency whereas the last takes care of fairness.

Kuppusamy and Kalaavathi [13] concentrated on data availability and data consistency issues in Mobile Adhoc networks. To attain the multipurpose, they planned two methods named as Adaptive Push and Pull Algorithm for Clusters and Cluster Based Data Consistency (CBDC) correspondingly. In order to accomplish this, it follows exact cluster head configuration (Fig. 2).

Guo and Yang [14] considered the effect of cache timeout and backup routes on faithful source routing. The analysis exposed that in case of heavy mobility situation, the backup routes have its impact on the toughness to mobility, whereas, timeout scheme has its importance on performance.

3 Proposed Method

Different parameters are measured using the proposed method. Those parameters are listed below.

3.1 *Overhead*

Overhead in routing is defined as quantity of routing data packets created in the network. The performance of routing protocol reduces as the overhead in routing improves this effect the quality of service in wireless mobile Adhoc networks.

3.2 *End-To-End Delay*

End-to-end delay can be defined as the ratio of difference in time between quantity of packets broadcasted and quantity of data packets received to the amount of time required to arrive at the target. To improve the performance of the network, delay must be reduced as much as possible.

$$T_y = \frac{1}{Q}\sum_{x=1}^{y}[R_y - S_x]/\, D_y \tag{1}$$

where, T_y = Time delay, R_y = amount of data packets obtained at yth node, S_x = amount of data packets dispatched at xth node, Q = quantity of nodes, D_y = . Time required to arrive at the target.

3.3 *Packet Drop*

Packet drop is defined as the amount of data packets not obtained successfully in the end node or destination.

$$P_D = T_x - R_y/T_x \tag{2}$$

where, P_D = Number of packets dropped during transmission, T_x = Amount of data packets sent from source node, R_y = Amount of data packets reached to destination node.

3.4 *Average Energy Utilization:*

Average Energy utilization is defined as the ratio of summation of energy utilization by each node to the overall nodes in the network. It defines each node energy utilization in the network. Mathematically it can be expressed as

$$Ex = \sum_{y=1}^{z} \mathrm{Ei} - \mathrm{Ez/Q}) \tag{3}$$

where, E_x = Average Energy utilization, E_i = Energy utilized by every node, E_z = Energy utilization by origin node. Q = Quantity of nodes.

3.5 *Packet Delivery Ratio (PDR)*

Packet delivery ratio is defined as quantity of data packets reached to destination without any failures. Mathematically it can be expressed as

$$Packet\ deliver\ ratio = \frac{Dr}{Ds} \tag{4}$$

where, Dr = Quantity of data Packets Received successfully, Ds = Quantity of data packets transmitted.

3.5.1 Procedure to Calculate PDR and Delay

Setep 1: Open the simulator window
Setep 2: Initialize the network
Setep 3: Choose the size of the network
Setep 4: Network with "n" number of nodes will be formulated and displayed
Setep 5: By using DSR protocol, data transmission was intiated in this step
Setep 6: Recognize the source node and destination node
Setep 7: Discover all possible routes availabe in the cache table
Setep 8: Start the route detection procedure using DSR. Such that origin node will discover route to the target node
Setep 9: The origin node will discover route to target node and then transmit the data packets to target node with the help of neighbor nodes
Setep 10: After completion of data broadcasting procedure, delete any one of the node which is in the way of data communication

Setep 11: After deleting the node from path, the data communication has been blocked since path is not available to transmit the data. The connection is broken

Setep 12: Execute the modified cache method in the command window

Setep 13: In the case of modified cache method, when link is broken down in the data communication path, all adjacent nodes to the origin node will modify the cache table, and delete all the faulty routes that are available in the cache table and discover fresh routes to target node

Setep 14: The origin node will locate the most excellent probable route to target node after link broken down by using modified cache mechanism

Setep 15: The origin node will execute the data communication to the target node

Setep 16: Calculate and plot the graph between PDR (vs.) Time, by using Packet deliver ratio $= \frac{Dr}{Ds}$
where, Dr = Quantity of data Packets Received successfully, Ds = Quantity of data packets transmitted

Setep 17: Calculate and plot the graph between Delay (vs.) Time, by using below formula $T_y = \frac{1}{Q} \sum_{x=1}^{y} [R_y - S_x] / D_y$
Where T_y = Time delay, R_y = amount of data packets obtained at yth node, S_x = amount of data packets dispatched at xth node, Q = quantity of nodes, D_y = . Time required to arrive at the target

Setep 18: Calculate and plot the graph between overhead (vs.) Time

Setep 19: Analyze the above three parameters for DSR and updated cache DSR

Setep 10: End of the process.

4 Results

In Fig. 3, the packet delivery ratio for dynamic source routing protocol and cache-modified dynamic source routing protocol methods are calculated using Eq. (4). Packet delivery ratio for dynamic source routing protocol is "0.55" and Packet delivery ratio for cache-modified dynamic source routing protocol is "0.95."

In Fig. 4, the overhead for dynamic source routing protocol and cache-modified dynamic source routing protocol methods are calculated. Overhead for dynamic source routing protocol is "0.977" and overhead for cache-modified dynamic source routing protocol is "0.741."

In Fig. 5, the delay needed to transmit data packets from origin node to target node is calculated using Eq. (3) for dynamic source routing protocol and cache-modified dynamic source routing protocol methods. Average packet delay for dynamic source routing protocol is "0.053" and Average packet delay for cache-modified dynamic source routing protocol is "0.020."

Packet delivery ratio (PDR) is the amount of data packets successfully reached to target node from origin node. The red line in Fig. 6 represents packet delivery ratio

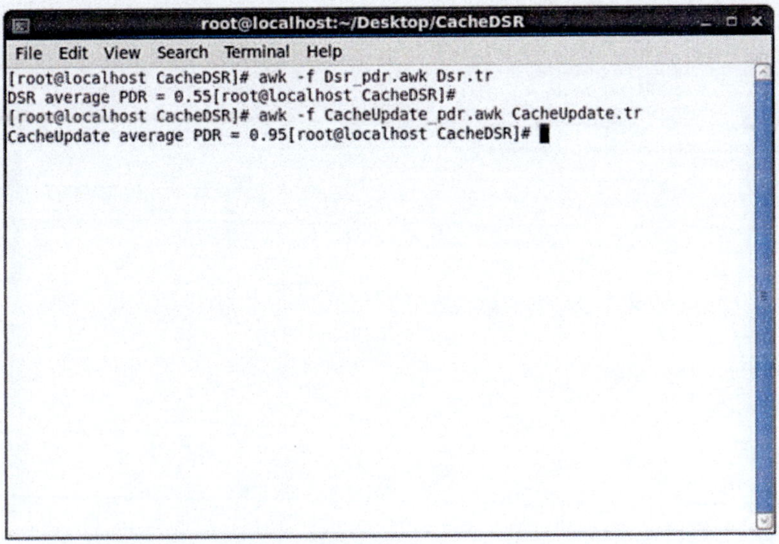

Fig. 3 Comparison of PDR in DSR and UDSR

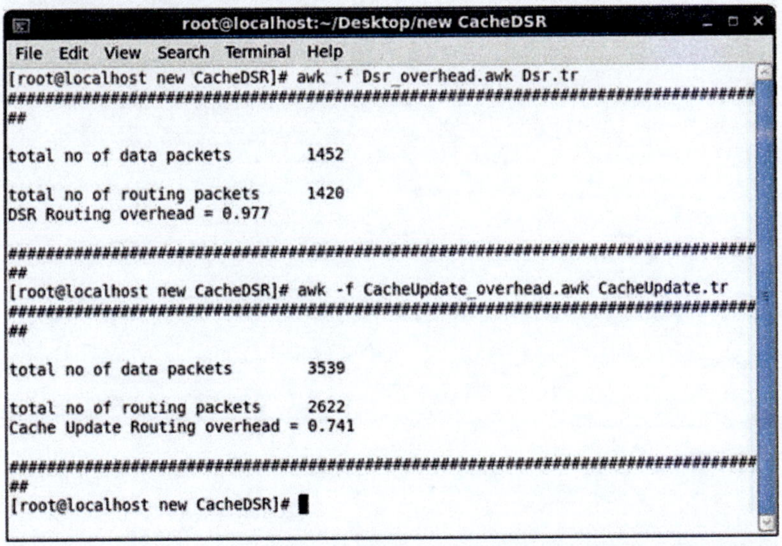

Fig. 4 Comparison of overhead in DSR and UDSR

for dynamic source routing protocol, and green line represents packet delivery ratio for cache-modified dynamic source routing protocol. In Fig. 6, x- axis represents time, and y-axis represents packet delivery ratio.

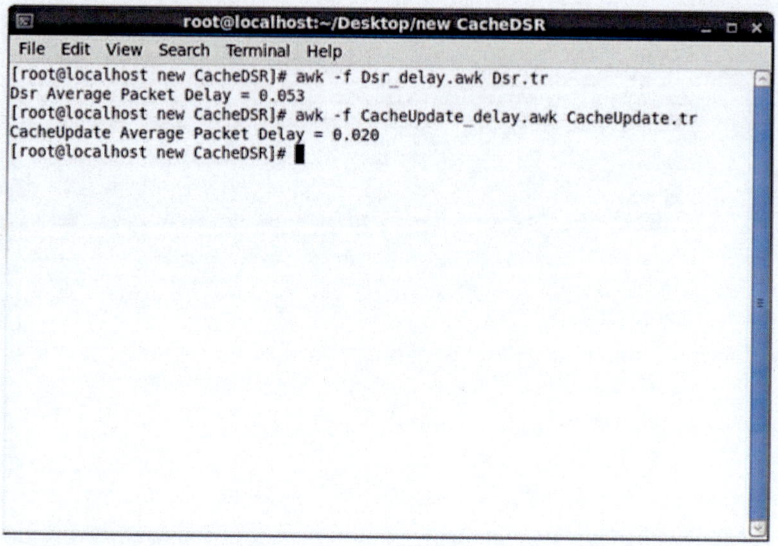

Fig. 5 Comparison of Delay for DSR and UDSR

Fig. 6 PDR Comparison for DSR and UDSR

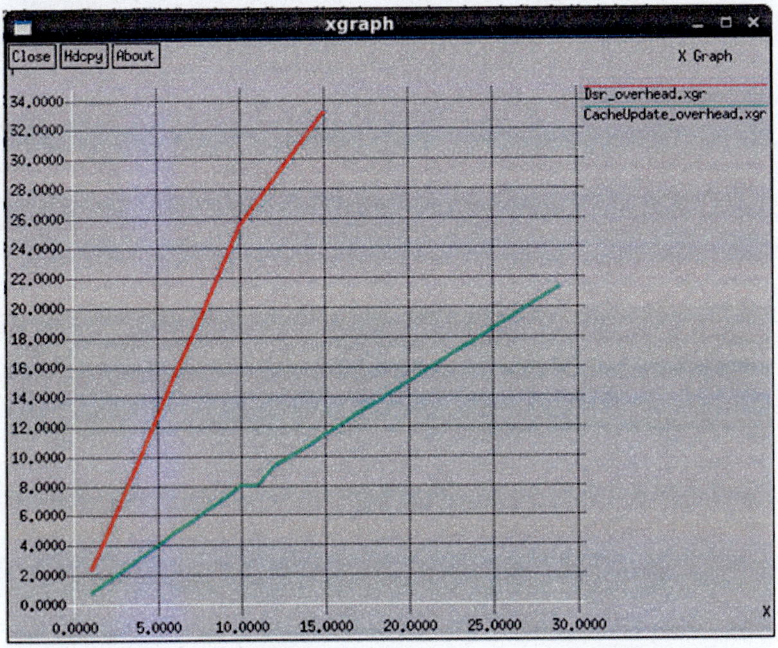

Fig. 7 Overhead Comparison for DSR and UDSR

Overhead in routing is defined as number of routing information packets produced in the network. The performance of routing protocol decreases as the overhead in routing increases, this results in the quality of service in wireless mobile Adhoc networks. The red line shown in Fig. 7 represents overhead for dynamic source routing protocol, and green line represents overhead for cache-modified dynamic source routing protocol. In Fig. 7, x-axis represents time, and y-axis represents overhead.

Delay in data transmission in wireless mobile Adhoc networks is defined as the ratio of difference in time between number of packets broadcasted and number of data packets received to the total time needed to reach the target. To increase the performance of the network, delay must be minimized. The red line shown in Fig. 8 represents delay for dynamic source routing protocol, and green line represents delay for cache-modified dynamic source routing protocol. In Fig. 8, x- axis represents time and y- axis represents delay.

5 Conclusion

In this paper, a new and efficient version of dynamic source routing protocol named cache-modified dynamic source routing protocol was proposed. By integrating cache modification with dynamic source routing protocol, the performance parameters are

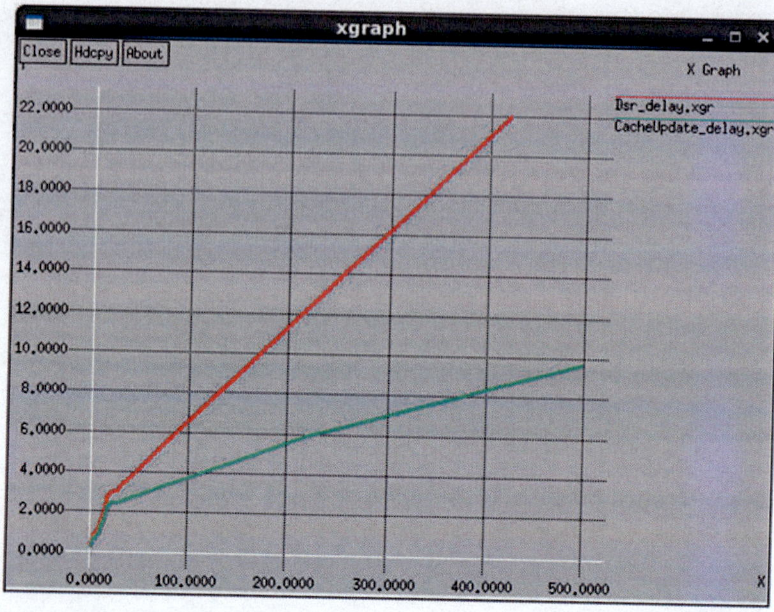

Fig. 8 Delay Comparison for DSR and UDSR

improved. In the results, the proposed method is compared with the existing DSR protocol. Packet delivery ratio for existing DSR protocol is 0.55; this can be improved to 0.95 with the proposed method. Routing overhead for existing method is 0.977; this can be reduced to 0.741 with the projected method. Average Packet delay for dynamic source routing protocol is 0.053; this can be reduced to 0.020 with the cache-modified dynamic source routing protocol. In the results, the projected method was compared with the dynamic source routing protocol, and the experimental results show a significant improvement in performance in terms of packet delivery ratio, routing overhead, average packet delay. It can be concluded that the proposed algorithm can be more useful to improve the performance of real-time wireless mobile ad hoc networks.

References

1. Clausen T, Jacquet P, Viennot L, Hansen G (2002) Optimizing route discovery in reactive protocols for ad hoc networks, vol 8, pp 1–8
2. Pathak GR, Patil SH, Rana AD, Suralkar YN (2014) Mathematical model for routing protocol performance in NS2: Comparing DSR, AODV and DSDV as example. In: Wireless computing and networking (GCWCN), 2014 IEEE global conference on. IEEE, pp 184–188
3. Barve A, Kini A, Ekbote O, Abraham J (2016) Optimization of DSR routing protocol in MANET using passive clustering. In: Proceedings of the 2nd international conference on communication control and intelligent systems (CCIS). pp 23–27

4. Tambuwal AB, Noor RM, Michael O (2013) Improvement of DSR routing protocol using TTL-based scheme in mobile ad hoc networks. In: Proceedings of the IEEE international conference on RFID-technologies and applications (RFID-TA). pp 1–6
5. Agrawal P, Sarkeja A (2017) A survey on optimum neighbour discovery techniques in mobile ad-hoc networks. Int J Comput Appl 162(12):1–4
6. Gupta AK, Sadawarti H, Verma AK (2011) A review of routing protocols for mobile ad hoc networks. SEAS Trans Commun 10(11):331–340. ISSN: 1109–2742
7. Satyanarayana P, Narmadha R (2019) Implementation of wireless sensor network based indoor air quality monitoring system using GSM. Int J Eng Adv Technol 8(6):4885–4889
8. Leung R, Liu J, Poon E, Chan A, Li B (2001) MP-DSR: A QoS-Aware multi-path dynamic source routing protocol for wireless ad-hoc networks. In: Proceedings of the 26th Annual IEEE conference on local computer networks (LCN 2001). pp 132–141
9. Poonia R et al (2011) DSR routing protocol in wireless ad-hoc networks: drop analysis. Int J Comput Appl (0975–8887) 14–(7)
10. Johnson D (2003) The dynamic source routing protocol for mobile ad hoc networks (DSR) IETF Internet Draft, draft-ietf-manet-dsr-09.txt
11. Satyanarayana P, Aruna VBKL, Kumar PNVS (2018) International boundary scanning and ship surveillance system. International conference on electrical, electronics, communication computer technologies and optimization techniques, ICEECCOT 2017. pp 212–215
12. Gupta AK, Sadawarti H (2009) Secure routing techniques for MANETs. Int J Comput Theory Eng 1 (4):456–460
13. Satyanarayana P, Narmadha R (2019) Real time implementation of air quality monitoring system using iot and ubidots. Int J Recent Technol Eng 8(3):7680–7685
14. Barati M, Atefi K, Khosravi F, Daftari YA (2012) Performance evaluation of energy consumption for AODV and DSR routing protocols in MANET. In: Computer & Information Science (ICCIS), 2012 International conference on (vol 2, pp 636–642). IEEE

Chapter 25 Design of Fault Tolerant Single RAM-Based Parallel Real Fast Fourier Transform Architectures Using Error Correction Codes and Parseval Checks

Rajasekhar Turaka, B. Koteswar rao, and M. Nageswara Rao

1 Introduction

Technology advancement makes the electronic systems exposed to errors. Scaling [1] of device is one such advancement that can provide the designer to involve more number of components (transistors) within less area. Decreasing gate length makes the transistor more sensitive to minor variations (errors). The errors may damage the system permanently or temporarily. Soft errors [2] are a type of temporary faults that affect the system to produce erroneous output for an operation time. For instance, errors from the environment through particles such as alpha rays due to cosmic radiation usually referred as radiation-induced faults are a type of soft errors. These do not harm the system but halts the actual operation. In order to resist the effect of these errors, mechanisms such as incorporating new process of fabricating devices, i.e., SOI (silicon on insulator) is preferred, whereas the system level errors can be avoided by adding extra circuitry (redundant hardware) to the actual one for its protection. The traditional N-Modular redundancy (NMR) [3] replicates the module to protect by N times. The voting mechanism among all modules identifies the error presence. More overhead for protection is its main drawback.

The less overhead consumption technique to protect parallel blocks is developed in [4], where the parallel filters are protected by conventional ECC scheme. It is suitable when the number of parallel blocks is large. One more scheme based on

R. Turaka (✉)
ECE, School of Engineering, Nalla Narasimha Reddy Education Society's Group of Institutions, Narapally, Hyderabad, Telangana, India

B. Koteswar rao
ECE, Marri Laxman Reddy Institute of Technology & Management, JNTUH, Hyderabad, Telangana, India

M. Nageswara Rao
ECE, DVR&Dr HS MIC College of Technology, Kanchikacherla, Andhra Pradesh, India
e-mail: nagesh.malisetti@gmail.com

S. Dawn et al. (eds.), *Smart and Intelligent Systems*, Algorithms for Intelligent Systems,
https://doi.org/10.1007/978-981-16-2109-3_25

Parseval theorem is observed in [5]; through this the overhead is less compared to the ECC scheme. Another improvised scheme that combines the below discussed ones is proposed in [6], having the name as "Parity-SOS-ECC" scheme. It provides the same fault tolerance with reduced hardware compared to either ECC technique or Parseval check. The technique that uses special encoder and decoder as convolutional encoder and viterbi decoder, instead of SOS, checks with parity FFT for fault identification and correction [7]. This new technique requires less hardware than the parity-SOS-ECC technique with multi-bit fault correction.

Fast Fourier Transform (FFT) is an efficient and fast algorithm used to compute the traditional Discrete Fourier Transform (DFT) and Inverse DFT (IDFT) operations. It is one of the crucial blocks in various systems, employed to process the time domain as well as frequency domain signals. There are numerous architectures for realizing FFT computations, which are mainly classified into three types, namely, Parallel, Pipelined, and Memory-based architectures. Memory-based architectures are having the advantage of low latency over pipelined ones at the cost of more hardware. Real-valued Fast Fourier Transform (RFFT) architecture [8] is designed to compute real-valued signals especially, as most of the signals in nature are real.

Although we can compute FFT of real signals using complex-valued FFT by simply equating imaginary part to zero while processing, but it may take very less number of complex adder and multipliers out of the available ones which results in wastage of hardware and increasing computational complexity. This reason finds importance of the real-valued FFT. Novel memory-based architecture for real signals proposed in [9] contains a different stage partition strategy to reduce the number of computations. Various algorithms to process real signals are seen in [10]. Processing element (PE) plays a vital role in the FFT architecture. The operation of RFFT can be improved by replacing the normal adder and multipliers with high speed architectures. Also, the optimized memory usage can reduce the hardware complexity of FFT [11].

In this paper, the fault tolerant single RAM-based parallel RFFTs is proposed in which the RFFT architecture is considered that utilizes the single RAM unit and its PE with carry-look-ahead adder (CLA) and Vedic multiplier. The technique to make it fault tolerant is a parity-SOS-ECC technique.

The rest of this paper is organized as follows. In Sect. 2, the parity-SOS-ECC fault tolerant technique for parallel RFFTs is presented. In Sect. 3, RFFT architecture with CLA adder and Vedic multiplier for PE calculation is described. In Sect. 4, experimental results and schematic are presented, and also comparison between existing and proposed system on two different Xilinx FPGA devices, i.e., on Virtex 6 and Virtex 7, is shown in Table 1. Conclusion is proposed in Sect. 5.

2 Parity-SOS-ECC Technique

In this paper, the protection of parallel RFFTs by a fault tolerant technique formed by the fusion of conventional error correction code (ECC) scheme and a Parseval check

(SOS) is discussed. The ECC scheme that can identify double faults and correct single fault, i.e., hamming code technique, is used here. Let us consider the protection of "K" parallel FFTs using ECC scheme alone that needs the FFT module to replicate "1 + $\log_2$K" times. Using Parseval check scheme, "k" number of Parseval (SOS) checks and one extra (parity) FFT are required. The hardware needed to implement SOS check is very less compared to FFT module implementation. Therefore, hardware got reduced to that of ECC. Further, to improve the overhead, a new scheme that takes the advantage of combining the two schemes mentioned earlier, i.e., parity-SOS-ECC, is seen. The brief explanation of it with block diagram is seen in Fig. 1.

To protect four parallel RFFTs, three Parseval checks are used based on Hamming code concept. The inputs to the four actual RFFTs are y1, y2, y3, and y4. The corresponding transformed results are Y1, Y2, Y3, and Y4, respectively. And, for the SOS blocks, the linear combination of inputs (y5, y6, y7) and transformed results.

$$\text{i.e.,}\quad \begin{aligned} Y5 &= Y1 + Y2 + Y3 \\ Y6 &= Y1 + Y2 + Y4 \\ Y7 &= Y1 + Y3 + Y4 \end{aligned}$$

are fed. The SOS block inside consists of magnitude square units, accumulators, and a comparator depicted in Fig. 2. The comparison result indicates the error presence.

For example, if RFFT1 is effected by an error, then P1, P2, and P3 bit combination will set to "111," then the error corrected by the extra RFFT and remaining unaffected RFFTs is.

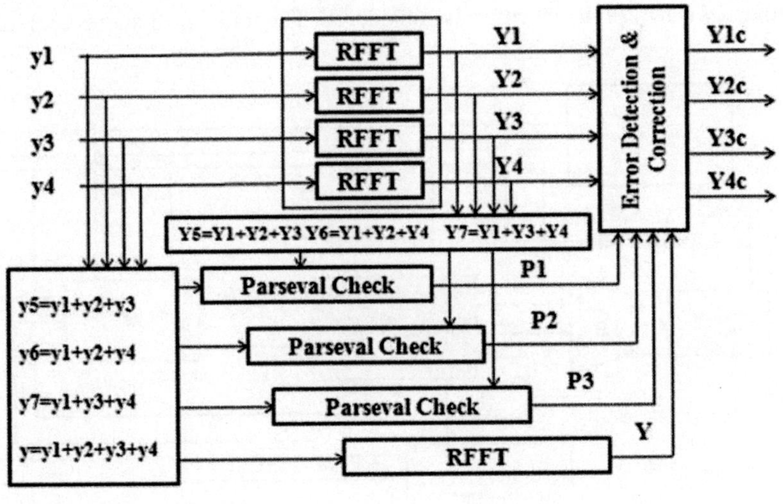

Fig. 1 Parity SOS fault tolerant parallel RFFTs

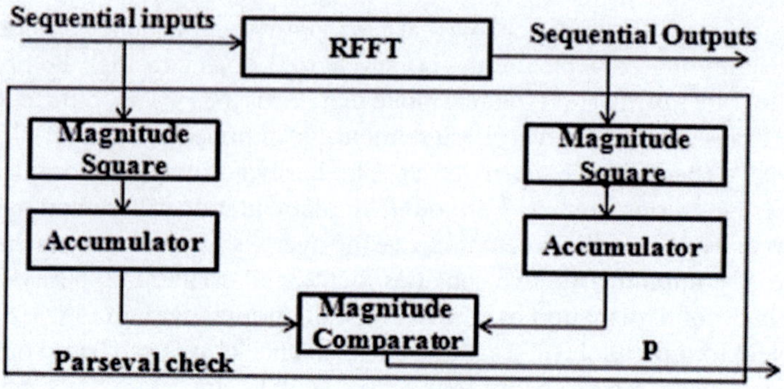

Fig. 2 Parseval check (SOS) block

$$Y1c = Y - Y2 - Y3 - Y4$$

Similarly, an error identification and rectification can be done for the FFTs as well as SOS blocks when they get affected by error.

3 Single RAM RFFT Architecture with CLA Adder and Vedic Multiplier

The Block diagram of High Speed RFFT with a single dedicated RAM named a Dual port RAM is shown in Fig. 3. Here, a 32 point input is fed and a radix-2 Decimation-in-frequency (DIF) FFT algorithm is considered. The number of stages included are

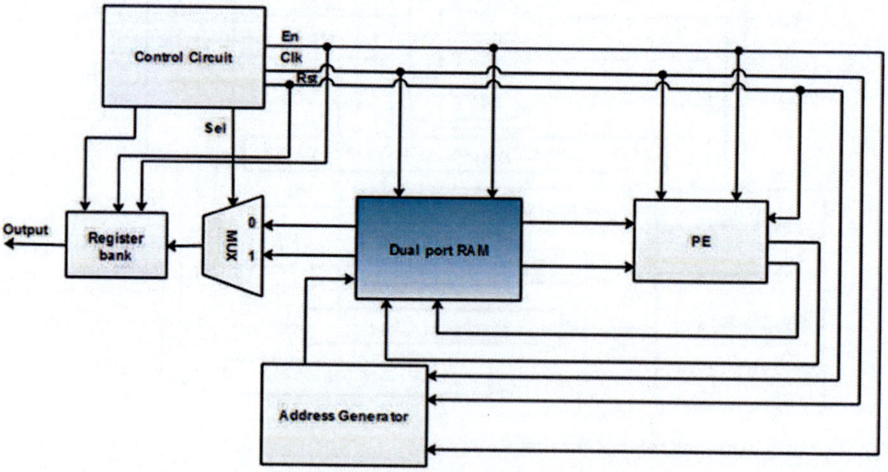

Fig. 3 Block diagram of single RAM-based RFFT

five (no. of stages = $\log_2 32 = 5$). This architecture comprises various blocks such as Dual port RAM, Multiplexer, address generator, control circuit, and a register bank.

The control circuit produces the control signals such as clock, reset, selection, and enable to the other blocks. From the address generator block, the input value is stored in Dual port RAM (DPRAM). The stored data is fed to the processing element (PE). The result of PE is also stored in DPRAM and fed to PE again until the last stage of FFT algorithm is reached. Based on the select line of MUX, PE result gets stored in register bank. When the value of enable shows 1 (en = 1), DPRAM output is transferred to the register bank, which gives the final phase output.

The processing element (PE) used in this RFFT is shown in the Fig. 4, where $\times$ 0 and $\times$ 1 are inputs to it. In this paper, efficient use of PE leads to less complexity and better operation speed.

Inside PE, multiplication of butterfly unit result with twiddle factor values stored in twiddle ROM, which are extracted using MATLAB tool, uses CLA adders and Vedic multipliers.

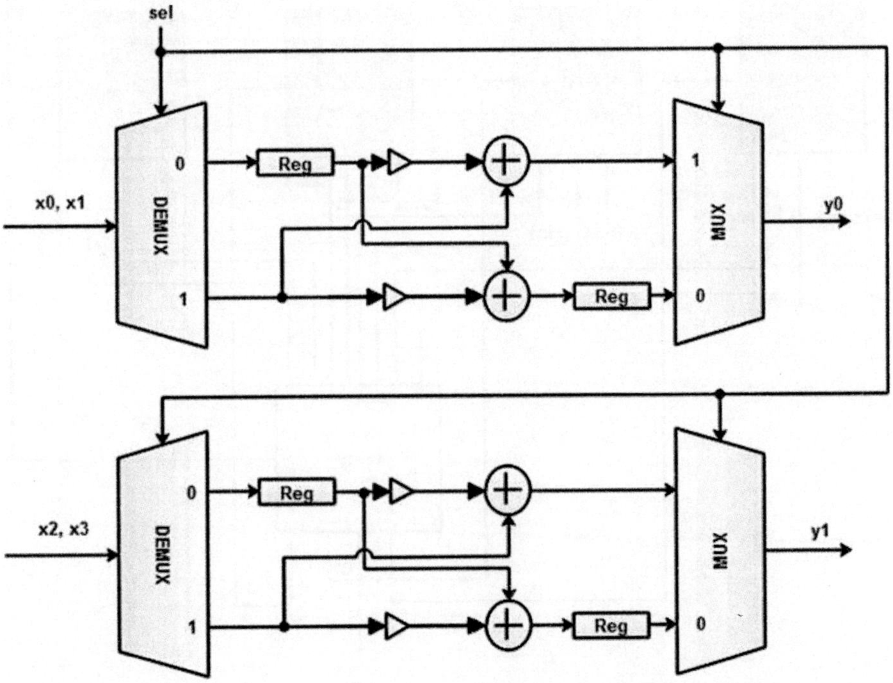

Fig. 4 PE of RFFT

3.1 CLA Adders

In the processing element, a carry look-ahead adder (CLA) has been used instead of a normal adder. The main advantage of using this adder is to minimize the delay due to carry propagation from one adder to the other in an n-bit adder.

In this, the carry generation takes place in parallel with sum computation and fed that calculated carry and propagation terms to the next stage. Data processing techniques with fast arithmetic operations have the requirement of CLA adders in it. The sum is performed by the full adders. In this paper, 16-bit adders are used. The 16-bit addition using 4-bit CLA adders is depicted in Fig. 5.

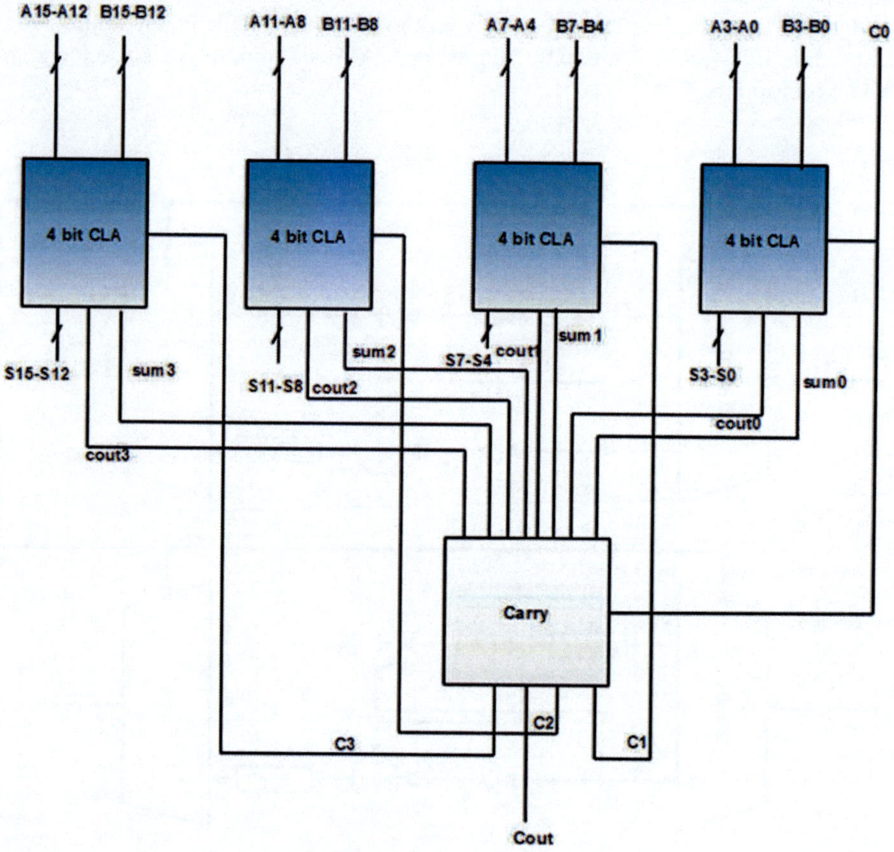

Fig. 5 16-bit addition using 4-bit CLA

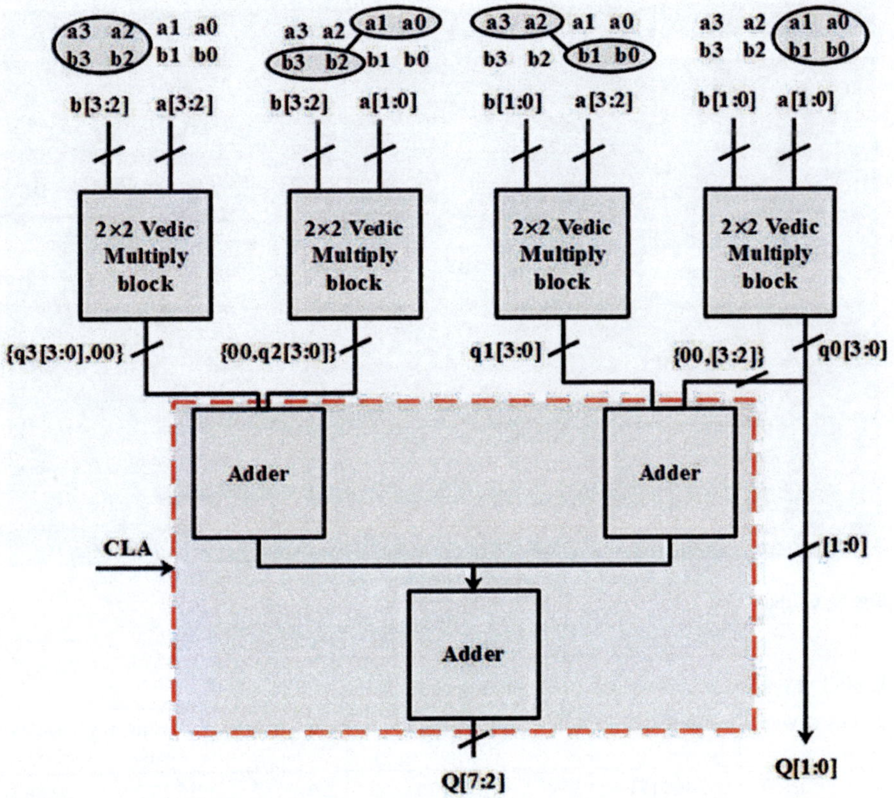

Fig. 6 16-bit vedic multiplier architecture

3.2 Vedic Multiplier

The multiplier used in the PE of RFFT is a Vedic multiplier. It improves the speed of the multiplication operation than the normal multiplier. For the applications with high speed of operation, Vedic multiplier finds its importance. The 16-bit multipliers used in this work are shown in Fig. 6.

4 Results and Discussion

In this section, the performance of the design is analyzed in various FPGA devices, such as Virtex 6 and 7, by using Xilinx ISE software tool. The parity-SOS-ECC protected single RAM-based parallel RFFTs are designed using the Verilog Hardware Description Language (HDL), and the simulation results shown in Fig. 7 have been verified in the ISIM simulator.

Fig. 7 Output waveform of Fault Tolerant Single RAM-based parallel RFFTs

Table 1 Comparison between existed and proposed architectures

FPGA device	Architecture	Slice LUTs	Flip-flops	Slice registers	Frequency (MHz)	Delay (ns)
XC6VCX75T Virtex 6	Existed [7]	359	406	251	373.32	0.664
	Fault tolerant RFFTS	368	307	226	428.97	0.676
XC7V3X30T Virtex 7	Existed [7]	354	406	242	423.25	0.580
	Fault tolerant RFFTS	325	307	218	956.02	0.525

The performance parameters, such as SliceLUTs, slice registers, flip-flops, frequency, and delay for two Xilinx FPGA devices, are listed in Table 1, and it is observed that the numbers of Slice LUTs, slice registers, flip-flops are reduced and frequency of operation is increased along with double fault identification and single fault correction.

5 Conclusion

In this paper, a fault tolerant single RAM-based Parallel Real Fast Fourier Transform Architecture using Error Correction Codes and Parseval Checks is proposed. This architecture can be used in various fields such as digital communications and signal processing devices. The results show that the proposed fault tolerant architecture performs better in terms of area usage and delay as compared with existing architecture. The speed of operation is improved for this architecure with the help of using Vedic multiplier and CLA.

References

1. International Technology Roadmap for Semiconductors 2008 Update http://www.itrs.net/Links/2008ITRS/Home2008.htm
2. Baumann R (2005) Soft errors in advanced computer systems. IEEE Des Test Comput 22(3):258–266
3. Koren I, Su S Reliability analysis of N-modular redundancy systems with intermittent and permanent faults. IEEE Trans Comput C-28
4. Gao Z et al (2015) Fault tolerant parallel filters based on error correction codes. IEEE Trans Very Large Scale Integr (VLSI) Syst 23(2), pp 384–387
5. Reviriego P, Bleakley CJ, Maestro JA (2012) A novel concurrent error detection technique for the fast fourier transform. In: Proc. ISSC, Maynooth, Ireland, pp 1–5
6. Gao Z et al (2016) Fault tolerant parallel FFTs using error correction codes and parseval checks. IEEE Trans Very Large Scale Int (VLSI) Syst 24(2):769–773
7. NagaSaiLakshmi B, RajaSekhar T (2018) Design of low- area, power fault tolerant parallel FFTs using trellis codes. Int J Eng Technol 7:2338–2343. https://doi.org/10.14419/ijet.v7i4.14106
8. Ayinala M, Brown M, Parhi KK (2012) Pipelined parallel FFT architectures via folding transformation. IEEE Trans VLSI Syst 20:1068–1081
9. Ma Z, Yin X, Yu F (2015) A novel memory-based FFT architecture for real-valued signals based on a Radix-2 Decimation-In-Frequency Algorithm. IEEE Trans Circuits Syst II Express Briefs 62(9):876–880
10. Sorensen HV, Jones DL, Heideman M, Burrus CS (1987) Real-valued fast Fourier transform algorithms. IEEE Trans Acoust Speech Signal Process ASSP-35(6):849–863
11. Rajasekhar Turaka M (2019) Satya Sai Ram "low power VLSI implementation of real fast fourier transform with DRAM-VM-CLA." Microprocess Microsyst 69:92–100

Chapter 26
Feasibility Study of Floating Solar–Hydro Hybrid System with IoT

Satya Vamsi Gudimella, Sandhya Thotakura, and Srichandan Kondamudi

1 Introduction

The optimization of the power plant structure shows that a combination of more than one renewable technology is needed to ensure continuous off-grid energy generation even under extreme conditions [1]. The hybrid power station consists of a pumped-storage hydropower plant, photovoltaic cells, and wind turbines. With immense energy demand and a shortage of wasteland for solar photovoltaic plants in cities, India can harness solar energy through floating PV plant technology for renewable energy generation. Mittal, D. et al. also estimate the power generated by the two plants and the amount of water saved from evaporation, and the reduction of CO2 emissions [2].

While implementing any systems, long-term data is necessary. Bhandari et al. [3] presented a practical guide for implementing different techniques in various locations. Of the four off-grid PV systems installed by the authors for village electrification in Nepal, one was further hybridized with wind and hydropower sources. Floating solar, which is installing photovoltaic systems on water bodies, is a nascent and fast-growing PV deployment option with global market potential for terawatts. Liu et al. provide an overview of floating solar technology and highlight some of the test bed's learning and testing results [4].

Thotakura S et al. have presented a case study on calculating the efficiency of a megawatt-scale grid-connected rooftop solar photovoltaic plant mounted on an educational institution's building roofs located in the coastal regions of Andhra Pradesh. This research has become a motivation in testing solar power at various locations all around [5].

S. V. Gudimella (✉) · S. Thotakura · S. Kondamudi
Department of Electrical, Electronics and Communication Engineering, Gitam Institute of Technology GITAM, Visakhapatnam 530045, India

S. Kondamudi
e-mail: sthotaku@gitam.edu

S. Dawn et al. (eds.), *Smart and Intelligent Systems*, Algorithms for Intelligent Systems,
https://doi.org/10.1007/978-981-16-2109-3_27

In [6], Lake of Jablanica was taken as an example. Jablanica Lake is an artificial accumulation lake on the Neretva River with 13 sq. km within the Jablanica hydroelectric power plant with 180 MW of installed power. And a floating photovoltaic plant with a capacity of 30 MW could be installed on a surface of less than 3% of the total area of HPP Jablanica accumulation. The deployment of floating solar photovoltaic installations progresses, with different designs starting to appear in various marine environments [7].

Pedroso D.A. et al. address the production and regulation of a micro-generation hybrid power system. The hydro device consists of a self-excited induction generator coupled to an unregulated primary unit [8].

Mohamed et al. presented the effectiveness and superiority of the hybrid metaheuristic technique is a grey wolf optimizer and cuckoo search (HGWOCS) algorithm for solving the short-term hydrothermal (STHS) problem [9]. Fortaleza et al. have been motivated to include IoT, hybrid power systems [10].

In hybrid photovoltaic–hydropower projects, photovoltaic panels are inserted into the hydropower plant in two ways: construction of photovoltaic panels on the downstream face of the dam, an alternative only available in some plants, or floating photovoltaic panels on the hydropower reservoir water surface.

Some of the advantages of floating solar panels on hydropower reservoirs:

1. Property usage can be reduced.
2. The panel's productivity may increase due to the cooling effect of the water below.
3. Water quality improves considerably.
4. Reservoir areas are free from shading, which enhances PV generation.

Often known as the Internet of Things or the Industrial Internet, the Internet of Things is a new technology model envisaged as a vast network of computers and sensors capable of communicating.

Some of the benefits of IoT in the power industry are real-time control, maintenance assistance, and fault position efficiently tracking for generation, transmission, substation, delivery, electricity, and other power grid facets.

Therefore, this paper's contribution is towards investigating the achievability and confinements of using IoT-based algorithm for estimating the power delivering from floating PV systems and hydro power plant.

2 Hybrid PV–Hydro System

2.1 Hydropower Plant

The region under study is the Srisailam dam, constructed across the Krishna River in the Kurnool district of Andhra Pradesh s in India's southern part. Around 16.05° and 17.50° N and 78.50° and 80.05° E, the dam is situated. In 1984, this multipurpose

Fig. 1 Srisailam Reservoir (licensed under CC BY-NC-SA 2.0)

dam was designed to increase food production in India and generate electricity. It has a 616 square kilometer reservoir (238 sq mi). The project has an approximate live capacity of 178.74 Tmcft at its maximum reservoir depth of 885 feet (270 m) of MSL. The minimum drawdown level (MDDL) of the reservoir shall be 705 feet (215 m) MSL from its river sluice gates, and the resulting dead storage shall be 3.42 Tmcft. The left bank underground power station contains six × 150 megawatts (200,000 hp) reversible Francis-pump turbines for pumped-storage operation (each Turbine can pump 200 cum). And the right bank semi-underground power station houses seven × 110 megawatts (150,000 hp) Francis-turbine generators (Fig. 1).

2.2 *Floating PV System*

Floating PV systems is one of the steadily rising technologies in the green energy market. The key advantage of floating PV plants is that they do not take up much ground, save for the minimal surfaces required for electrical cabinets and grid connections. There are more lightweight than land-based plants, easier to maintain, and simpler to construct and decommission. The key argument is that there are no set structures like the foundations used by a land-based farm so that their installation can be entirely reversible. Partial exposure of the basins can minimize water evaporation. This outcome depends on the atmospheric conditions and the proportion of

Fig. 2 Floating PV system (licensed under CC BY-NC-ND 2.0)

the area covered. So, floating PV is an encouraging future aspect in the field of power generation (Fig. 2).

The basic floating PV system consists of floating polymer, docking system, submerged cables, and PV cells. The floating polymer facilitates the mounting of solar cells on the water's surface; the docking mechanism helps preserve PV panels' location, and for tapping power, submerged cables are preferred.

3 IoT-Based Controller for Hybrid Floating PV and Hydropower

Hydropower is one of the promising prospects in the field of electricity. But, hydropower's total capability is hampered due to some of the factors like availability and variability. And the same applies to Solar PV systems also. PV continues to grow, and businesses continue to expand. The power industry is searching for new places for the construction of electricity scaling of PV plants. Thanks to the fast combination of the two technologies, there is a strong interest in constructing PV plants coupled with HPP. Floating PV plants provide a further increase in the variety of solar and hydropower.

Load scheduling is one of the issues in integrating the floating PV system and hydro system. A trail on automation-controlled load scheduling is shown in Fig. 3.

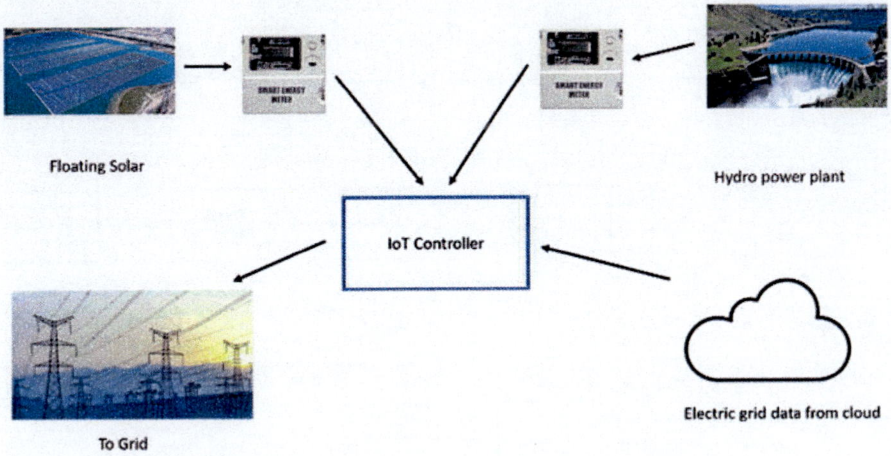

Fig. 3 Mind map of IoT-based controller for hybrid floating PV-hydropower

The basic idea is that the amount of instantaneous solar power is sensed, and it is compared with the grid's cloud power demand. If the power demand required matches the total solar energy generated, then the hydropower will rest. Suppose the demand is much more than the floating solar's instantaneous capacity. In such a case, the comparator will send a signal to the Arduino board, where the Arduino is programmed to control the position of the penstock's valve. Based on this idea, the Arduino will take control of the value-based solar energy availability (Fig. 4).

3.1 Estimated Model

Figure 5 represents the estimated model of IoT-based hybrid floating PV and hydropower system. Components used in prototype are as follows:

ESP8266:

The ESP8266 is a low-cost Wi-Fi microchip, with a full TCP/IP stack and microcontroller capability, produced by Espressif Systems in Shanghai, China. This small module allows microcontrollers to connect to a Wi-Fi network and make simple TCP/IP connections using Hayes-style commands. In this prototype, it is used for getting the information of solar power generated from floating solar panels which are sent to the Internet.

Transistor:

A transistor is a semiconductor device used to amplify or switch electronic signals and electrical power. It is composed of semiconductor material usually with at least three terminals for connection to an external circuit. A transistor in this model is used

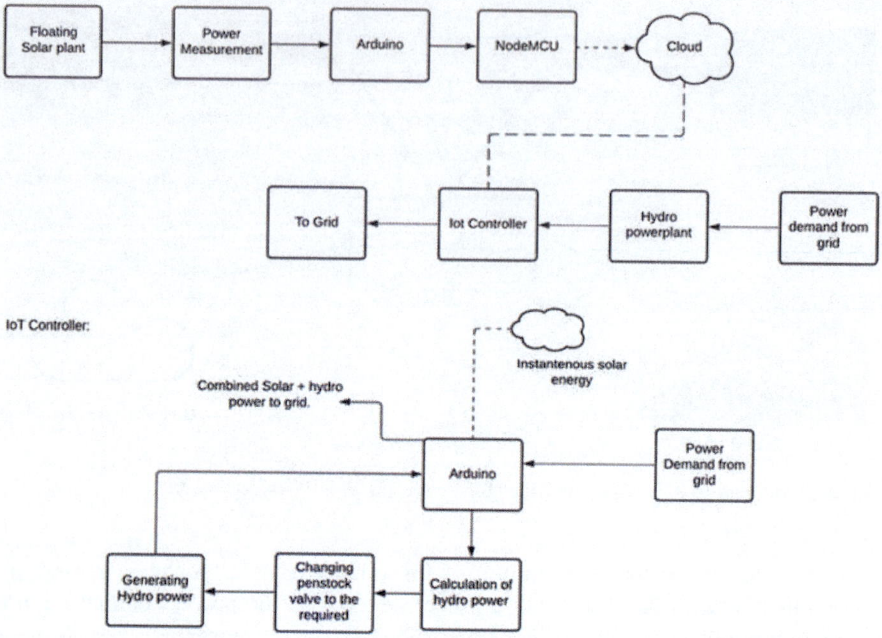

Fig. 4 Operational principle of the IoT controller

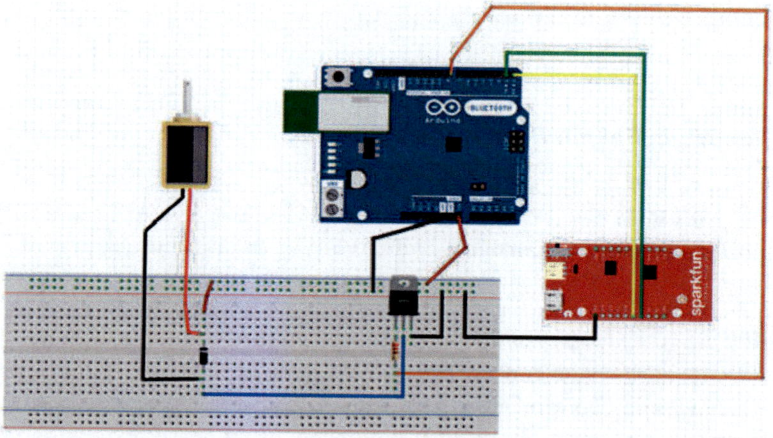

Fig. 5 Estimated model

because the stepper motor for rotation requires more current so the NPN transistor which consists of the Darlington pair is connected.

Diode:

A diode is a two-terminal electronic component that conducts current primarily in one direction (asymmetric conductance); it has low (ideally zero) resistance in one direction, and high (ideally infinite) resistance in the other. It is used for protecting motor from back emf.

Stepper motor:

A stepper motor, also known as a step motor or stepping motor, is a brushless DC electric motor that divides a full rotation into a number of equal steps. The motor's position can then be commanded to move and hold at one of these steps without any position sensor for feedback (an open-loop controller), as long as the motor is carefully sized to the application in respect to torque and speed. It is used in our prototype for changing degrees of rotation in steps for different power inputs.

Arduino:

It is a microcontroller to which many shields and components can be attached to perform different operations. It is used as the IoT controller for our work. It is connected to ESP8266 which helps in getting power data. The program fed in the microcontroller calculates the power to be generated from the hydropower plant and then generates a signal which rotates the valve connected to the stepper motor to a required value.

3.2 *Flow Diagram*

The cloud data is taken on a timely basis, and the data is transferred to the controller. This data will tell the power plant about demand power. Next, the data from the floating solar power plant is measured and forwarded to the controller. The control station will estimate the total power produced by the hybrid plant. If the power requirement is more than the solar energy capacity, then the signal is sent to the hydropower governor to open the valve. If the power requirement is less than or equal to solar power, then hydropower will be a hot reserve to the grid. Total control will be an automation process (Fig. 6).

4 Case Study

For understating the proposed model's feasibility, data of the Srisailam power plant is considered. Solar irradiance data has been collected from the NREL website at the location.

Figure 7 presents the solar direct normal irradiation [Wh/m^2] for the year 2020.

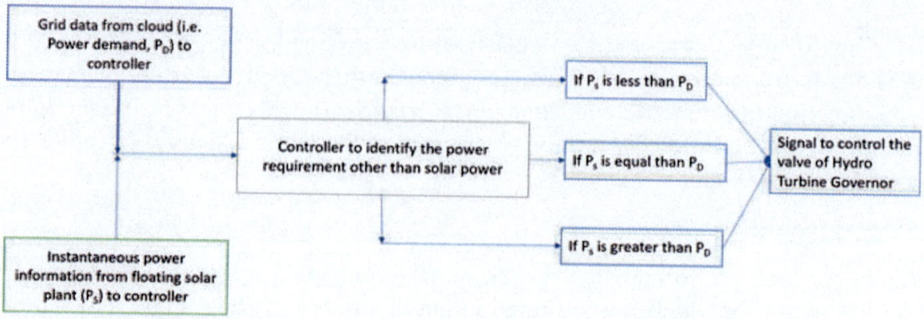

Fig. 6 Flow diagram of the model

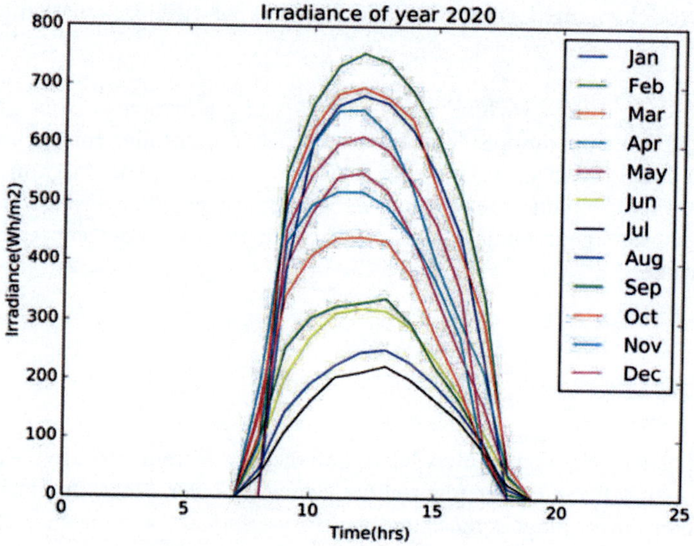

Fig. 7 Solar direct irradiation at Srisailam reservoir

Based on the solar direct irradiation characteristics, the noticeable solar energy at the location is available in the time range of 9 am to 4 pm. And in between 11.30 am and 12.00 pm, maximum irradiation is observed for all months, and an average of 500 Wh/m^2 has been observed.

For the analysis, solar panels 1STH-250-WH [10] and Sunny Boy 3000TL inverters [11] were selected. Table 1 lists their parameters.

Module per string can be evaluated using the following formula:

$$Number\ of\ Module\ per\ string\ (N_m) = \frac{V_{\text{max, MPPT}} + V_{\text{min, MPPT}}}{2V_{\text{module}}} \quad (1)$$

Table 1 Parameters of 1STH-250-WH and Sunny Boy 3000TL

1STH-250-WH		Sunny Boy 3000TL	
Parameter	Value	Parameter	Value
Maximum power of the panel	250 W	Maximum AC power	3000 VA
Panel Width	990.0 mm	Maximum DC power	3200 W
Panel length	1652.0 mm	Maximum DC current	15 A
Maximum working voltage	30.7	Maximum DC voltage	750 V
Maximum working current	8.15	Minimum MPPT voltage	175 V
Open circuit voltage	37.3	Maximum MPPT voltage	500 V
Short circuit current	8.66	Efficiency	97%
Temperature coefficient	−0.482%/°C	Maximum output current	16 A
Efficiency	15.2%		

$V_{\text{max, MPTT}}$—maximum MPPT voltage.
$V_{\text{min, MPTT}}$—the minimum MPPT voltage in the inverter.
V_{module}—the maximum power voltage of the module.

This gives the parameters of the solar panel and inverter a number per string of 11 modules. From the parameters in Table 1, the number of strings can be estimated based on the following formula:

$$Number\ of\ strings\ N_s = \frac{P}{Module\ per\ string * P_{module}} \quad (2)$$

where P is the power requirement of the plant. In this paper, for the basic review, a 250 MW floating PV power plant is considered. On calculating the value, the number of strings needed for the 250 MW plant is 90909 strings, which says that the number of panels needed is 1000,000.

The number of inverters required can be obtained using

$$Number\ of\ inverters = \frac{N_m N_s P_{module}}{R_{DC-AC} * P_{inverter}} \quad (3)$$

After calculation, the approximated number of inverter need is 83,333.25.

The area required to mount the panels, which depends on the slope (α) with which the panels are placed. The distance between the panels can be calculated using

$$D_{space} = panel\ width * [cos\alpha + sin\alpha * \tan(latitude + 23.5^0] \quad (4)$$

$$Area\ = N_m * N_s * D_{space} * Length\ of\ the\ panel \quad (5)$$

The spacing between the panels is 10.94 mts and the required surface area is 175.039825 km^2.

The total surface area of the Srisailam reservoir is 616 km^2, so there is more scope for keeping PV panels for more solar power capacity.

Figures 8 and 9 illustrate the relationship between the AC system output and the solar radiation and plane of solar radiation. In the month of march, the 559.93670654 kWh is the highest amount of AC output generated. The annual AC output observed in the system is 5824.76086425 kWh. In Fig. 10, the maximum DC array output is 622.08349609 kWh, and it is also observed in the month of March. The annual DC array output obtained is 6480.66885377 kWh. On the basis of the statistics, floating solar can be a promising add-on to the hydropower plant. Also, detailed analytics can explore more hidden points in the feasibility analysis of the system.

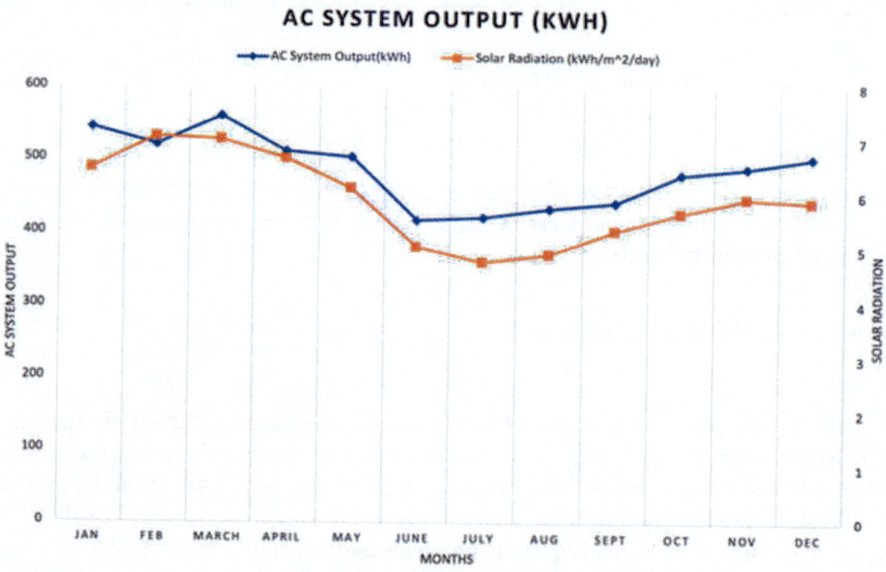

Fig. 8 AC system output based on solar irradiation at Srisailam reservoir

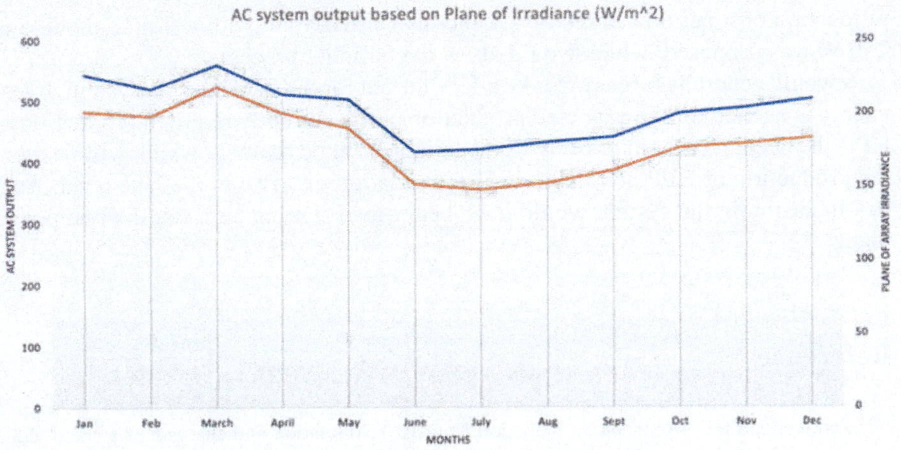

Fig. 9 AC system output based on plane of solar irradiation at Srisailam reservoir

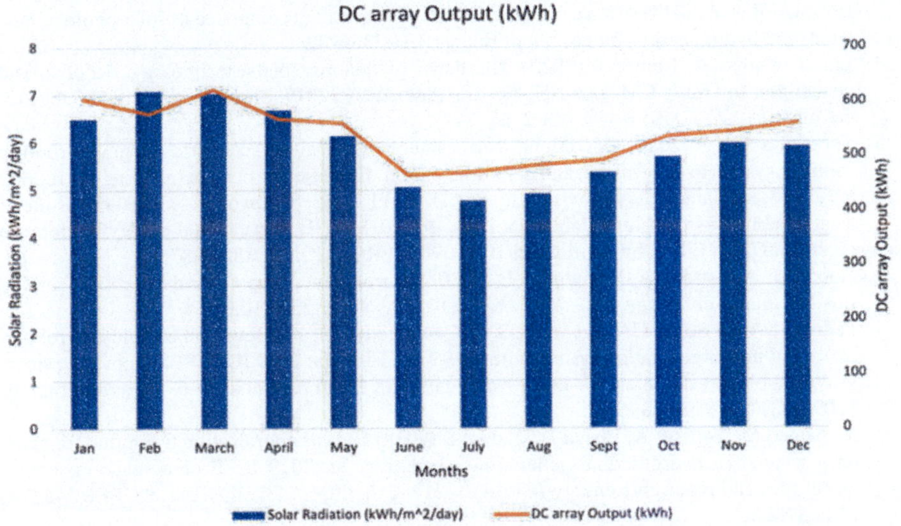

Fig. 10 DC array output based on solar irradiation at Srisailam reservoir

5 Conclusions

This paper is a primary attempt to propose the possible use of a floating PV system on the Srisailam reservoir. IoT-based automation is proposed for a better load dispatch. It is shown that on a surface of 175 m^2, there could be a solar plant with a nominal power of 250 MW. An area of 28.4% of the lake surface can produce 250 MW of power. Despite this technology's high cost, a plant of this type could improve production cost per unit of energy. A modular installation with a simple module of 250 W was proposed, which would allow the building process.

It would generally be easy to take all PV output for the transmission system. If less energy is needed for the expected production or the currently required production of HPP, the necessary number of hydro generators will be omitted, which will result in the production of HPP and increased accumulation or a rise in the lake level. And, IoT inclusion in the system would have better load sharing between the two power plants.

References

1. Petrakopoulou F, Robinson A, Loizidou M (2016) Simulation and analysis of a stand-alone solar-wind and pumped-storage hydropower plant. Energy 96:676–683
2. Mittal D, Saxena BK, Rao KVS (2017) Floating solar photovoltaic systems: an overview and their feasibility at Kota in Rajasthan. In: 2017 international conference on circuit, power and computing technologies (ICCPCT) 1–7. https://doi.org/10.1109/ICCPCT.2017.8074182
3. Bhandari B et al (2014) A novel off-grid hybrid power system comprised of solar photovoltaic, wind, and hydro energy sources. Appl Energy 133:236–242
4. Liu H, Kumar A, Reindl T (2020) The dawn of floating solar—technology, benefits, and challenges. In: Wang CM, Lim SH, Tay ZY (eds) WCFS2019. Springer, pp 373–383. https://doi.org/10.1007/978-981-13-8743-2_21
5. Thotakura S et al (2020) Operational performance of megawatt-scale grid integrated rooftop solar PV system in tropical wet and dry climates of India. Case Studies Thermal Eng 18:100602
6. Pašalić S, Akšamović A, Avdaković S (2018) Floating photovoltaic plants on artificial accumulations—example of Jablanica Lake. In: 2018 IEEE international energy conference (ENERGYCON), 1–6. https://doi.org/10.1109/ENERGYCON.2018.8398765
7. Hooper T, Armstrong A, Vlaswinkel B (2020) Environmental impacts and benefits of marine floating solar. Sol Energy. https://doi.org/10.1016/j.solener.2020.10.010
8. Pedroso DA, de Souza CA, Scherer LG, de Camargo RF (2019) Development and control of a hydro-PV power generation system with BESS and ELC. In: 2019 IEEE PES innovative smart grid technologies conference—Latin America (ISGT Latin America), 1–6. https://doi.org/10.1109/ISGT-LA.2019.8895340
9. Mohamed M, Youssef A, Ebeed M, Kamel S (2019) Hybrid optimization technique for short term wind-solar-hydrothermal generation scheduling. In: 2019 IEEE conference on power electronics and renewable energy (CPERE), 212–216. https://doi.org/10.1109/CPERE45374.2019.8980237
10. http://www.solarhub.com/product-catalog/pv-modules/2315-1STH-250-WH-1Soltech
11. http://files.sma.de/dl/15330/SB5000TL-21-DEN1551-V20web.pdf

Chapter 27
A Rectangular SIW MIMO Antenna for IoT Devices and Ultra-Wideband Applications

N. Kartheek Ram Reddy, K. Sneha, B. Alekhya, and E. Prathyusha

1 Introduction

Substrate-Integrated Waveguide technique (SIW) has gained much attention in the field of high-density integration in various subsystems. It has created many ways to compound all passive and active elements upon a similar substrate like the amplifiers, couplers, filters, and oscillators. The rows were narrowly spaced with metallic vias between two planes on the adjoining walls of a filmy rectangular-shaped waveguide occupied with dielectric. The SIW has different attributes such as lossless, high power efficiency, best integration, and fabrication with planar technology [1].

The efficiency of the spectrum is improved by using MIMO technology, which provides greater isolation among the elements around wider band frequencies. Recently, there is a rapid increase in devices that can operate in various frequencies from 2 to 7 GHz for applications like WiMAX, WLAN, ZigBee, Bluetooth, and UWB. Hence for applications like LTE, UTMS, etc., a quad element MIMO antenna with a rectangular SPLIT-RING RESONATOR (SRR) and an altered slotted ground plane was designed.

In this paper, we present a SIW structure penetrated on the ground by using metallic vias perpendicularly, which improves the operating frequency up to 6.7 GHz, achieving an efficiency of 95% and around 4 dBi gain.The diversity performance of antenna is best and suitable for practical systems. The single element structure is shown in Fig. 1. The dimensions of the proposed antenna are W = 45 mm, L = 2.5 mm, W1 = 3 mm, W2 = 10 mm, W3 = 21 mm, L1 = 19 mm, L2 = 5.2 mm, L3 = 12 mm, L4 = 5 mm, L5 = 16 mm, d = 1 mm, g = 0.4 mm, s = 0.5 mm, and c = 0.5 mm [2].

N. K. Ram Reddy · K. Sneha (✉) · B. Alekhya · E. Prathyusha
Department of ECE, Velagapudi Ramakrishna Siddhartha Engineering College, Velagapudi, India

S. Dawn et al. (eds.), *Smart and Intelligent Systems*, Algorithms for Intelligent Systems,
https://doi.org/10.1007/978-981-16-2109-3_28

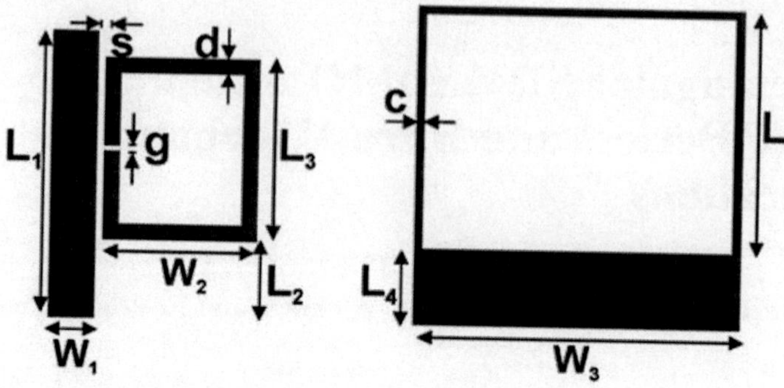

Fig. 1 Attributes of a single element. **a** Front view, **b** Back view

2 Antenna Design

The antenna has been fabricated using FR4 substrate with a thickness of 1.6 mm, $\varepsilon_r = 4.4$, and $\tan\delta = 0.02$. The Ansoft High-Frequency Structure Simulator (HFSS) is used for performing simulations [2].

A. *Four-Element Antenna*

A four-element MIMO antenna of various elements with an area of $0.027\lambda^2$ are placed alternatively and equally in a quadrate substrate with a separation of 90° as shown in Fig. 2.

B. *Design of SIW four-element Antenna*

The frequency range of the antenna is increased by using the below structure. The design consists of metallic vias on the substrate. Twelve metallic vias are constructed

Fig. 2 Four-port MIMO antenna design

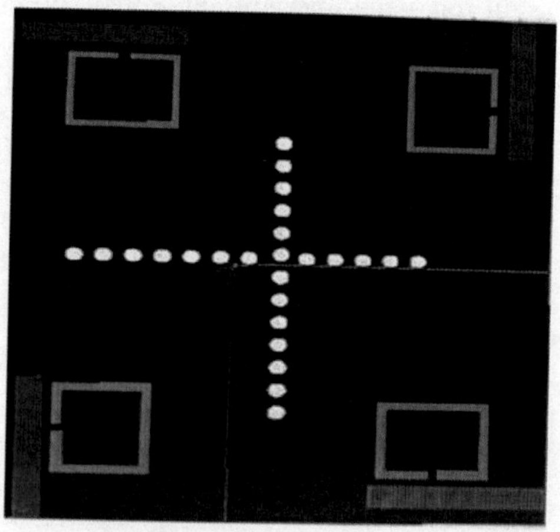

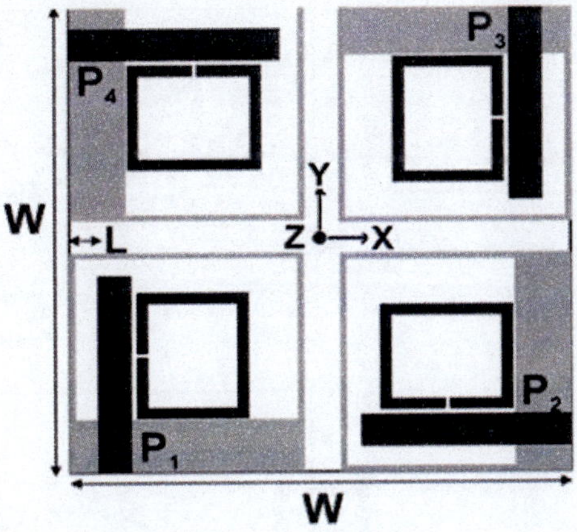

Fig. 3 Design of four-port SIW MIMO antenna

using a cylindrical structure throughout the axis perpendicularly with a radius of 1 mm each as shown in Fig. 3.

The height of each cylindrical via is 1.6 mm and they are placed with space in between them. This structure improved the operating frequency up to 6.7 GHz with an S11 of –10 dB.

3 Results

The evolution stages of the structure from a single element to quad element and quad element with SIW are shown in the Fig. 4.

In graph 4a, we can observe that the efficient return loss is obtained at –30 dB for the single element.

In Fig. 4b, the returns loss of the antenna at each port is clearly shown which is < –5 dB, which is generating a good efficiency. Overcoming the above results, the SIW MIMO antenna yielded an excellent efficiency at all the ports which is –40 dB, and hence, by comparing the S_{11} values we can prove that the SIW MIMO antenna is the most efficient one. For this value of S_{11}, we are getting a better result with all the values below –10 dB [3].

By observing all the results, we can say that the s-parameters of the quad element SIW antenna are more efficient and the radiation patterns are satisfactory. The radiation patterns are given for 2, 4, and 6 GHz.

4 Fabricated Antenna

The fabricated structure of the antenna is clearly shown in the below images with the S11 measurement result (Fig. 5).

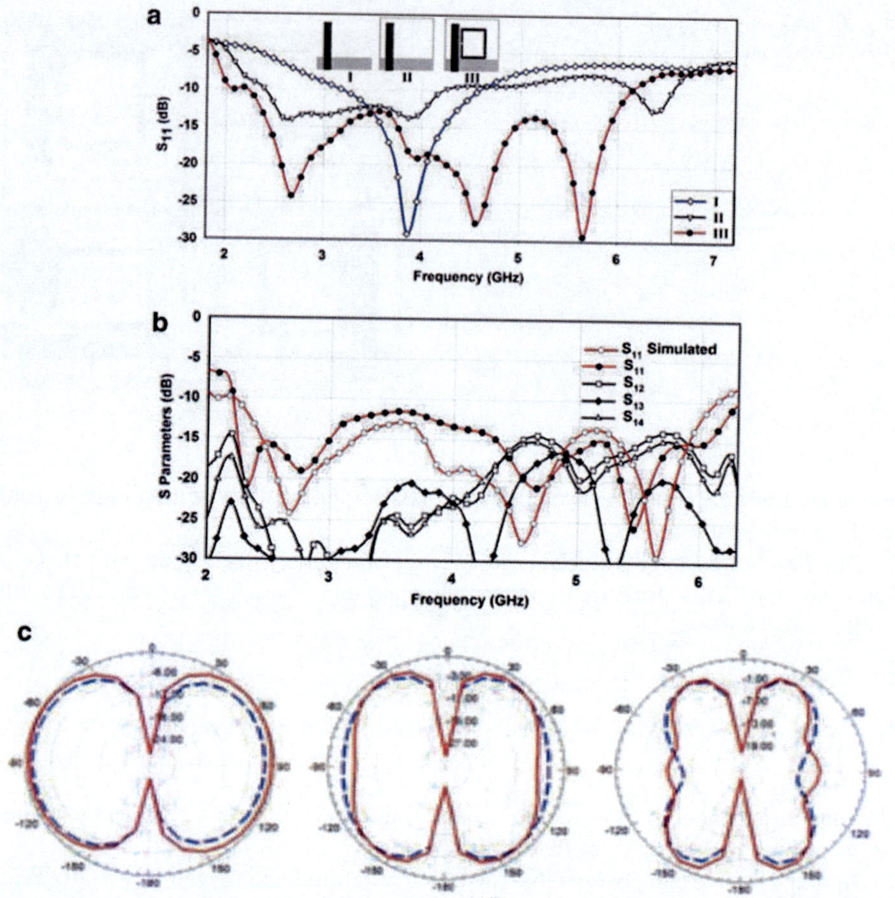

Fig. 4 **a** *S*11 characteristics of single-port antenna, **b** Quad element MIMO antenna S-parameters, **c** Quad element SIW MIMO antenna radiation patterns

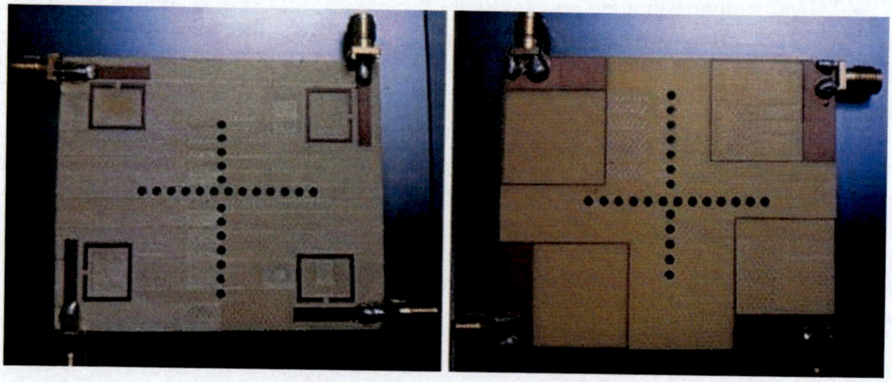

Fig. 5 Front and back view of the manufactured structure

The real view of the proposed antenna 1is shown in the above image and the vias are in a way penetrating through the ground which is clearly visible on both sides. The feeds and rectangle structures are made using copper material. The measurement of return loss is done using VNA obtaining a value of −40 dB, which is similar to the simulated value [4] (Fig. 6).

This clearly enhances the real-time operations and advantage of the proposed antenna. The ECC is measured and the acquired value is <0.25 is shown in the Fig. 7.

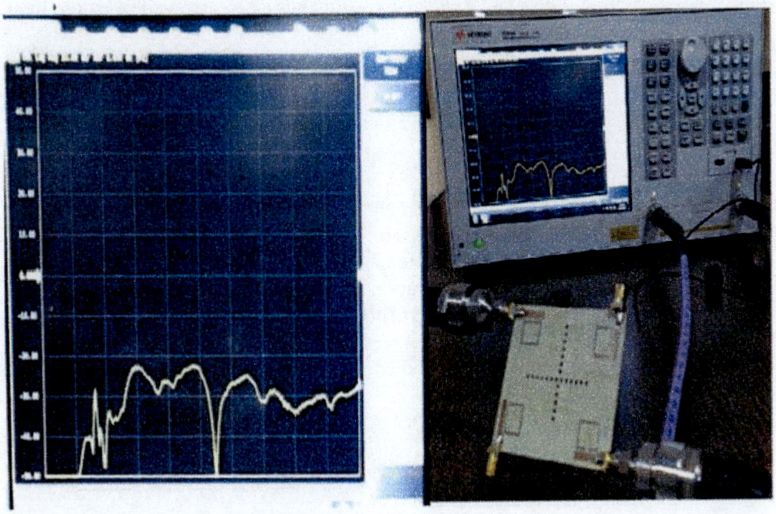

Fig. 6 Return loss measurement of the fabricated antenna using VNA

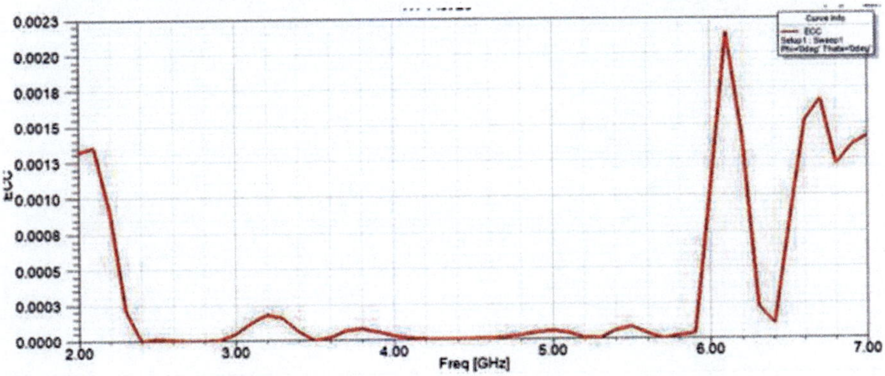

Fig. 7 Simulated ECC graph

5 Conclusion

By comparing the simulation and measurement values, we can say that the proposed antenna works most effectively for the IoT devices and UWB applications by yielding the S11 up to −30 dB, the overall efficiency is 95%, and a gain of 4 dBi. The value of ECC is $\leq$0.25.

Acknowledgements The authors are thankful to Dr. A. V. Ratna Prasad, Principal, for constant encouragement. They are also grateful to Dr. P. V. Subbaiah, the Head of the Department of Velagapudi Ramakrishna Siddhartha Engineering College for making avail of all facilities.

References

1. Vinesh PV, Anitha R (2016) A quad element slotted ground wideband antenna for MIMO applications. IEEE Trans Antennas Propag 10
2. Moradi A, Khalily M Wide band MIMO antenna system for WiFi/LTE access point applications. IEEE Antennas Wireless Propag Lett Common Elements 13:1601–04
3. Sonkki M, Dandekar KR Wide band planar four element linear antenna array. IEEE Antennas Wireless Propag Lett 13:1663–66
4. Li F, Chu QX (2012) A compact wide band MIMO antenna with two novel bents. IEEE Trans Aantennas Propag 60(2):125–28

Chapter 28
Cloud-Connected Smart Energy Meter with Remote Monitoring and Control

Vanka Saritha, Anil Kumar Karra, Shaik Khader Zelani, and Ch. Prasanth

1 Introduction

Electrical utilities are facing a lot of losses as a result of power theft, inadequate utilization of energy, and unpaid bills. Apart from these, there are other losses like transmission losses which cannot be avoided. The main problem with the existing electricity meter is that the consumers are not known about their consumption on a time to time basis. Besides this, electricity board officials have to go to each house in order to generate the bills, which is a time-consuming process. Smart energy meter is a two-way communication device, which establishes communication between meter and substation. The energy consumed data is directed to the cloud and the mobile app of the consumer.

It is also a tamper-proof device so there is no chance of power theft. On nonpayment of bills, there is no need for the lineman to go near the meter to remove the fuse or cut the line. All this is done at the substation itself by a simple connect and disconnect feature. Overload indication is also available so that the board officials will know about overload usage without manual inspection.

An IoT-Based Energy Meter Reading, energy theft Detection and Disconnection that uses PLC modem and optimization of Power is proposed in [1]. This work described the design based on PIC18F46k22 microcontroller and putting into practice of energy meter using IoT concept. This system enables the KSEB office for

V. Saritha (✉) · Ch. Prasanth
Department of E.C.E, Velagapudi Ramakrishna Siddhartha Engineering College, Vijayawada, Andhra Pradesh, India
e-mail: vsaritha@vrsiddhartha.ac.in

A. Kumar Karra
Department of E.C.E, VFSTR, Vadlamudi, Andhra Pradesh, India

S. K. Zelani
Department of E.C.E, Narasaraopeta Institute of Technology, Narasaraopeta, Andhra Pradesh, India

S. Dawn et al. (eds.), *Smart and Intelligent Systems*, Algorithms for Intelligent Systems,
https://doi.org/10.1007/978-981-16-2109-3_29

collecting the bill and power theft detection data robotically deprived of applying the manpower. The data that is collected at the customer sites is shared to the office through the web and GSM [2–5]. Intended for advanced metering system and billing System, GSM-based Smart Energy Meters by making use of Arduino are developed in [6–8, 9]. These works presented a smart energy meter for automatic and superior metering and billing system. GSM or ZigBee technologies are used for the functioning of smart energy meters [10–19].

This paper presents the novel idea of a smart energy meter using IoT and Arduino for automatic metering and billing, where overload indication and meter tampering information will be sent to the substation.

2 Proposed Design

2.1 Hardware Implementation:

The proposed smart energy meter consists of the hardware components, which are shown in Fig. 1, i.e., Block diagram of proposed Energy Meter.

The smart energy meter consists of the following blocks shown in Fig. 1. The Arduino UNO is the main component of the block. The Wi-Fi module used here is the Node MCU component. The current and voltage are computed with the help of the potential transformer and current sensor-ACS712. The cloud platform used is the AllThingsTalkMaker and the mobile app is the Talk Maker app. The smart energy meter comprises the following hardware components. Arduino UNO is the heart of the smart energy meter. It is used for the computation of power (Watts) and energy (KWH), which is performed by the microcontroller ATMEGA328P. The power supply is 230 V AC, which is fed to the potential transformer. The potential transformer is a step-down transformer that steps down from 230 V toward 9 V AC. The conversion unit consists of a full-wave rectifier, that converts AC to DC and a capacitor is used before giving it to Arduino analog input pins. The current sensor ACS712A, which is Hall-Effect-based meter is considered for the measurement of RMS current consumed by the load. IR LED and photodiode are used for detection of the meter tampering.

Relay module is employed for meter connection and disconnection on nonpayment of bills. The values computed are directed to the cloud platform with the help of the NodeMCU and are also displayed on the LCD display screen.

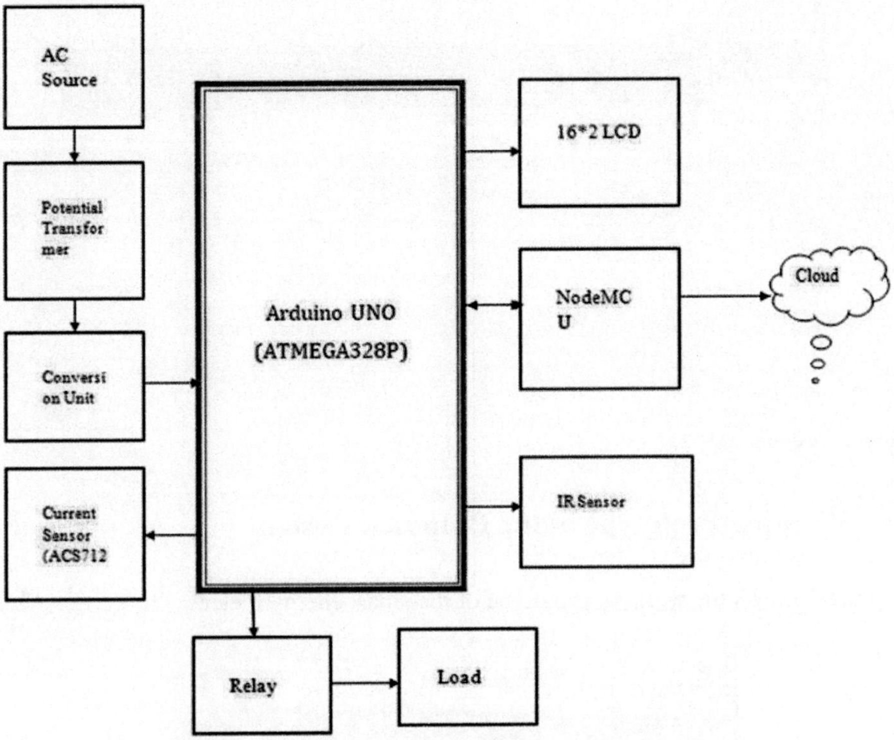

Fig. 1 Block diagram of the recommended energy meter

2.2 Software Implementation

Smart energy meter uses the following software:

Arduino IDE,

AllThingsTalkMaker.

Arduino IDE:

The (IDE) comprises a code editor with functions like cutting, pasting, examining and swapping text, indenting, brace matching, syntax underlining, and offers simple one-click mechanisms for compiling and uploading programs to a board. A program that is written with the IDE is named as a sketch. Figure 2 shows the sketch.

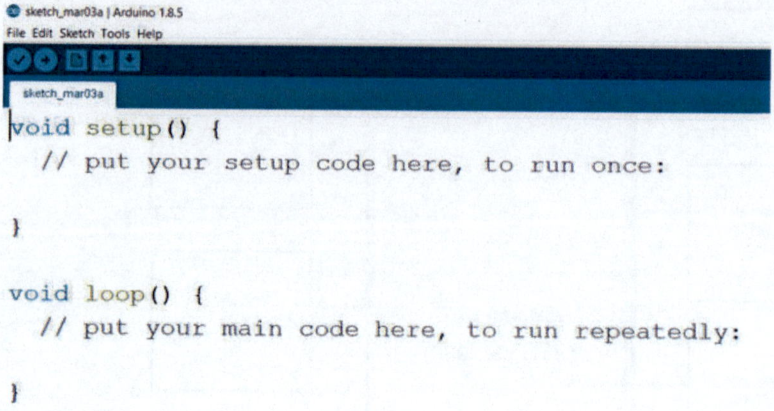

Fig. 2 Arduino IDE

3 Working Prototype of the Proposed Design

Figure 3 shows the working prototype of the smart energy meter.

Fig. 3 Working prototype of smart energy meter

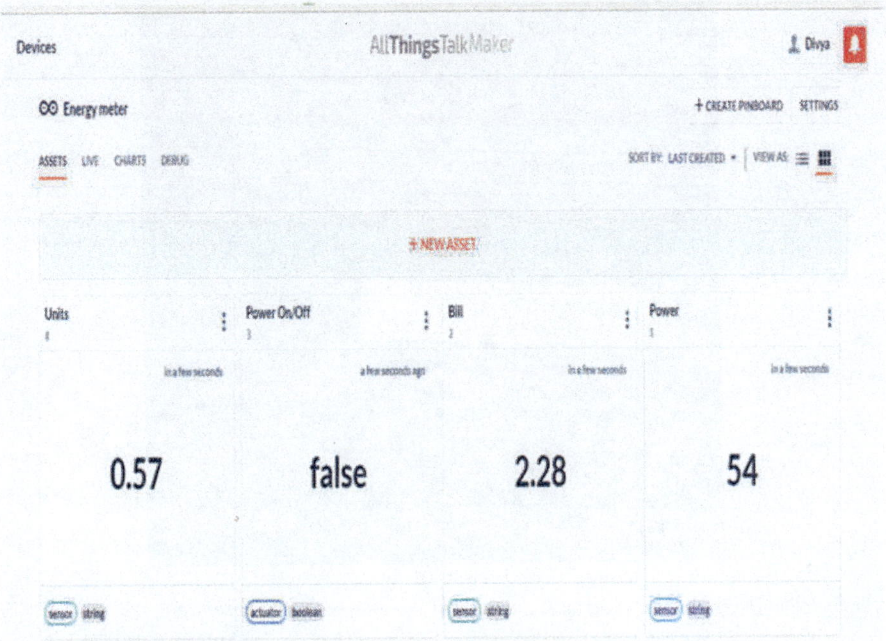

Fig. 4 Result of the window on cloud

4 Results

The following results are observed in the cloud platform AllThingsTalkMaker, App and LCD, respectively. The results on the cloud are shown in Fig. 4.

The output screen using LCD display is shown in Fig. 5 (Fig. 6).

Figure 7 shows the values displayed in the app window regarding the units and bill.

Figure 8 shows the output for the 200 W bulb with an overload indication.

Figure 9 shows that the meter is not overloaded with a 60 W bulb.

5 Conclusion

This work has reviewed the currently available flaws in the existing energy meter and has identified a key area, i.e., IoT that can provide a solution to these problems. An effort is made to develop a practical model of "IoT Based Smart Energy Meter". This proposed model is used to compute the consumption of energy of the household and make the energy unit reading to be manageable. By using this smart energy meter,

Fig. 5 Display on LCD

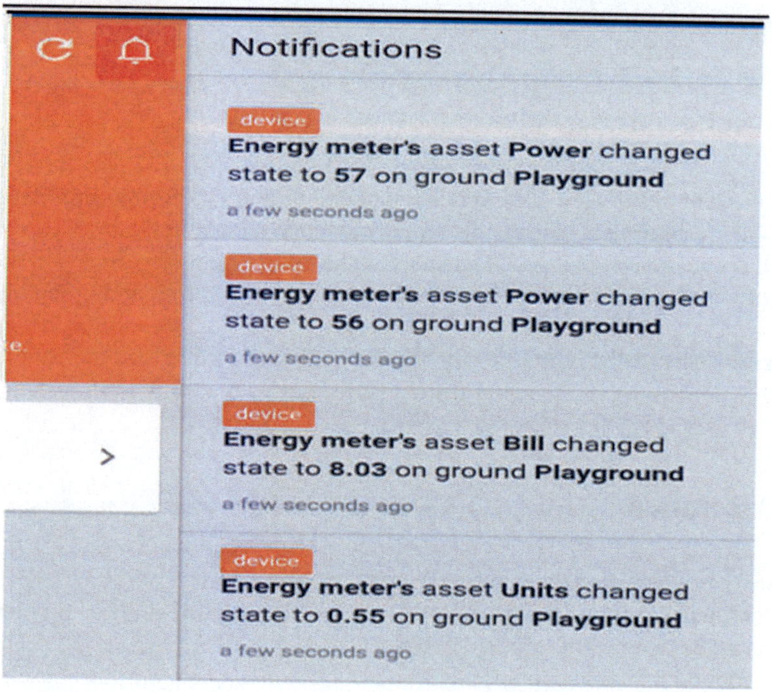

Fig. 6 Notifications in App

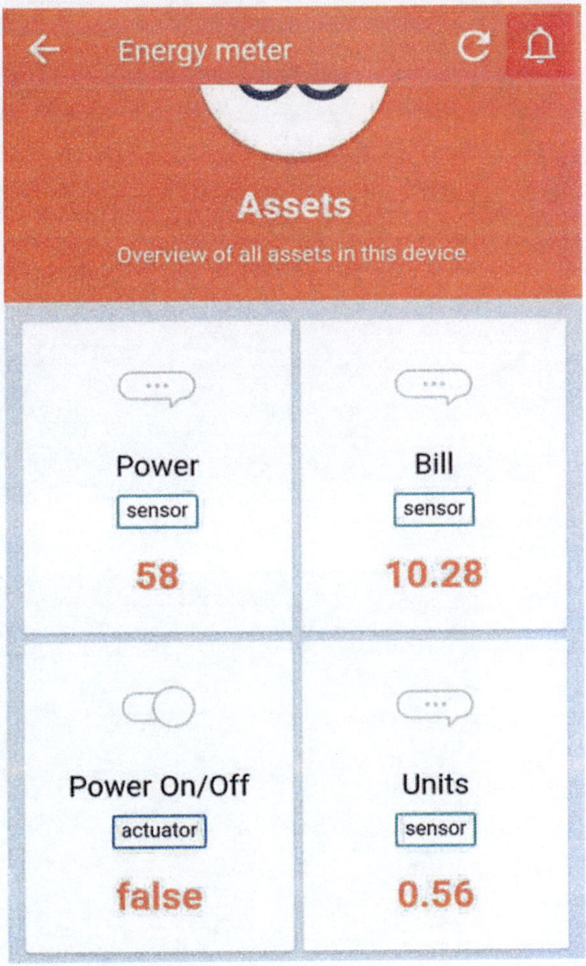

Fig. 7 Displayed values in app

the power, voltage, and current are calculated, they are displayed on the LCD screen. It can monitor the customer about the power consumption and also provides overload indication whenever an overload occurs. If any tampering occurs, it can be detected by using an IR sensor. It can also provide a billing facility. If a bill is not paid by the customer, meter connection and disconnection are done through a relay module.

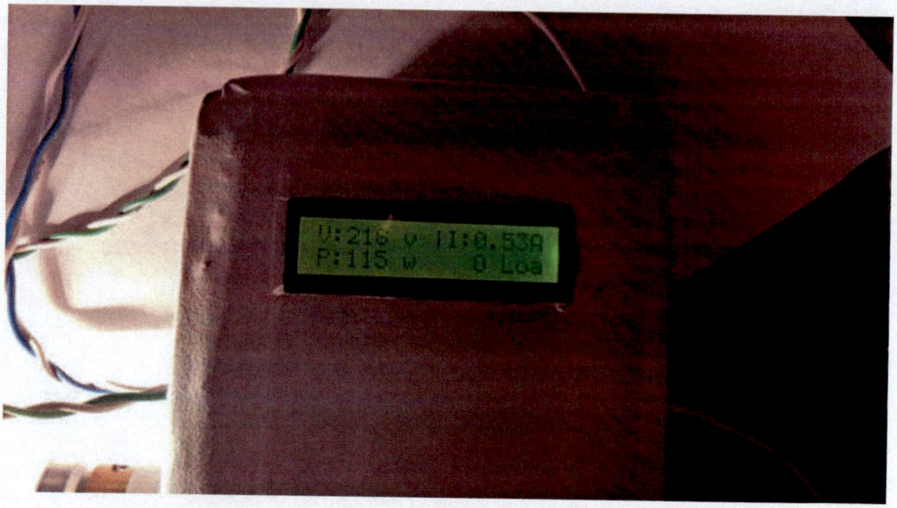

Fig. 8 Output for 200 W bulb overload indication

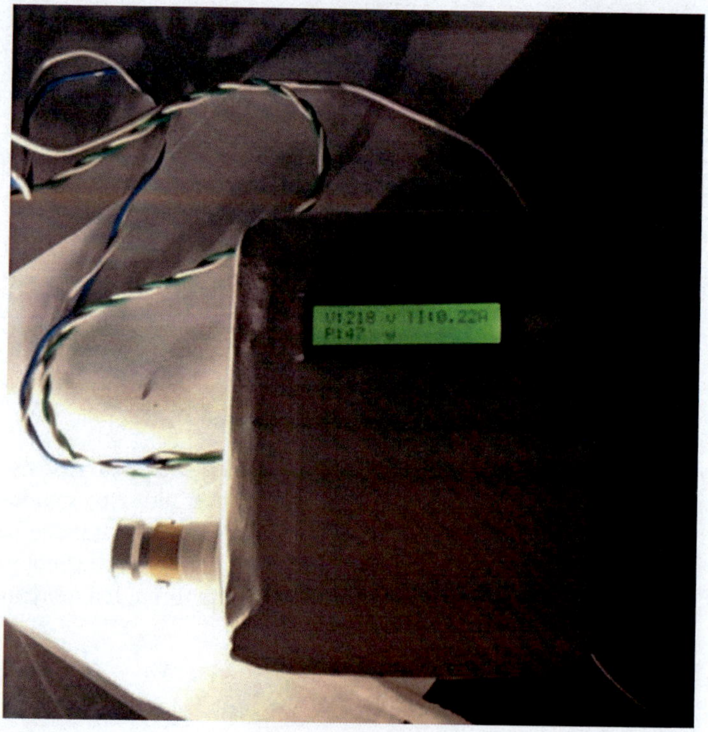

Fig. 9 Output for 60 W Bulb with no overload indication

References

1. DarshanIyer N, Radhakrishna Rao KA (2015) IoT based electricity energy meter reading, theft detection and disconnection using plc modem and power optimization. Int J Adv Res Electric Electron Instrum Eng (IJAREEIE) 4(7)
2. Talwar PD, Kulkarni SB (2016) IoT based energy meter reading. Int J Recent Trends Eng Res (IJRTER) 02(06)
3. Thanth D, AmolBorole (2015) Wi-Fi based smart energy meter. Int J Res Comput Commun Technol (IJRCCT) 4(4)
4. Sahani B, AkibjavedTamboli TR, Pisal RS (2017) IoT based smart energy meter. Int Res Eng Technol (IRJET) e-ISSN: 2395-0056 04(04)
5. Ann Mathews J, Varghese J, Raju J, Daley L, Beena AO (2017) Intelligent energy meter with power theft detection. Int J Current Eng Sci Res (IJCESR) 4(4)
6. Rastogi S, Sharma M, Varshney P (2016) Internet of things based smart electricity meters. Int J Comput Appl (0975–8887) 133(8)
7. Bhavani R, Alagammal S Design and implementation of gsm based smart energy meter (sem) for home applications. Int J Latest Trends Eng Technol
8. Preethi V, Harish G Design and implementation of smart energy meter. 8(1):431–439
9. Rahman Md.M, Jannat NE, Islam Md.O, Salakin Md.S (2015) Arduino and gsm based smart energy meter for advanced metering and billing system. In: 2nd international conference on electrical engineering and information & communication technology (ICEEICT) 2015
10. Chen D, Kalra S, Irwin D, Shenoy P, Albrecht J (2015) Preventing occupancy detection from smart meters. IEEE Trans Smart Grid
11. Xie J, Hong T, Stroud J (2015) Long-term retail energy forecasting with consideration of residential customer attrition. IEEE Trans Smart Grid
12. Rahman Md.M, Jannat NE, Islam Md.O, Salakin Md.S (2015) Arduino and GSM based smart energy meter for advanced metering and billing system. In: IEEE 2015 Ref 8–1463–666–2/15 & 2nd international conference on electrical engineering and information & communication technology (ICEEICT)
13. Singh M, Sanduja Er.V (2015) Minimizing electricity theft by internet of things. Int J Adv Res Comput Commun Eng 4(8)
14. Haijun L, Lingmin L, Xianyi L (2010) A novel preprocessing approach for digital meter reading based on computer vision. In: Proceedings of the third international symposium on computer science and computational technology (ISCSCT ’10) Jiaozuo, P. R. China, pp 308–311. ISBN 978–9525726–10–7
15. Naveenkumar, Jagadeesh SK (2016) Smart energy meter. Int J Tech Res Appl 4(3):126–129
16. Kate PG, Rana JR (2015) ZigBee based monitoring theft detection and automatic electricity meter reading. In: International conference on energy systems and applications (ICESA 2015)
17. Prashanthi GL, Prasad KV (2014) Wireless power meter and intimation system using GSM and Zigbee networks. J Electron Commun Eng 9(6), Ver. 1
18. Bansal A, Goyal V (2017) Real time electricity monitoring using smart energy meter in a smart LAN based network. Int J Electron Electric Comput Syst IJEECS, ISSN 6(5):2348–3117
19. Rahman Md.M, Jannat NE, Islam Md.O, Salakin Md.S (2015) Arduino and GSM based smart energy meter for advanced metering and billing system. In: International conferenced on electrical engineering and information & communication technology (ICEEICT)

Chapter 29
IBM Watson Assistant and Node-RED-Based Movie Ticketing Bot Design

Arpita Ghosh

1 Introduction

The chatbots [1–6] or software-based conversational agents are gaining popularity. There are several available chatbots, among them, a companion for dementia patients was introduced by the Russian technology company, endurance. Casper introduced chatbots for the insomniac people to get through the night. The chatbot is named Insomnobot 3000 [7]. The nonprofit organization UNICEF uses chatbots for helping people living in developing nations raise their voice regarding their most urgent needs. The bot called U-Report [8] emphasizes largescale data gathering via polls. A chatbot called Med What [9] is used to make the medical diagnoses faster. Ticket booking system for the cinema halls still uses human agents for answering the repetitive queries of the customers and guide them to book their tickets. In this system, the customer has to wait in a queue for an indefinite time based on the number of customers already present in the queue before him/her. Even per customer time requirement is also not fixed that is why the delay during the peak hours is unpredictable and even in busy schedule it depends upon the efficiency of the agents. Moreover, it has a fixed operation time and customer has to be physically present. The proposed system with a movie ticketing bot eliminates the problems associated with onsite movie ticket booking. They can book their movie tickets from anywhere using internet-enabled devices. It is easy to access, easy to get the information related to the show timing, movie list, seat categories, seat availability, etc. Moreover, in this system different add-on order facility has been also incorporated along with the movie ticket booking. They can even choose the mode of payment. The movie ticketing bot is nothing but a simple chatbot which has all the information related to the name or list of movies, their show timings, seat availability details, prices, etc. It can be trained with all the necessary details for booking a movie ticket to select the seat category, select different add-ons to choose the payment modes while paying for the

A. Ghosh (✉)
RCC Institute of Information Technology, Kolkata 700015, India

S. Dawn et al. (eds.), *Smart and Intelligent Systems*, Algorithms for Intelligent Systems,
https://doi.org/10.1007/978-981-16-2109-3_30

tickets. It is a kind of question–answer session initiation between the trained chatbot and the customer.

The purpose of designing this movie ticketing bot is mainly to save the time from the customer point of view as well as the service provider end. Even during the peak hours of day, weekends, and holidays, this movie ticketing bot can work effortlessly without any exhaustion. It reduces the chance of confusion while placing the order and customers can repetitively check the desired query responses. Even it can handle more number of customers within a time span. The cost of the service provider end can be reduced by reducing the number of human agents. Moreover, it provides easy access as it can be used from any electronic gadgets such as mobile phones, tabs, laptops, desktops with internet connectivity.

2 Theoretical Analysis

A movie ticketing chatbot using the Watson Assistant [10] is designed in this work. The designed chatbot is capable of showing the list of different movies in the multiplex. Here, there are three categories of movies mainly—Hindi movies, English movies, and regional movies. Under the Hindi movies category, three movies are enlisted such as Three idiots, Border, and Swades. Similarly for English movies, ET and Step Up 2 movies are included in the list. Among the regional type Bengali, Tamil, and Gujarati options are there. After viewing the list of the movies with categories, customer can select a specific movie to know the details such as show timing, price during the weekdays as well as during weekends. Three seat categories are included such as silver class, gold class, and the platinum class. While checking the details of a specific movie by clicking on it the customer can get the idea of the price on weekdays and weekends as per the seat category as well. The availability of the total number of the seats as per category can be also viewed while clicking on the specific movie. There is another option for the customer to choose different add-ons such as popcorn, ice-cream, and cold drinks. By clicking on the specific add-on they can also check the price details. Finally, the chatbot is capable of booking the ticket for the specific movie and timing with preferred seat category along with the add-ons. Different payment options are also included as a part of the chatbot functionality, so that customer can choose among the different payment modes.

3 Block Diagram

The overall architecture of the movie ticketing chatbot development is shown in the block diagram below (Fig. 1). Mainly, the Watson Assistant is required to develop the functionalities of the chatbot using different entities, intents, and dialog. Watson Assistant is a conversation AI platform. It provides customers straightforward, fast, as well as accurate answers to their queries, across any application or device. It reduces

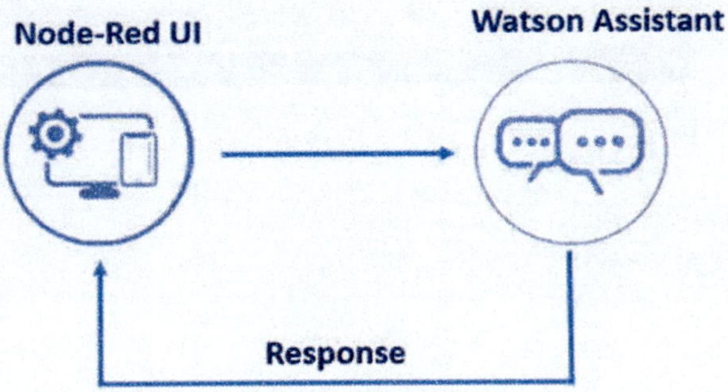

Fig. 1 Block diagram

the burden on the customer care executives or agents and handles the customer traffic in a smooth way without any disruption even during the peak times. The exhaustion due to repetitive queries can be handled and provides customer support 24 × 7. Watson Assistant is smarter than a conventional chatbot, where human emotion mimicking creates misunderstanding, as it is capable of searching for an answer from a knowledge base. It knows when to ask for clarity and when to forward the query of users to a human. It is possible to deploy it in any cloud or on-premises environment.

Node-RED [11], developed by the Emerging Technology Services team of IBM now a part of JS foundation, is a flow-based programming tool. The flow-based programming is a way of describing an application's behavior as a network of black-boxes, or "nodes". Each node with a well-defined purpose can take some data, processes it and passes it on. The flow of data among the nodes is due to the network. Node-RED comprises of a Node.js-based runtime so that a web browser can be pointed to access the flow editor. It is possible to create applications within the browser by dragging nodes from palette to the workspace and they can be wired together. A single click is required to deploy back to the runtime where it is run. Final integration of Watson Assistant has been done with the Node-RED UI for generating the URL.

4 Software Part

The core part of this work is to build the chatbot for booking movie ticket. The software part is mainly designed with the Watson Assistant. The intents and then the entities are created for the chatbot. Figure 2 shows the created intents #booking, #enquiry, and #greetings.

In Fig. 3, the generated entities are shown in the screenshot such as @add_on, @booking, @email, @enquiry, @movies, @payment, @seat_availability, @seat_category, and @show_time. The associated dialog is created and in each step the outputs are checked. The dialog part is shown in Fig. 4.

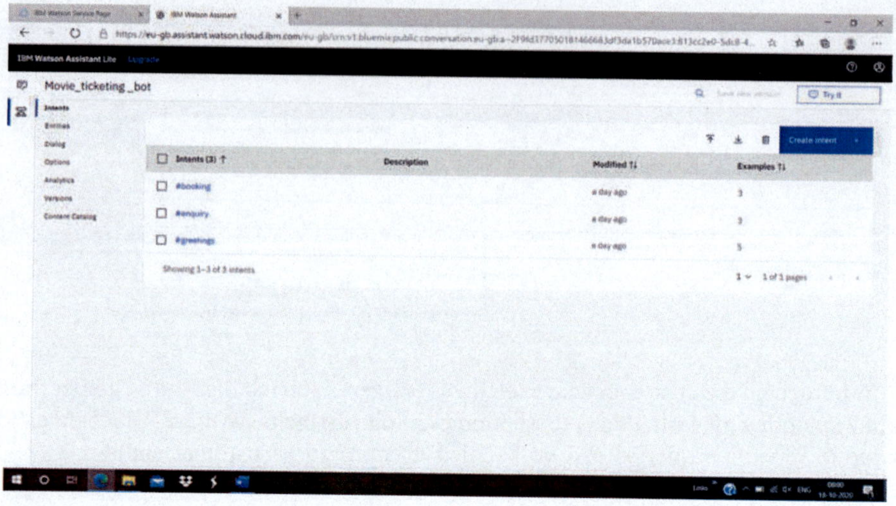

Fig. 2 Intent creation

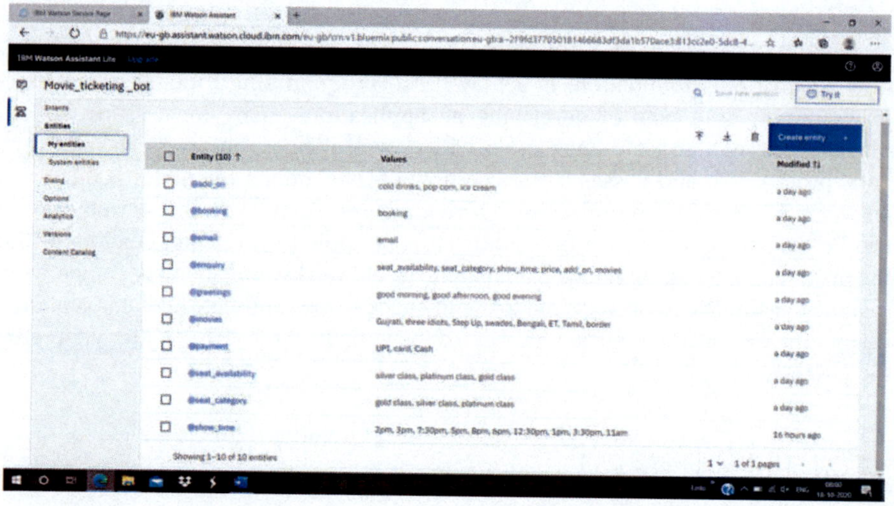

Fig. 3 Creation of Entities

The nodes generated in the dialog part are welcome, greetings, enquiry, booking, and anything else. The child nodes created under node welcome is mail id, under enquiry it is movies. The individual responses are configured as per the requirements. After successful completion of dialog generation part Node-RED instance is created. The Node-RED instance is then integrated with the Watson Assistant in this case the Movie_ticketing_bot. The successful integration generates the preview link, which can be accessed from any internet-enabled device for booking tickets as shown in Fig. 5.

The flowchart of the complete process of developing the movie ticketing chatbot is shown in the flowchart below in Fig. 6.

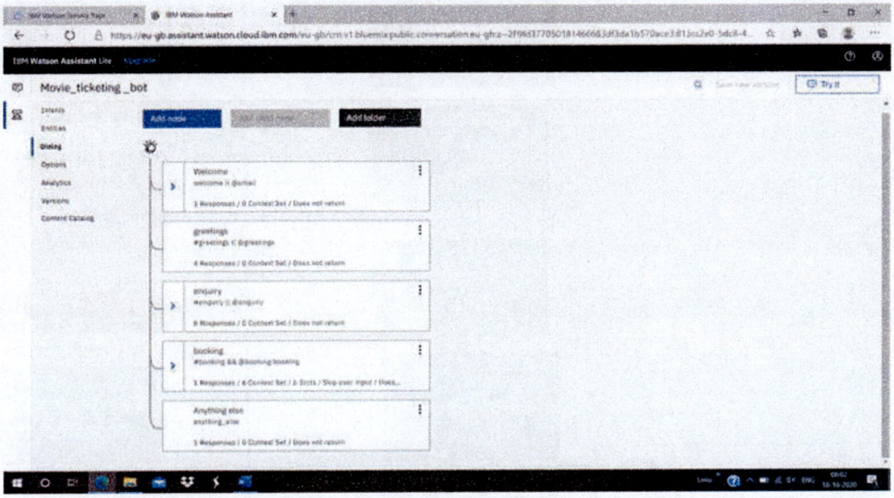

Fig. 4 Defining dialogs

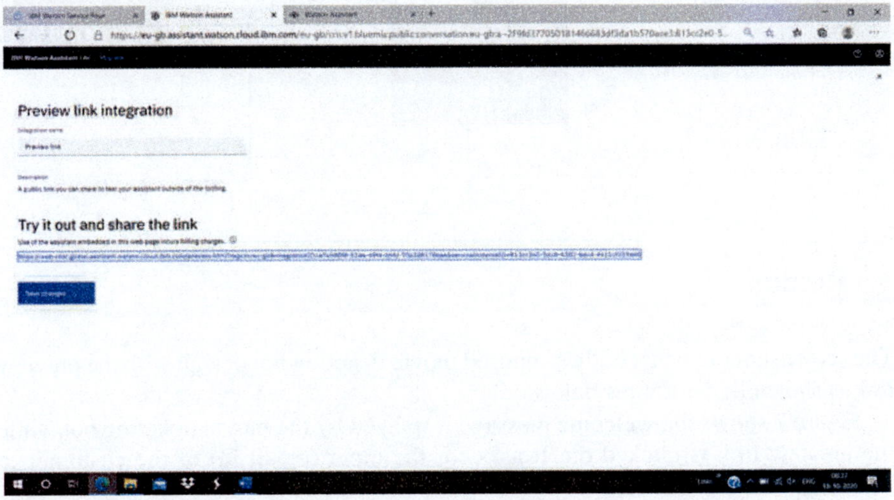

Fig. 5 Preview link generation after successful integration

Fig. 6 Flowchart of the entire movie ticketing bot design process

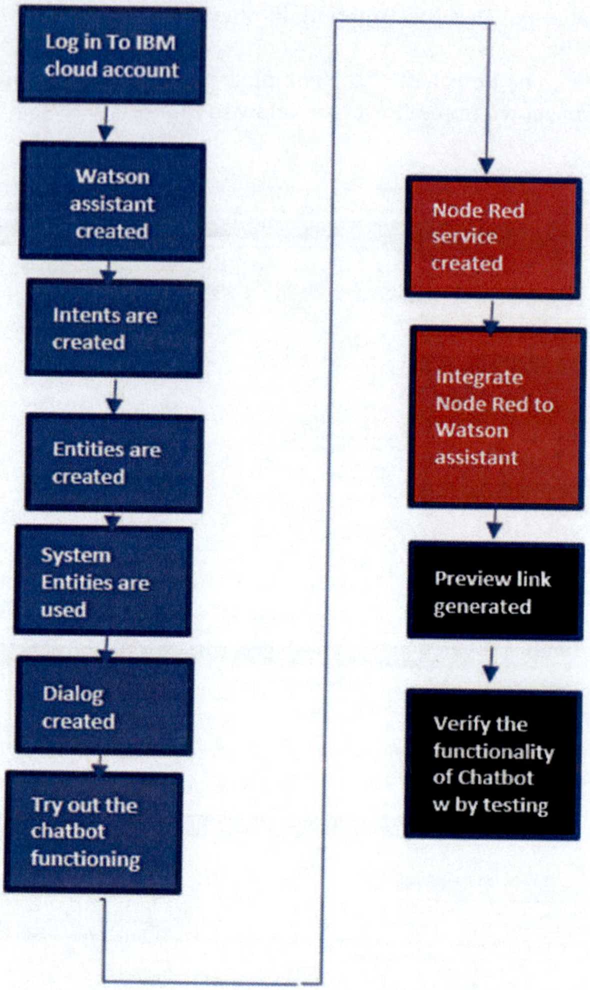

5 Results

The screenshot of results of the proposed movie ticketing bot design with the preview link is shown in the figures below.

Figure 7 shows the welcome message displayed by the movie ticketing bot while the preview link is clicked on. It asks for the input (email id) of the customer to initiate the conversation flow.

As a response, whenever the customer enters the email id it shows the next message. Figure 8 also shows the continuous flow of queries and responses between customer and the chatbot and as a response chatbot greets the customer and shows

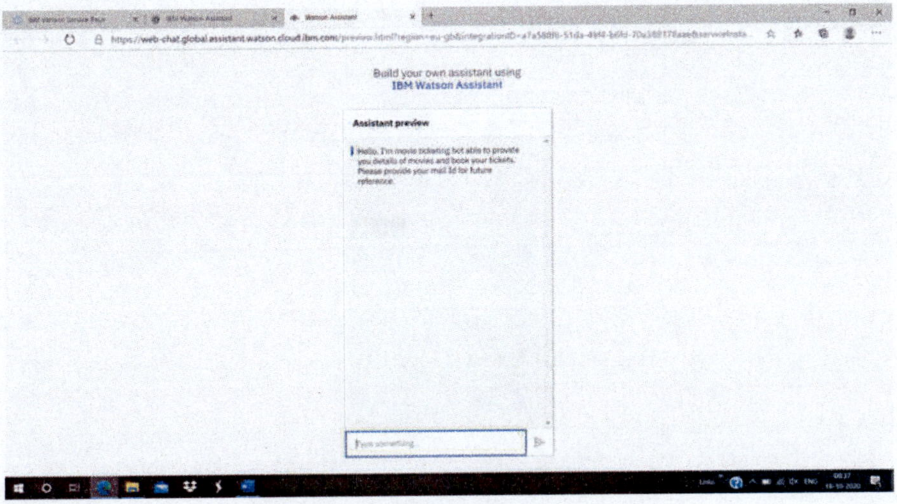

Fig. 7 Welcome message upon opening the chatbot link

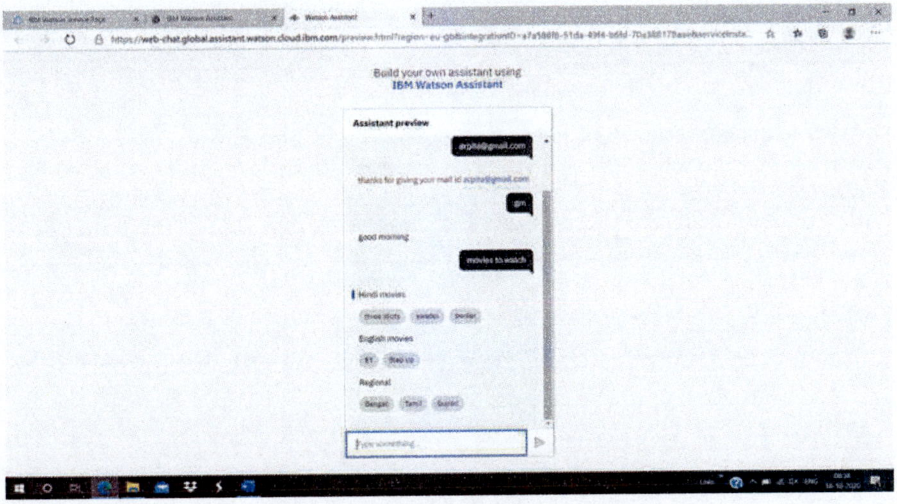

Fig. 8 Continuous flow of responses of the movie ticketing bot and customer conversation

the list of movies. Whenever the customer clicks on the specific movie, seat price, timing, and availability is displayed by the bot (shown in Fig. 9).

Finally, the order or booking of the tickets with movie name, timing, number of tickets, seat category, add-on, and payment mode is stored by the bot and the ticket is booked (shown in Fig. 10).

Result of customized web chat is shown in Fig. 11.

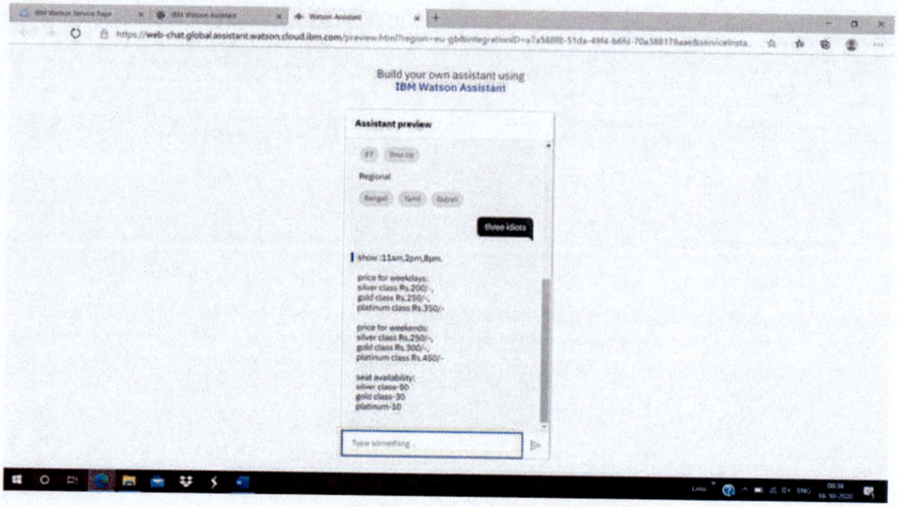

Fig. 9 Responses of the movie ticketing bot upon customer query

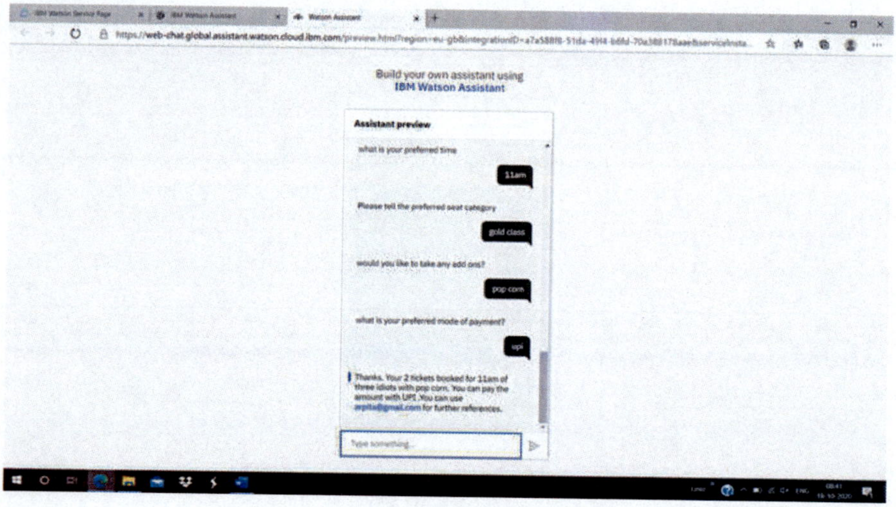

Fig. 10 Order completion message by movie ticketing bot

6 Advantage and Disadvantage

The advantages of the designed chatbot is it reduces human effort. It reduces the time constraint from the service provider point of view. It can handle the customer traffic smoothly during peak hours as well. It is easy to use and the customer can decide by viewing the list of movies, their show timings, seat category, and seat availability. They can even add some add-ons to their ticket booking as per their

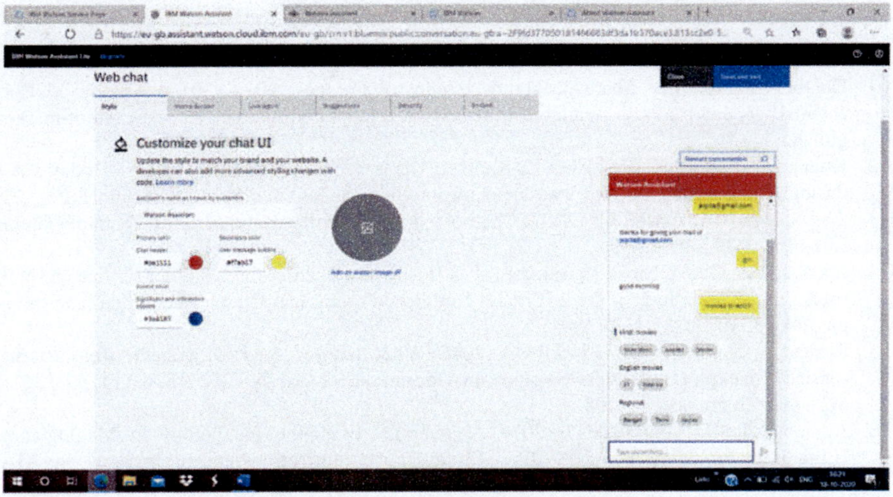

Fig. 11 Customization options available for web chat

choice by checking the list of available add-ons and their respective prices. The provision of payment options selection is also available to the customer.

7 Conclusion and Future Scope

The presented work describes the steps to develop the movie ticketing bot system with IBM Watson Assistant and Node-RED UI service. The screenshots of the steps involving the stage-by-stage development of the chatbot is provided. The Node-RED service has been created and finally the integration of the skill (movie_ticketing_bot) is successfully done to get the preview link. The successful execution of the complete procedure can be verified by the screenshots of the ultimate movie ticketing bot, responding to the customer in a proper manner provided in the result section.

As a future scope to this work, many features can be added to make the movie ticketing bot more convenient to use. The cancelation of ticket option can be added to make it more user-friendly. The genre or overview of the movie can be added in the future to help the customer decide which movie to see. Voice integration of the chatbot would have been another milestone to achieve by using the text-to-voice converter. Autoupdatation facility of the tickets on seating class basis as per the customer request can be added as another feature. Reminder facility with SMS and the email id can be also integrated to the system.

Acknowledgements The author is thankful to Mr. Hemant Kumar Gahlot from Smartbridge and also IBM for all the support.

References

1. Chatbot Market Size Share & Trends | Industry Forecast 2022 (2016) Arizton advisory & intelligence. [online] Available https://www.arizton.com/market-reports/chatbot-market-global.
2. Khanna A, Pandey B, Vashishta K, Kalia K, Bhale P, Das T (2015) A study of today's A.I. through chatbots and rediscovery of machine intelligence. Int J u- e-Serv Sci Technol 8:277–284
3. Abu Shawar BA, Atwell ES (2007) Chatbots: are they really useful? J Lang Technol Comput Linguist 22:29–49
4. Xu A, Liu Z, Guo Y, Sinha V, Akkiraju R (2017) A new chatbot for customer service on social media. In: Proceedings of the 2017 CHI conference on human factors in computing systems, pp 3506–3510. ACM, New York
5. Følstad A, Nordheim CB, Bjørkli CA (2018) What makes users trust a chatbot for customer service? An exploratory interview study. In: Bodrunova SS (ed) INSCI 2018, vol 11193. LNCS. Springer, Cham, pp 194–208
6. Adamopoulou E, Moussiades L (2020) An overview of chatbot technology. In: Maglogiannis I, Iliadis L, Pimenidis E (eds) Artificial intelligence applications and innovations. In: AIAI 2020. IFIP advances in information and communication technology, vol 584. Springer, Cham
7. Insomnobot 3000. https://insomnobot3000.com/?source=post_page-
8. U-Report.https://www.unicef.org/innovation/ureportCOVID19
9. Med What. https://medwhat.com
10. IBM Watson. https://www.ibm.com/watson
11. Node-RED . https://nodered.orgBaldonado. Chang M, C-CK, Gravano L, Paepcke A (1997) The Stanford digital library metadata architecture. Int J Digit Libr 1:108–121

Chapter 30
A Comparative Analysis of IC and RCC MPPT Techniques for High-Power PV Systems

Pankaj Sahu and Rajiv Dey

1 Introduction

With the increase of huge demand for energy, and due to the fast and continuous depletion of conventional resources sustainable energy resources demand is increasing with time. Among all these sustainable energy resources, solar energy is preferred widely and which are capable to substitute these conventional energy sources. It is one of the most sustainable sources of renewable energy and is harvested using photovoltaic systems; these systems are environment-friendly. This green energy harvesting also suffers some drawbacks, like change in the level of solar and change in temperature degrades the power transfer to load. Also, in partly cloudy weather conditions, its performance becomes poor. Due to which the power produced by solar panels becomes non-uniform with time and hence each and every time the MPP varies [1]. PV array can transfer maximum power to any utility only at a fixed set of values of voltage and current, which is called Maximum Power Point (MPP).

The per watt production pricing of solar energy is very high, due to the limitations of material efficiency and losses, hence to improve its efficiency DC/DC converters and charge controllers are used. These DC/DC converters have many returns as, they provide isolation, effective in tracking maximum power point, and hence maximum power extraction. Also, their control can be made simple. Power electronic converter systems are used to deliver maximum power and the switching of these converters is controlled by the duty-cycle generated by the MPPT algorithm. Among, two available choices for charge controlling; (Maximum Power Point Tracking (MPPT) is preferred in Photovoltaic systems since a typical Pulse Width Modulation (PWM) controller can only have control on the voltage of a PV array, but not on current, whereas, MPPT controller has a capability to regulate both (Fig. 1).

There were many algorithms that have been presented in the literature of MPPT such as, Perturb and Observe (P&O) [1–12], Incremental Conductance (IC) [12–16],

P. Sahu (✉) · R. Dey
SoET, BML Munjal University, Gurugram, India

S. Dawn et al. (eds.), *Smart and Intelligent Systems*, Algorithms for Intelligent Systems,
https://doi.org/10.1007/978-981-16-2109-3_31

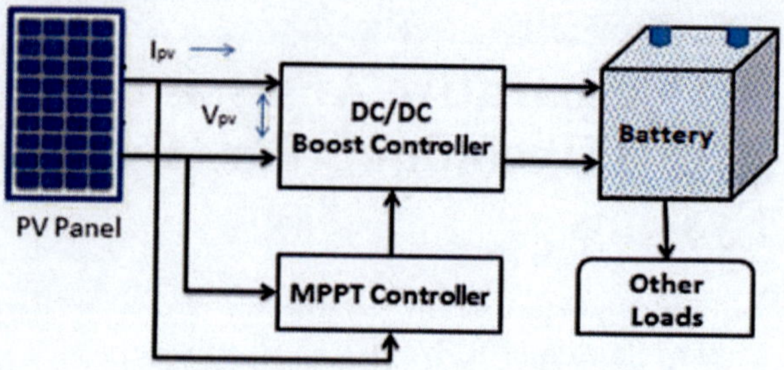

Fig. 1 Block diagram of PV system

Hill Climbing (HC) [17, 18], Fractional Open-Circuit Voltage (FOCV), Fractional Short-Circuit Current (FSCC) [19–25], Fuzzy Logic [26], Particle Swarm Optimization (PSO) [27], Ripple Correlation Control (RCC) [28–32] with their modified versions like fixed and variable step size P&O and IC, etc. The main shortcomings of conventional algorithms like HC, P&O, and IC are they are incapable of tracking MPP under varying environmental conditions, i.e., change in irradiance or temperature. The other major drawbacks are being slow tracking in terms of speed, poor convergence rate, with high amount of oscillations in steady state. Therefore, for tracking MPP under varying environmental circumstances, conventional methods should be improved or essentially be combined with other methods for better performance [4, 12, 30]. All later modified technique improves drawbacks of previous, but still, there is a need for a more accurate method for MPPT. It gives inspiration to the research community to find some better algorithms and techniques to improve this problem.

"Ripple Correlation Control (RCC) [28, 29] is a nonlinear MPPT technique, which provides a way to solve this problem of steady-state ripples, which are engendered due to the use of power electronic elements. As P&O, IC, and hill climbing, all these techniques do not bother about these ripples and ponder these ripples as exterior disturbances, which in turn make output oscillatory around MPP, which in turn creates oscillations in output current and voltage, and hence in their product power. These oscillations are harmful to the load and for system's stability (Figs. 2 and 3).

There are many alternative strategies that also evolved, like Soft Computing methods, optimization techniques to solve multiple MPP point problems. "RCC technique emerged and has been well accepted by the researchers, due to its ripple removal characteristics, ease of implementation, and robustness properties. This work presents a simulation analysis of IC and RCC MPPT techniques with boost converter for ripple-free high-power requirements and speedy tracking requirements for all types of environmental conditions" [33].

Fig. 2 P–V and I-V curves

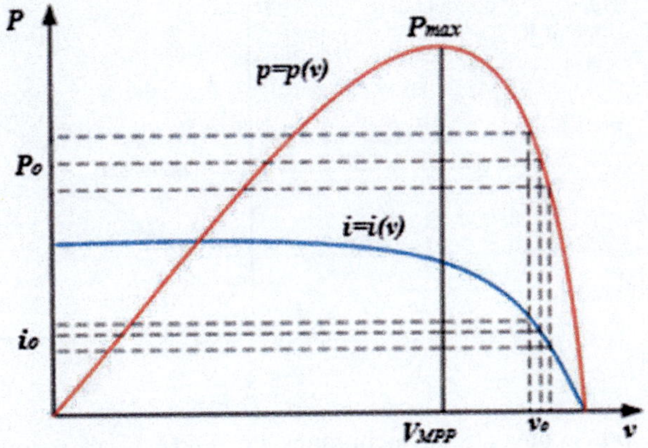

Fig. 3 I-V and P–V curves of the PV panel used in simulation

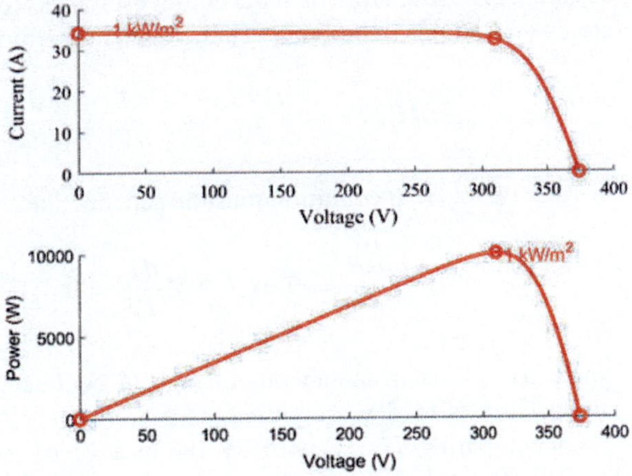

2 Background

This section describes a detailed overview of both MPPT techniques used for comparison, i.e., IC and RCC, along with their formulations.

2.1 *Incremental Conductance (IC)*

Incremental Conductance is also known as, conventional MPPT technique. It is better than Perturb and Observe (P&O) to locate the MPP [11–16]. "In this technique association of voltage and power is utilized to find out the derivative of power with respect to voltage for zero condition. The main weakness of this technique is that it cannot reduce the computation time, also complexity along with the cost is too high." Like,

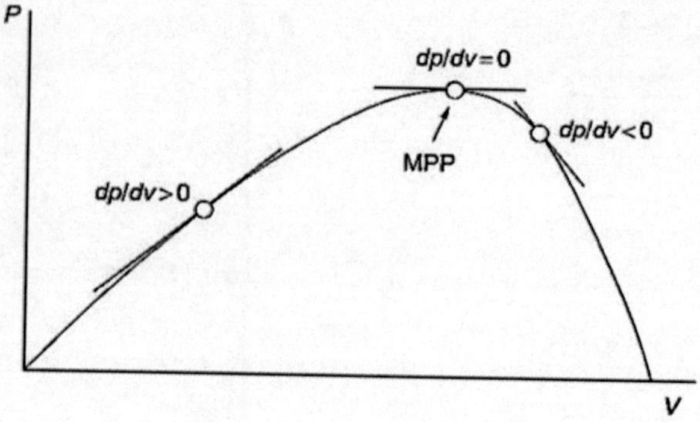

Fig. 4 P–V conductance slope of IC

other conventional techniques, i.e., P&O, IC also cannot discriminate external perturbations with inherently generated ripples, which are produced by power electronic circuitry. As a result of which the output oscillates around MPP to attain MPP, which create oscillations in output, even there are no perturbations outside (Fig. 4)."

$$dP/dV = \frac{d(VI)}{dV} = I\frac{dI}{dV} + V\frac{dI}{dV} = I + V\frac{dI}{dV}. \tag{1}$$

At MPP, $\frac{dP}{dV} = 0$ condition must be satisfied, i.e.,

$$\frac{dP}{dV} = 0 \Rightarrow I + V\frac{dI}{dV} = 0 \Rightarrow -\frac{I}{V} = \frac{dI}{dV} \tag{2}$$

where dI/dV is instantaneous change in conductance and I/V is instantaneous conductance of the PV array used.

On comparing dI/dV with I/V, the location of the operating point with respect to MPP can be discovered. Equation 3 gives the location of the operating point for all possible values of dP/dV.

$$\begin{cases} \frac{dP}{dV} > 0,\ for\ V < V_{MPP} \\ \frac{dP}{dV} = 0,\ for\ V = V_{MPP} \\ \frac{dP}{dV} \langle 0,\ for\ V \rangle V_{MPP} \end{cases} \tag{3}$$

2.2 Ripple Correlation Control (RCC)

"It is a nonlinear control technique, which is pertinent to power electronic circuits. It basically utilizes ripple content of voltage or current with power and performs correlation of this with ripple components of power, for effective control of switching [33]. The RCC provides solutions to problems for the systems having non-linearity, with less complex implementation. Ease of implementation, with fast computation time, are some motivations to use this technique. This technique does not require any exterior perturbation, as required in conventional MPPT techniques like HC, P&O, and IC. RCC is a technique that is used to compute the duty cycle of the switching element, through which maximum power can be obtained. The major benefit of RCC is, it utilizes inherent ripples that appear due to the involvement of power electronic elements and cannot become a cause at the output, like in conventional MPPT techniques. By the means of correlation of time-based derivative of power and voltage or current, the RCC attempts to check whether the correlation resultant is as shown in Eq. 4 [33];

$$\frac{dppV}{dt}\frac{dvpV}{dt} > 0\, when\; V_{pv} < V_M, \tag{4}$$

$$\frac{dppV}{dt}\frac{dvpV}{dt} < 0\, when\; V_{pv} > V_M, \tag{5}$$

$$\frac{dppV}{dt}\frac{dvpV}{dt} = 0\, when\; V_{pv} = V_M. \tag{6}$$

The basic equation which is used to find the ripple components is given as

$$\tilde{x}\ (t) = x(t) - \bar{x}\ (t) \tag{7}$$

where $x(t)$ is a quantity that can be either power, voltage, or current and comprises both moving average component as well as the ripple. $\tilde{x}\ (t)$ is the ripple component and $\bar{x}\ (t)$ is the average component of either power, voltage, or current. Ripple content can be found by using Eq. 7.

In Fig. 5, the architecture of the RCC MPPT block is shown. The objective of this RCC block is to find out the optimum duty cycle to regulate the switching of the MOSFET through the gate terminal of the boost converter. This architecture has low-pass filter implementation, as in [32, 33] and differs from [28, 29], where high-pass filter implementation has been done. In this architecture, error is the product of

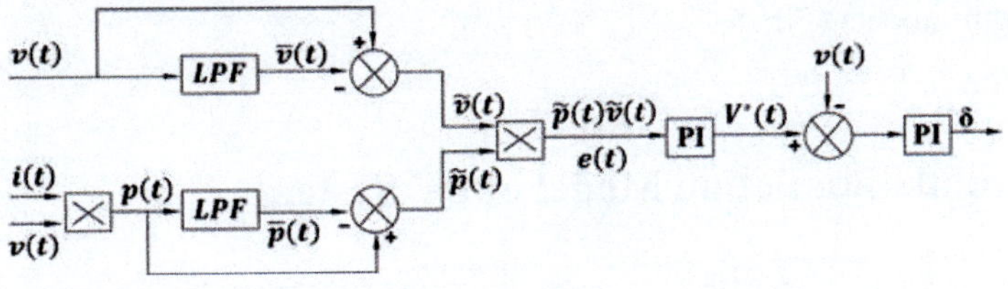

Fig. 5 RCC architecture block [32]

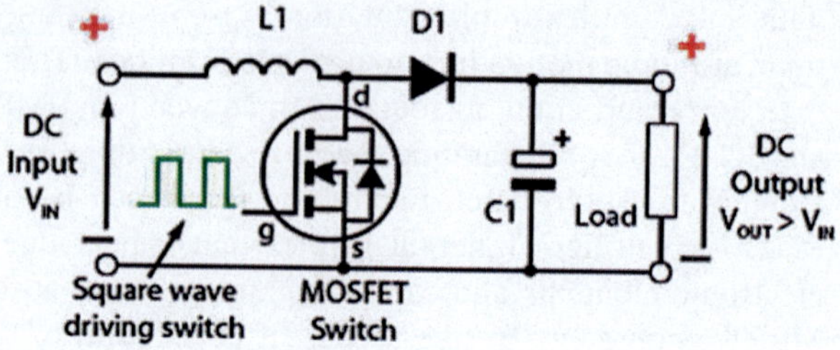

Fig. 6 Electrical equivalent of DC/DC Boost Converter

the ripple components of both the voltage and power. The final equation of the error $e(t)$ in RCC can be expressed as

$$e(t) = \tilde{p}\,(t) \times \tilde{v}\,(t) = \tilde{v}^2(t)\left[\frac{dp(t)}{dv(t)}\right] + \tilde{v}^2(t)\,\tilde{i}\,(t) \tag{8}$$

where $v(t)$ is PV voltage, $i(t)$ is PV current, $p(t)$ is PV power, $\bar{v}\,(t)$ is filtered DC voltage, $\bar{p}\,(t)$ is filtered DC power, $\tilde{v}\,(t)$ is voltage ripple, $\tilde{p}\,(t)$ is power ripple, $V^*(t)$ is reference voltage obtained through the first Proportional–Integral (PI) controller, and δ is optimal duty cycle obtained through the second Proportional–Integral (PI) controller. Since error is directly proportional to the magnitude of dp/dv as the avg. value $\tilde{v}^2\,(t)\,\tilde{i}\,(t)$ is zero over a cycle, $e(t)$ represents the distance from the MPP.

If o/p is $==$ Left of MPP $\rightarrow$ Avg. value of error is $+$ ve.
If o/p is $==$ Right of MPP $\rightarrow$ Avg. value of error is $-$ ve.
If o/p is $==$ At MPP $\rightarrow$ Avg. value of error is 0.

2.3 DC–DC Converters

The DC–DC voltage converter is a power electronic device, which is used for conversion of one level of DC to some other level DC, with the regulation in current. There are three distinguished types of converters available, in accordance with the requirements, these are, buck–boost converter, buck and boost converters. The main function of the boost converter is to increase the voltage in accordance with the need for reduction in the current. In this work, a boost converter is utilized, due to high voltage and power requirements (Fig. 6).

3 Simulink Simulation Model of PV System

Figure 7 shows the photovoltaic system consisting of a PV panel that is connected to a boost converter with a subsystem for IC and RCC MPPT controller selection, for

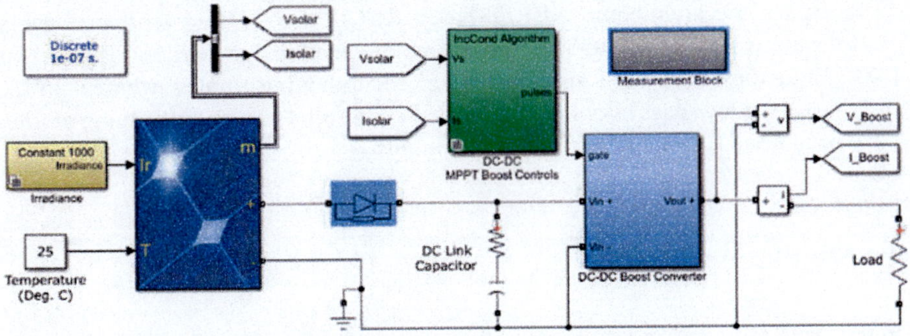

Fig. 7 Photovoltaic system model used for simulation

Table 1 Simulation parameter specifications

S. no	Parameter name	Value
1	Number of parallel strings	4
2	Series-connected modules per string	10
3	Open-circuit voltage	$V_{oc} = 37.4$ V
4	Short-circuit current	$I_{sc} = 8.55$ A
5	Series resistance of PV model	$R_s = 0.2351\ \Omega$
6	Parallel resistance of PV model	$R_p = 287\ \Omega$
8	DC link capacitor	$C_{dc} = 100\ \mu$F
Boost converter parameters		
9	Capacitance	$C = 120\ \mu$F
10	Inductance	$L = 0.01$ H
11	Switching frequency	$f_{sw} = 10^6$
12	Sample time	5e-6 s
RCC block parameters		
13	Time constant of low pass filters	7.5 ms
14	First PI controller (within MPPT block)	$K_p = 200$ $K_I = 5.5$
15	Second PI controller (outside MPPT block)	$K_p = $ 2e-9 $K_I = -0.009$

450 V output requirements, with a load of 20Ω resistance. In simulation Simulink, inbuilt PV panel Trina Solar TSM-250PA05.15 has been used. All specifications used for simulation are tabulated in Table 1.

In Fig. 5, the architecture of LPF-based RCC [33] has been presented, which controls the switching of MOSFET. RCC implemented in this work is different from [28–32], as they have used high-pass filters for removing ripples, whereas in [32] they have used low-pass filters. In this work, PI controllers along with tuning of filters have been done for optimum ripple-free outputs.

4 Simulation Results

In this work, simulation results are obtained for both IC and RCC MPPT for various levels of solar irradiance, i.e., 800 W/m^2, 1000 W/m^2, and for step response as shown in Fig. 8. This step response represents the partly cloudy weather condition, as the irradiance is not constant, when weather is cloudy, the irradiance level gets changes. This is for representation of variable irradiance conditions, i.e., cloudy weather conditions. Most of the MPPT techniques fail under variable environmental conditions. Simulation results for both IC and RCC MPPT are shown in voltage, current, and power for different solar irradiances. For comparison purposes outputs of both the MPPT techniques are shown in a single graph representation.

First, the simulations have been performed for steady-state weather conditions, i.e., at 800 W/m^2 and 1000 W/m^2. Then, after simulation results have been obtained for partly cloudy weather conditions, i.e., step irradiance. In this work, standard temperature has been taken for all simulations, which is 25 °C. All the results have been enlarged within Simulink itself, to show the oscillations properly.

In Figs. 9, 10, 11, 12, 13 and 14, outputs of IC and RCC MPPT techniques for voltage, current, and power have been presented for 800 W/m^2 and 1000 W/m^2, which are obtained by simulation using the model shown in Fig. 7. These results show the

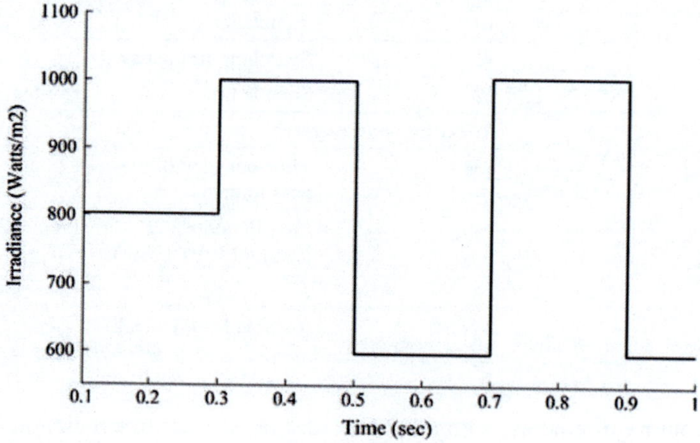

Fig. 8 Step response used in simulation

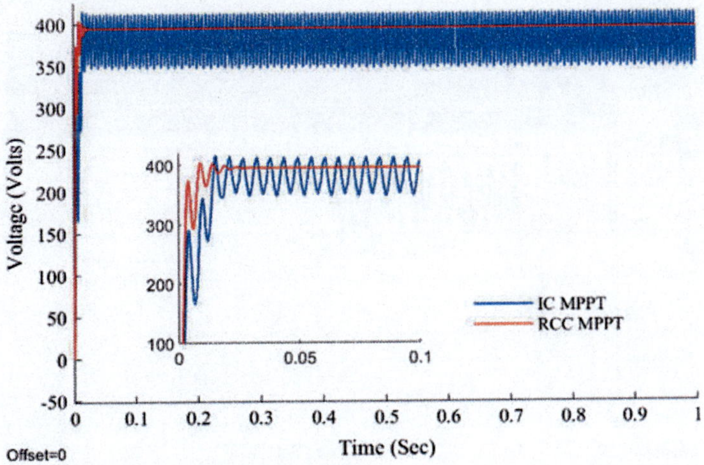

Fig. 9 Voltage output of IC and RCC at 800 W/m^2

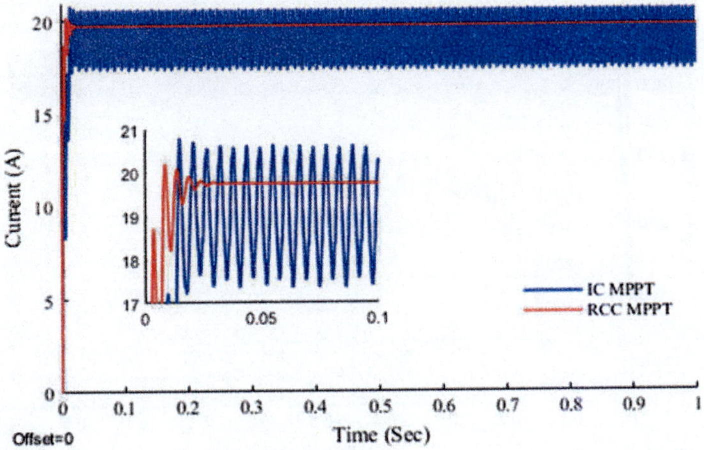

Fig. 10 Current output of IC and RCC at 800 W/m^2

steady-state environment condition. From these figures, it can be seen that all the outputs of the IC technique contain ripples or oscillations, which appear mainly due to inherently produced ripples. Conventional techniques like P&O and IC take these inherently produced ripples as external perturbations and tend to shift MPP each and every time. These inherent ripples are appearing due to the use of power electronics converters. Figures 11 and 14, i.e., power outputs at constant irradiance, clearly show the delayed response of IC, compared to the RCC technique.

Similarly, Figs. 15, 16 and 17 show the outputs of IC and RCC MPPT techniques for cloudy weather conditions, i.e., for step change in irradiance. From the figures,

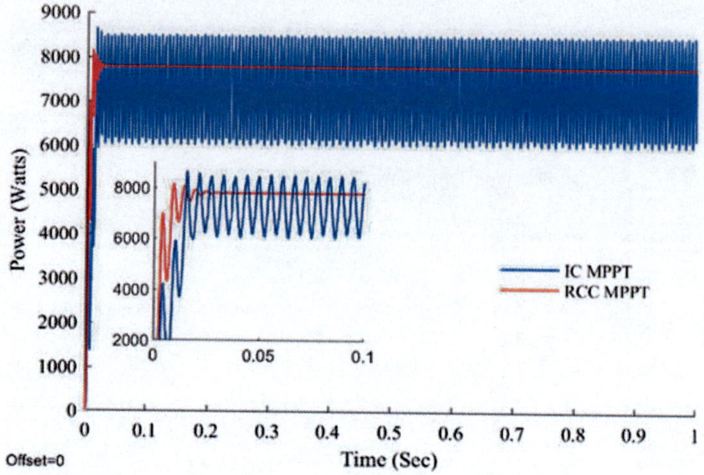

Fig. 11 Power output of IC and RCC at 800 W/m²

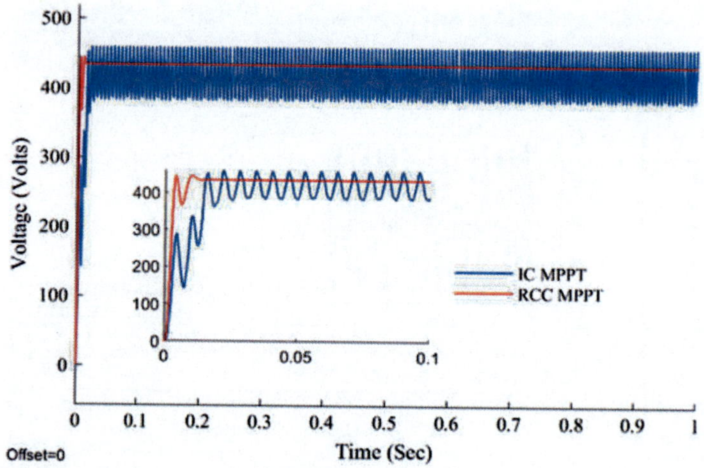

Fig. 12 Voltage output of IC and RCC at 1000 W/m²

it can be analyzed that outputs of the IC technique contain ripples or oscillations, which is the known problem of the IC technique, whereas outputs of the RCC MPPT are free from such ripple contents. From the simulations, it is observed that the IC technique suffers from slow tracking, when irradiance is changed, the output takes some time delay to follow the input, whereas, RCC MPPT is fast-tracking technique. Figure 17 at variable irradiance clearly shows fewer oscillations in power output in RCC MPPT as compared to IC MPPT, during transitions.

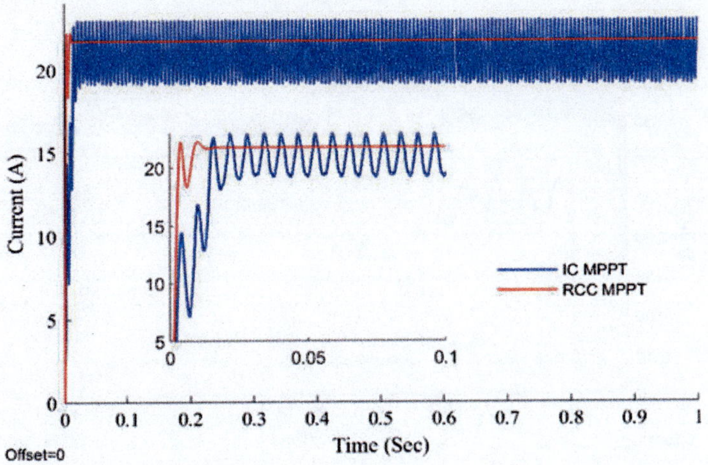

Fig. 13 Current output of IC and RCC at 1000 W/m^2

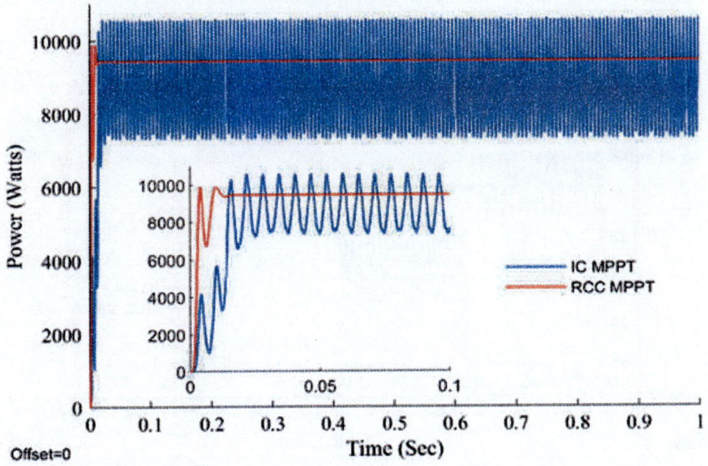

Fig. 14 Power output of IC and RCC at 1000 W/m^2

5 Conclusion

In this work, a comparative simulation analysis of Incremental Conductance (IC) with Ripple Correlation Control (RCC) MPPT techniques have been presented. Photovoltaic array along with the boost converter and its MPPT control have been implemented with both the MPPT techniques. The system has been designed for 450 V, 10 kW power output, at 1000 W/m^2 irradiations, 25 °C temperature (STC), which is attained through DC/DC converter, i.e., boost converter. From these figures, it can be seen that all the outputs of the IC technique contain ripples or oscillations,

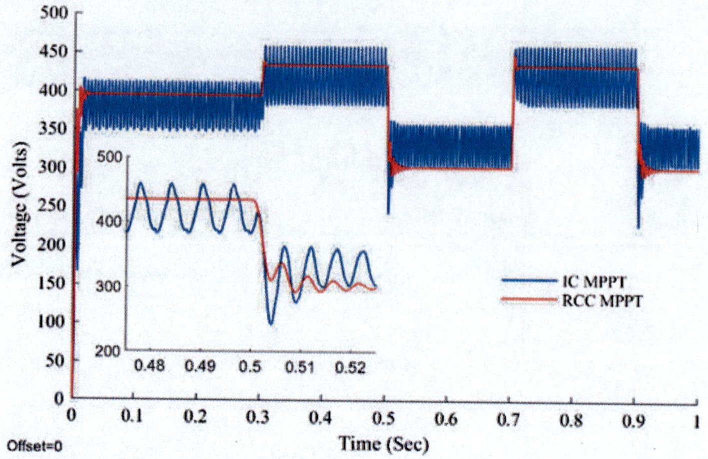

Fig. 15 Voltage output of IC and RCC for step irradiance

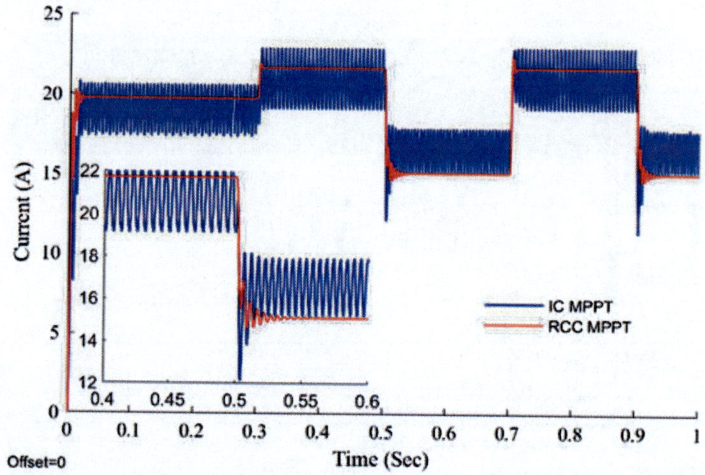

Fig. 16 Current output of IC and RCC for step irradiance

which appear mainly due to inherently produced ripples. Conventional techniques like P&O and IC take these inherently produced ripples as external perturbations and tend to shift MPP each and every time. Basically, these ripples are part of the system, due to power electronics circuitry involved. Also, IC shows poor tracking performance, whereas RCC removes all ripples from voltage, current, and power outputs, with good tracking performance, as RCC utilizes these internally generated ripples for stabilization of the system. RCC techniques perform better in steady-state conditions, but at transient conditions, it also suffers from oscillation problem, which can be further improved by using some suitable adaptive control strategy.

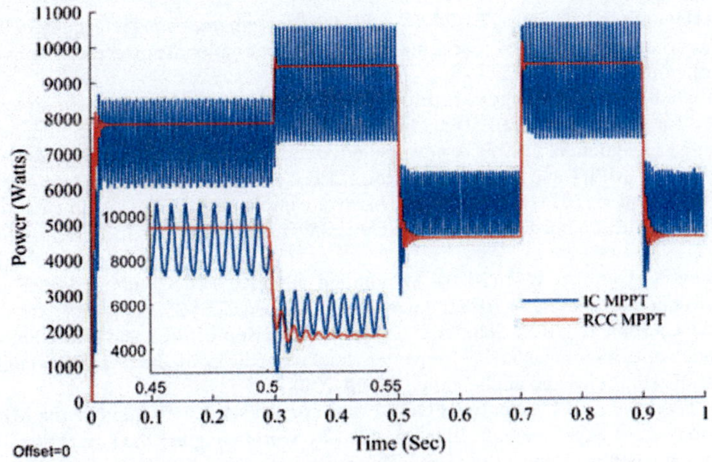

Fig. 17 Power output of IC and RCC for step irradiance

References

1. Nicola F, Giovanni P, Giovanni S (2005) Optimization of perturb and observe maximum power point tracking method. IEEE Trans Power Electron 20(4):963–973
2. Ashish P, Nivedita D, Ashok Kumar M (2008) High-performance algorithms for drift avoidance and fast tracking in solar mppt system. IEEE Trans Energy Convert 23:681–9
3. Abdelsalam AK, Massoud AM, Ahmed S, Enjeti Prasad N (2011) High-performance adaptive perturb and observe MPPT technique for photovoltaic-based micro grids. IEEE Trans Power Electron 26(4):1010–21
4. Mohammed A, Elgendy B, Zahawi, Atkinson David J (2012) Assessment of perturb and observe MPPT algorithm implementation techniques for PV pumping applications. IEEE Trans Sustain Energy 3(1):21–33
5. Viorel Banu I et al (2013) Comparative analysis of the perturb-and-observe and incremental conductance MPPT methods. In: 8TH IEEE international symposium on advanced topics in electrical engineering (ATEE)
6. Aureliano Gomes de BM, Luigi GJr, Poltronieri SL, e Melo Guilherme de A, Carlos Alberto C (2013) Evaluation of the main MPPT techniques for photovoltaic applications. IEEE Trans Indus Electron 60(3):1156–657
7. Ishaque Zainal SK, George L (2014) The performance of perturb and observe and incremental conductance maximum power point tracking method under dynamic weather conditions. Appl Energy 119:228–36
8. Sathish Kumar K, Mahesh Kumar M (2014) A novel adaptive PAO MPPT algorithm considering sudden changes in the irradiance. IEEE Trans Energy Convert 29(3):602–10
9. Elgendy MA, Bashar Z, Atkinson David J (2015) Operating characteristics of the PAO algorithm at high perturbation frequencies for standalone PV systems. IEEE Trans Energy Convert 30(1):189–98
10. Jubaer A, Zainal S (2015) An improved perturb and observe (PAO) maximum power point tracking (MPPT) algorithm for higher efficiency. Appl Energy 150:97–108
11. Salman S, AI X, Wu Z (2018) Design of a P-&-O algorithm based MPPT charge controller for a stand-alone 200W PV system. Springer Protection and Control of Modern Power Systems
12. Hanif Moin RR (2014) Design, testing and comparison of PAO, IC and VSSIR MPPT technique. In: Proceedings of the 3rd international conference on renewable energy research and applications, pp 19–22

13. HoBaeHyunSu LJ, Hyung CB (2006) Advanced incremental conductance MPPT algorithm with a variable step size. In. Proceedings of the 12th international power electronics and motion control conference. EPEPEMC
14. Liu Shanxu F, Fei D, Bangyin L, Liu, Yong Kang A (2008) Variable step size INC MPPT method for PV systems. IEEE Trans Ind Electron 55(7):2622–8
15. Jiyong L, Honghua W (2009) A novel stand-alone PV generation system based on variable step size INC MPPT and SVPWM control. IEEE Conf-IPEMC 2009, pp 2155–60
16. Azadeh S, Saad M (2011) Simulation and hardware implementation of incremental conductance MPPT with direct control method using Cuk converter. IEEE Trans Ind Electron 58(4):1154–1161
17. Weidong X, Dunford WG (2004) A modified adaptive Hill Climbing MPPT method for photovoltaic power systems. IEEE Power Electron Spec Conf 35:1957–63
18. Fangrui L, Yong K, Yu Z, Shanxu D (2008) Comparison of PAO and Hill Climbing MPPT methods for grid-connected PV converter. In: Proceedings of the 3rd IEEE conference on industrial electronics and applications, ICIEA 2008
19. Frezzetti A et al (2014) Adaptive FOCV-based control scheme to improve the MPP tracking performance: an experimental validation. In: 19th world congress the international federation of automatic control Cape Town, South Africa
20. Das P (2015) Maximum power tracking based open circuit voltage method for PV system. In: 5th international conference on advances in energy research, ICAER 2015
21. El Mentaly L et al (2017) Comparison between HC, FOCV and TG MPPT algorithms for PV solar systems using buck converter. International conference on wireless technologies, embedded and intelligent systems (WITS)
22. Bharath K et al (2017) Design and implementation of improved fractional open circuit voltage based maximum power point tracking algorithm for photovoltaic applications. Int J Renew Energy Res 7(3)
23. Baimel D et al (2019) Improved fractional open circuit voltage MPPT methods for PV systems. MDPI Electronics 2019
24. Nadeem A et al (2020) Online fractional open-circuit voltage maximum output power algorithm for photovoltaic modules. IET Renew Power Gener 14(2):188–198
25. Ganesh Moorthy J et al (2020) Performance analysis of solar PV based DC optimizer distributed system with simplified MPPT method. Springer
26. Ammirrul Atiqi Mohd ZM, Amran Mohd RM, Azura Che S, Nasrudin Abd R (2014) Development of adaptive perturb and observe-fuzzy control maximum power point tracking for photovoltaic boost dc–dc converter. IET Renew Power Gener 8(2):183–94
27. Mao M et al (2017) Maximum power point tracking for cascaded PV-converter modules using two-stage particle swarm optimization. Nat Sci Rep Article number 9381
28. Midya P et al (1996) Dynamic maximum power point tracker for photovoltaic applications. IEEE Power Electron Spec Conf
29. Krein PT (1999) Ripple correlation control, with some applications. IEEE international symposium on circuits and systems (ISCAS)
30. Khanna R et al (2014) Maximum power point tracking using model reference adaptive control. IEEE Trans Power Electron
31. Costabeber A et al (2014) Convergence analysis and tuning of a sliding-mode ripple-correlation MPPT. IEEE Trans Energy Convers
32. Satish R et al (2016) A maximum power point tracking technique based on ripple correlation control for single-phase single-stage grid connected photovoltaic system. Science Direct, Energy Proc
33. Sahu P (2020) Ripple correlation control maximum power point tracking for battery operated PV systems: a comparative analysis. IEEE international IOT, electronics and mechatronics conference (IEMTRONICS). Vancouver, BC, Canada, 1–6. https://doi.org/10.1109/IEMTRONICS51293.2020.9216414

Chapter 31
Development of Comprehensive Modelling and Simulation of Photovoltaic Module

S. Bhanu Prakash and Gagan Singh

1 Introduction

With the growth in power demand for industries and commercial uses, and also due to limited sources of fossil fuels the share of renewable power resources is growing in distribution systems. Renewable power resources mostly like solar PV power play a vital role in power generation. The choice of approximating the performance of the PV module with variant irradiance and temperature is very significant in developing a detailed model of the PV module. The non-linear behaviour of I-V characteristics requires the alteration of constraints by using experimental data. Some searchers propose a simplified model in which some constraints cannot be evaluated. For example, some authors do not consider both series and shunt resistance in the PV cell circuit model. One author considers that the shunt resistance value is very high while developing the PV module [1]. The equivalent PV cell circuit model by including the series resistance and by neglecting the shunt resistance has been modeled in [2]. Other authors considered these two internal parameters to have very significant role and need to be estimated appropriately. The key purpose of this paper is to develop a detailed model for single diode PV cell, which can be useful for those who works on the mathematical modelling of PV module.

S. Bhanu Prakash (✉) · G. Singh
Department of EE, DIT University, Dehradun, Uttarakhand, India
e-mail: sb.prakash@dituniversity.edu.in

G. Singh
e-mail: gagan.singh@dituniversity.edu.in

S. Dawn et al. (eds.), *Smart and Intelligent Systems*, Algorithms for Intelligent Systems,
https://doi.org/10.1007/978-981-16-2109-3_32

2 Modelling of PV Cell

An effective and exact model which fits the mathematical point to I-V curve for single exponential model and double exponential model for power electronic applications is described in [3–5]. In this paper, the detailed modelling of the single diode circuit model is described. Figure 1 illustrates the single diode PV cell equivalent circuit by neglecting both internal series and shunt resistance. By applying KVL we can get an expression for output current I:

$$I = I_{PH} - I_D \tag{1}$$

where I_{PH}—Photocurrent, I_D—Diode current and I—Module output current.

Diode Current

$$I_D = I_{DS}\left[\left(e^{\frac{qV_o}{CkT_A N_S}} - 1\right)\right] \tag{2}$$

where I_{DS}—Dark saturation current, q—Electron charge (1.602 X 10^{-19}) C, V_o—Voltage applied across the diode, C—Ideality factor, k—Boltzmann constant (1.38 X 10^{-23}) Joule/Kelvin, T_A—Absolute temperature in kelvin and V_T—Thermal voltage.

$$I_D = I_{DS}\left[\left(e^{\frac{V_o}{CV_T N_S}} - 1\right)\right] \tag{3}$$

where $V_T = \frac{kT_A}{q}$; V_T is approximately 25.856 mV at 300 K and N_s—series-connected PV cells.

In Fig. 2, the PV cell equivalent circuit with both internal resistances are shown. By applying KVL, the output current is obtained as

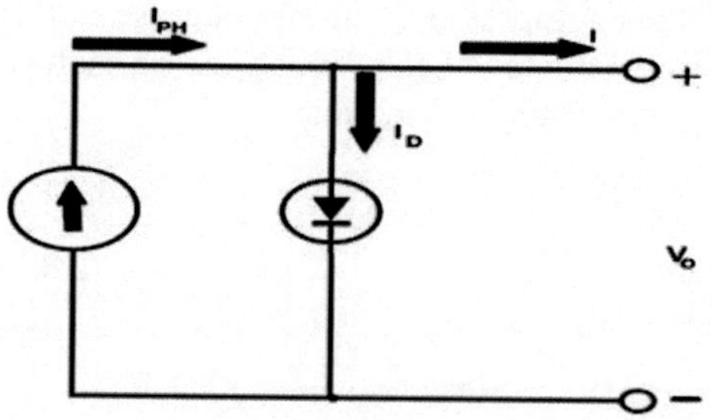

Fig. 1 PV Cell equivalent circuit model: ideal case

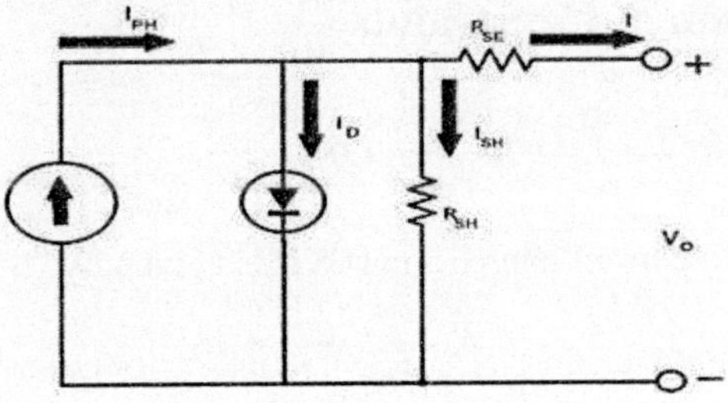

Fig. 2 PV Cell equivalent circuit model: Practical Case

$$I = I_{PH} - I_D - I_{SH}$$

$$I = I_{PH} - I_{DS}\left[\left(e^{\frac{V_o + IR_{SE}}{CV_T N_S}} - 1\right)\right] - \frac{V_o + IR_{SE}}{R_{SH}} \tag{4}$$

where I_{SH}—current through a shunt resistor, R_{SE}—Series resistance and R_{SH}—Shunt resistance.

Photo current

$$I_{PH} = (I_{SC} + \gamma_{SC}\Delta T)\frac{G_i}{G_{SC}} \tag{5}$$

where G_i—Solar irradiance W/m^2, G_{SC}—Solar irradiance (at STC) = 1000 W/m^2 $\Delta T = T_A - T_{A,ref}$(Kelvin), $T_{A,ref}$—cell temperature at STC - 25 + 273 = 298 (Kelvin)γ_{SC}—coefficient temperature of short-circuit current A/k, I_{SC}—Photo current at STC and V_{Go}—band gap energy (eV), (Silicon has 1.12 eV).

Saturation current

$$I_{DS} = I_{RS,ref}\left[\frac{T_A}{T_{A,ref}}\right]^3\left[e^{\left(\frac{qV_{Go}\left[\frac{1}{T_{A,ref}} - \frac{1}{T_A}\right]}{Ck}\right)}\right] \tag{6}$$

$$I_{RS,ref} = \frac{I_{SC}}{\left(e^{\frac{qV_o}{CN_s KT_A}} - 1\right)} \tag{7}$$

3 Determination of Constraints

3.1 Determination of I_{PH} and I_{DS}

From Fig. 1, the load current expression at STC is given by

$$I = I_{PH,ref} - I_{DS,r}\left[\left(e^{\frac{V_o+IR_{SE}}{CV_T N_S}} - 1\right)\right] \tag{8}$$

This shows that without knowing the $I_{PH,r}$, the output current cannot be determined. Under short-circuit conditions but the following is valid only in ideal cases:

$$I_{SC,ref} = I_{PH,r} - I_{DS,r}\left[\left(e^{\frac{0}{CV_T N_S}} - 1\right)\right] = I_{PH,r} \tag{9}$$

$$I_{SC,r} \approx I_{PH,r} \tag{10}$$

This demonstrates that the short-circuit current is approximately proportional to the photocurrent at STC.

3.2 Determination of R_{SE} and R_{SH}

For accurate model, it is always necessary to estimate the values of both internal resistances. Series and shunt resistances of PV cells can be determined using a suitable iterative method for single and double exponential model [6]. R_{SE} and R_{SH} must be determined when P_{MPP} is equal to the experimental value $P_{MPP,\text{exp}}$ at STC. Therefore, the equation can be obtained as

$$P_{MPP,ref} = V_{MPP,ref} I_{MPP,ref}$$

$$P_{MPP,ref} = I_{PH} - I_{DS}\left[\left(e^{\frac{V_{MPP,ref}+I_{MPP,ref}R_{SE}}{CV_T N_S}} - 1\right)\right] - \frac{V_{MPP,ref} + I_{MPP,ref}R_{SE}}{R_{SH}} \tag{11}$$

By using Eq. (11), it is easy to determine both series and shunt resistance by iteration method. The iteration starts from $R_{SE} = 0$ further which must vary to match the maximum power point to experimental value. There will be only one pair that satisfies the state. Table 1 shows the datasheet of model PWX 500–50 W PV module.

Table 1 Component specifications of PWX 500–50 W PV module

Constraint	Values
P_{MPP}(W)	50
V_{MPP}(V)	17.2
I_{MPP}(A)	2.90
$V_{O.C.}$(V)	21.6
$I_{O.C.}$(A)	3.10
γ_{SC}(K^0)	0.0013
N_s	36

4 Simulation Model and Results

With the help of Eqs. (4), (5), (6) and (7), the overall simulation of the PV module is developed. Figure 3, 4, 5, 6, 7 and 8 represent the simulation models of photocurrent, saturation current, reverses saturation current, output current and overall model of PV cell.

Figure 9, 10, 11, 12, 13, 14, 15 and 16 represent the I–V and P–V curves of

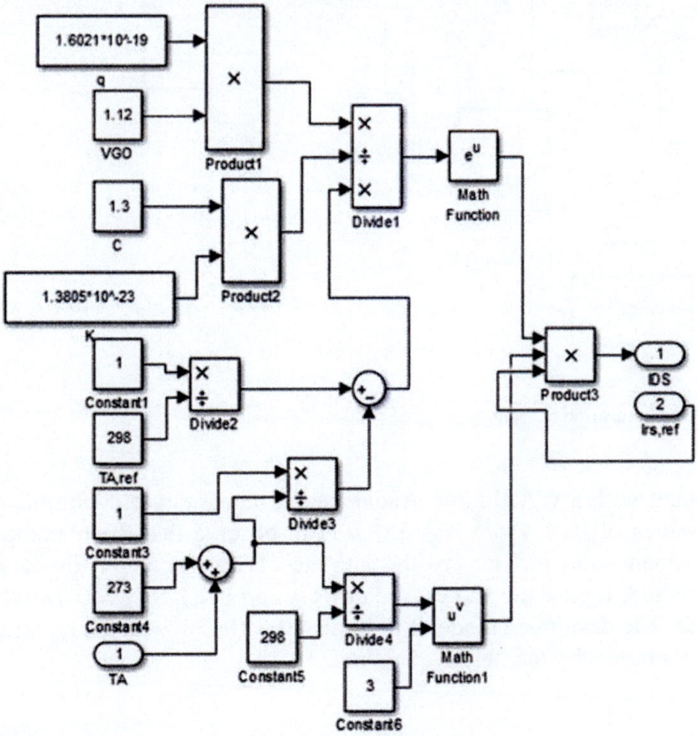

Fig. 3 Saturation current

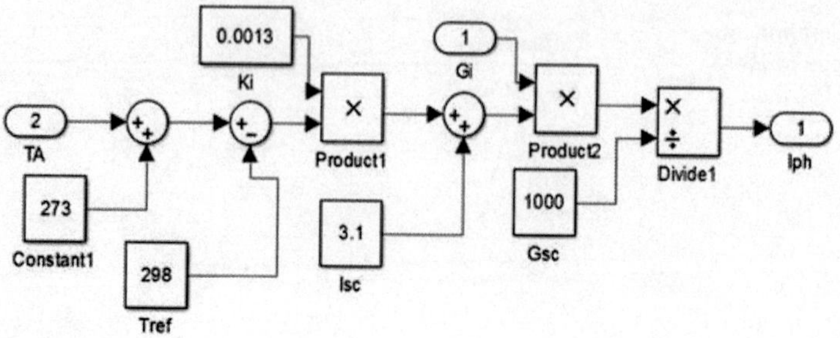

Fig. 4 Photo current

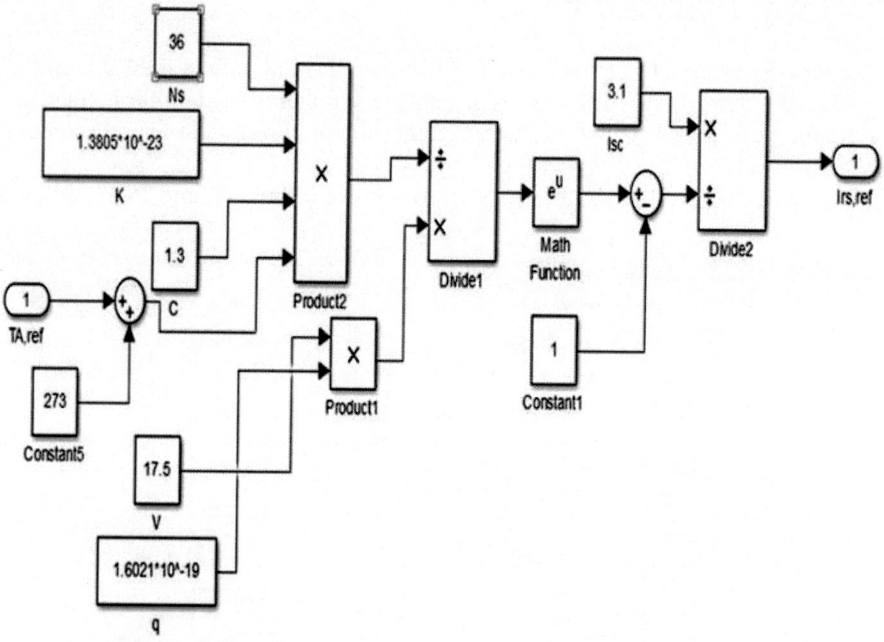

Fig. 5 Reverse saturation current

the PV model with a variation of irradiance and temperature conditions and with different values of R_{SE}. From Fig. 15, we can observe that the maximum power point experiment value provided by the data sheet is approximately the same. By the iterative method, it gave the values of pair (R_{SE} and R_{SH}), $R_{SE} = 0.071\ \Omega$ and $R_{SH} = 333.37\ \Omega$. The developed model is of utmost illustration with the preferred model with values attained by the iterative method.

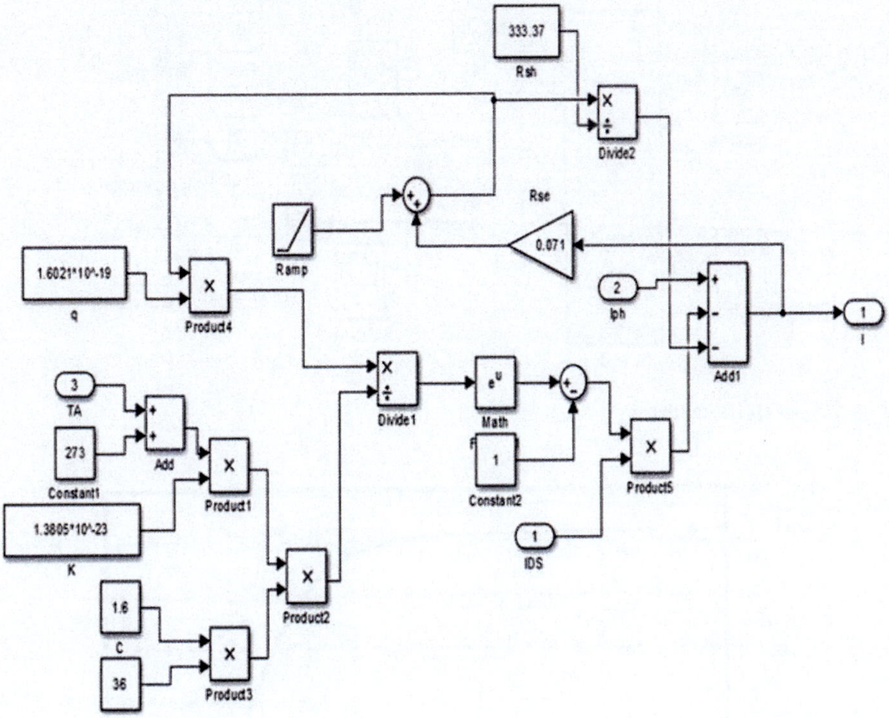

Fig. 6 Output current

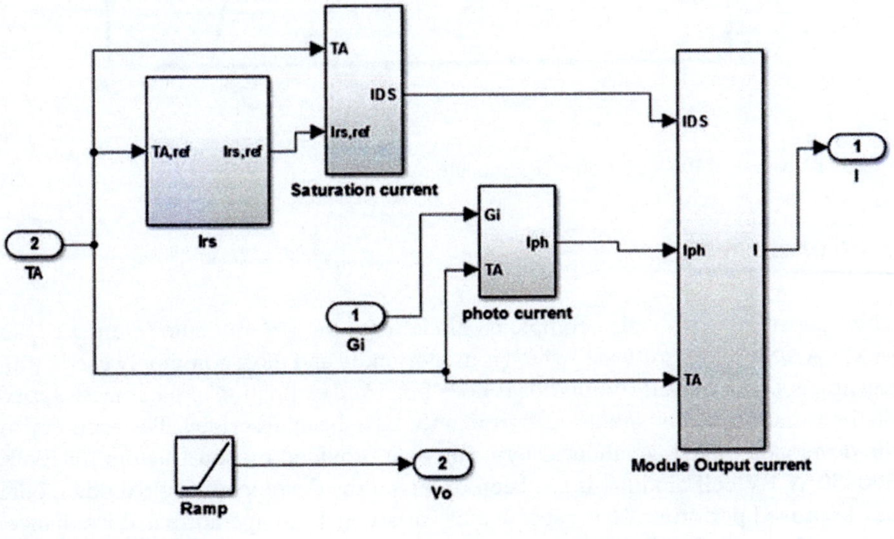

Fig. 7 Subsystem model

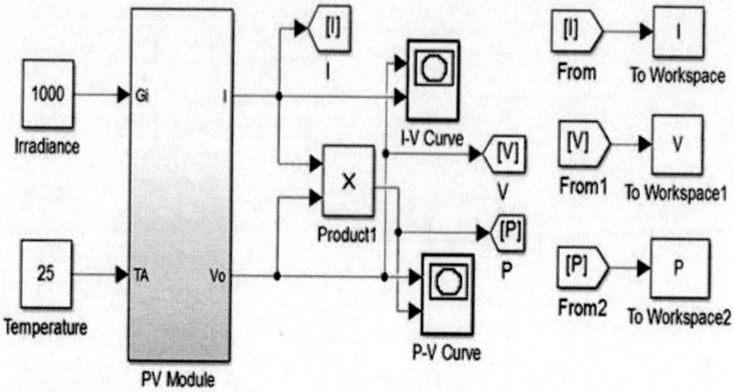

Fig. 8 Overall PV model

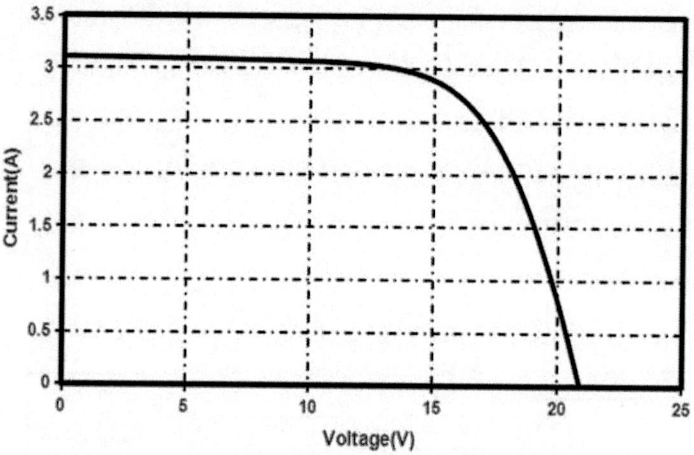

Fig. 9 I-V Curve of PWX 500–50 W PV module

5 Conclusion

This paper presents the complete modelling of PV module simulated in MATLAB/Simulink software, which is user-friendly and most commonly used by all researchers. The overall PV module is developed and estimation of parameters based on the variation of temperature and irradiance have been described. The accuracy of the developed model is validated by using data provided by constructors for PWX 500–50 W PV cell module. It has been observed the developed single diode model has improved performance irrespective of variations in temperature and irradiance. This model can be used for the study of any type of PV cell model provided by the manufacturer under standard test conditions (STC).

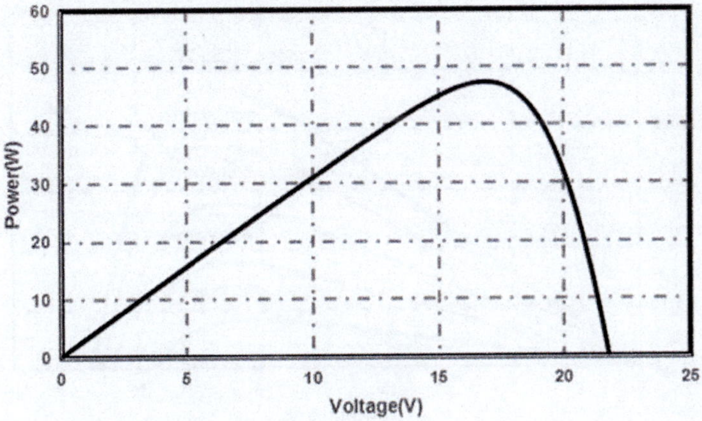

Fig. 10 P–V Curve of PWX 500–50 W PV module

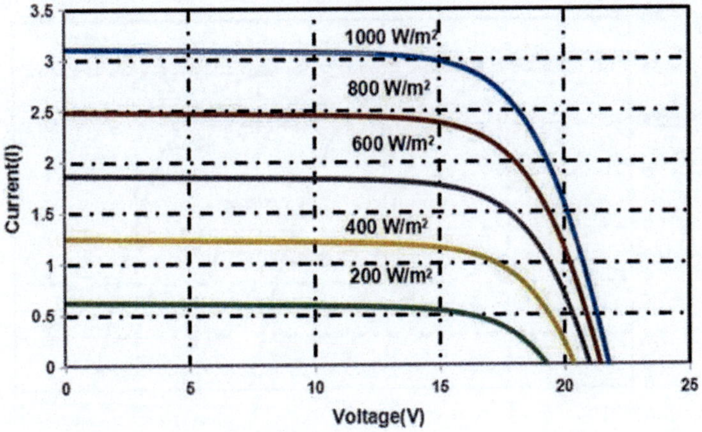

Fig. 11 I-V Curve with variation of irradiance

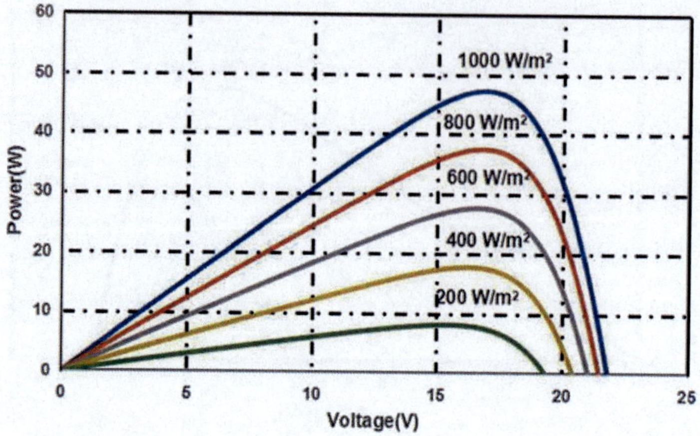

Fig. 12 P–V Curves with variation of irradiance

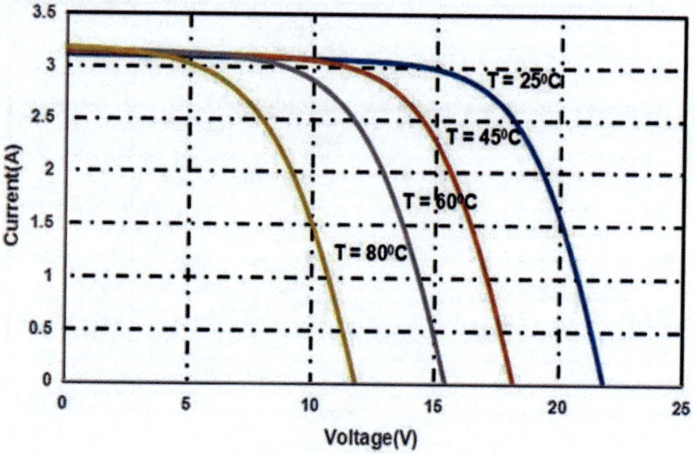

Fig. 13 I-V Curves with variation of temperature

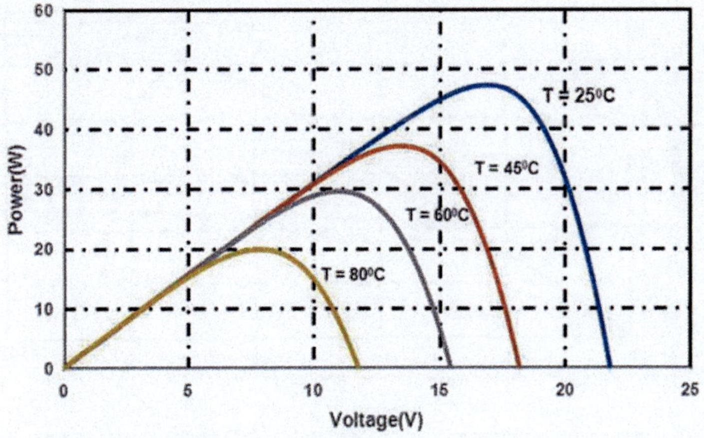

Fig. 14 P–V Curves with variation of temperature

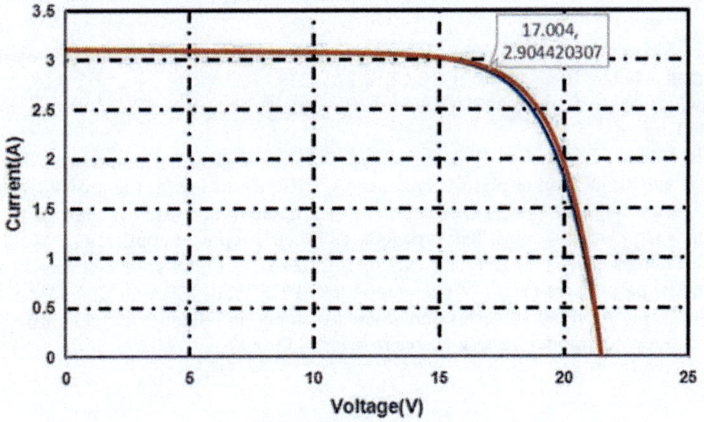

Fig. 15 I-V Curves with different values of R_{SE} ($R_{SE} = 0\ \Omega$ and $R_{SE} = 0.55\ \Omega$)

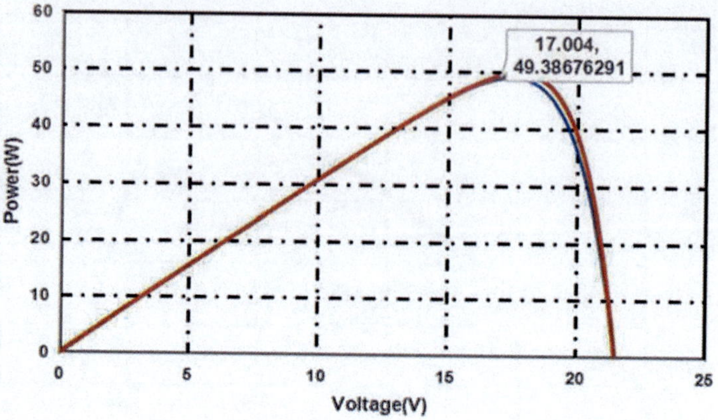

Fig. 16 P–V Curves with different values of R_{SE} ($R_{SE} = 0\ \Omega$ and $R_{SE} = 0.55\ \Omega$)

References

1. Atlas IH, Sharaff AM (1992) A fuzzy logic power tracking controller for a photovoltaic energy conversion scheme. Energy Power Syst Res 227–238
2. Walker G (2001) Evaluating MPPT converter topologies using a MATLAB PV model. J Electric Electron Eng
3. Gradella Villalva M, Rafael Gazoli J, Ruppert Filho E (2009) Comprehensive approach to modeling and simulation of photovoltaic arrays. IEEE Trans Power Electron 24:1198–1208
4. Gow JA, Manning CD (1999) Development of a photovoltaic array model for use in power-electronics simulation studies. IEE proceedings Electrical power applications 146:193–200
5. Chan D, Phang J (1987) Analytical methods for the extraction of solar cell single- and double-diode model parameters from I-V characteristics. IEEE Trans Electron Dev 34(2):286–293
6. Singh R (2011) Analysis of series and shunt resistance in silicon solar cells using single and double exponential model. Emerg Mater Res 1:33–38

Chapter 32
Analysis and Prediction of Crime Using Machine Learning Techniques

S. Srinivasulu Raju, G. Narasimha Swamy, M. Rejoice Angelina, M. Sai Snehitha, M. Sai Chandana, and M. Priya Mythili

1 Introduction

The economic growth of a nation is unfavorably influenced by ever-rising crime every day. It is one of our society's most significant challenges and reducing the incidence of crime has become an incredibly important task. Therefore, recognizing numerous causes, frequency relationships of crimes and thus evaluating optimized ways to minimize crime rates are very relevant. The above dilemma made me go for a research report about how it was easier to solve a crime case. It has been demonstrated across several documents and cases that machine learning and data science will make the job easier and faster. Mittal and Patidar [1] conducted a research and analyzed that Naïve Bayes classifier performs faster than any other classifier nonetheless, it does not produce the desired accuracy. Furthermore, they said that machine learning algorithm requires adequate dataset to produce good results. Kharde and Sonawane [2] took a survey and found out that SVM and Naïve Bayes algorithm provide better outcomes. Lexicon-based approach is useful in many applications. According to their study, bigram model predicts better accuracy. Rout et al. [3] proposed a model with supervised learning with bigram and unigram as part of the features. The Multinomial Naive Bayes classifier produces the best outcome. The prevalence of cancer and deaths in both men and women was correlated with adult height as calculated by genetics, and one-third of the six cancer risk differences are accounted for by childhood nutrition [4–6]. Early breast cancer diagnosis and detection will reduce the death risk and provide the means for timely care. Copious research has been done in the field of diabetes detection and prediction using machine learning models and image classification. They mostly consist of using SVM, which is a prominent classification model—mostly because it gave superlative accuracy in

S. Srinivasulu Raju · G. Narasimha Swamy · M. Rejoice Angelina (✉) · M. Sai Snehitha · M. Sai Chandana · M. Priya Mythili
Department of EIE, VR Siddhartha Engineering College, Vijayawada, India

S. Dawn et al. (eds.), *Smart and Intelligent Systems*, Algorithms for Intelligent Systems,
https://doi.org/10.1007/978-981-16-2109-3_33

all the models apart from the neural network models like Artificial Neural Network (ANN) and Convolution Neural Network (CNN). Voluminous features have been used for cancer prediction and different researchers have extensively used different parameters and shown surprisingly accurate results. In [7], the authors present an automatic keyword extraction method based on Bi-LSTM RNN. This aims to automatically generate keywords from reviews and based on these keywords to aim to detect the fake reviews present [8]. The goal of the author is to decide whether or not a piece of news is solely based on its content, thus addressing the issue from a purely deep learning point of view by RNN and LSTM. Kaur and Bawa [9] demonstrate the Consumer reviews of Amazon products, datasets are available in world datasets. It is in CSV format and contains 24 columns where 3 columns are important to classify the requirements, i.e., category, primary category and reviews.text and reviews.title. This dataset contains almost 33,332 reviews. References [10, 11] focused on the performance of different requirement prioritization techniques such as numerical assignment, binary search tree, analytic hierarchy process, cumulative voting, planning game, B-tree and value-oriented prioritization. The performance of these seven techniques was assessed based on the criteria—time complexity, fault tolerance. As per the conclusion, there is still additional work required to enhance the effectiveness of requirement prioritization techniques in terms of time complexity, complexity and fault tolerance. In Priya [12], the sentiment analysis may be used for this classification. Sentiment analysis involves unstructured text feedback on product reviews, events, so on, from all feedback shared by dissimilar users and classifies comments as positive, negative or neutral opinions in different categories. This study compares the performance of various ML algorithms in the SA of Twitter data. Zhang et al. [13] online product feedback and other user-generated content related analysis is a meaningful topic of study in its various applications. The limited range of functionality induced by too short text length is limited to conventional feature-based approaches. Bidirectional Short-Term (Bi-LSTM) Memory, a neural network that removes automated text features commonly used for data and prediction processing. Our study will describe how the emotional polarity of product feedback is defined by Bi-LSTM. In [14], the authors aim to detect spammer groups that are those referees who work collectively composing fake reviews to organize or demote some goal inventions. The steps used in this are frequent pattern mining, computing spam indicator values and ranking using SVM rank. In Korde [15], SA also known as OM is a technique utilized to bring collective opinions of a particular feature by reviews dataset. Other users' views contribute to the decision-making procedure. This article discusses various approaches for evaluating sentiments. Semantic-based techniques, ML, NN, and syntactic strategies of their strength differ. While there is a hybrid approach, the main concept is to incorporate two or more approaches to improve their accuracy. A context in which sentimental research is carried out using recommended feature reduction techniques is suggested. Word embedding is a method that produces low-dimensional vector representations of words. An SVM classifier is used in feature reduction technique. The framework analyses user opinions via the ML method and offers a recommendation system that encourages user decision-making [15]. In [16, 17], the authors aim to detect fake reviews using product-related features. A CNN model is used here to capture product-related review characteristics and a classifier is defined based on the composition characteristics of the product word.

The goal of this work is to predict crime using the characteristics present in the dataset. The dataset is taken from the official websites. Using python as the heart, we can predict the type of crime that will occur in a specific region with the aid of the machine learning algorithm. Machine learning algorithms are used to analyze and predict the crime rate at specified areas by using the provided data from kaggle dataset. Classification algorithms are used to analyze the data, and visualization tools are utilized to envision the data.

2 Categories of Analytical Models Algorithms

The prediction and analysis of crime is a very logical method that will assist the authorities concerned with different patterns of crime. The clustering algorithm is used to figure out how often in a given region a specific form of crime has occurred. It also helps to evaluate the trend and statistical analysis of crime details and also Pandas is used to read and analyze the datasets. In [18, 19], random forest is used to forecast future crimes and predict the area by using the decision trees. K-means clustering is an unsupervised learning procedure based on the principle of Euclidean distance, i.e., data points are completely scattered over an area, after feeding into the algorithm using clusters between similar points. K-means clustering algorithm is employed to build clusters of the dataset obtained from the kaggle depends upon the crime rate.

2.1 Naive Bayes

The Naive Bayes classifier is the most common algorithm, which is used for sentiment analysis. The Bayes theorem is the basis of this algorithm. Naïve Bayes is the fastest and straight forward algorithm [5]. We have used Bernoulli Naive Bayes from scikit-learn library. In this, the weight of each word is equal to 1, if present or 0 if not [20, 21].

2.2 K-means Clustering Algorithm

K-means is an unsupervised machine learning algorithm, which makes inference from datasets using only input data without labeled findings. The K-means attempts to clearly group together identical data points and to uncover the trends behind them. In [22], K implies searches for a fixed number (k) of clusters in the dataset

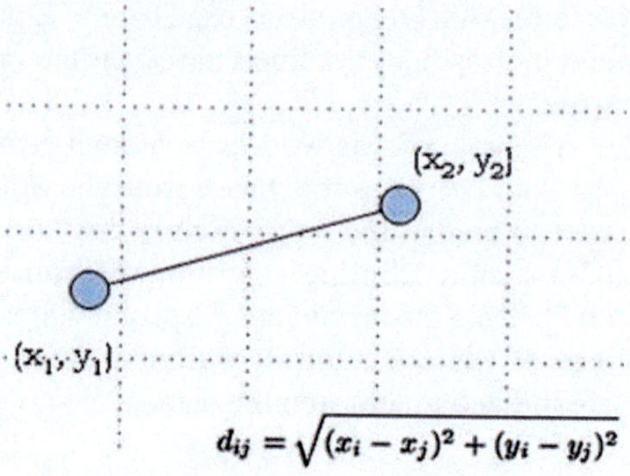

Fig. 1 Euclidean distance between two points

to accomplish this goal. Because of such parallels, the clusters are a series of data points. The number K is the number of centers in the dataset that are needed: a centroid is an imaginary or actual position in the middle of clusters.

The first randomly chosen centroids are used as start points for each cluster, and then iterative calculations are made to refine the centroid positions. Then assign each data point to the nearest center using Euclidean distance or simply put the distance between the centroid and the point. For each centroid, calculate the mean value for all points and this mean value would become the new centroids. These new points are put through the same steps mentioned above producing yet another set of centroid. This process repeats until the position of the centroids converges. When it converges it means they are accurately grouped.

2.3 *Random Forest*

When we are confused in taking a decision we often use a technique wherein we take an average of the decisions of many people, then use that average decision as a result. In [23, 24], these methods are also called ensemble techniques. Random forest classifier uses the same technique to calculate the result—it takes the average/majority voting of many predictions made by the decision trees and then represents that as a new model. It is quite basic but a very powerful method, considering its ability to give such accurate results.

Figure 2 depicts the representation of a random forest classifier. This method is also known as an ensemble method because it is a collection of decision tree forests.

Random forest is an ensemble learning method consisting of several decision trees used for classification as well as regression. Random forest is a kind of Bagging Algorithm or Bootstrap Algorithm. The fundamental goal is not to draw on individual decision treasures but to merge multiple decision trees to decide the ultimate efficiency.

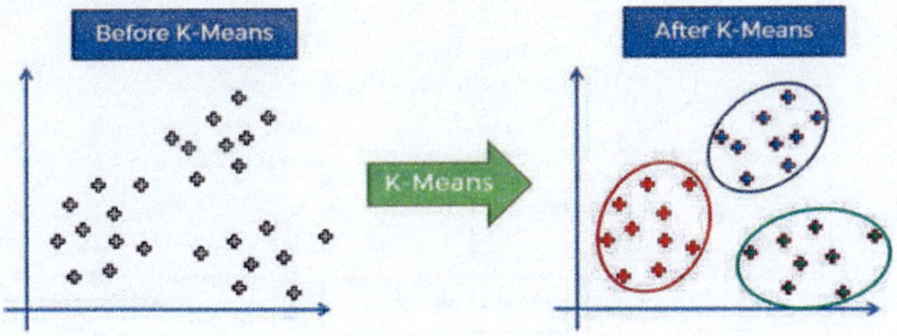

Fig. 2 The working of K-means clustering

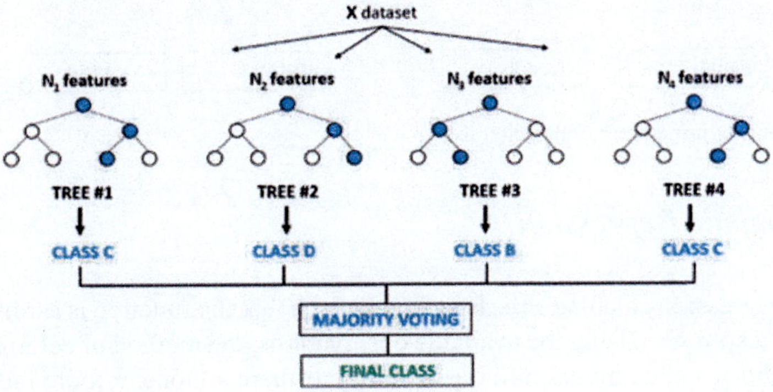

Fig. 3 Random forest diagram

2.4 Decision Tree

Decision trees are a type of supervised learning method so this algorithm requires training of data before implementation. It is very similar to text classification: given a set of documents (e.g., represented as TFIDF vectors) together with their labels, the algorithm can determine how much each word correlates with a particular label. For example, if the word "good" frequently appears in data labeled it as positive, whereas the word "bad" will be labeled as negative. By integrating all these observations, it builds a model that can assign a label to any data. In [25], decision tree classifier uses an actual decision tree to take out useful information from data. It creates a set of questions in the nodes and classifies the data on the basis of those questions hence creating a set of information or useful bifurcation, which one cannot see directly. Decision tree is a method that uses tree-like decision model and its implications. It is a system like a matrix in which each node in the inner is a test, each branch represents the output, and each leaf node represents the class label.

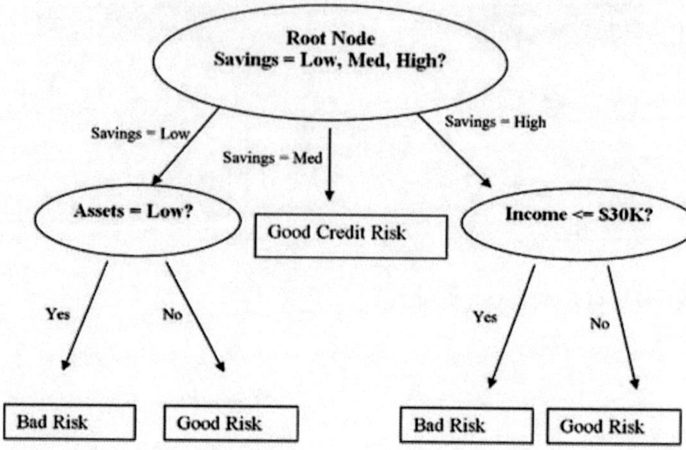

Fig. 4 Working principle of a decision tree

2.5 *Logistic Regression*

Logistic regression calculates the best weight such that the function is as similar as to all actual responses. Using the available observations, the method of determining the best weights is called model training or fitting. If there are one or more independent variables, it is used to evaluate the output. The output value is in binary form.

2.6 *Support Vector Machines*

Support Vector Machines is another common method used for classification. To provide complete separation of data points, a hyperplane is constructed in a high-dimensional space. This is the reason why the SVM is often named the maximum margin classifier. The hyperplane describes certain instances that are closed to the plane, which are referred to as support vectors. We used kernel equals to "rbf" for better accuracy. In [26], the whole idea behind support vectors is that to separate two categories of data sets, a hyperplane is used. Vectors, which are the farthest points in the categories, support that hyperplane hence making it easier to find the optimal hyperplane. Figure 5a depicts how SVM works. When the data are non-linearly separable, the hyperplane needs to be high dimensional and hence k-SVM is used, which uses Gaussian surfaces as hyperplanes, where k stands for the kernel.

In random forest, we extract K random points from the training dataset and then create decision trees associated with these K data points. Make each tree to forecast the output value for the data point for a new data point and assign the average of all the projected output values to the new data point. Tree-based learning algorithms are known to be one of the easiest and most commonly used supervised learning

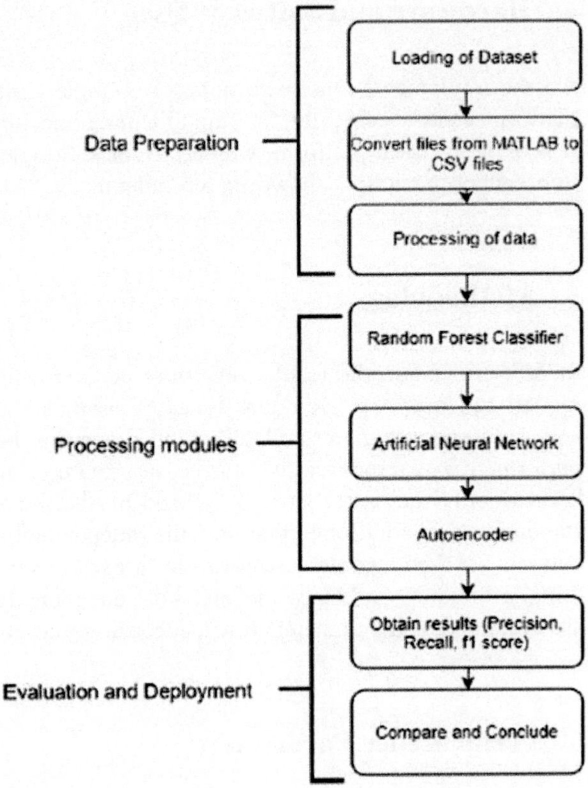

Fig. 5 System architecture

techniques. These methods develop predictive models with high precision, stability and ease of analysis. Unlike linear structures, non-linear interactions are very well mapped out.

3 Dataset

The data were collected from kaggle data center for classifying and detecting the crime rate. Data files obtained in MATLAB format are converted to comma-separated values (.csv). Datasets used in the study are normal baseline and used for segmentation to train and test the machine learning algorithms.

4 Hardware and Software Used

All the implementations were done on Google Colaboratory, which is an online platform used specifically for implementing machine learning and data science. It is a cloud-based platform, which provides Google's hardware, GPU and TPUs. Subsequent paragraphs, however, are indented.

5 Methodology

In order to obtain the results regarding accuracy, precision and anomalies, three models are used, i.e., Random Forest Classification, Artificial Neural Networks and Auto encoder. After collecting and preparing the data, feature extraction was performed, i.e., removing the unnecessary features and data from the data set and keeping only the ones that will be used by the models. Next, all the features are standardized in addition to that only the independent features are standardized. The datasets are then modeled using various algorithms, i.e., Random Forest Classification, Artificial Neural Networks and Auto encoders. In order to do this, the datasets are evaluated, and parameters tuned accordingly to give better results.

6 Performance Parameters

6.1 Confusion Matrix

As shown in the table, it is a matrix of the actual class and expected class outcomes for a classification problem.

True Positive (TP): Forecast the quantity is positive,

True Negative (TN): Forecast the quantity is negative, also it is true.

False Positive (FP): Forecast the quantity is positive, also it is false.

False Negative (FN): Forecast the quantity is negative, also it is false and it is true.

Based on the survey, the performance parameters to be considered in this work are as follows.

6.2 Accuracy

It is one of the metrics used to evaluate the trained model. It is calculated by finding the ratio of predicated output dependent variable and actual dependent variable.

$$\text{Accuracy} = \frac{\text{TP} + \text{TN}}{\text{TP} + \text{FP} + \text{TN} + \text{FN}}$$

6.3 Precision

The ratio of relevant samples classified as positive to recovered samples classified as positive is precision.

$$\text{Precision} = \frac{\text{TP}}{\text{TP} + \text{FP}}$$

6.4 Recall

The ratio of the relevant samples labeled as positive to the relevant positive samples in the test collection is recall.

$$\text{Recall} = \frac{\text{TP}}{\text{TP} + \text{FN}}$$

6.5 F1-Support

It is the harmonic mean of precision and recall.

$$\text{F1} - \text{support} = \frac{2 * \text{precision} * \text{recall}}{\text{precision} + \text{recall}}$$

6.6 Time Complexity

It is the time required to execute the algorithm.

7 Results and Discussions

All machine learning classifier algorithms are implemented on the breast cancer dataset. During comparison work, we stick up to the following rules:

1. Evaluation of the model for classifier is compulsorily done to check whether output is satisfactory or not?

2. Performance accuracy of the model is understood as "how correctly and incorrectly instances are classified out of the total number of instances". The result is expressed as a percentage and is shown in Table 1.

3. Use of confusion matrix for describing the performance of the model. It has some attributes like False Positive (FP), False Negative (FN), True Positives (TP) and True Negatives (TN).

4. Total sample count is sum of all FP, FN, TP and TN.

5. Use of Accuracy parameter to record correctness of the model, which is TP + TN / total sample count.

6. Precision is taken as TP + FP / total sample count.

7. Recall is taken as TP + FN / total sample count.

8. F1 Score is harmonic mean of Precision and Recall as they both deal with relevance. It is the best value at 1 and worst at value 0.

9. Area under the curve (AUC) shows the ability of the model to distinguish between FN and FP classes and is used as a synopsis of receiver operator characteristic (ROC) curve that narrates about the performance of binary classification algorithm.

The dataset had different kinds of data types like strings, integer and floats, which had to be dealt with. While pre-processing the data, some of the datasets are allocated for training and rest of the datasets are for testing purpose. To form the number of Clusters in K means algorithm, Elbow method is used and plotted the scatter plot of a single crime in the region. After finding the number of clusters or K, we had to randomly select centroids on that clusters and then find the Euclidean distance between the centroids and all the points in that cluster so as to produce the new points.

The above considerations used to measure the performance of the model are shown in Table 2. Graphical representations of the results are shown in Fig. 6.

From the table, Naive Bayes classifier has the maximum accuracy while decision trees have the least one. We performed the experiments using various classifiers. We used 20% of the training dataset for validation of our models to check against overfitting, i.e., 21,465 for training and 5398 for testing. We trained the classifier with this dataset. The classifier is used in the rest of 20% of the tweets to predict the

Table 1 Confusion matrix

	Predicted class		
Actual class		P	N
	P	True Positives (TP)	False Negatives (FN)
	N	False Positives (FP)	True Negatives (TN)

Table 2 Experimental results

Algorithm	Accuracy %	Precision	Recall	F1-score	ROC-AUC score
LR	82.1	70.7	85.7	74	69.7
RF	81.5	67.4	26.8	38.1	56.7
SVM	78.4	59.9	40.5	44.3	49.0
NB	83.1	74.2	74.4	66.9	63.1
DT	73.4	62.1	85.7	74	69.7
XGB	78.1	69.2	78.5	69.2	74.3

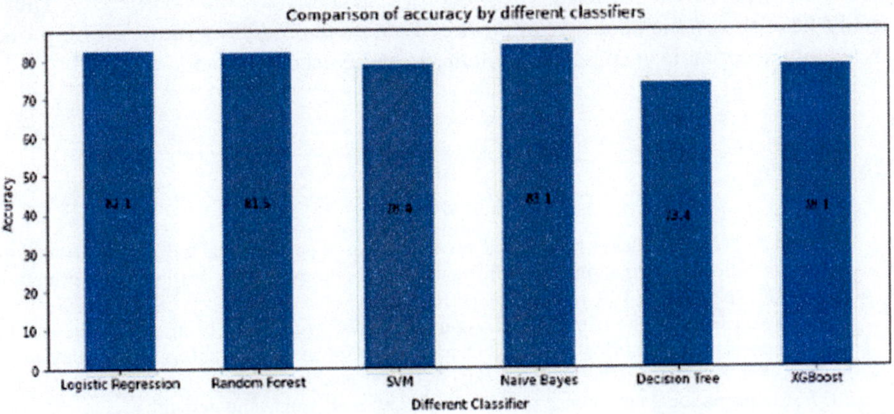

Fig. 6 Accuracy of the algorithms

accuracy. The training and test data we used contain instances divided into positive, negative and neutral. Additionally, we calculated the accuracy score from Scikit Learn Library along with F-Score. After evaluating the models, we get an accuracy score of 83.1% for Naïve Bayes algorithm and the second best of 82.1% from logistic regression algorithm while using random forest, the model predicted the score of 81.5%.

8 Conclusions

Using machine learning technology, it has become easy to figure out the relationships and trends between different data. The work in this project specifically focuses on predicting the type of crime that may occur if we know where it has occurred. The clustered results made it possible to recognize areas in our country that are vulnerable to crime and can be used to warn the authorities of the need to take precautionary measures in advance. Different offenses require different penalties, and it is simple to complete this application. We have developed a model using a training dataset that has undergone data cleaning and data transformation using the principle of machine learning. Despite the strong assumptions made about the independence between the features, the Naive Bayes Classifier provides relatively good results with an accuracy of 83.1%. Basically, it outflanks all the other classifiers. The proposed models collect the dataset of the crime from kaggle and then refine the datasets and features are extracted from these datasets, which are supposed to be analyzed and predicted. Finally, features are analyzed from these texts and are classified as positive, negative and neutral values. By increasing the size of the dataset in stages, it has been shown that accuracy improves as the size of the training dataset increases. The findings have given that there are limitations of existing ML algorithms used to identify and prioritize requirements in terms of scalability, dependency and complexity. These results have shown that, because of its effectiveness in predicting future crime-prone areas with acceptable accuracy, this method can be used very well.

References

1. Agarwal S, Yadav L, Thakur MK, Crime prediction based on statistical model. Department of Computer Science Engineering and Information Technology, Jaypee Institue of Information Technology, Noida, 201309, India
2. Arietta SM, Efros AA, Ramamoorthi R, Agarwala MA City forensic: using visual elements to predict non-visual city attributes. EECS Department, University of California, Berkeley
3. Joshi A, Sai Sabitha A, Choudhury T (2017) Crime analysis using K-means clustering. In: 2017 3rd international conference on computational intelligence and networks
4. Yu L-C, Chen T-Y, Lin Y-L (2017) Using machine learning to assist crime prevention. In: 2017 6th IIAI international congress on advanced applied informatics (IIAI-AAI)
5. Drew J, Moore T Automatic identification of replicated criminal websites using combined clustering. Computer Science and Engineering Department, Southern Methodist University, Dallas, TX, USA
6. Sarvari H, Abozinadah E, Mbaziira A Constructing and analyzing criminal networks. Damon McCoy George Mason University
7. Lin Y, Zhu T, Wu H, Zhang J, Wang X, Zhou A. (2014) Towards onlineanti-opinion spam: spotting fake reviews from the review sequence. In 2014 IEEE/ACMinternational conference on advances in social networks analysis and mining (ASONAM2014), pp 261–264. IEEE
8. Wang CC, Day MY, Chen CC, Liou JW (2018) Detecting spammingreviews using long short-term memory recurrent neural network framework. In: Proceedings of the 2nd international conference on e-commerce, e-business and e-government, pp 16–20
9. Makrehchi M, Aghababaei S (2016) Mining social media content for crime prediction. In: 2016 IEEE/WIC/ACM international conference on web intelligence (WI)
10. Gerber MS, Al Boni Md (2016) Area-specific crime prediction models. In: 2016 15th IEEE international conference on machine learning and applications (ICMLA)
11. Brantingham PL, Glasser U, Ester M, Tayebi MA (2014) CRIMETRACER: activity space based crime location prediction. In: 2014 IEEE/ACM international conference on advances in social networks analysis and mining (ASONAM)
12. Priya KS (2020) A comparative sentiment analysis of sentence embedding using machine learning techniques. In: 2020 6th international conference on advanced computing and communication systems (ICACCS).
13. Zhang K, Song W, Liu L, Zhao X, Du C (2019) Bidirectional long short-term memory for sentiment analysis of Chinese product reviews. In: 2019 IEEE 9th international conference on electronics information and emergency communication (ICEIEC)

14. Mukherjee A, Liu B, Wang J, Glance N, Jindal N (2011) Detecting groupreview spam. In: Proceedings of the 20th international conference companion on WorldWide Web, pp 93–94
15. Shelke PPP, Korde AN (2020) Support vector machine based word embedding and feature reduction for sentiment analysis-a study. In: 2020 fourth international conference on computing methodologies and communication (ICCMC).
16. https://www.analyticsvidhya.com/blog/2017/12/fundamentals-of-deep-learning-introduction-to-lstm/
17. Sun C, Du Q, Tian G (2016) Exploiting product related review features for fake reviewdetection. Math Probl Eng 201
18. https://ujjwalkarn.me/2016/08/09/quick-intro-neural-networks/
19. http://www.cs.bham.ac.uk/~jxb/INC/l11.pdf
20. https://www.saedsayad.com/support_vector_machine.html [svm]
21. https://en.wikipedia.org/wiki/Long_short-term_memory
22. https://www.cse.unsw.edu.au/~cs9417ml/MLP2/BackPropagation.html [mlp]
23. https://www.analyticsvidhya.com/blog/2017/12/introduction-to-recurrent-neural-networks/
24. https://en.wikipedia.org/wiki/Recurrent_neural_network
25. https://www.geeksforgeeks.org/introduction-to-recurrent-neural-network/
26. https://ayearofai.com/rohan-lenny-3-recurrent-neural-networks-10300100899b

Chapter 33
Short-Term Power Forecasting for Renewable Energy Sources Using Genetics-Based Harmony Search Algorithm

Rejo Roy, Albert John Varghese, and S. R. Awasthi

1 Introduction

A major problem affecting mankind globally is greenhouse gas, which is rapidly increasing due to the emission of carbon dioxide in atmosphere. The atmospheric concentration of carbon dioxide has risen to 425 parts per million in 2019 [1]. Generation of power from renewable energy aids to reduce greenhouse gases and safeguards environment [2]. These issues can be survived if environmentally friendly power turns out to be effectively versatile and is used for a bigger scope. It has been estimated that by 2022, 18.9% of power consumption in India will be provided by 175 GW of renewable energy [3].

The supply of power from renewable resources such as solar and wind is on a global rise. The variable nature of renewable power resources leads the power grid into a vulnerable condition [4]. The prediction of solar energy is important as it is variable in nature due to weather conditions and day and night cycles [5]. It is important to accurately forecast PV power output for the reliable operation of the power industry [6, 7]. It plays a very significant role in integrating conventional power plants with solar power plants [8] in the power grid.

Artificial neural network is a field where the system is made intelligent and it is made to work like a human brain in decision-making capabilities. Forecasting is the best field where artificial neural networks are applied. The central unit of a neural network works like a human brain and it is connected by a large number of processing elements called neurons that work in parallel to resolve any problem [9]. Each neuron consists of an associated weight factor that is summed up to achieve an internal level

R. Roy (✉) · A. J. Varghese
Department of Electrical and Electronics Engineering, RNTU, Bhopal, India

S. R. Awasthi
RNTU, Bhopal, India

S. Dawn et al. (eds.), *Smart and Intelligent Systems*, Algorithms for Intelligent Systems,
https://doi.org/10.1007/978-981-16-2109-3_34

of activity [10]. Artificial intelligence method is used to learn the relationship between the generated output power and predicted weather conditions [11].

Forecasting is an approach in which using the present and past data, future data are estimated so that the quantity forecasted is not unpredictable but an anticipated one. Forecasting can be classified into short-term, medium-term and long-term forecasting based on the time intervals it deals with. The forecasting approaches helps in minimizing the error in decisions made and hence in turn reduce the unpredictable nature of renewable energy sources. Forecasting approach when used in renewable energy source power output forecasting, helps in providing a vital input to state load dispatch centers for proper scheduling of power by various energy sources [12]. For many years, ANN is used to achieve optimized results of problems that are observed in various disciplines and sectors [13]. It is very difficult to propose an accurate forecast system because of climatic variables of renewable energy sources, seasonality, daily variability, etc. causing the process of forecasting to be a challenging task.

Forecasting is done widely using statistical methods or artificial intelligence techniques. In this study, artificial intelligence technique uses artificial neural networks. These methods are capable of dealing with non-linearity. The developed algorithm can classify data and form a relationship with output values. In this paper, Sect. 2 deals with the various steps in forecasting. The application of algorithm to forecast solar PV power at Bhilai is explained in Sect. 3. The results and discussion are provided in Sect. 4. Finally, the conclusion of the paper is provided.

2 Methodology

This paper introduces an evolutionary hybrid optimization algorithm as a tool to train an ANN used for solar PV power forecasting [14] on a short-term basis [15]. The purpose of this work is to simulate a Short-Term Power Forecasting model with the help of HSA and GA. The flowchart for genetics-based harmonic search model is shown in Fig. 1.

GA introduces the principle of survival of the fittest in nature into optimizing parameters of coding populations [16]. The genetic algorithm selects a number of individuals through a certain choice of strategies for breeding in pairs and matches the best individual in the group to breed the next individual generation [17]. Genetic algorithm is an effective method to explore an entire search space. But it is relatively poor to find the precise local optimal solutions in convergence regions. Hence, some additional operators are introduced to get the improved predictive power of ANN.

Harmony search (HS) is a natural search process by a musician for a better state of harmony forming an array of harmonies within a possible range of notes from the musical performance process. Harmony search algorithm helps to solve the optimization problem of non-linear models [18]. The search range and adaptability can be enlarged by making use of a modified search algorithm for increasing the convergence speed of the forecast model [19].

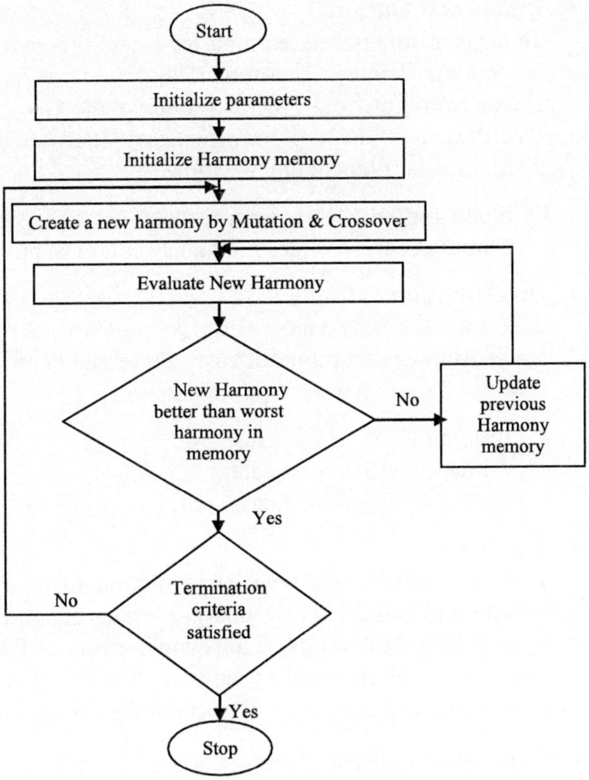

Fig. 1 Flowchart for GA + HSA

For forecasting, two or more methods can be combined instead of using a single method to get better results [20]. In the genetics-based harmony search algorithm, some positive traits from both algorithms are combined together, i.e.,

- Harmony search algorithm tries to find a pleasing harmony by imitating musician's improvisation process.
- Genetic algorithm creates a solution of high quality by depending upon biologically inspired selection process such as mutation and cross over.

Genetics-based harmony search algorithm is used so that simplicity and search efficiency of harmony search algorithm and the ability to not be trapped in local optimal solution from the genetic algorithm are combined. Here the ANN is designed using the "nntraintool" in MATLAB using Bayesian Regularization and the tuning of the weights and biases is done using genetics-based harmony search algorithm, i.e., a combination of harmony search and genetic algorithm.

Some of the terms that need discussion are as given below, these terms are used repeatedly and the description below gives clarity into its use and how the ANN functions.

- **Inputs and Outputs**
  To achieve supervised learning for a neural network, it is needed that inputs and outputs are defined. The numerical values for input and output for a period of 1 year were got from PVGIS (Photovoltaic Geographical Information System). PVGIS is an online tool that provides typical meteorological year (TMY) data for a virtual solar plant at any location.
  - Inputs for the ANN are hourly solar irradiance and temperature data for a year.
  - Output or target values are hourly solar PV power for a year.
- **Data Division**
  8760 (i.e., 365*24) data are used for the purpose of training. The values selected are as follows ("nntraintool" uses these values by default). These values can be changed based on user requirements.
  - Training (70% of the data)
  - Validation (15% of the data)
  - Testing (15% of the data)
- Training
  Training is done using Bayesian regulation method in MATLAB. This updates the weights and bias using Levenberg–Marquardt optimization. But here it minimizes a combination of weights and squared errors and determines the combination to produce a well-generalized network. The weights and bias are further adjusted using genetics-based harmony search algorithm.
- Termination Criteria

The neural network stops training if any of the following conditions are met:

- Maximum iterations are reached (User-defined)
- Mean square error value is reached (User-defined)
- Minimum error gradient value of 1e-7 is reached (As defined in "nntraintool")
- Mu function reaches maximum value of 1e10 (As defined in "nntraintool")

3 Application of Developed Methodology

The forecasting application is applied for Bhilai (Latitude—21.220 N and Longitude—81.380E). There are two inputs, 49 hidden neurons and one output. The transfer function used in the hidden layer is tan-sigmoid and in the output layer is linear [1]. The value of 49 neurons was reached by trial and error, any increase or decrease from the value increases the error. The structure of the neural network used is shown in Fig. 2.

The artificial neural network is achieved using Bayesian Regularization. The weights of the neurons are optimized using the combination of a heuristic algorithm, i.e., harmony search algorithm and an evolutionary algorithm, i.e., genetic algorithm

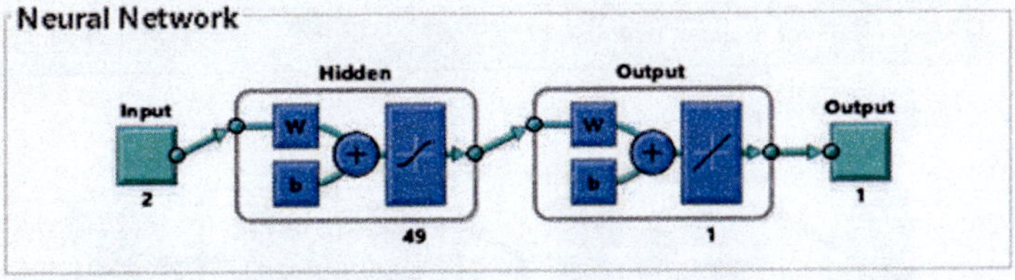

Fig. 2 Neural network structure for short-term power forecasting

Table 1 Parameters of neural network training

Parameters	GA + HSA model
Execution time (minutes)	26:07
Error (MSE)	0.99951
Terminated due to	Defined error reached

so that the weight of the neurons can be optimized to improve the performance in terms of reduced error and lesser execution time.

The parameters defined by the user for the execution of the ANN are:

- Maximum iterations—20,000
- Inputs and outputs—hourly data (8760) from PVGIS
- Number of neurons—49
- Mean square error—1
- Data initialization—random
- Weight and bias tuning—genetics-based harmony search algorithm

The tuning of ANN was completed as the mean square error (MSE) value below 1 is reached after completing 15,372 iterations. The values of different parameters after the end of training are listed in Table 1.

The training status obtained from 'nntraintool' is shown in Fig. 3. Training status gives information about execution time, iterations, error values, it also helps in forming various plots related to status after training the neural network. It also displays the criterion fulfilled for termination of the training process.

4 Results and Discussion

The research work focuses mainly on the short-term forecasting of solar PV power. Hourly forecasting of 6 h ahead is done for solar PV power. The forecasting model uses artificial intelligence-based artificial neural networks tuned using genetics-based harmony search algorithm.

The graph of Actual solar PV power and forecasted solar PV power obtained is shown in Fig. 4. The exact overlapping is not visible but the algorithm used is able to learn the trends and pattern of the actual values very closely. Actual v/s forecasted

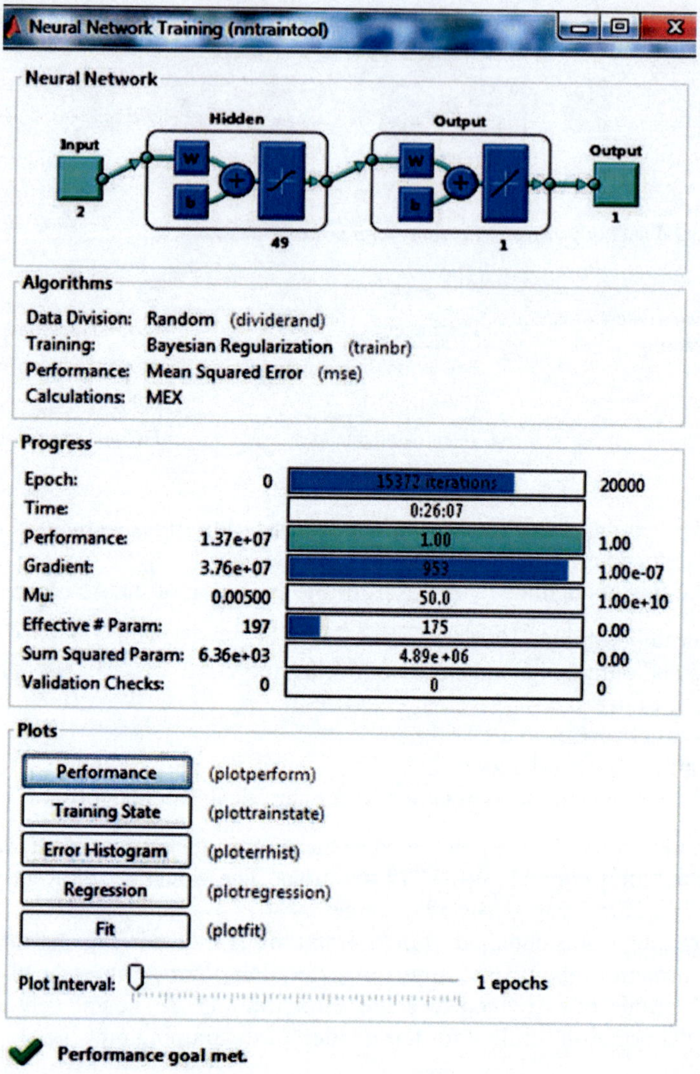

Fig. 3 Neural network training status

value is plotted on the same graph in order to show how the neural network is able to adapt itself to the actual values. Here red graph shows the forecasted values and the blue graph shows the actual values.

A small portion of the graph (from 1080 to 1220 h) in Fig. 4 has been zoomed and shown in Fig. 5 to have a clear view of the superimposition of the actual PV power plot on the forecasted PV plot. The exact overlapping of actual and forecasted values shows a single colored graph for most of the portion.

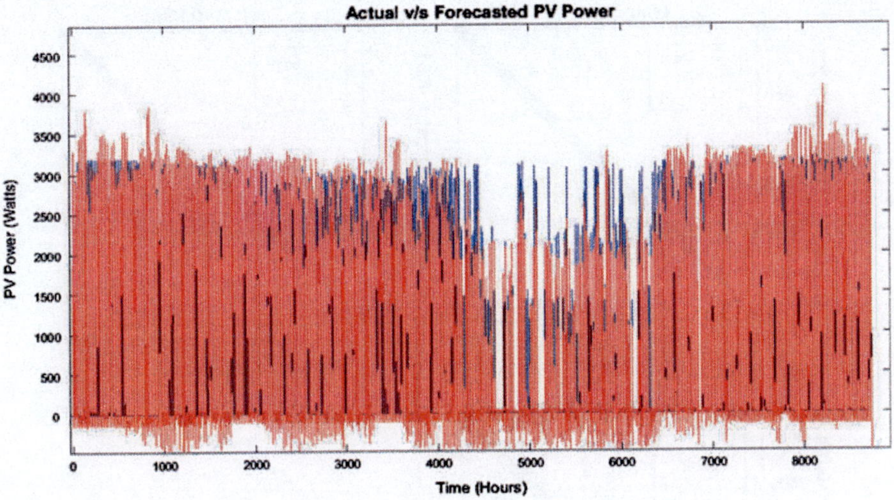

Fig. 4 Variation in actual and forecasted solar PV power

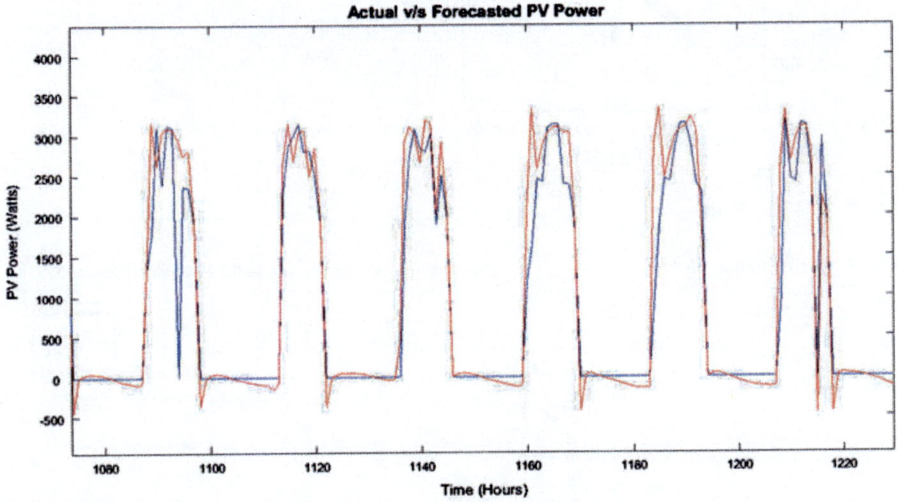

Fig. 5 Exploded view of variation in actual and forecasted solar PV power

The regression plots obtained for the artificial neural network tuned using harmonic search algorithm simulated in MATLAB are shown in Fig. 6. It is a linear graph plotted between the target value and output value. If all the points lie on the 450 lines or the value of slope or R is in the range of 1, it means the neural network is properly trained.

The performance plots obtained for the artificial neural network tuned using harmonic search algorithm simulated in MATLAB is shown in Fig. 7. It is a graph

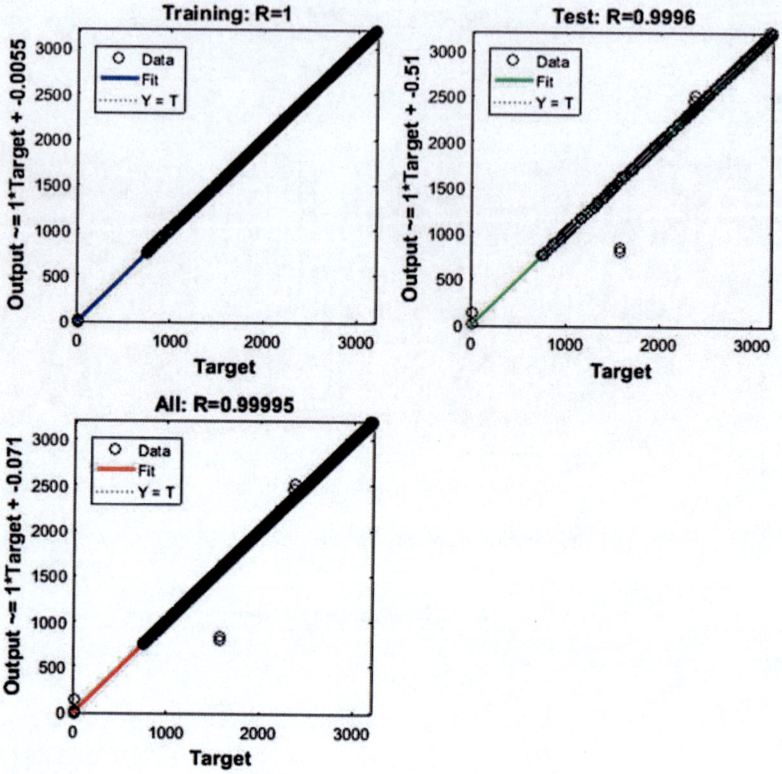

Fig. 6 Regression plot

Fig. 7 Performance plot

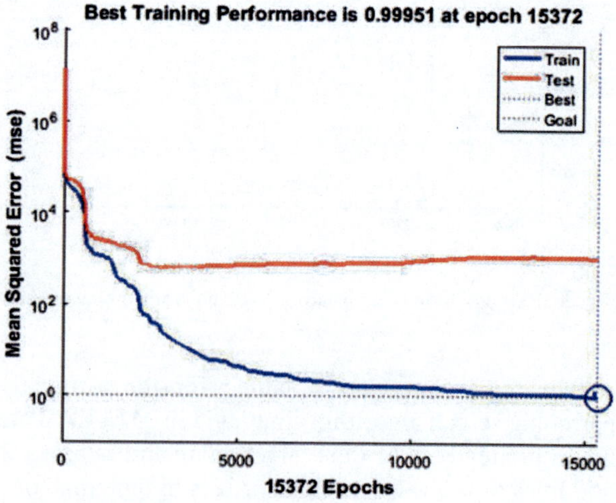

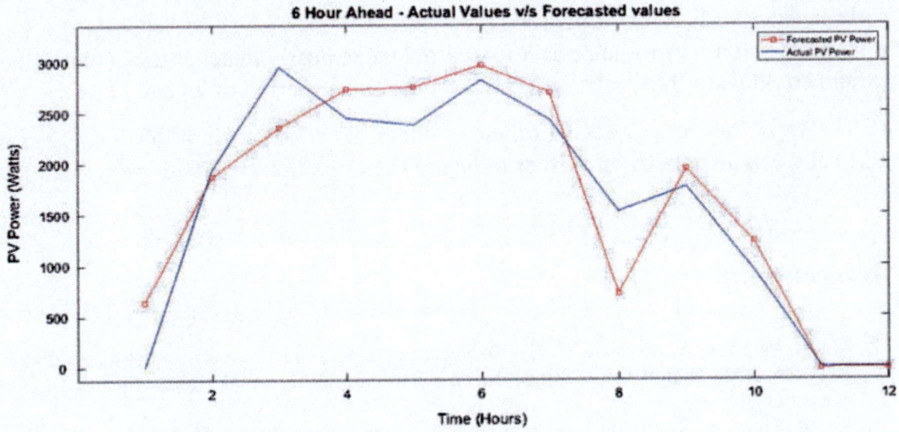

Fig. 8 Forecasted hourly values for 6 h ahead

which show errors during each iteration. This gives an idea about the change in error with each iteration.

4.1 *Six Hours ahead forecasting*

Using the ANN, the next 6 h values can be forecasted. Hereafter the completion of the training, six values of irradiance were fed to the algorithm. The values from 0 to 6 h are for the values of solar PV power for the irradiance values fed. The values from 6 to 12 h are the forecasted value based on its training. The graph showing the values for solar PV power based on solar irradiance values up to 6 h and 6 h ahead of forecasted solar PV power is shown in Fig. 8.

5 Conclusion

Forecasting was done using the artificial intelligence technique, i.e., ANN tuned using genetics-based harmony search algorithm. The work focuses mainly on short-term forecasting of solar PV power. Artificial neural networks are trained using annual hourly data. Training was done using Bayesian Regularization and tuning of weights and bias was done using genetics-based harmony search algorithm. The algorithm was run for 20,000 iterations and a training goal of reducing mean square error up to 1.

- It gave an execution time (i.e., 26.07 min) lower than the interval (i.e., 1 h ahead) for which forecasting is done.

- It gave an MSE value of 0.999
- It gives better performance as it forecasted value nearly matches the actual values and could learn the trends of power change over a period of a year.

The work can be extended further by going for a longer duration (i.e., 12 h or 18 h) and can also focus on further reducing the execution time and error.

References

1. Roy R, Varghese AJ, Awasthi SR (2020) Recent developments in solar forecasting—a review. BITCON 2020-National conference on new horizons in electrical engineering to Combat current challenges
2. Roy R, Vinothini N (2018) Comparative analysis between wind and solar forecasting methods using artificial neural networks and fuzzy logic. J Sci Eng Technol 5
3. Nial FH, Niavand H (2017) Impact of renewable energy consumption on economics in India
4. Alanazi M, Mahoor M, Khodaei A (2017) Two-stage hybrid day-ahead solar forecasting. In: North American power symposium
5. Patel A, Varghese AJ (2017) Simulation of converter based solar PV module, for prediction of output solar power generation using NARX neural network. Int J Sci Res Dev 5
6. Shi J, Lee W-J, Liu Y, Yang Y, Wang P (2012) Forecasting power output of photovoltaic systems based on weather classification and support vector machines. IEEE Trans Indus Appl 48:1064–1069
7. Vanderstar G, Musilek P, Nassif A (2018) Solar forecasting using remote solar monitoring stations and artificial neural networks. IEEE Canadian conference on electrical & computer engineering
8. Srivastava R, Tiwari AN, Giri VK (2018) Forecasting of solar radiation in India using various ANN models. IEEE Uttar Pradesh section international conference on electrical, electronics and computer engineering, 1–6
9. Singh VP, Vaibhav K, Chaturvedi DK (2012) Solar power forecasting modeling using soft computing approach. Nirma University international conference on engineering Ahmedabad
10. Ghanbarzadeh A, Noghrehabadi AR, Assareh E, Behrang MA (2009) Solar radiation forecasting based on meteorological data using artificial neural networks. In: IEEE international conference on industrial informatics, 227–231
11. Abuella M, Chowdhury B (2015) Solar power forecasting using artificial neural networks. In: IEEE North American power symposium
12. Roy R, Varghese AJ, Awasthi SR (2020) ANN optimization for short term forecasting of solar PV power. Anusandhan—RNTU J. X
13. Sharma G, Pandey A, Chaudhary P (2016) Prediction of output solar power generation using neural network time series method. In: 3rd international conference on electrical, electronics, engineering trends, communication, optimization and sciences, 806–808
14. Caputo D, Grimaccia F, Mussetta M, Zich RE (2010) Photovoltaic plants predictive model by means of ANN trained by a hybrid evolutionary algorithm. IEEE, 1–6
15. Snegirev AD, Eroshenko SA, Valiev RT, Khalyasmaa AI (2017) Algorithmic realization of short-term solar power plant output forecasting. IEEE 228–231
16. Li J, Wang R, Zhang T (2016) Wind speed prediction using a cooperative coevolution genetic algorithm based on back propagation neural network. IEEE congress on evolutionary computation, 4578–4583
17. Yang S-X (2018) Neural network forecast under the organic hybrid model of genetic algorithm and particle swarm algorithm. In: International conference on wavelet analysis and pattern recognition IEEE, 254–258

18. Sun W, Wang J, Chang H (2012) Forecasting annual power generation using a harmony search algorithm-based joint parameters optimization combination model. Energies 5:3948–3971
19. Jiao Y, Wu J, Tan Q-K, Tan Z-F, Wang G (2017) An optimization model and modified harmony search algorithm for microgrid planning with ESS. Discrete Dyn Nat Soc 1–11
20. Ehab Elattar E, Salah Elsayed K, Tamer Ahmed F (2021) Hybrid local general regression neural network and harmony search algorithm for electricity price forecasting. IEEE Access, 2044–2054

Chapter 34
Investigation and Implementation of Low Profile Patch Beam Steering Antenna for Vehicular Applications

Ch. Raghavendra, M. Neelaveni Ammal, K. Krishna Sai, and V. S. N. Pranav

1 Introduction

The ambitious goals set for fifth generation communication systems as well as for communication systems requires the adoption of low profile antennas are resonant devices, which operate efficiently over a relatively common to design high-frequency band. There is much more important for steering of main beam in mobile and wireless communications, because of which they can reduce the interference, noise ratio and subsequently multipath fading. Beam steering also exhibits certain advantages such as they can low power and can travel longer distances [1].

Beam steering and pattern reconfiguration for phased array can be achieved by phased array antennas, but the disadvantage of phased array antenna is they need more heavy structures that are more number of antenna elements. As they are of high cost and occupy more space such that they are not preferable in practical applications. So, we have to find such a method that obtains beam steering and pattern reconfiguration that is suitable in the present scenario. Then comes into account of single element antenna which are of low cost and occupies less place and provides beam steering and pattern re-configurability [2]. Pattern reconfigurability can be obtained by varying the current distribution of the antennas and so we use a rectangular square loop but beam steering of the antenna is limited by giving a single port so we use antenna that we fed them by having four ports and sometimes diodes and extra circuits are also used. Beam steering is nothing but changing the direction. To steer the beam in nine different directions, circular patch with excitation of four ports and having diodes is used. For producing beam steering here, we are using circular patches with different

Ch. Raghavendra (✉) · K. K. Sai · V. S. N. Pranav
Department of ECE, VRSEC, Vijayawada, India
e-mail: raghi.2u@vrsiddhartha.ac.in

M. N. Ammal
Department of ECE, SRMIST, Chennai, India
e-mail: neelavem@srmist.edu.in

S. Dawn et al. (eds.), *Smart and Intelligent Systems*, Algorithms for Intelligent Systems,
https://doi.org/10.1007/978-981-16-2109-3_35

Fig. 1 Beam steering

excitations with a coaxial probe (Fig. 1). Cylindrical slots are placed in the antenna and the rectangular patches are placed and the fractal antenna is used [3].

2 Beam Steering Design

In this proposed design, the antenna frequency is taken as 2.4 GHz. The frequency range of antenna for vehicular applications is 1.7–3.78 GHz. The selected operating frequency f = 2.4 GHz [3].

$$\lambda = c/f$$

$$\lambda = 3 \times 10^8/2.45 \times 10^9$$

$$\lambda = 0.125\,\text{cm}.$$

On the ground plane with dimension 100 mm × 100 mm, Taconic TLP-5 substrate of thickness and with permittivity 2.2 is placed, and on that, a square patch of 76 mm × 76 mm is etched [4] (Fig. 2).

Four right-angle-shaped slots with dimensions are cut (formed) in the vicinity of feeding points for achieving 50Ω impedance match The center metal section of 25 mm × 35 mm is removed and replaced with the rectangular fractal antenna. The basic fractal antenna is used and up to the second iteration process has been implemented [5].

The feeding used here is coaxial cable feeding, which is of three layers, where the inner layer is conducting layer for which pec material is assigned, and the outermost layer is assigned by Teflon, and the bottom of the cylinder is closed with a circle, which is assigned with excitation. Now, these feeding ports act as switches to operate different cases.

The iteration factor is 1/3. Hence in the first iteration process, L/3 × W/3 rectangular region is removed and in the second iteration process, L/9 × W/9 rectangular

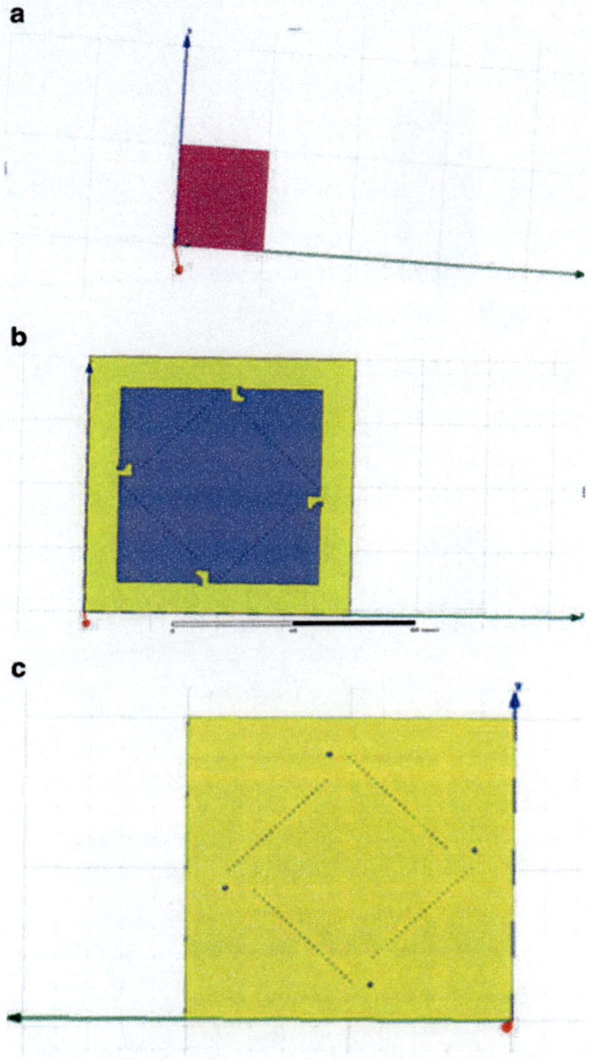

Fig. 2 **a** Basic patch, **b** Front view, **c** Back view, **d** Coaxial cable feeding, **e** Coaxial feeding top view, **f** Inserting fractal antenna on the center metal section, **g** Second iteration of fractal antenna, **h** Lumped port feeding for fractal antenna

region is removed (Table 1). The iteration process involves removing metal surfaces with a particular iteration factor. For this particular fractal antenna, lumped port feeding technique is used (Fig. 3).

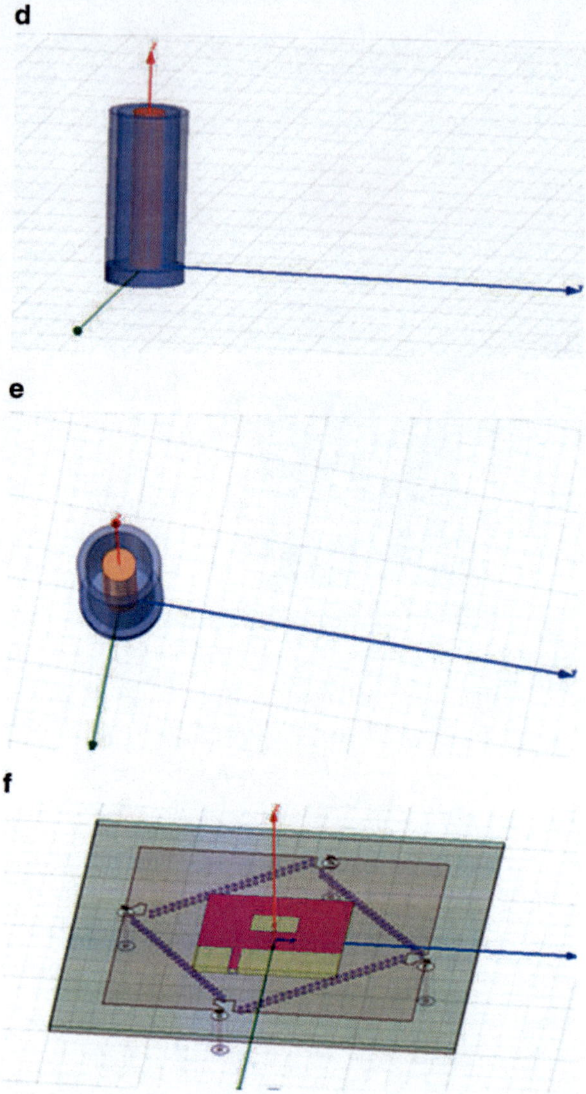

Fig. 2 (continued)

3 Simulated Results

The S11 response shows the designed antenna operates at 2.45 GHz and has a return loss of 14 dB (Figs. 4, 5, 6 and 7).

g

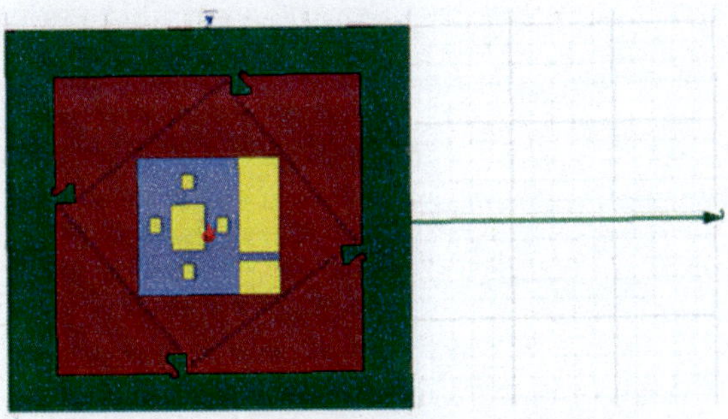

h

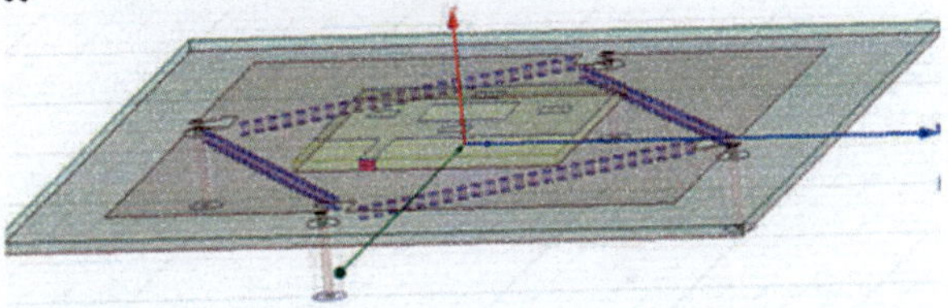

Fig. 2 (continued)

Table 1 Radiation patterns of the proposed antenna for single port excitation and multi-port excitation.

Port	Type of beam	Gain	θ_{max}	Φ_{max}
A	Tilted beam	6.1dBi	30^0	45^0
B	Tilted beam	6.1dBi	30^0	135
C	Tilted beam	6.1dBi	30^0	225^0
D	Tilted beam	6.1dBi	30^0	315^0
A&B	Tilted beam	6.1dBi	30^0	90^0
B&C	Tilted beam	6.1dBi	30^0	180^0
C&D	Tilted beam	6.1dBi	30^0	270^0
D&A	Tilted beam	6.1dBi	30^0	0^0
A&C	Twin tilted beam	5.8dBi	$\pm 36^0$	45^0 & 225^0
B&D	Twin tilted beam	5.8dBi	$\pm 36^0$	135^0 & 315^0

4 Fabricated Antenna

The base antenna is fabricated using Taconic TLP-5 substrate and the center metal section fractal antenna is fabricated using FR4 substrate and the diagonal via are soldered in order to restrict current flow to the corners.

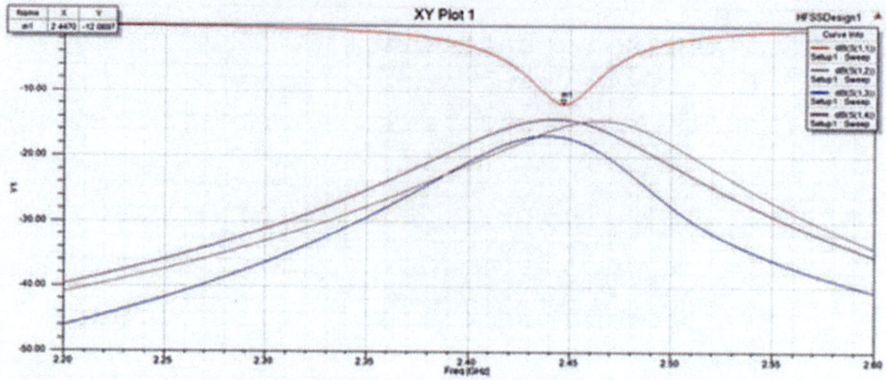

Fig. 3 S11 for designed antenna

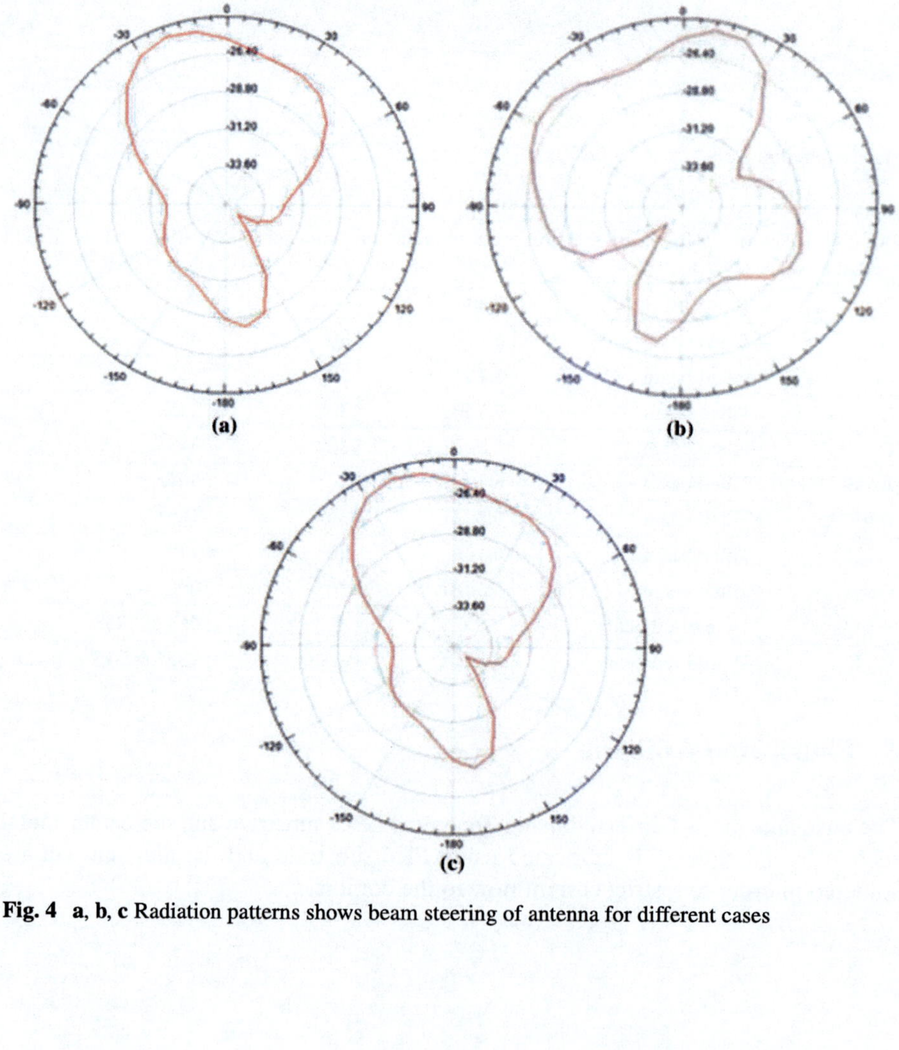

Fig. 4 **a**, **b**, **c** Radiation patterns shows beam steering of antenna for different cases

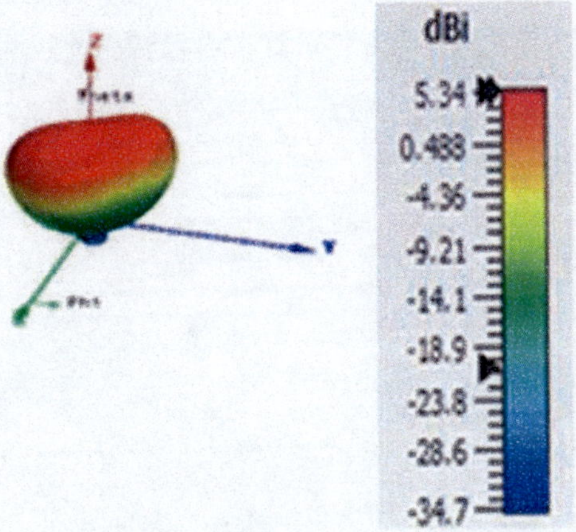

(a)

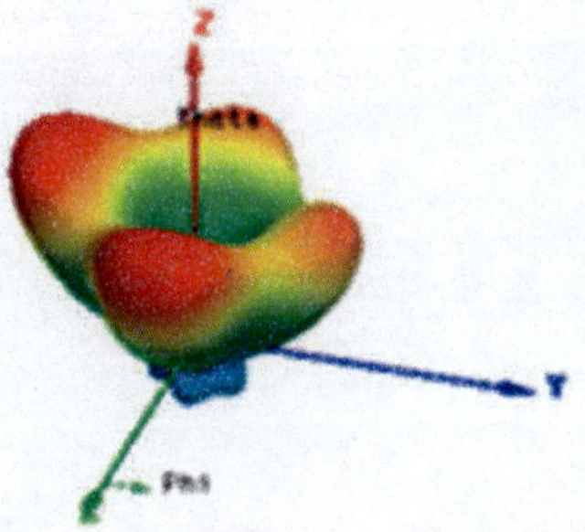

(b)

Fig. 5 **a**, **b** 3D polar plot of the radiation pattern of antenna for different cases

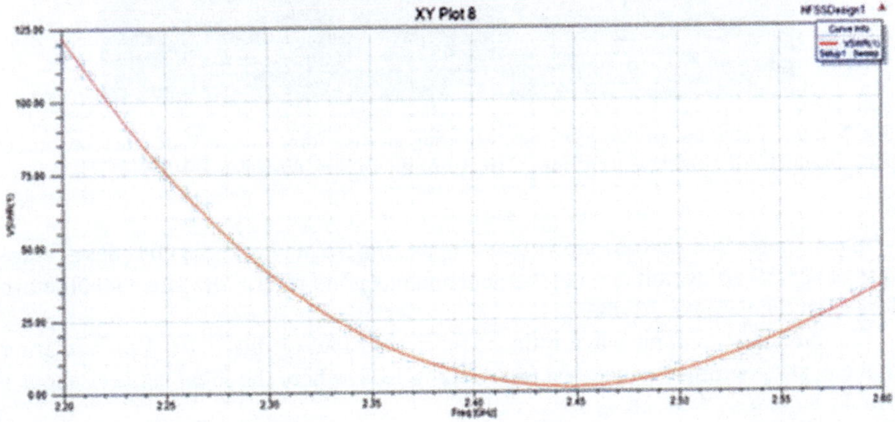

Fig. 6 VSWR for the designed antenna

a

b

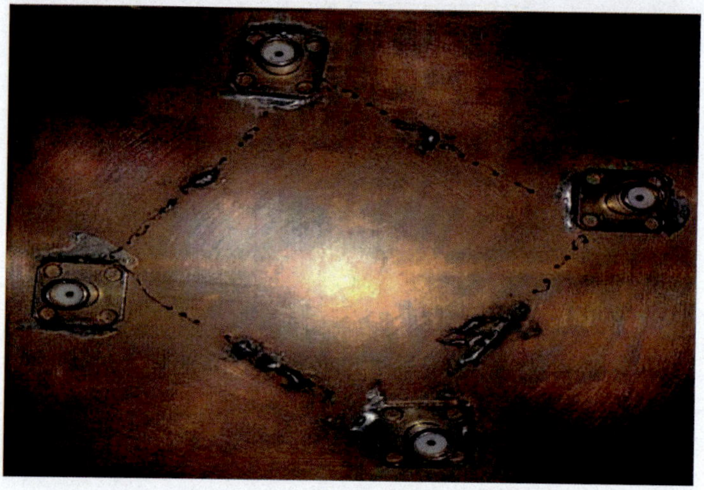

Fig. 7 **a** Fabricated low profile patch beam-steering antenna front view, **b** Fabricated low profile patch beam-steering antenna back view, **c** Measured return loss, **d** Measured VSWR

S parameter (return loss) shows that at operating frequency 2.908 GHz has a return loss of -28.50 db, which depicts that antenna has good return loss since ideal return loss should be above -10 db.

The voltage standing wave ratio of fabricated low profile patch beam-steering antenna at operating frequency 2.908 GHz is 1.08, where the ideal value is about 1 (Table 2).

c

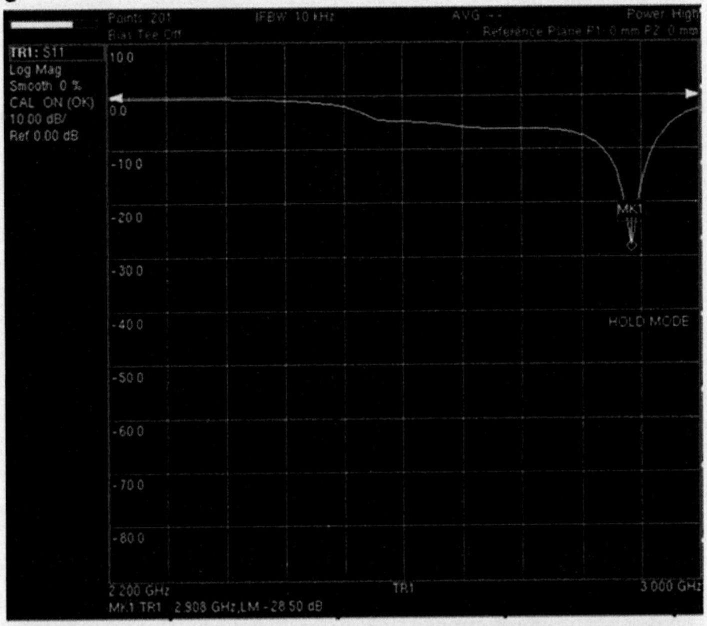

d

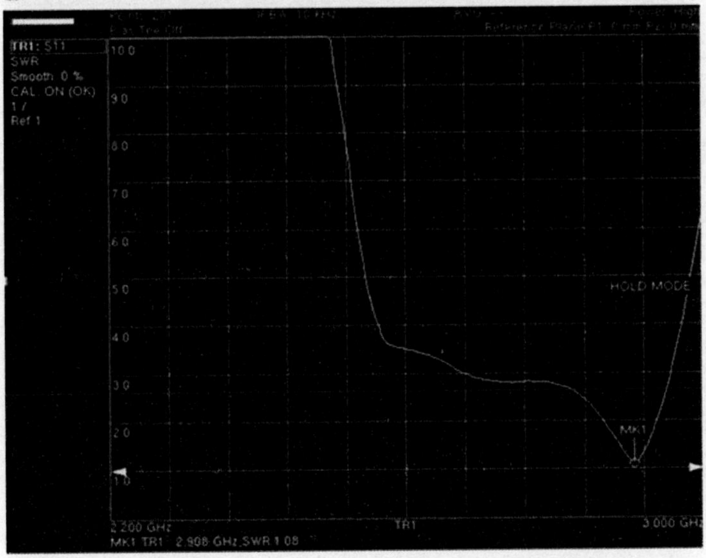

Fig. 7 (continued)

Table 2 Comparison results

Antenna characteristics	Operating frequency (GHz)	Return loss (S11 dB)	VSWR
Simulated	2.45	−14	1
Fabricated	3.0 to 5.7	−28.5	1.08

5 Conclusion

The proposed low-profile patch beam-steering antenna that can steer up to 12 beams is simulated and fabricated in this major project. The characteristics of the proposed antenna are simulated using the HFSS tool. The proposed antenna operates at 2.4 GHz frequency. It can steer up to 12 beams with high gains and also provides a good return loss. Further rectangular fractal antenna up to the second iteration is placed at the center metal section, this has increased gain of the beams formed. The antenna is used for vehicular applications. It provides signal strength at weak signal zones by connecting base stations nearby. The various parameters are measured and compared with fabricated results. The number of iterations can be increased for rectangular fractal antenna, which makes it furthermore low profiles. Also, rectangular fractal antenna can be replaced with Sierpinski carpet fractal antenna, Koch snowflakes and Sierpinski gaskets and many more.

References

1. Nieh CM, Wei C, Lin J (2015) Concurrent detection of vibration and distance using unmodulated CW doppler vibration radar with an adaptive beam-steering antenna. IEEE Trans Microw Theory Tech 63(6):2069–2078
2. Nor MZM, Rahim SKA, Sabran MI, Soh PJ, Vandenbosch GAE (2013) Dual-band, switched-beam, reconfigurable antenna for WLAN applications. IEEE Antennas Wirel Propag Lett 12:1500–1503
3. Pozar DM (1986) Finite phased arrays of rectangular microstrip patches. IEEE Trans Antennas Propag 34(5):658–665
4. Mehta A, Mirshekar-Syahkal D (2004) Spiral antenna with adaptive radiation pattern under electronic control. In: Proceedings of IEEE Antennas propagation society symposium 1. Monterey, CA, USA, pp 843–846
5. Mehta A, Mirshekar-Syahkal D, Nakano H (2006) Beam adaptive single arm rectangular spiral antenna with switches. IEE Proc Microws Antennas Propag 153(1):13–18

Chapter 35
Updated Review on the Classification of Target Tracking Algorithms in Wireless Sensor Networks

Urvashi Saraswat and Anita Yadav

1 Introduction

Wireless Sensor Network (WSN) is an emerging area for research innovations because of its advancements in the cost-effective embedded processors and the technology of wireless transmission. Various areas of applications [1–4], where WSN plays a crucial role, are healthcare [5, 6], military [7], agriculture [8, 9], smart houses [10], robotics [11]. Target tracking is widely used in many real-life applications like environment [12] and wildlife monitoring [13, 14], forest fire detection [15, 16], intruding movements at border [17], etc. A single node or a collaboration of multiple sensors can be used for performing target tracking, where the use of a single node might lead to power loss while multiple sensors would lead to efficient results and save energy. There have been different taxonomies proposed for target tracking algorithms, but there is still no standard classification given to it yet. Some algorithms focus on energy efficiency [18], mobility, security [19], and so on. The paper surveys the existing target tracking techniques and then compares and showcases the discussions on these techniques. Finally, the paper concludes with open issues in target tracking to assist in future research directions. This paper establishes, outlines, and compares some target tracking algorithms that are used in wireless sensor networks based on a specific classification.

1.1 Different Categories of Target Tracking Algorithms

Target tracking is one of the most promising applications in WSNs, which has been given widespread attention due to its involvement in various applications like robotics

U. Saraswat (✉) · A. Yadav
Department of Computer Science and Engineering, Harcourt Butler Technical University, Kanpur, Uttar Pradesh 208002, India

S. Dawn et al. (eds.), *Smart and Intelligent Systems*, Algorithms for Intelligent Systems,
https://doi.org/10.1007/978-981-16-2109-3_36

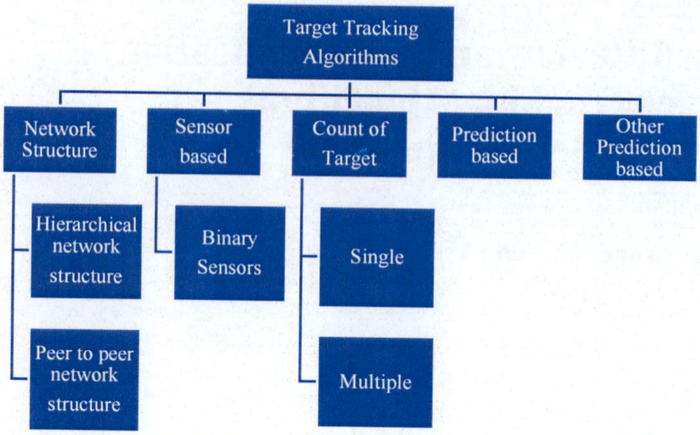

Fig. 1 Different categories of algorithms in target tracking in WSN [23, 24, 27, 28]

[11], human–computer interaction, traffic monitoring [20], intelligent transportation [21], etc. With the implementation of target tracking algorithms in various applications, it enables to detect and monitor the movement of the target be it a person or a vehicle, or any other object intended to be looked upon. The recent implementations of different target tracking-based applications can be seen mainly based on network structure, types of sensors used, type of target, recovery, prediction, and energy efficiency. In this paper, the taxonomy of different target tracking approaches is categorized under network structure, sensor type, count of a target, and prediction strategy as given in Fig. 1.

2 Target Tracking Algorithms for Network Structure

In [22], the author has defined the classification of target tracking algorithms on network structure in four structure types: cluster-based [23], tree-based [24], leader-based, cluster prediction, and tree prediction structures. These five structure types described by the authors use prediction strategies so that they consume less energy and diminish the probability of losing the location of a target. Concurrently, authors in [1–3, 25] have classified the network structure in target tracking into two categories: hierarchical [26, 27] and peer to peer [28]. These categories of network structure are defined with embedded filter-based consensus.

2.1 Target Tracking Algorithms for Hierarchical Network Structure

The hierarchical network structure is the approach of target tracking where the entire methodology is organized in further four categories: tree-based [29], cluster-based [30], hybrid [31, 33], and activation methods [31, 34]. In hierarchical network structures, the communication is performed using multi-hop radio-connectivity between the sensor nodes deployed in the region.

2.2 Tree-Based Hierarchical Network Structure in Target Tracking Algorithms

Tree-based target tracking algorithms are those network structures where nodes are organized in a hierarchical tree-like structure [29]. Here, the sensor nodes have vertices as nodes and edges as the link between nodes to help direct communication. The information is captured and broadcasted by the leaf node. This transmission occurs till the captured information reaches the sink node.

Scalable Tracking Using Networked Sensors (STUN) using Drain and Balance (DAB) Tree: STUN was introduced using the DAB Tree approach and was proposed in [35] for efficient message-pruning. It forms a tree structure to support a large target count and then utilizes target motion patterns to produce message-pruning hierarchies. The STUN approach uses the hierarchy to document the information about the presence of targets [35]. This method of DAB creates an entire tree in a bottom-up fashion where each DAB step has a subset of sensor nodes that integrate into balanced sub-trees. The drawback of STUN is that the DAB cannot replicate the physical sensor network like a logical tree thus, the multiple communication hops exist at the edge of the tree.

Dynamic Convoy Tree-based Collaboration: In the Dynamic Convoy Tree-based collaboration, the network is like a convoy tree-like structure where the main concentration is on the network layer domain [36, 37]. There is a lot of heavy message exchange and broadcasts that happen during the communication. These broadcasts are not much required when the target is moving in a high [38]. The consumed energy increases when there is an increase observed in the number of nodes or any expansion in the monitoring area. There are certain strategies to perform optimization for energy consumption but are not suitable for practical implementation [36].

Deviation Avoidance Tree: The DAT [25] considers each sensor node as a single sub-tree. It is a two-stage process where the first phase is responsible for updates, followed by a query phase. This update phase is revised whenever the target changes its position. During this phase, the target moves from one sensor location to another sensor location. Depending on the series of queries, a substantial amount of energy is consumed, which is counted as one of the drawbacks of the DAT approach.

Zone-based Deviation Avoidance Tree: Similar to that of DAT approach, Z-DAT is also proposed for reducing the update costs [39]. The drawback of energy consumption in DAT is somewhat improved in Z-DAT where the update cost is reduced but lacks to update the query cost [40]. In this approach, the whole coverage is divided into smaller square regions, which individually uses DAT approach, after which these areas are merged. To overcome the failure of reducing the query cost update, Query Cost Reduction (QCR) is designed, which reduces the total update and the query by balancing the tree obtained by DAT or Z-DAT.

Dynamic Object Tracking: The DOT approach proposed in [41] reports the tracking of target to the mobile source. In the process of finding the target, the source releases a request to the sensor nodes to which the neighbor of the target node acknowledges with a reply. The concept of beacon nodes is used by this algorithm where every node ignores the beacon nodes if the target passes through its sensing range. A Gabriel Graph is used by DOT that creates faces in the graph structure area [42]. With the help of these graphs, the area is divided into different regions and the neighbors are formed on the basis of the Gabriel edges.

2.3 Cluster-Based Hierarchical Network Structure in Target Tracking Algorithms

In cluster-based hierarchical network structures, the organization of all sensors is done in form of cluster-based tree [24] and the election of cluster head determines the node, which will be responsible for collecting the target information. This category algorithms are categorized in three types: Static Clustering [43, 44], Dynamic Clustering [3, 45–48], and Space–time Clustering.

In Static Clustering, the coverage area network and the present nodes are all static. The architecture in this type is simple and it is known in advance how to form the network. The advantage of this approach is that the range of communication is higher than their sensing range. Concurrently, the drawback is that it can deal only with constant parameters like speed, whereas varying speed is not considered at all.

In Dynamic Clustering [3, 45, 49, 50], there are many advantages when compared with the static approaches. Here, the nodes are not left useless as each of them can be a part of more than one cluster. Therefore, this approach is considered more energy-efficient and much likely for a large WSN. The only drawback associated with dynamic clustering is it is loosely defined and misses the target recovery procedure. Dynamic Clustering is much efficient for local sensor-based collaboration as the cluster which is formed can change dynamically. Also, there is no consideration of how efficiently the data is sent to the sink [51]. The different types of dynamic clustering approaches are ISDQ (Information-Driven Sensor Querying) [52], DELTA (Distributed Event Localization and Tracking) [47], RARE [48].

Information-Driven Sensor Querying: ISDQ is a dynamic clustering approach that allows sensors to reach an active state when “interesting” events occur around to report [52]. Here, only specific parts of the network with useful information are balanced by the communication cost and are made active. The prime idea behind *information-driven* approach is to conclude the sensor collaboration on the basis of information as well as the constraints of cost and resource consumption. This approach is mainly built on the work of *direct-diffusion routing* method, which successfully deploys for the adhoc sensor networks.

Distributed Event Localization and Tracking: DELTA is the approach that tracks the mobile target moving at a constant speed and which dynamically forms a cluster and selects the CH. The only drawback of this approach is about handling only targets moving with constant speed [47].

RARE-Area (Reduced Area Reporting) and RARE-node (Reduction of Active node REduncancy): RARE algorithm is used in two categories: RARE-AREA and RARE-NODE. The primary purpose of RARE algorithm is to reduce the number of nodes that are involved in the process of tracking. In the RARE-AREA, this algorithm is capable of reducing the number of sensors that are involved in the process of tracking, while in the RARE-NODE algorithm, the redundancy of the data is decreased in the network [48]. The RARE-AREA algorithm mainly consists of two operations. First, it limits the sensors that participate in the process of tracking. Second, it regulates the amount of data that have to be transmitted to the cluster head. When the RARE-NODE algorithm runs on a node without focusing on the redundancy, it first performs check on the neighboring sensors in the vicinity of the sensing range [48].

Dynamic Space–Time Clustering (DSTC): Another type of cluster structure defined under space–time clustering is the Dynamic Space–Time Clustering (DSTC) where the sensor networks use a Closest Point Approach (CPA) technique [49, 52–54]. CPA is defined as the local maximum, which is identified in the time of arrival of an acoustic signal at a sensing node. The main concept used here is based on the neighboring space–time, a defined event and a dynamic window. The main role in this algorithm is performed by the clusters; therefore, the important measure is the average number of clusters present in the region, instead of the total number of nodes present in the sensing region [53].

2.4 Hybrid Methods in Hierarchical Network Structure in Target Tracking Algorithms

Hybrid clustering consists of algorithms like Hierarchical Prediction Strategy (HPS) [30, 55], Dynamic Clustering for Acoustic Tracking (DCAT) [33], Distributed Predictive Tracking (DPT), and Adaptive Multi-feature Framework for-MSPF (AMF-MSPF) [33]. Hierarchical prediction strategy (HPS): HPS was proposed by Wang et al. [55], which also refers to the cluster structure of tracking and further implementing a real-time-based target tracking system. The basic idea of HPS is to sustain efficiency in network management and perform real-time data routing as shown in Fig. 2. There is a periodic change in CHs to balance the energy consumption over

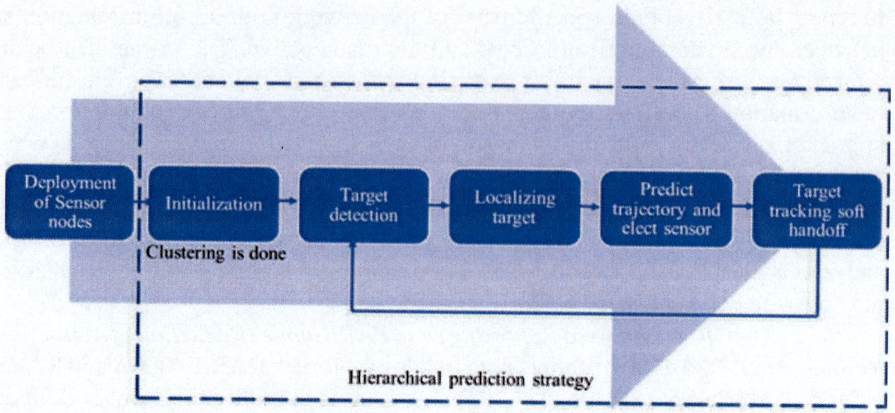

Fig. 2 Organization of hierarchical prediction strategy (HPS) for wireless sensor networks [55]

the sensors, which extends the lifetime of the network.

The process of detection is followed by localizing the target with the implementation of soft handoff scheme; the CH of the selected cluster is responsible to estimate the next location of the target as well as that cluster will be responsible to comprise of the predicted location [30].

Dynamic Clustering for Acoustic Tracking (DCAT): Chen et al. proposed DCAT, the main purpose is to perform tracking for single target. Voronoi diagrams are used by the CHs to assemble an appropriate number of sensor nodes and actively form clusters [56]. There is only one CH that turns into active state on detecting the signal strength, which is more than the predetermined threshold. Now, the CH requests for the sensors in its local area to join the cluster by releasing a broadcast packet. Now the sensor estimates the probabilistic distance between itself and the target, depending on which it decides to reply to the CH or not.

Distributed Predictive Tracking (DPT): In DPT, the approach for tracking uses cluster-based architecture for maintaining scalability and robustness, which yields to the accuracy of the tracking mechanism [56]. Here, several sensors are present in different areas, like on the border or some at specific distance from the border and non-border areas. These sensors present in the border sense actively all the time to detect all the targets, which enter the sensing region. On the contrary, the non-border sensors stay in hibernation until its CH asks to sense. This method has a key design guideline for efficient energy, where the assumption is that most of the sensor nodes are in hibernation at a certain time.

Adaptive Multi-feature Framework for-MSPF (AMF-MSPF): The AMF-MSPF [33] is a hybrid method based on embedding the meanshift (MS) into the particle filter (PF) methodology, where the MS optimization takes over the expensive and imprecise particle validation activity of PF method. This optimization decreases the count of particle by identifying the local nodes for each particle. This leads to the increase in the accuracy of the state of every particle that results in the reduction of the need for large particle count. This algorithm is able to process in real time.

2.5 Activation-Based Methods in Hierarchical Network Structure in Target Tracking Algorithms

The main idea behind these algorithms is the more active sensors are involved in the network, the better is the quality of the tracking method. The different tracking algorithms defined in the literature [51, 57, 58] of activation method are Naïve activation-based tracking (NA), Randomized Activation (RA), Selective activation (SA) and Duty Cycle activation (DA).

Naïve activation-based tracking (NA): The NA tracking scheme is also called direct communication-based tracking and is the simplest approach where all nodes present are active all the time. Each node keeps performing tracking operations until it exhausts its energy and becomes dead [51]. There is a new method defined in [58] for conserving energy. This strategy is considered the most secure and offers the best tracking results. However, it offers the worst energy efficiency because of which it is rarely used.

Randomized Activation (RA): The RA strategy is based on randomized scheduling [59], where every node is active with some possibility. If any node is compromised and makes itself the target tracking node using its random method, then this type of strategy becomes vulnerable. In such cases, such nodes can release wrong information related to the target to the base station.

Selective activation based (SA): In the SA-based prediction, there is only a small subset of all nodes, which is used for tracking at a given time. This algorithm is able to predict the "next" position of the target and uses this prediction to select in which set of nodes should be in tracking mode. The nodes apart from the active mode are under communication mode [51, 57, 58]. These nodes can easily switch to the tracking mode whenever are alerted by the signals of tracking nodes.

Duty Cycle activation (DA): The DA is a type of activation-based algorithm where the whole sensor network keeps turning on and off in a specific period, following a regular duty cycle [60]. The important feature of this activation is that it can be jointly used with any other activation strategy for target tracking.

3 Target Tracking Algorithms for Peer-To-Peer Network Structure

Another defined architecture for target tracking algorithms is the Peer-to-Peer (P2P) WSN, which ensures the sensors of the desired estimations. This strategy establishes single-hop communication between the neighboring nodes. The peer-to-peer network structures are also known as point-to-point networks and sometimes even referred to as flat networks [28]. They possess single hop communication between the deployed sensor nodes. They define the algorithms on the basis of using the embedded filter. They perform target tracking either as centralized or decentralized sensor networks.

3.1 Embedded Filter-Based Peer-To-Peer Network Structure

The beginning of each iteration in the algorithm consists of a single communication step, where the information is exchanged between the sensors and their neighbors. After this, an update step is performed where each sensor utilizes the information from the first step for refining its local estimate.

Distributed Kalman Filter (DKF): The DKF [61] algorithm comprises a network of micro-Kalman filters that are embedded with a high pass and high gain consensus filter. These filters are used to estimate the global information grant by utilizing the local as well as the neighboring information.

Distributed Scalable Sigma-Point Kalman Filter (DSPKF): Unlike the decentralized sigma point Kalman filter [62], DSPKF is entirely distributed and is applicable for large-scale networks. Each node present in the network communicates only with its neighbor. This node estimates the global average information contribution by the local as well as the neighbor node's information using a dynamic consensus strategy.

4 Target Tracking Algorithms as Per the Sensor Type

Sensor nodes in target tracking algorithms can be either or ordinary or binary, where binary sensors are widely used. These binary sensors generate information of the target in one-bit information (either 0 or 1). The low cost and less energy consumed are the two important factors, which make binary sensors very common in use. There are even some drawbacks associated with the binary sensors where they lack the accuracy in estimating the location of target, its direction or even the velocity.

Distributed sensor activation algorithm (DSA^2): A distributed sensor activation algorithm (DSA^2) was proposed to track the mobile targets, which possess high acceleration [51]. In this algorithm, each sensor node is activated with a probability of target detection. On the other hand, the algorithm is also considered less efficient as there is no target recovery model. The authors in [63] have even reported the tracking issues when it comes to catch a group of targets using binary sensor nodes. Hull and Cir algorithms were even proposed to predict the trajectory of the target group. The former is quite expensive as well as complex in its use and the latter lacks in accuracy.

5 Target Tracking Algorithms as Per the Count of Target

The target tracking methods defined as per the count of the target fall into two categories, i.e. single and multiple targets. In Single target tracking, energy consumed is less, hence, making the tracking method efficient. The algorithms in [64–68] are based on a single target tracking mechanism. While chasing the target, the traffic generated in the network is low, which maintains the efficiency of the algorithms using single target tracking. Multiple target tracking is challenging and complex because of the speed and direction variations of the targets. The approaches in [51] use multiple target tracking mechanisms. Here, a sensor node can obtain more than one measurement, thus making it difficult to map the observation with the right target.

6 Other Prediction-Based Target Tracking Algorithms

Target tracking approaches are also categorized with respect to the estimation of future movement and position of the target(s). *Dual Prediction Reporting (DPR)*: In the DPR, the sensor node with a target in its territory has to predict the movements of the target for the upcoming reporting interval [69]. This algorithm reduces the count of long distance-based transmissions between the sensing node and the base station. This minimization is done with a suitable overhead to lower the energy consumption. While the sensor node predicts the movements, the base station also performs the same predictions using the same target movement history. The transmission performed here is single hop, between the neighbor sensors in which target movement history is exchanged. Therefore, this makes it more energy-efficient as compared to when there is multi-hop transmission performed.

Prediction-based energy saving scheme (PES): The PES consists of three mechanisms: the prediction model mechanism, wakeup mechanism, and the recovery mechanism. This scheme reduces the sampling frequency and the node count and stabilizes the overhead, which occurs because of missing targets [70]. The wakeup mechanism is based on heuristics where energy and performance are the main factors. These factors are considered to decide which sensor node has to be activated. This algorithm considers that if any sensor node is not performing target tracking, it should remain in sleep mode as long as possible.

Prediction-based energy-efficient target tracking protocol (PET): In a prediction-based energy-efficient target tracking protocol (PET), it identifies the path moved by the target and utilizes the different patterns moved by the target, which saves energy for target tracking in sensor networks [66]. A linear predictor is used to estimate the next location of the target. This protocol reduces the messages, which were exchanged between the beacon and the sensors and trackers, by simplifying the beacon's computation.

Predictive-based tracking technique using sequential pattern (PTSP): PTSP algorithm is used to predict the future movements of the target using the least number of sensor nodes. There are two stages involved: Generation of sequence pattern followed by tracking and monitoring of target. In the pattern generation stage, a prediction model is formed using a wide data that are collected from the sensor network, which produces the inherited behavioral patterns of movements by the target [71].

Exponential distributed predictive tracking (EDPT): The EDPT algorithm is considered suitable to perform tracking using reduced computation complexity [72]. It can predict the location of the target without any noise state or target matrix. This algorithm is considered to serve with higher precision as well as lower computation complexity. The protocol of EDPT is best suitable for target tracking and estimating the position of the target with lower computational complexity.

Prediction-based optimistic object tracking (POOT): Hsu et al. proposed in [65] the POOT strategy for tracking mobile targets using face routing and linear movement prediction mechanism. Two other schemes were proposed by the authors: Time-efficient object recovery scheme (TORS) and communication-efficient object recovery scheme (CORS) for improvement in the target recovery phase. TORS is designed to locate the lost target and wakes up all the sensor nodes in a circular area. CORS is designed to use minimum communication for locating the lost target using the direction of path and distance. This algorithm focuses on integrating and maintaining the tracking information to obtain better performance results concerning the communication cost.

7 Discussion and Conclusion

Target tracking has become popular in recent years because of its wide range of applications such as in the field of military, wildlife, environment, healthcare, etc. The use of target tracking in wireless sensor networks in the different application mediums encouraged us to present this survey. There are many reasons behind hampering the tracking approaches, which often lead to the loss of target, like—obstacles in the network, communication failures, abrupt modifications in the velocity or direction of the target, low energy resources or battery-driven sensor nodes (faster energy depletion), inaccurate target estimation (location), etc. Algorithms possessing these disadvantages may result in the loss of target might not implement any recovery mechanisms for finding the missed target. However, tree-based algorithms are a good choice for tracking but they possess a high overhead cost.

Algorithms like FOTP and POOT yield more accurate results as they integrate face structure along with the prediction strategy and utilize recovery mechanisms in case of target loss. However, while treating large-scale networks this recovery process might take a long time in the extraction of target position (exact).

In this paper, we have categorized the different target tracking algorithms that are currently used in wireless sensor networks. Our survey analyses that most of these algorithms focus on parameters like energy efficiency or the accuracy, and not on the security.

8 Open Research and Future Directions

In target tracking, many open research issues are into call, where maximum issues lie under the mobile target(s). Table 1 summarizes the target tracking algorithms in this survey and focuses either on accuracy or fault-tolerant, and not on the security aspects of it. As depicted from the survey, it can be seen that the number of targets or even the difference in the speed or direction of the targets can vary the results. Therefore, the future research direction for this problem is to coin a target tracking algorithm, which is accurate in results without getting affected by the change in speed or direction. Another most important factor about algorithms in WSNs is about security. Thus, security along with target tracking can turn beneficial for many research fields. Maximum algorithms covered in the above analysis focus on maintaining energy with accuracy, thus, make headways to propose a target tracking algorithm focusing on security.

Table 1 Comparison of target tracking algorithms

Target Tracking algorithms	Performance		Security		
	Accuracy in tracking	Fault tolerant	Confidentiality	Authentication	Integrity
HPS [30]	No	Yes	No	No	No
STUN [35]	No	Yes	No	No	No
DCTC [37]	No	Yes	No	No	No
DAT [25]	No	No	No	No	No
DELTA [47]	No	Yes	No	No	No
RARE [48]	No	No	No	No	No
DSA^2 [51]	Yes	No	No	No	No
DCAT [33]	Yes	Yes	No	No	No
DPT [57]	No	Yes	No	No	No
DKF [62]	Yes	No	No	No	No
DSKPF [62]	Yes	No	No	No	No
DPR [69]	Yes	No	No	No	No
PET [66]	Yes	Yes	No	No	No
EDPT [72]	Yes	Yes	No	No	No
POOT [65]	No	Yes	No	No	No

References

1. Bhatti S, Jie X (2009) Survey of target tracking protocols using wireless sensor network. In: IEEE Fifth international conference on wireless and mobile communications, pp 110–115
2. Fayyaz M (2011) Classification of object tracking techniques in wireless sensor networks. Wirel Sens Netw 3(4):121
3. Li J, Yan Z (2010) Target tracking in wireless sensor networks. wireless sensor networks: application-centric design, pp 1–20
4. Nandhini M, Sarma Dhulipala VR (2012) Energy-efficient target tracking algorithms in wireless sensor networks: an overview. Int J Comput Sci Technol 3(1):66–71
5. Al Ameen M, Jingwei L, Kyungsup K (2012) Security and privacy issues in wireless sensor networks for healthcare applications. J Med Syst 36(1):93–101
6. Alemdar H, Ersoy C (2010) Wireless sensor networks for healthcare: a survey. Comput Netw 54(15):2688–2710
7. Diamond Scott M, Marion GC (2007) Application of wireless sensor network to military information integration. In: 5th IEEE international conference on industrial informatics, 1:317–322
8. Srbinovska M, Gavrovski C, Dimcev V, Krkoleva A, Borozan V (2015) Environmental parameters monitoring in precision agriculture using wireless sensor networks. J Clean Prod 88:297–307
9. Sahota Hn, Ratnesh K, Ahmed K, Jing H (2010) An energy-efficient wireless sensor network for precision agriculture. In: The IEEE symposium on computers and communications, pp 347–350
10. Li M, Lin H-J (2014) Design and implementation of smart home control systems based on wireless sensor networks and power line communications. IEEE Trans Industr Electron 62(7):4430–4442
11. Onate JMB, Dario JMC, Nancy del Rocio VE (2017) Tracking objects using artificial neural networks and wireless connection for robotics. J Telecommun Electron Comput Eng (JTEC) 9(1–3):161–164
12. BenSaleh, MS, Raoudha S, Yessine HK, Mohamed A (2020) Wireless sensor network design methodologies: a survey. J SensS
13. Othman MF, Khairunnisa S (2012) Wireless sensor network applications: a study in environment monitoring system. Procedia Eng 41:1204–1210
14. Wamuyu PK (2017) A conceptual framework for implementing a WSN based cattle recovery system in case of cattle rustling in kenya. Technologies 5(3):54
15. Jan MA, Priyadarsi N, Xiangjian H, Ren PL (2018) A Sybil attack detection scheme for a forest wildfire monitoring application. Futur Gener Comput Syst 80:613–626
16. Tsao S-L, Huang C-H (2011) A survey of energy efficient MAC protocols for IEEE 802.11 WLAN. Comput Commun 34(1):54–67
17. Yang T, Dejun Mu, Wei Hu, Zhang HuiXiang (2014) Energy-efficient border intrusion detection using wireless sensors network. EURASIP J Wirel Commun Netw 2014(1):46
18. Demigha O, Hidouci W-K, Ahmed T (2012) On energy efficiency in collaborative target tracking in wireless sensor network: a review. IEEE Commun Surv & Tutor 15(3):1210–1222
19. Oracevic A, Suat O (2014) Secure and reliable prediction based target tracking for wireless sensor networks. In: 5th international conference on intelligent systems. Modell Simul 646–651
20. Tian B, Qingming Y, Yuan G, Kunfeng W, Ye L (2011) Video processing techniques for traffic flow monitoring: a survey. In: 14th international IEEE conference on intelligent transportation systems (ITSC), pp 1103–1108
21. Tubaishat M, Zhuang P, Qi Qi, Shang Yi (2009) Wireless sensor networks in intelligent transportation systems. Wirel Commun Mob Comput 9(3):287–302
22. Naderan M, Mehdi D, Hossein P (2009) Mobile object tracking techniques in wireless sensor networks. In: 2009 international conference on ultra modern telecommunications & workshops, pp 1–8

23. Feng J, Shi X, Zhang J (2018) Dynamic cluster heads selection and data aggregation for efficient target monitoring and tracking in wireless sensor networks. Int J Distrib Sens Netw 14(6):1550147718783179
24. Chen M-X, Che-Chen H, Wen-Yen W (2010) Dynamic object tracking tree in wireless sensor network. EURASIP J Wirel Commun Netw 1:386319
25. Lin C-Y, Wen-Chih P, Yu-Chee T (2006) Efficient in-network moving object tracking in wireless sensor networks. IEEE Trans Mob Comput 5(8):1044–1056
26. Chamberland J-F, Venugopal VV (2004) Asymptotic results for decentralized detection in power constrained wireless sensor networks. IEEE J Sel Areas Commun 22(6):1007–1015
27. Oh S, Shankar S, Luca S (2005) A hierarchical multiple-target tracking algorithm for sensor networks. In: Proceedings of the 2005 IEEE international conference on robotics and automation, pp 2197–2202
28. Wang X, Wang S, Bi D-W, Ma J-J (2007) Distributed peer-to-peer target tracking in wireless sensor networks. Sensors 7(6):1001–1027
29. Long T, Liang Z, Liu Q (2019) Advanced technology of high-resolution radar: target detection, tracking, imaging, and recognition. Sci China Inf Sci 62(4):40301
30. Wang Z-B, Wang Z, Chen H-L, Li J-F, Li H-B, Shen J (2013) HierTrack: an energy-efficient cluster-based target tracking system forwireless sensor networks. J Zhejiang Univ Sci C 14(6):395–406
31. Zhou Y, Li JX, Wang DL (2012) Target tracking in wireless sensor networks using adaptive measurement quantization. Sci China Inf Sci 55(4):827–838
32. Oiwa D, Shinji F, Yuji I, Boonserm K, Tsuyoshi N, Manas KB (2017) Tracking with extraction of moving object under moving camera environment. Procedia Comput Sci 112:1479–1487
33. Khattak AS, Gulistan R, Nadeem A (2018) Adaptive framework for multi-feature hybrid object tracking. Appl Sci 8(11):2294
34. Park J, Wonsang H, Wook B, Chang-Hun L, Tae-Il K, Muhammad MS, Kwang-Soo K (2011) Pan/tilt camera control for vision tracking system based on the robot motion and vision information. IFAC Proc 44(1) 3165–3170
35. Kung H-T, Dario V (2003) Efficient location tracking using sensor networks. In: IEEE wireless communications and networking WCNC, vol 3, pp 1954–1961
36. Luo J, Wang J, Huazhong Xu, Hanqing Lu (2016) Real-time people counting for indoor scenes. Signal Process 124:27–35
37. Zhang W, Cao G (2004) DCTC: dynamic convoy tree-based collaboration for target tracking in sensor networks. IEEE Trans Wireless Commun 3(5):1689–1701
38. Jeeva S, Sivabalakrishnan M (2015) Survey on background modeling and foreground detection for real time video surveillance. Procedia Comput Sci 50:566–571
39. Lin Chih-Yu, Yu-Chee Tseng (2004) Structures for in-network moving object tracking in wireless sensor networks. In: First international conference on broadband networks, pp 718–727
40. Yi C, Cui L, Luo C (2015) Moving target tracking algorithm based on improved optical flow technology. Open Autom Control Syst J 7(1)
41. Reboul L, Kieffer M, Piet-Lahanier H, Reynaud S (2019) Cooperative guidance of a fleet of UAVs for multi-target discovery and tracking in presence of obstacles using a set membership approach. IFAC-PapersOnLine 52(12):340–345
42. Xu C, Daqing H, Jianye L (2019) Target location of unmanned aerial vehicles based on the electro-optical stabilization and tracking platform. Measurement 147:106848
43. Darabkh KA, Ismail SS, Al-Shurman M, Jafar IF, Alkhader E, Al-Mistarihi MF (2012) Performance evaluation of selective and adaptive heads clustering algorithms over wireless sensor networks. J Netw Comput Appl 35(6):2068–2080
44. Teng J, Snoussi H, Richard C (2012) Prediction-based cluster management for target tracking in wireless sensor networks. Wirel Commun Mob Comput 12(9):797–812
45. Chen W-P, Hou JC, Sha L (2004) Dynamic clustering for acoustic target tracking in wireless sensor networks. IEEE Trans Mob Comput 3(3):258–271
46. Huang K-F, Hsing-Hsien W, Wei-Jie W, Ying-Hong W (2012) A dynamic tracking mechanism for mobile target in wireless sensor networks. In: International symposium on intelligent signal processing and communications systems, pp 822–826

47. Wälchli M, Piotr S, Michael M, Torsten B (2007) Distributed event localization and tracking with wireless sensors. In: International conference on wired/wireless internet communications. Springer, Berlin, Heidelberg, pp 247–258
48. Olule E, Guojun W, Minyi G, Mianxiong D (2007) Rare: an energy-efficient target tracking protocol for wireless sensor networks. In: International conference on parallel processing workshops (ICPPW), pp 76–76
49. Shang L, Kang Z, Zhengguo C, Dan G, Maolin H (2014) An energy-efficient collaborative target tracking framework in distributed wireless sensor networks. Int J Distrib Sens Netw 10(7):396109
50. Phoha S, Koch J, Grele E, Griffin C, Madan B (2005) Space-time coordinated distributed sensing algorithms for resource efficient narrowband target localization and tracking. Int J Distrib Sens Netw 1(1):81–99
51. Chen J, Cao K, Li K, Sun Y (2011) Distributed sensor activation algorithm for target tracking with binary sensor networks. Clust Comput 14(1):55–64
52. Zhao F, Shin J, Reich J (2002) Information-driven dynamic sensor collaboration. IEEE Signal Process Mag 19(2):61–72
53. Phoha S, Noah J, David F, Richard B (2003) Sensor network based localization and target tracking through hybridization in the operational domains of beamforming and dynamic space-time clustering. In: GLOBECOM'03. IEEE global telecommunications conference (IEEE Cat No 03CH37489), vol 5, pp 2952–2956
54. Friedlander D, Christopher G, Noah J, Shashi P, Richard RB (2003) Dynamic agent classification and tracking using an ad hoc mobile acoustic sensor network. EURASIP J Adv Signal Process 4:819145
55. Wang Z, Hongbin L, Xingfa S, Xice S, Zhi W (2008) Tracking and predicting moving targets in hierarchical sensor networks. In IEEE international conference on networking, sensing and control, pp 1169–1173
56. Wang Z, Wei L, Zhi W, Junchao M, Honglong C (2013) A hybrid cluster-based target tracking protocol for wireless sensor networks. Int J Distrib Sens Netw 9(3):494863
57. Yang H, Biplab S (2003) A protocol for tracking mobile targets using sensor networks. In: Proceedings of the first IEEE international workshop on sensor network protocols and applications, pp 71–81
58. Pattem S, Sameera P, Bhaskar K (2003) Energy-quality tradeoffs for target tracking in wireless sensor networks. In: Information processing in sensor networks. Springer, Berlin, Heidelberg, pp 32–46
59. Zhou W, Weiren S, Xiaogang W, Kai W (2012) Adaptive sensor activation algorithm for target tracking in wireless sensor networks. Int J Distrib Sens Netw 8(6):515906
60. Díaz MO, Kin KL (2010) Randomized scheduling algorithm for data aggregation in wireless sensor networks. In: European wireless conference (EW), pp 42–48
61. Olfati-Saber R (2007) Distributed Kalman filtering for sensor networks. In: 46th IEEE conference on decision and control, pp 5492–5498
62. Zhou Y, Jianxun L (2009) Distributed sigma-point Kalman filtering for sensor networks: dynamic consensus approach. In: IEEE international conference on systems, man and cybernetics, pp 5178–5183
63. Cao D, Jin B, Das SK, Cao J (2010) On collaborative tracking of a target group using binary proximity sensors. J Parallel Distrib Comput 70(8):825–838
64. Kong J-I, Jin-Woo K, Doo-Seop E (2014) Energy-aware distributed clustering algorithm for improving network performance in WSNs. Int J Distrib Sens Netw 10(3):670962
65. Hsu JM, Chao CC, Chia CL (2011) Short-term prediction-based optimistic object tracking strategy in wireless sensor networks. In: IEEE fifth international conference on innovative mobile and internet services in ubiquitous computing, pp 78–85
66. Bhuiyan MZA, Guo-Jun W, Li Z, Yong P (2010) Prediction-based energy-efficient target tracking protocol in wireless sensor networks. J Cent South Univ Technol 17(2):340–348
67. Deldar F, Mohammad HY (2011) Designing a prediction-based clustering algorithm for target tracking in wireless sensor networks. In: International symposium on computer networks and distributed systems (CNDS), pp 199–203

68. Jiang B, Ravindran B, Cho H (2012) Probability-based prediction and sleep scheduling for energy-efficient target tracking in sensor networks. IEEE Trans Mob Comput 12(4):735–747
69. Xu Y, Julian W, Lee W-C (2004) Dual prediction-based reporting for object tracking sensor networks. In: IEEE the first annual international conference on mobile and ubiquitous systems: networking and services MOBIQUITOUS, pp 154–163
70. Xu Y, Julian W, Wang-Chien L (2004) Prediction-based strategies for energy saving in object tracking sensor networks. In IEEE international conference on mobile data management, Proceedings, pp 346–357
71. Samarah S, Al-Hajri M, Boukerche A (2010) A predictive energy-efficient technique to support object-tracking sensor networks. IEEE Trans Veh Technol 60(2):656–663
72. Xue L, Liu Z, Guan X (2011) Prediction-based protocol for mobile target tracking in wireless sensor networks. J Syst Eng Electron 22(2):347–352

Chapter 36
Performance Evaluation of Various Traditional Controllers in Automatic Generation Control of Multi-Area System with Multi-Type Generation Units

CH. Naga Sai Kalyan and Chintalapudi V. Suresh

1 Introduction

The rapid industrialization in developing countries like India, urbanization in people's lifestyle tends to depend more on electric power, which leads to faster growth in power system. However, the quick growth in the power system results in few uncertainties of individual power outages and disruptions which in turn affects the country's economy. Apart from this, the major issue facing the power sector is establishing the equilibrium with demand and supply of real power, failing will lead to more fluctuations in frequency and sometimes may resulting in system instability. To avoid this, there is a need for an automatic generation controller (AGC), which continuously monitors the disruptions in system behaviour under varying load conditions. AGC is an ancillary service comprised of primary and secondary control loops. Small deviations in the system performance can be governed by the primary regulator. If the deviations are greater than the capability of the governor's dead band, then a secondary regulator comes into action.

From the past few years, secondary regulator design for AGC of power systems is the trend in recent research. In [1], various AGC strategies on different test system models are consolidated in an extensive manner. AGC of conventional generation sources and a combination of conventional- and non-conventional-based generation sources had also been reported in the literature [2]. Irrespective of the power system models, most of the researchers are majorly concentrated on the design of AGC controllers utilizing the mechanism of soft computing techniques [3]. Various controllers like PI/PID [4] and degree of freedom (DOF) controllers, fractional order (FO) [5], combination of traditional and FO based, intelligent fuzzy and neural network [6], cascade of fuzzy and FO-based controllers [7] are proposed by the

CH. Naga Sai Kalyan (✉) · C. V. Suresh
EEE Department, Vasireddy Venkatadri Institute of Technology, Guntur, India

S. Dawn et al. (eds.), *Smart and Intelligent Systems*, Algorithms for Intelligent Systems,
https://doi.org/10.1007/978-981-16-2109-3_37

researchers in their work. In terms of design, conventional controllers of PI/PID are simpler when compared to other FO and intelligent controllers. Whereas the FO controllers involve additional parameters to optimize and the intelligent types require more approximations which definitely affects the controller robustness [8]. Moreover, optimization of conventional controllers imposes fewer burdens for computation and also gets optimal values in quick time, which attracts the researchers to implement traditional controllers in the present study of AGC.

However, the selection of optimization techniques also plays a vital role in controller performance. Different optimizations that are employed by the researchers are particle swarm optimization (PSO), dragonfly algorithm (DFA) [9], differential evolution (DE), grey wolf optimizer (GWO) [2], imperialist competitive algorithm (ICA) [7], artificial electric field (AEFA), ant lion optimizer (ALO) [10], DE-AEFA [11] algorithms, etc. Moreover, HAEFA is the latest hybrid optimization which proves its performance superiority and stated as best suitable for engineering optimization problems motivates the author to implement in this work.

The contributions of this paper are as follows:

(a) Dual area system is designed in MATLAB/Simulink.
(b) PIDN tuned with HAEFA algorithm is implemented as secondary regulator whose efficacy is deliberated with PI/PD and PID controllers.
(c) TCSC-SMES control mechanism is implemented for further performance enhancement.
(d) Boldness of the presented control approach is validated through sensitivity analysis.

2 Investigative Model

Equal capacities of dual area systems consisting of multi-type generation sources like thermal, hydro and gas units are considered for study in this work. The model of the system is rendered in Fig. 1 and is developed in MATLAB/Simulink domain, the necessary time and gain constant parameters are directly taken from [4]. System performance is examined upon subjugating a step load change of 1% (SLP) on only Area-1. Secondary controller engaged in this present work is traditional controllers, which are optimized with a new hybrid algorithm of HAEFA having the capability of preserving the reconciliation within the features of exploitation and exploration. Optimization is performed subjected to the minimization of error squared over the integral [12] function given in Eq. (1).

$$J = \int_{0}^{T} (\Delta f_1^2 + \Delta P_{tie12}^2 + \Delta f_2^2)dt \quad (1)$$

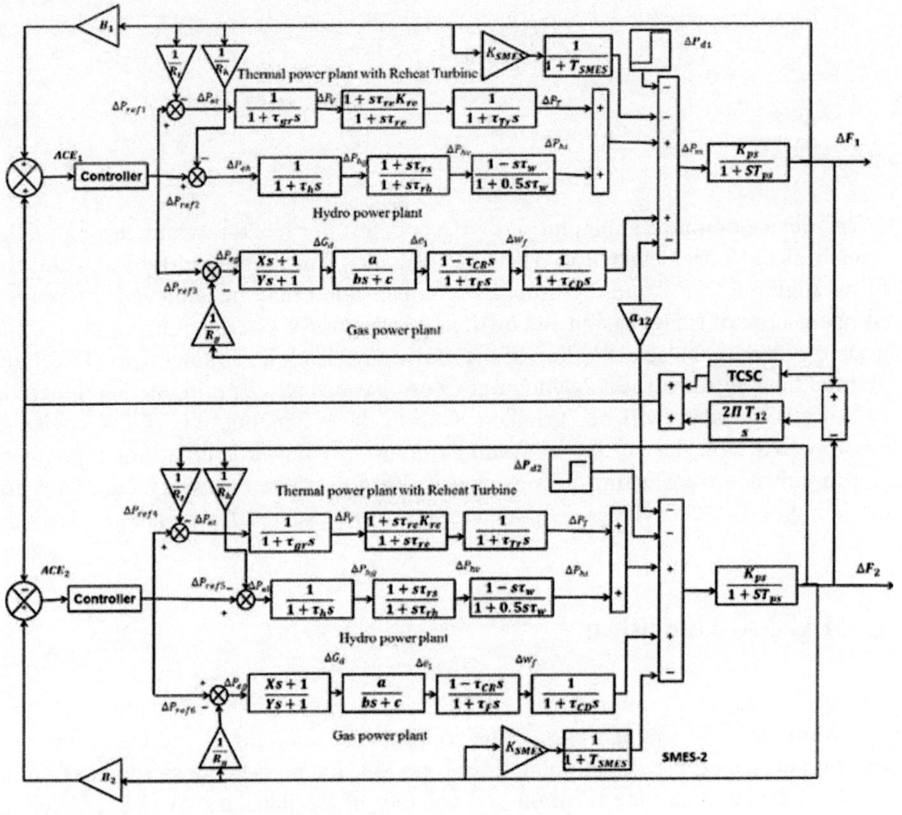

Fig. 1 Dual area system with multi-type generation units

3 Coordinated Strategy of SMES-TCSC

3.1 SMES

SMES works on the basic principle idea of charging the magnet build-up of semiconducting material of nearly zero resistance, which is stored at cryogenic temperature to behave as superconductors. The magnet charging will be done at off-peak durations and stores in the form of DC. That DC will be converted back to AC by making use of control and power conditioning units whenever necessary. Further, these devices having the tendency of damping out the inter-area oscillations thereby leading to an enhancement in power transfer feasibility. Moreover, SMES acts as a spinning reserve and can meet the load demand during the period of generator outages. The model of SMES is given in Eq. (2) and the parameters of $K_{SMES} = 1$ and $T_{SMES} = 0.98$ is employed in this work.

$$G_{SMES} = \frac{K_{SMES}}{1 + ST_{SMES}} \tag{2}$$

3.2 *TCSC*

TCSC device comprises the unit of thyristor controller reactor, which has a parallel branch of the series capacitor. The line reactance can be altered by adjusting the firing angle of TCSC and then the line compensation may be achieved. Moreover, compensation of transmission can be done continuously and dynamic control of the flow of power in the selected line of the meshed network be attained with TCSC for optimal flow. Further, the subsynchronous oscillations will be mitigated and thereby DC off-set voltages will be shrunken. Due to these benefits, the TCSC device is incorporated with the tie-line to eliminate inter-area power oscillations effectively and to enhance transferring power capacity (Fig. 2). The necessary equations for modeling of TCSC device in dual area system are considered from [13].

4 HAEFA Algorithm

HAEFA algorithm is proposed by [4] in 2020 and is applied to the study of LFC. In this algorithm, initially, the particles of the controller are initialized and the simulation will be performed for the stipulated time period. Later, the fitness function value will be calculated and the position and velocity of the particles will be updated as mentioned in [4]. After completion of the third iteration, the pairwise comparison will be performed among fitness values and the populations of the best fitness values are only entered into the second stage. By doing so, the total population that enters into a further stage becomes half and is efficient in locating the final best solution there by the computational burden on the algorithm will be relieved. This is the key most aspect of this HAEFA algorithm. In this work, the HAEFA algorithm is executed for a maximum of 50 iterations for a total of 100 populations. The mathematical analysis required to design this algorithm in (.m file) format are considered from [4] and the flowchart is depicted in Fig. 3.

Fig. 2 Single line diagram of the dual area system with TCSC and SMES

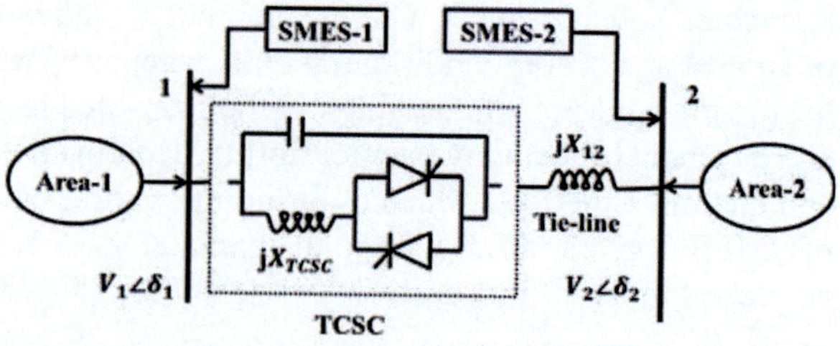

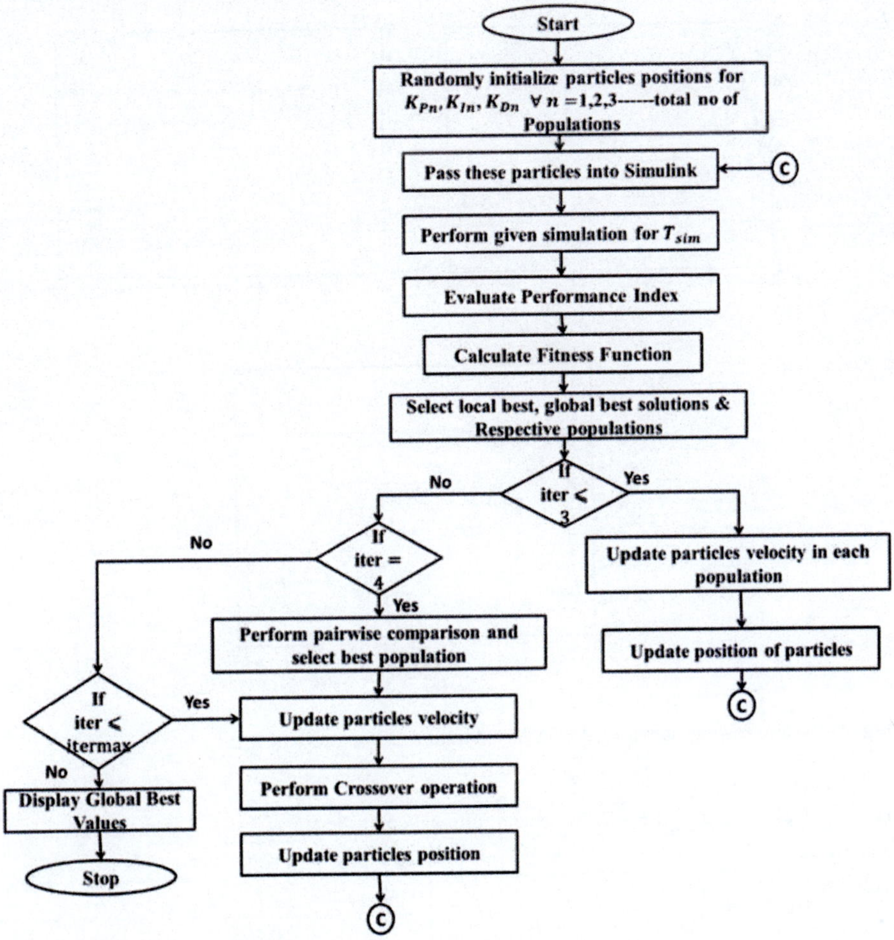

Fig. 3 HAEFA flowchart

5 Simulation Results

5.1 Evaluating the Performances of Several Traditional Controllers

The traditional controllers of PI/PD/PID and PIDN controllers are inserted in both the areas one after the other as secondary regulators and are optimized with a new HAEFA algorithm. The entire analysis is done by creating a disturbance in Area-1 of 1% SLP. To get the most comparative approach, the dynamical behaviour of the system responses are depicted in Fig. 4 and the respective controller parameters are noted in Table 1. The responses shown in Fig. 4 are analyzed in the settling time view

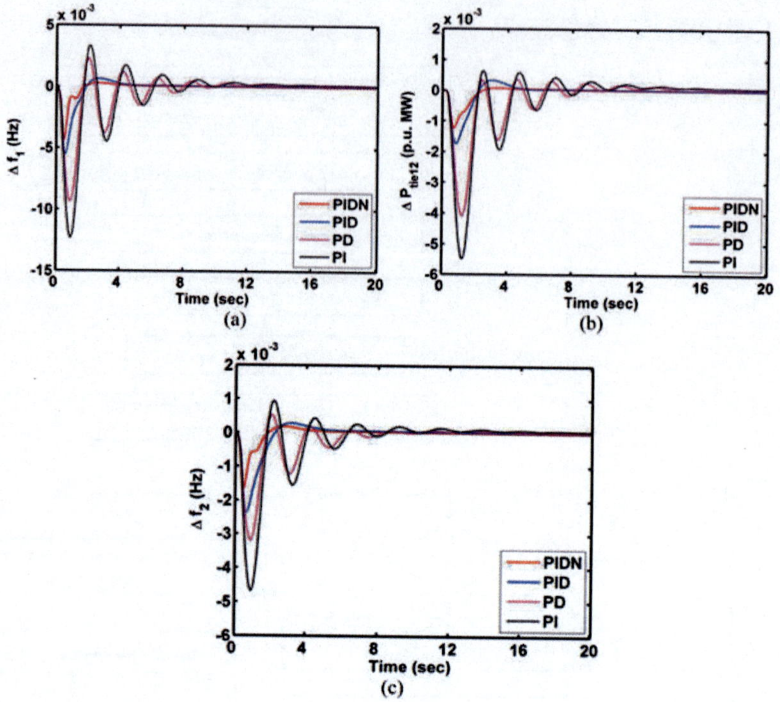

Fig. 4 System responses with HAEFA based traditional controllers. **a** Δf_1, **b** ΔP_{tie12}, **c** Δf_2

Table 1 Optimal gains of HAEFA-based traditional controllers

Parameter		K_P	K_I	K_D	N
Area-1	PI	1.778	0.097	-	-
	PD	1.950	-	0.405	-
	PID	1.146	0.932	0.369	-
	PIDN	1.127	0.694	0.337	147.69
Area-2	PI	1.106	0.054	-	-
	PD	1.913	-	0.255	-
	PID	1.141	0.800	0.455	-
	PIDN	1.156	0.964	0.779	149.32

and are noted in Table 2. Reviewing the responses and settling times, it is pointed out that the PIDN controller outrages other traditional controllers.

Table 2 Settling time of system responses

Control strategy	Settling time (Sec)			ISE * 10^{-3}
	ΔF1	ΔPtie,12	ΔF2	
PI	16.60	15.94	17.23	5.26
PD	14.76	15.71	15.45	1.215
PID	9.323	10.58	10.06	0.429
PIDN	6.643	7.52	6.97	0.125
PIDN with SMES only	5.609	7.17	5.146	0.00613
PIDN with TCSC-SMES	4.757	5.93	4.793	0.00414

5.2 *Evaluating the Effect of Coordinated SMES-TCSC Strategy on System Performance*

As the HAEFA-based PIDN controller is proven to be the best among other traditional regulators, it is continued to act as a secondary regulator. To further improve the behaviour of the system under disturbed conditions, SMES are stationed in two areas and the TCSC device is positioned with the tie-line. The responses are rendered in Fig. 5. It is clear that the TCSC-SMES coordinated scheme highly regulates the

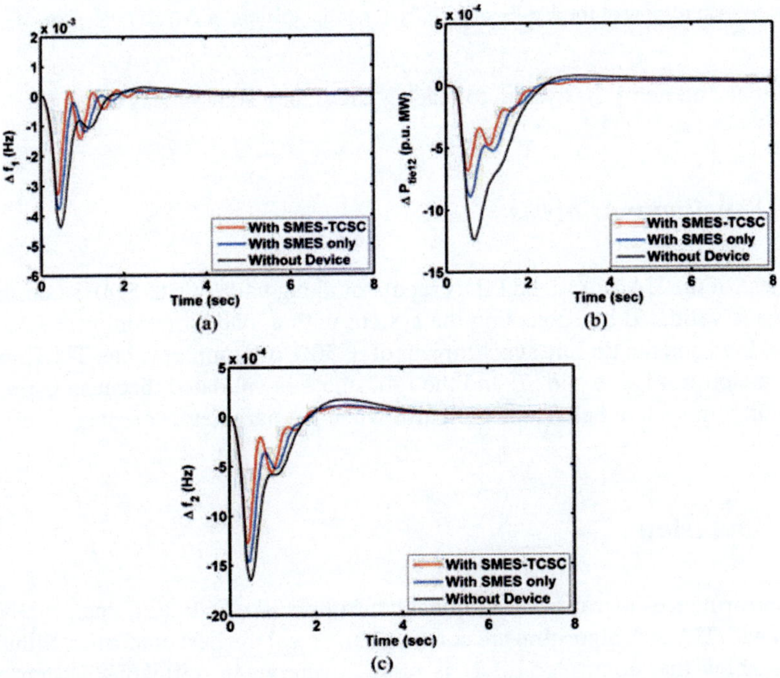

Fig. 5 System responses under SMES-TCSC strategy. **a** Δf_1, **b** ΔP_{tie12}, **c** Δf_2

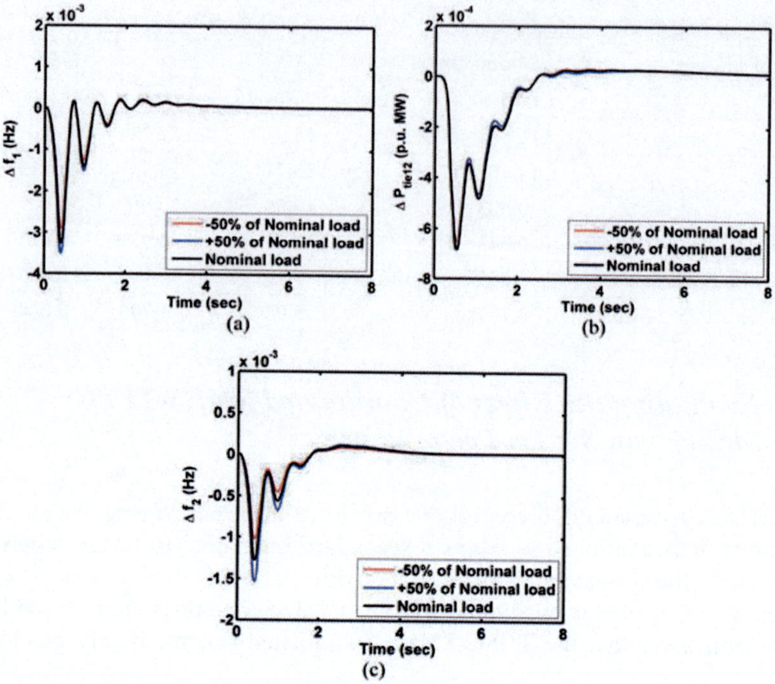

Fig. 6 System responses for $\pm$ 50% of load from normal loading. **a** Δf_1, **b** ΔP_{tie12}, **c** Δf_2

deviations and helps the system to reaches the steady state in less duration.

5.3 *Sensitivity Analysis*

Boldness of the HAEFA tuned PIDN regulator along with TCSC-SMES coordinated scheme is validated by subjecting the system with a load variation of $\pm$ 50% from normal load and the tie-line synchronizer of $\pm$ 50% of nominal value. The responses are depicted in Figs. 6 and 7, and the robustness is validated, because there is not much drag in system behaviour even also when the parameters are largely affected.

6 Conclusion

The performances of various traditional controllers like PI/PD/PID and PIDN optimized with HAEFA algorithm are compared to reveal the best controller. Simulation results show that controller PIDN is more sovereign in mitigating system variations under normal and abnormal conditions. The coordinated control approach of TCSC-SMES is implemented to enhance the system performance further. Sensitivity

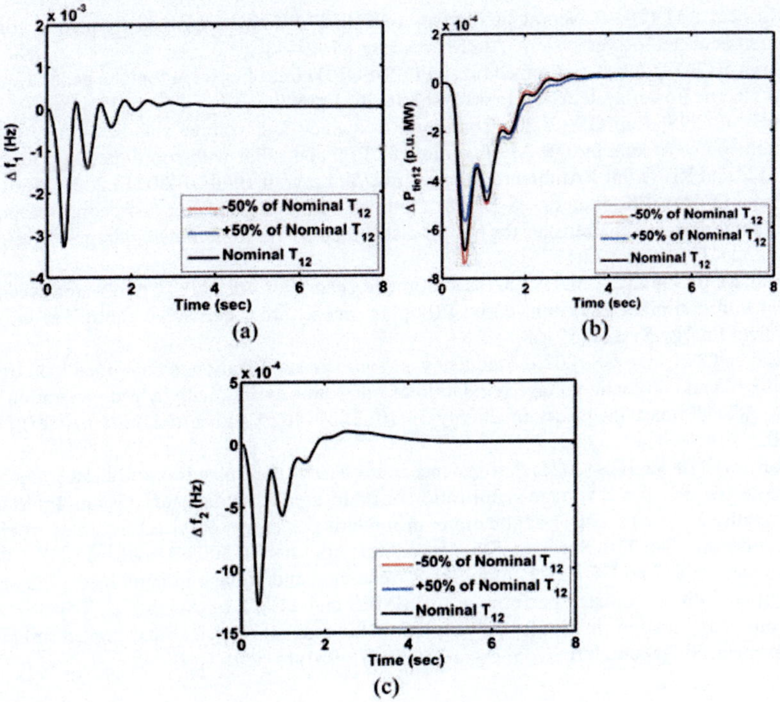

Fig. 7 System responses for ± 50% of T_{12} from normal loading. **a** Δf_1, **b** ΔP_{tie12}, **c** Δf_2

analysis exposes the robustness of the presented coordinated governing approach of TCSC-SMES along with HAEFA optimized PIDN controller.

References

1. Tungadio DH, Sun Y (2019) Load frequency controllers considering renewable energy integration in power system. Energy Rep 5:436–454
2. Sharma Y, Saikia LC (2015) Automatic generation control of a multi-area ST-thermal power system using grey wolf optimizer algorithm based classical controllers. Int J Electr Power Energy Syst 73:853–862
3. Kalyan CNS, Rao GS (2020) Performance comparison of various energy storage devices in combined LFC and AVR of multi area system with renewable energy integration. Int J Renew Energy Res 10(02):933–944
4. Kalyan CNS, Rao GS (2020) Coordinated SMES and TCSC damping controller for load frequency control of multi area power system with diverse sources. Int J Electr Eng Inform 12(04):747–769. https://doi.org/10.15676/ijeei.2020.12.4.4
5. Tasnin W, Saikia LC (2018) Comparative performance of different energy storage devices in AGC of multi-source system including geothermal power plant. J Renew Sustain Energy 10:024101

6. Kaseem AM (2010) Neural predictive controller of a two-area load frequency control for interconnected power system. Ain Shams Eng J 1:49–58
7. Arya Y (2019) A new optimized fuzzy FOPI-FOPD controller for automatic generation control of electric power systems. J Frankl Inst 356:5611–5629
8. Kalyan CNS, Rao GS (2020) Combined frequency and voltage stabilization of multi-area multisource system by DE-AEFA optimized PID controller with coordinated performance of IPFC and RFBs. Int J Ambient Energy. https://doi.org/10.1080/01430750.2020.1860130
9. Guha D, Roy PK, Banerjee S (2018) Optimal tuning of 3 degree-of-freedom proportional-integral-derivative controller for hybrid distributed power system using dragon fly algorithm. Comput Electr Eng 72:137–153
10. Raju M, Saikia LC, Sinha N (2016) Automatic generation control of a multi-area system using ant lion optimizer algorithm based PID plus second order derivative controller. Int J Electr Power Energy Syst 80:52–63
11. Kalyan CNS, Rao GS (2020) Frequency and voltage stabilization in combined load frequency control and automatic voltage regulation of multi area system with hybrid generation utilities by AC/DC links. Int J Sustain Energy 39(10):1009–1029. https://doi.org/10.1080/14786451.2020.1797740
12. Kalyan CNS, Rao GS (2021) Performance index based coordinated control strategy for simultaneous frequency and voltage stabilization of multi-area interconnected system. In: Singh AK, Tripathy M (eds) Control applications in modern power system. Lecture notes in electrical engineering, vol 710. Springer, Singapore. https://doi.org/10.1007/978-981-15-8815-0_4
13. Kalyan CNS, Rao GS (2021) Stabilizing frequency and voltage in combined LFC and AVR system with coordinated performance of SMES and TCSC. In: Singh AK, Tripathy M (eds) Control applications in modern power system. Lecture notes in electrical engineering, vol 710. Springer, Singapore. https://doi.org/10.1007/978-981-15-8815-0_4

Chapter 37
Conventional and Heuristic Optimization Techniques Comparison for Economic Load Dispatch

P. Sowmith, N. Vamsi Krishna, and B. Varunkumar

1 Introduction

Power system optimization problems involving economic dispatch (ED) involve nonlinear and convoluted characteristics including inequality and equality constraints. In order to trim the operating cost of electric energy, significant depletion in the operating cost as well as in the quantity of consumed fuel conventional methods are essential. The ED problem with valve-point loading and emission is represented as a nonlinear and nonconvex optimization problem having a set of equality and inequality constraints, and this makes the problem of identifying the global optimum by using conventional methods [1]. To solve this problem, many conventional methods have been proposed such as a mathematical approach, nonlinear programming [2], quadratic programming [3], and genetic algorithm [4]. The heuristic optimization techniques such as particle swarm optimization (PSO) and genetic algorithms are considered as realistic and powerful solutions schemes to obtain the global optimums in power system optimization problems [5] and are better alternatives to the conventional mathematical approaches.

2 Problem Formulation

The ED problem is one of the critical problems in power system planning and operation. Minimizing the total fuel cost of the generating units considering several operating constraints is called the classical ED problem. The ED problem for a given system is nothing but minimization of total fuel cost as defined by Eq. (1) under a set of operating constraints.

P. Sowmith (✉) · N. V. Krishna · B. Varunkumar
Department of Electrical and Electronics Engineering, V R Siddhartha Engineering College, Vijayawada, AP 520007, India

S. Dawn et al. (eds.), *Smart and Intelligent Systems*, Algorithms for Intelligent Systems,
https://doi.org/10.1007/978-981-16-2109-3_38

$$f = \sum_{i=1}^{N} \left(a_i P_i^2 + b_i P_i + c_i\right) \quad (1)$$

where f is the total fuel cost of generation in the system ($/hr), ai, b_i, c_i are the fuel cost coefficients of the ith generating unit, P_i is power generated by the ith unit and n is the number of thermal units.

The valve-point effect [6] with sinusoidal function is as follows:

$$f = \sum_{i=1}^{N} \left(a_i P_i^2 + b_i P_i + c_i + |d_i \sin\{e_i\left(P_i^{\min} - P_i\right)\}|\right) \quad (2)$$

where f is the total fuel cost of generation ($/hr) including valve-point loading, e_i, d_i are the fuel cost coefficients of the ith generating unit reflecting the valve-point effect.

The emission dispatch problem can be described as the optimization (minimization) of total amount of emission release defined by

$$f = \sum_{i=1}^{N} \left(\alpha_i p_i^2 + \beta_i p_i + \delta_i\right) \quad (3)$$

2.1 Constraints

The cost is minimized with the following generator capacities and active power balance constraints as

$$p_{i,\min} \leq p_i \leq p_{i,\max} \quad (4)$$

$$\sum_{i=1}^{N} P_i = P_D + P_L \quad (5)$$

where $p_{i,min}$,$p_{i,max}$ are the minimum and maximum power generation by i_{th} unit, respectively, P_D is the total power demand and P_L is the total transmission loss.

The transmission loss P_L can be calculated by using the B matrix technique and is defined as

$$P_L = \sum_{i=1}^{N} \sum_{j=1}^{N} P_i B_{ij} P_j \quad (6)$$

where B_{ij} is the element of loss coefficient matrix B.

In this paper, the aim of the ED problem is to minimize the total fuel cost by allocating appropriate outputs of all thermal generating units and to meet the required constraints. The efficient and reliable PSO-based solutions provides minimum emission, minimum cost. The proposed PSO algorithm is applied to a test system with a 6-generator system, 10-generator system, and 14-generator system to present its effectiveness.

3 Methodology

3.1 Conventional Techniques

Lambda iteration method is the conventional method that minimizes the cost of generating the power at any demand for more number of units. It is more accurate and incremental cost curves of all units are stored in memory. Unique mapping is required from a value of lambda to each generator's MW output PG(λ), for any choice of lambda the generators collectively produce a total MW output, lambda iteration with generation limits are taken into account when calculating PG(λ). Again, we continue the iterations until the conversation condition is satisfied with maximum limits that will always cause λ to either increase or remain the same. The computational framework of lambda iteration method is shown in Fig. 1.

The optimization toolbox provides the functions which satisfy constraints that minimize or maximize the objective function to find the parameters. In this toolbox, it includes solvers for nonlinear programming (NLP), linear programming (LP), quadratic programming (QP), genetic algorithm(GA) [7], etc.… with the equality and inequality constraints and these are defined with the function and matrices. By using optimization toolbox, we can find the optimal solution, it defines the problem quickly and modifies speedily and this optimization toolbox's setup is shown in Fig. 2

3.2 Heuristic Techniques

3.2.1 Particle Swarm Optimization

Eberhort and Kennedy developed PSO which is an evolutionary optimization technique in 1995, through simulation of bird flocking in two-dimensional spaces. Comprehensive nonlinear problems are effectively resolved using PSO [8]. It is one of the classifications of the evolutionary computations, which is used to resolve the problems of optimization. A random sample of values is considered in the solution space and is grouped as particles. Each particle is associated with position and

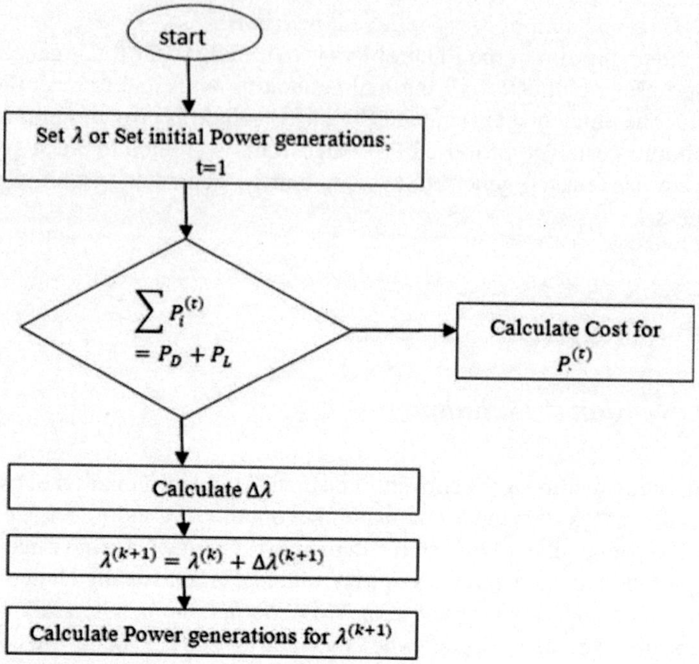

Fig. 1 Computational framework of lambda iteration method

velocity, which are stochastically adjusted according to the chronological best position for the particle itself and the region particle best position at the end of every iteration. An optimal or near-optimal solution is obtained to the natural moment of the particles, which can generate good quality solutions with a reduced amount of computational time and more constant convergence characteristics. Most of the analytical methods fail to converge in the above case. The flowchart of PSO is shown in Fig. 3. The velocities and positions are kept up to date for all the particles using the below equations, respectively.

$$V_{ij}{}^{K+1} = w * V_{ij}{}^{k} + c_1 * r_1 * \left(Pbest_{ij}{}^{k} - X_{ij}{}^{k}\right) + c2 * r2 * \left(Gbest_{j}{}^{k} - X_{ij}{}^{k}\right) \quad (7)$$

$$X_{ij}^{k+1} = X_{ij}^{k} + V_{ij}^{k+1} \quad (8)$$

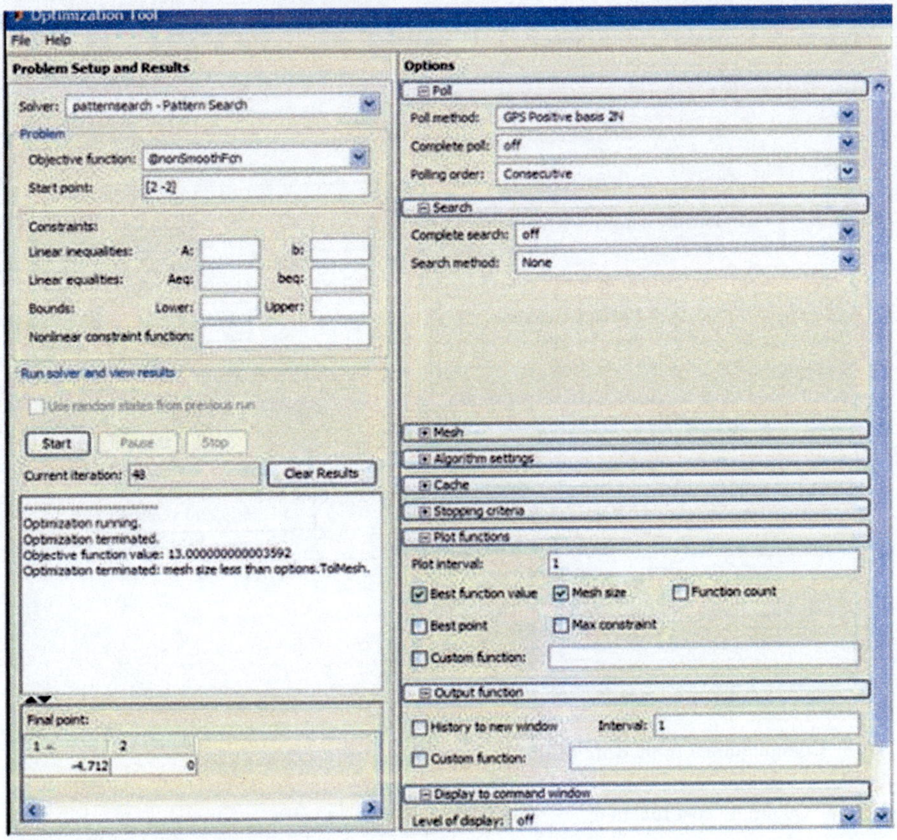

Fig. 2 Optimization toolbox setup

4 Simulation Results

Test System I: A system with six generating units having quadratic cost is considered as a test system, whose data is taken from Appendix-I, and 2.834 P.U is the maximum demand set. Loss coefficients are also considered in the test system (Tables 1 and 2).

Case 1: Quadratic cost function,

Case 2: Quadratic cost function including losses.

Test System II: A ten generating units system with a quadratic cost function, valve-point loading effect, and emission is considered as a test system whose data is taken from Appendix-II [9], and 2000 MW is the maximum demand set. Loss coefficients are also considered in the test system (Figs. 4, 5, 6, 7 and Tables 3, 4, 5, 6, 7).

Case 1: Quadratic cost function,

Case 2: Quadratic cost function having losses,

Case 3: Quadratic cost function involving valve-point load effect,

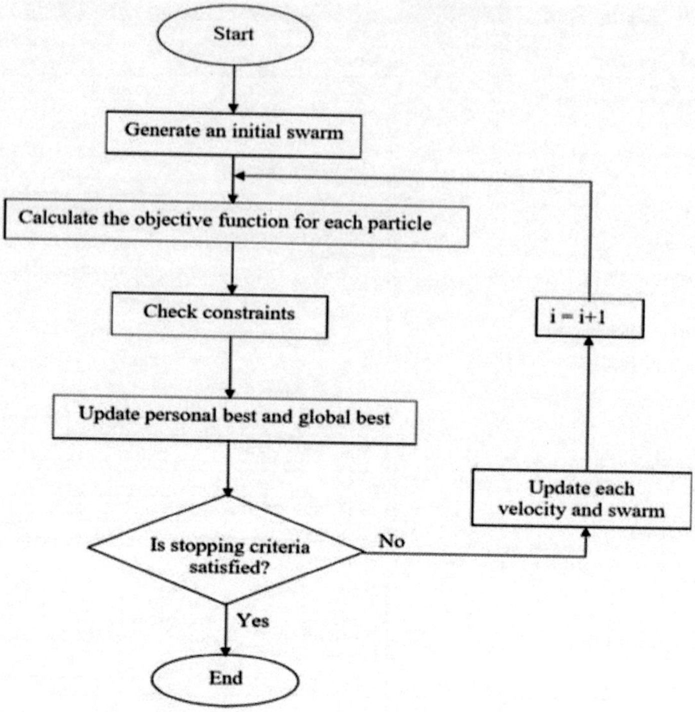

Fig. 3 Computational framework of PSO

Table 1 Quadratic cost function

Methods	Fuel cost ($/h)			Standard deviation	Computational time	Number of hits to get the best solution	Solution accuracy
	Minimum	Average	Maximum				
Lambda-iterative method	785.31	–	–	0.70661	2.00271	1	50
Nonlinear programming (Optimization toolbox)	767.6	–	–	0	0.725940	1	50
Quadratic programming (Optimization toolbox)	767.6	–	–	0	0.168912	1	50
Genetic algorithm (Optimization toolbox)	767.66	768.89	769.21	0	1.374535	5	13
Particle swarm optimization	767.6	–	–	0.69131	0.092191	1	50

Table 2 Quadratic cost function including losses

Methods	Fuel cost ($/h)			Standard deviation	Computational time	Number of hits to get the best solution	Solution accuracy
	Minimum	Average	Maximum				
Lambda-iterative method	777.27	–	–	0.7012	2.105946	1	50
Nonlinear programming (Optimization toolbox)	776.53	–	–	0	0.347736	1	50
Quadratic programming (Optimization toolbox)	787.37	–	–	0	0.190577	1	50
Genetic algorithm (Optimization toolbox)	777.28	779.83	780.14	0	1.87542	8	6
Particle swarm optimization	776.50	776.55	776.59	0.69117	0.350757	8	5

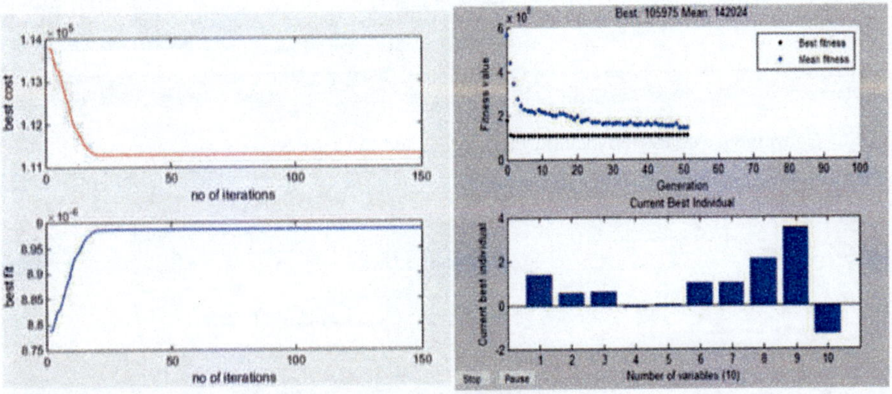

Fig. 4 Convergence characteristics for PSO case1and GA case1

Case 4: Quadratic cost function having valve-point load effect, transmission losses, and.

Case 5: Quadratic emission pollution function.

Test System III: A system with 14 generating units having quadratic cost, valve-point loading effect is considered as a test system, whose data is taken from Appendix III [10], and 2000 MW is the maximum demand set (Fig. 8 and Tables 8, 9, 10).

Case 1: Quadratic cost function,

Case 3: Quadratic cost function including valve-point load effect, and.

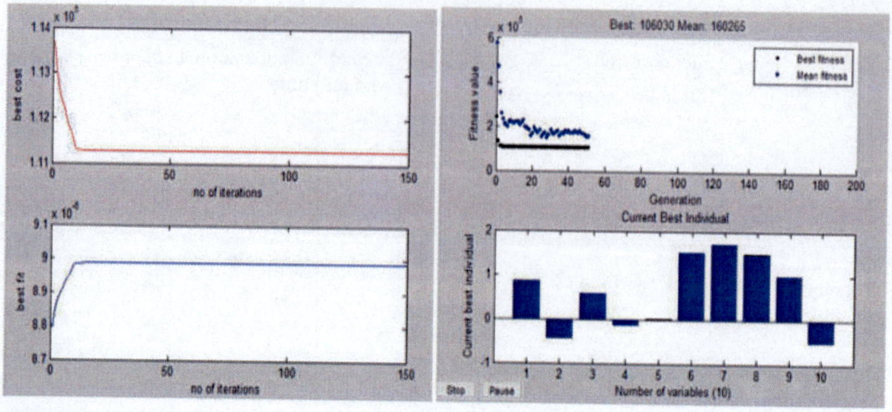

Fig. 5 Convergence characteristics for PSO case 2 and GA case 2

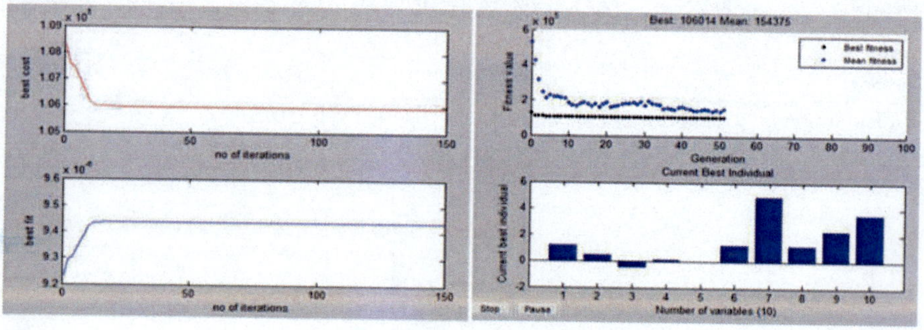

Fig. 6 Convergence characteristics for PSO case 3 and GA case 3

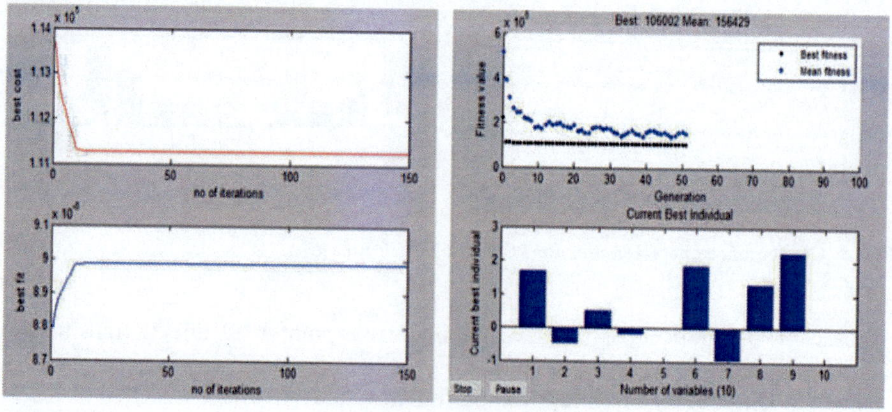

Fig. 7 Convergence characteristics for PSO case 4 and GA case 4

Table 3 Quadratic cost function

me Methods	Fuel cost ($/h)			Standard deviation	Computational time	Number of hits to get the best solution	Solution accuracy
	Minimum	Average	Maximum				
Lambda-iterative method	101,770	–	–	170.324	2.160205	1	50
Nonlinear programming (Optimization toolbox)	105,970	–	–	0	0.365913	1	50
Quadratic programming (Optimization toolbox)	105,970	–	–	0	0.360710	1	50
Genetic algorithm (Optimization toolbox)	105,975	106,017	106,057	0	1.67152	9	3
Particle swarm optimization	105,940	105,950	105,960	172.127	0.183922	6	5

Table 4 Quadratic cost function having losses

Methods	Fuel cost ($/h)			Standard deviation	Computational time	Number of hits to get the best solution	Solution accuracy
	Minimum	Average	Maximum				
Lambda-iterative method	99,863	–	–	169.8344	2.57947	1	50
Nonlinear programming (Optimization toolbox)	106,420	–	–	0	0.582501	1	50
Quadratic programming (Optimization toolbox)	98,691	–	–	0	0.400528	1	50
Genetic algorithm (Optimization toolbox)	106,030	106,035	106,110	0	1.31148	6	5
Particle swarm optimization	111,260	111,290	111,410	168.9855	0.988243	8	18

Table 5 Quadratic cost function involving valve-point load effect

Methods	Fuel cost ($/h)			Standard deviation	Computational time	Number of hits to get the best solution	Solution accuracy
	Minimum	Average	Maximum				
Lambda-iterative method	101,780	–	–	170.3244	1.99421	1	50
Nonlinear programming (Optimization toolbox)	105,970	–	–	0	0.605952	1	50
Quadratic programming (Optimization toolbox)	95,790	–	–	0	0.59271	1	50
Genetic algorithm (Optimization toolbox)	106,014	–	–	0	1.994211	14	8
Particle swarm optimization	105,960		105,980	172.75	0.429707	4	24

Table 6 Quadratic cost function having valve-point load effect, transmission losses

Methods	Fuel cost ($/h)			Standard deviation	Computational time	Number of hits to get the best solution	Solution accuracy
	Minimum	Average	Maximum				
Lambda-iterative method	99,876	–	–	169.83	4.859278	1	50
Nonlinear programming (Optimization toolbox)	106,430	–	–	0	1.022102	1	50
Quadratic programming (Optimization toolbox)	98,691	–	–	0	0.190826	1	50
Genetic algorithm (Optimization toolbox)	106,002	–	–	0	1.359737	11	6
Particle swarm optimization	111,280	111,300	111,430	168.76	2.412035	6	18

Table 7 Quadratic emission pollution function

Methods	Fuel cost ($/h)			Standard deviation	Computational time	Number of hits to get the best solution	Solution accuracy
	Minimum	Average	Maximum				
Lambda-iterative method	3074.9	–	–	113.5127	1.618242	1	50
Nonlinear programming (Optimization toolbox))	3568.1	–	–	0	0.396814	1	50
Quadratic programming (Optimization toolbox)	3568.1	–	–	0	0.162112	1	50
Genetic algorithm (Optimization toolbox)	3573.41	–	–	0	1.60505	24	4
Particle swarm optimization	3560	3563.6	3570.7	0	0.242381	13	8

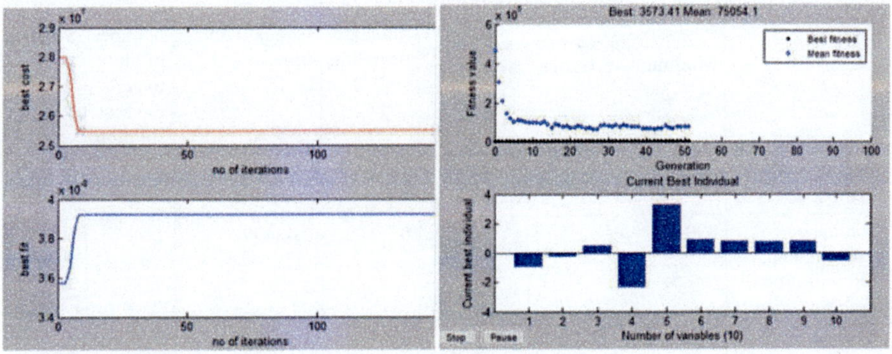

Fig. 8 Convergence characteristics for PSO case 5 and GA case 5

Case 5: Quadratic emission pollution function.

5 Conclusion

The PSO algorithm is applied to the ED problem considering typical test systems in the literature. In order to satisfy the inequality constraints, a position adjustment strategy is also considered in the PSO framework. The degree of freedom is reduced

Table 8 Quadratic cost function

Methods	Fuel cost ($/h)			Standard deviation	Computational time	Number of hits to get the best solution	Solution accuracy
	Minimum	Average	Maximum				
Lambda-iterative method	7459.5	–	–	60.6051	1.83570	1	50
Nonlinear programming (Optimization toolbox)	9690.2	–	–	0	0.364933	1	50
Quadratic programming (Optimization toolbox)	8738.9	–	–	0	0.160447	1	50
Genetic algorithm (Optimization toolbox)	8741.07	8753.24	8779.12	0	1.6893	8	2
Particle swarm optimization	8738.9	8767.3	8861.2	69.343	0.155448	14	6

Table 9 Quadratic cost function including valve-point load effect

Methods	Fuel cost ($/h)			Standard deviation	Computational time	Number of hits to get the best solution	Solution accuracy
	Minimum	Average	Maximum				
Lambda-iterative method	7727.5	–	–	60.6051	2.032561	1	50
Nonlinear programming (Optimization toolbox)	8904.1	–	–	0	0.424998	1	50
Quadratic Programming (Optimization toolbox)	7216.9	–	–	0	0.170697	1	50
Genetic algorithm (Optimization toolbox)	8905.3	8921.01	8940.10	0	1.804601	5	2
Particle swarm optimization	8904.1	8918.0.1	8948.21	69.135	0.533403	5	8

Table 10 Quadratic emission pollution function

Methods	Fuel cost ($/h)			Standard deviation	Computational time	Number of hits to get the best solution	Solution accuracy
	Minimum	Average	Maximum				
Lambda-iterative method	1906.6	–	–	37.319	2.145452	1	50
Nonlinear programming (Optimization toolbox)	2533.5	–	–	0	0.378345	1	50
Quadratic programming (Optimization toolbox)	2742.7	–	–	0	0.160873	1	50
Genetic algorithm (Optimization toolbox)	2746.26	2778.15	2800.26	0	1.990196	17	2
Particle swarm optimization	2742.7	–	2744.2	54.656	0.1522299	3	26

by one to resolve the equality constraint in the ED problem. In order to preserve the dynamic process of the PSO, the strategies for handling constraints are appropriately devised. The PSO has provided the global solution satisfying the constraints with valve-point effects, the PSO has also provided the global solution with a high probability for 6-, 10-, 14-generator system.

Appendix-I: Test System I: 6-generator System Data

See (Tables 11 and 12).

Table 11 Cost coefficients

System	a($/PU)	b ($/PU)	c ($)	Pmin	Pmax
1	37.5	200	0	0.5	2
2	175	175	0	0.2	0.8
3	83.4	325	0	0.1	0.35
4	250	300	0	0.1	0.3
5	625	100	0	0.15	0.5
6	250	300	0	0.12	0.40

Table 12 Loss coefficients

0.0017	0.0002	0.1522	−0.0001	−0.0005	−0.0002
0.0012	0.0014	0.0009	0.0001	−0.0006	−0.0001
0.0007	0.0009	0.0031	0	−0.0010	−0.0006
−0.0001	0.0001	0	0.0024	−0.0006	−0.0008
−0.0005	−0.0006	−0.0010	−0.0006	0.0129	−0.0002
−0.0002	−0.0001	−0.0006	−0.0008	−0.0002	0.0150

Table 13 Cost coefficients

System	a($/MW^2)	b ($/MW)	c ($)	d	e	f	g	h	Pmin	Pmax
1	0.12951	40.5407	1000.403	33	0.0174	360.0012	−3.9864	0.04702	10	55
2	0.10908	39.5804	950.606	25	0.0178	350.0056	−3.9524	0.04652	20	80
3	0.12511	36.5104	900.705	32	0.0162	330.0056	−3.9023	0.04652	47	120
4	0.12111	39.5104	800.705	30	0.0168	330.0056	−3.9023	0.04652	20	130
5	0.15247	38.5390	756.799	30	0.0148	13.8593	0.3277	0.00420	50	160
6	0.10587	46.1592	451.325	20	0.0163	13.8593	0.3277	0.00420	70	240
7	0.03546	38.3055	1243.531	20	0.0152	40.2669	−0.5455	0.00680	60	300
8	0.02803	40.3965	1049.998	30	0.0128	40.2669	−0.5455	0.00680	70	340
9	0.02111	36.3278	1658.569	60	0.0136	42.8955	−0.5112	0.00460	135	470
10	0.01799	38.2704	1356.659	40	0.0141	42.8955	−0.5112	0.00460	150	470

Appendix-II: Test System II: 10-Generator System Data

See (Tables 13 and 14).

Appendix-III: Test System III: 14-Generator System Data

See (Table 15).

Table 14 Loss coefficients

0.000049	0.000014	0.000015	0.000015	0.000016	0.000017	0.000017	0.000018	0.000019	0.000020
0.000014	0.000045	0.000016	0.000016	0.000017	0.000015	0.000015	0.000016	0.000018	0.000018
0.000015	0.000016	0.000039	0.000010	0.000012	0.000012	0.000014	0.000014	0.000016	0.000016
0.000015	0.000016	0.000010	0.000040	0.000014	0.000010	0.000011	0.000012	0.000014	0.000015
0.000016	0.000017	0.000012	0.000014	0.000035	0.000011	0.000013	0.000013	0.000015	0.000016
0.000017	0.000015	0.000012	0.000010	0.000011	0.000036	0.000012	0.000012	0.000014	0.000015
0.000017	0.000015	0.000014	0.000011	0.000013	0.000012	0.000038	0.000016	0.000016	0.000016
0.000018	0.000016	0.000014	0.000011	0.000013	0.000012	0.000016	0.000040	0.000015	0.000016
0.000019	0.000018	0.000016	0.000014	0.000015	0.000014	0.000016	0.000018	0.000042	0.000019
0.000020	0.000018	0.000016	0.000015	0.000016	0.000015	0.000016	0.000016	0.000019	0.000044

Table 15 Cost coefficients

System	a($/MW^2)	b ($/MW)	c ($)	d	e	f	g	h	Pmin	Pmax
1	0.0050	1.89	150	300	0.0350	0.016	−1.50	23.333	150	455
2	0.0055	2	115	200	0.420	0.031	−1.82	21.022	150	455
3	0.006	3.5	40	200	0.420	0.013	−1.24	22.050	20	130
4	0.0050	3	122	150	0.0630	0.012	−1.35	22.983	20	130
5	0.0050	3.05	125	150	0.0630	0.020	−1.90	21.313	150	470
6	0.0070	2.75	120	150	0.0630	0.007	0.805	21.900	135	460
7	0.0070	3.45	70	150	0.0630	0.015	−1.40	23.001	135	465
8	0.0070	3.45	70	150	0.0630	0.018	−1.80	24.003	60	300
9	0.0050	2.45	130	150	0.0630	0.019	−2.00	24.121	25	162
10	0.0050	2.45	130	100	0.0840	0.012	−1.36	22.990	25	160
11	0.0055	2.35	135	100	0.0840	0.033	−2.10	27.010	25	80
12	0.0045	1.60	200	100	0.0840	0.018	−1.80	25.101	20	80
13	0.0070	3.45	70	100	0.0840	0.018	−1.81	24.313	25	85
14	0.0060	3.89	45	100	0.0840	0.030	−1.92	7.119	15	55

References

1. Sowmith P, Madhusudhanrao R, Dr Gouthamkumar N (2020) Optimal scheduling of hydrothermal plant using particle swarm optimization. Lecture notes in mechanical engineering. Springer, Singapore. (14 March 2020)
2. Non linear programming. https://www.mathworks.com/discovery/nonlinear-programming.html
3. Dr Govindaraj T, Thendral R (2014) Multi objective economic emission load dispatch using quadratic programming. Int J Innov Res Electr Electron Instrum Control Eng 2(1). (January 2014)
4. Gajendra S, Swarnkar K (2014) Economic load dispatch by genetic algorithm in power system. Int J Sci Eng Technol Res (IJSETR) 3(8). (August 2014)
5. Lee KY, El-Sharkawi MA (eds) (2002) Modern heuristic optimization techniques with applications to power systems: IEEE power engineering society (02TP160)
6. Mandal KK, Chakraborty N (2008) Effect of control parameters on differential evolution based combined economic emission dispatch with valve-point loading and transmission loss. (Jadavpur University)
7. Walters DC, Sheble GB (1993) Genetic algorithm solution of economic dispatch with the valve point loading. IEEE Trans Power Syst 8:1325–1332
8. Park J-B, Lee K-S, Shin J-R, Lee KY (2005) A particle swarm optimization for economic dispatch with nonsmooth cost functions, member. IEEE. (Fellow, IEEE)
9. Wood AJ, Wallenberg BF (1984) Power generation, operation, and control. Wiley, New York
10. Güvença U, Sönmezb Y, Dumanc S, Yörükerend N (2012) Combined economic and emission dispatch solution using gravitational search algorithm

Chapter 38
Effect of Au-Al Dual-Metal Gate on 3D Double-Gate Junctionless Transistor Performance

Achinta Baidya, Rajesh Saha, Amarnath Gaini, Chaitali Koley, Somen Debnath, and Subir Datta

1 Introduction

As device dimensions are scaled down drastically to meet the high speed and high functionality requirements, the conventional Metal–Oxide–Semiconductor (MOS) devices face performance degradation which is termed as Short Channel Effects (SCEs). To diminish the SCEs, MOS devices feel the requirement of a super shallow junction at source–channel and channel–drain junctions. The microelectronics industry experts persistently invested their knowledge to improve the technology to ease the fabrication process of extremely scaled devices. In recent years, novel devices like Fin-FET [1–3], Tunnel FET (TFET) [4–6], spin-based device [7–9], Silicon-On-Insulator (SOI) structures [10–13], carbon nanotubes [14–17] are

A. Baidya (✉)
Department of Electronics and Communication Engineering, Mizoram University, Aizawl 796004, India

R. Saha
Department of Electronics and Communication Engineering, Malaviya National Institute of Technology Jaipur, Jaipur 302017, India

A. Gaini
Department of Electronics and Communication Engineering, Marri Laxman Reddy Institute of Technology and Management, Hyderabad, India

C. Koley
Department of Electronics and Communication Engineering, National Institute of Technology, Aizawl, Mizoram 796012, India
e-mail: chaitali.ece@nitmz.ac.in

S. Debnath
Department of Information Technology, Mizoram University, Aizawl 796004, India

S. Datta
Department of Electrical Engineering, Mizoram University, Aizawl 796004, India

S. Dawn et al. (eds.), *Smart and Intelligent Systems*, Algorithms for Intelligent Systems,
https://doi.org/10.1007/978-981-16-2109-3_39

invented with the hope of solving the problems of extreme scaling. However, fabrication of ultra-shallow junction remains in the case of extremely scaled devices. As a solution to this problem, a device named as Junctionless Transistor (JLT) with no junction throughout the source–channel–drain was proposed in 2009 [18] and later fabricated in 2010 [19]. The absence of any junction and need of any doping gradient makes the device an attractable solution to the researchers. With existing MOS fabrication technology, the device shows the possibility of more scalability and cost reduction. To make the device fully compatible with the technology use, an immense amount of examination and analysis are required. Despite several advantages over inversion mode conventional MOSFET, JLT is not fully free from the SCEs. Even many studies have focused on the scaling of the JLT device [20–23]. In order to understand the tunability of the JLT further more studies is required. A 20 nm double-gate junctionless transistor was proposed and analog performance was analyzed with different gate work functions [24, 25]. Though JLT with low work function metal gate shows suitable analog performance, very low threshold voltage is very unrealistic for circuit applications. So, further optimization in terms of gate engineering is investigated in the present paper, and the effect of the gate metal work function is optimized through the combined use of high and low work function metal at the gate. Section 2 describes the device dimension and the simulation environment.

2 Device Dimension and Simulation Environment

A double-gate junctionless transistor structure with a 30 nm gate length is considered for the present study. The device acts like a gated resistor that controls the current flow through it by the MOS gate. Silicon channel and SiO_2 gate oxide are considered for the JLT device. As the silicon channel has to enter full depletion to stop the current flow from the source to drain, the channel should be low enough. Presently the thickness of the silicon bar is taken as 10 nm and a double gate will confirm the control over the channel. To optimize the effect of gate metal work function two different metals are used in both gates of the JLT structure. The length of the two metals in the gate is kept equal (=15 nm) for the study. To understand the behavior of JLT with multi-metal gate, we have chosen Gold (Au) and Aluminum (Al) for the present investigation. As the study shows that gate with low work function metal gives high on current and gives higher linearity at low power [24, 25], low work function material like Al (WF = 4.28) is suitable for gate metal. But it gives a very low threshold voltage, which brings difficulty for the application in circuits. Though high work function gate metal like Au (WF = 5.1) has different drawbacks, it helps to set a required threshold voltage of the device. So, a combination of high and low work function material at the gate may be the solution for device characteristics optimization.

The gate-engineered JLT 3D structures are prepared with the help of the 3D device structure editor of Synopsys Technology Computer-Aided Design (TCAD). Afterward, the structure is simulated for some physics models using Sdevice. Different models for mobility, recombination, high electric field effect and carrier concentration are used in Sdevice. The simulation result is analyzed and data is extracted through Svisual, inspect, and Tecplot tools. The simulated 3D structure of the JLT is shown in Fig. 1 and the details of the structure attributes are mentioned in Table 1. Further detailed discussion on the device performances is included in Sect. 3.

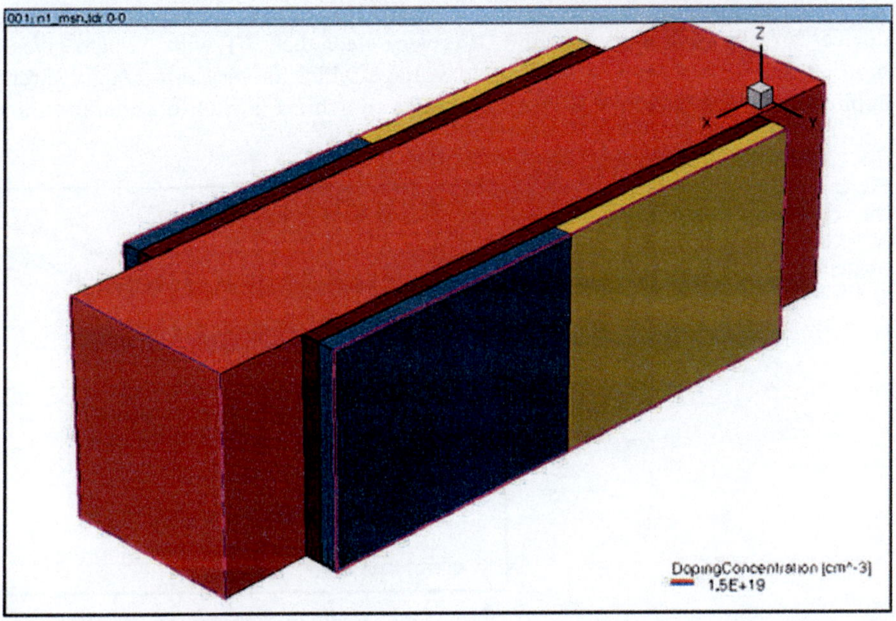

Fig. 1. 3D simulated structure of dual-metal double-gated junctionless transistor [24]

Table 1 Device structural attributes and values of the dual-metal double-gate junctionless transistor

Attributes	Values
Device gate length (l_g)	30 nm
Gate metal 1 length (lm1)	15 nm
Gate metal 2 length (lm2)	15 nm
Gate material	Gold, Aluminum
Channel width (w)	12 nm
Channel thickness (t_{si})	10 nm
Doping Concentration at channel (N_D)	1.5×10^{19} cm^{-3}
Doping material	Arsenic
Gate oxide thickness	1 nm

3 Results and Discussion

In this section, the JLT structure is studied with dual-gate metal and single-gate metal structures. In single-metal gate JLTs, Gold and Aluminum are used. Normally for MOS devices, high work function gate metal near to the source is useful for excitation of the electrons. In double-metal gate, a combination of Au (WF = 5.1) and Al (WF = 4.28) is used and characteristics are analyzed. The transfer characteristics of these devices are shown in Figs. 2 and 3. It is very clear that JLT with Al gate gives a high current with a very low threshold voltage which is very difficult for circuit applications. The device will face difficulties to achieve switch-off condition and

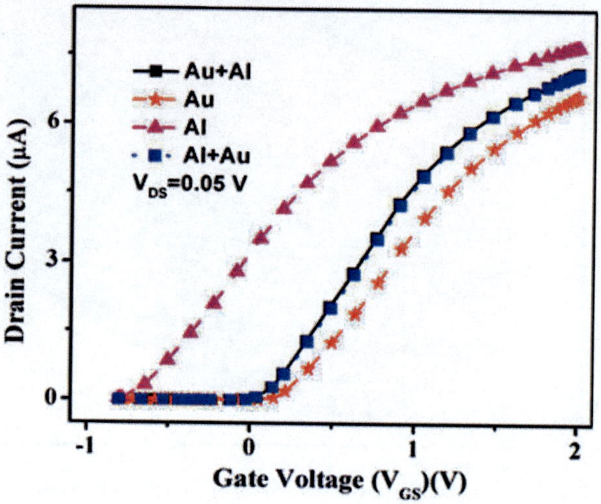

Fig. 2 Transfer characteristics (in linear scale) of JLT for single-metal and dual-metal gate with $V_{DS} = 0.05$ V

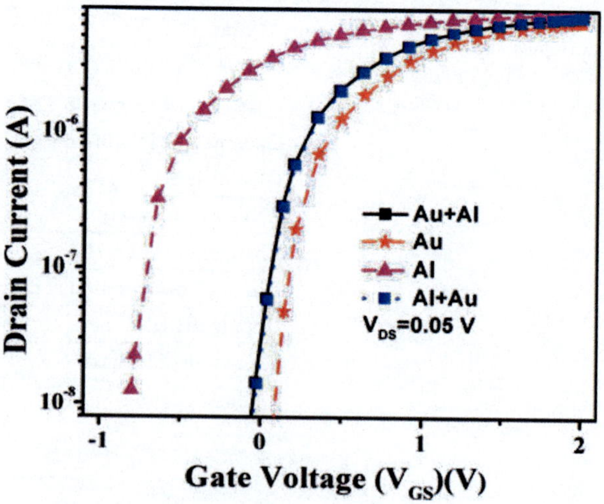

Fig. 3 Transfer characteristics (in log scale) of JLT for single-metal and dual-metal gate with $V_{DS} = 0.05$ V

further optimization is required to adjust the threshold voltage. Whereas, JLT with Au gate gives a suitable threshold voltage for the scaled JLT device but faces lower on current compared to JLT with Al.

Rest two JLT dual-metal gate structures have 15 nm of Au and 15 nm Al gate in both gates. The result shows that characteristics are almost similar. A small improvement is present in the case of Au-gate metal placed near to the source. Whereas both devices perform well compared to single-metal gate cases. The details of threshold voltage, on current and subthreshold swing are mentioned in Table 2. It is quite clear that Au-Al gate gave a better subthreshold swing too.

Transconductance of JLTs for single-metal and dual-metal gates are shown in Fig. 4. From the transconductance curves, it is clear that JLT with Au-Al gate gave the highest transconductance of 5.38 μS. If we interchange the position of the metals, a shallow change or decrement is observed in transconductance. The occurrence of maximum g_m value is different for different devices and it depends on its threshold voltage. Further Transconductance Generation Factor (TGF) is another important parameter for analog applications. The TGF for all the JLT devices is shown in Fig. 5. JLT with Au-Al gate gives better TGF than the single Al metal gate JLT and

Table 2 Performance of JLT for different channel doping concentrations

Gate material		Threshold voltage (V_{th}) (V)	"On" current at $V_{GS} = 2$ V (μA)	Maximum transconductance ($g_{m,max}$) (μS)	Subthreshold swing (mV/dec)
LM1	LM2				
Al	Al	−0.629	7.69	5.08	101
Au	Au	0.271	6.62	5.07	74.7
Al	Au	0.189	7.11	5.36	88
Au	Al	0.106	7.14	5.38	86.4

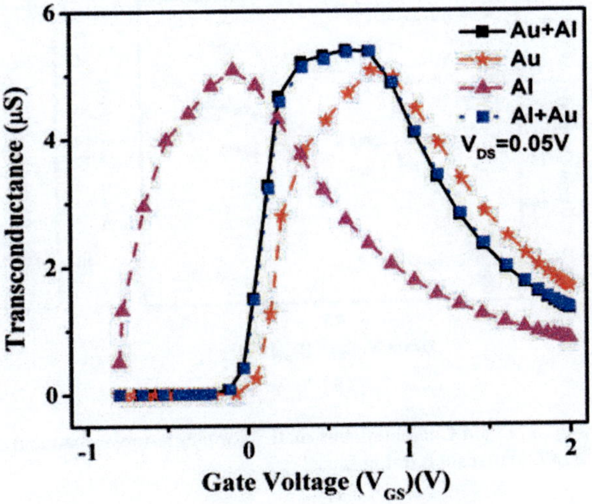

Fig. 4 Transconductance of JLT for single-metal and dual-metal gate with $V_{DS} = 0.05$ V

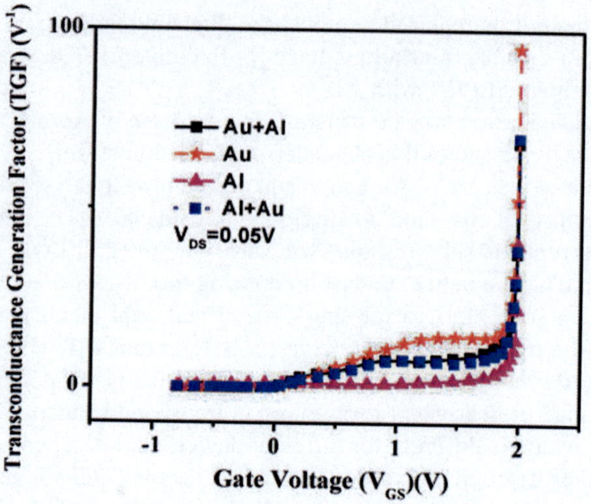

Fig. 5 Transconductance Generation Factor (TGF) of JLT for single-metal and dual-metal gate with $V_{DS} = 0.05$ V

comparatively less TGF than the Au-gated JLT.

The drain characteristics of JLTs for single-metal gate and double-metal gate with $V_{GS} = 1$ V is shown in Fig. 6. All the devices show a good on–off ratio. The drain characteristics comparison in Fig. 7 shows that JLT with Al gate has the highest drain current. It also shows higher drain conductance which means the change in drain current with the change in drain voltage is more in the case of Al gate compared to Au gate. The performance of the Au-Al gate is moderate when compared to these two. That means the influence of drain in short length device dimension is less in

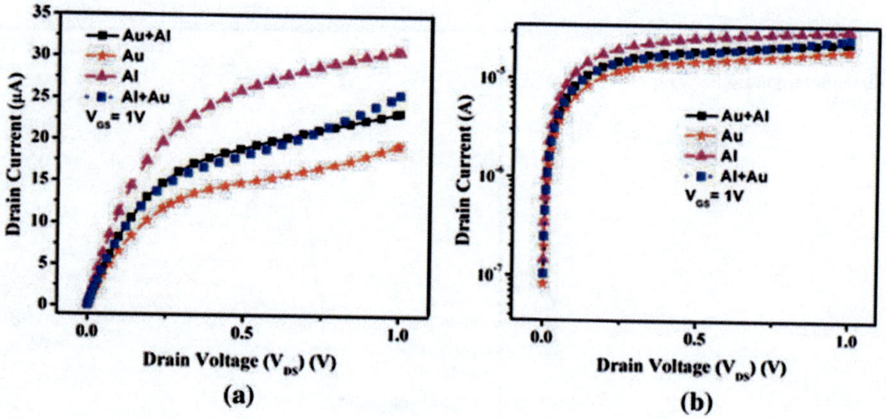

Fig. 6 Drain Characteristics of JLT for single-metal gate and double-metal gate with $V_{GS} = 1$ V **a** in linear scale **b** in log scale

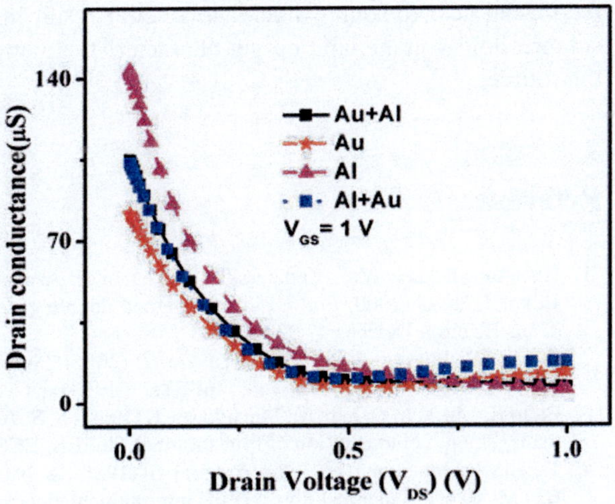

Fig. 7 Drain conductance of JLT for different channel doping with $V_{GS} = 1$ V

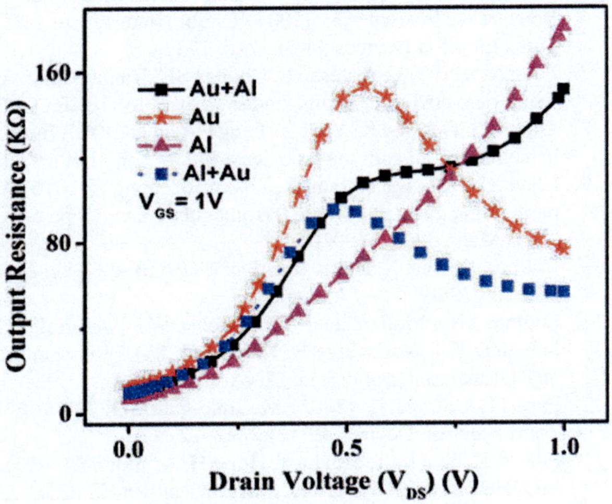

Fig. 8 Output resistance of JLT for different channel doping with $V_{GS} = 1$ V

case of Au-Al gate compare to Al gate. Even the output resistance in Fig. 8 shows similar performance for the double-metal JLTs.

4 Conclusion

The paper has emphasized the use of dual metal at the double gates of JLT. A combination of high and low work function metal is chosen for the JLT structure to incorporate the advantages of both. The device with dual metal gave optimized characteristics of JLT compared to a single-metal gate of either high or low work function. JLT with

Au placed near to source side gave a better result in threshold voltage, on current, subthreshold voltage, and output characteristics with respect to the Al placed near the source.

References

1. Hisamoto D, Lee W-C, Kedzierski J, Takeuchi H, Asano K, Kuo C, Anderson E, King T-J, Bokor J, Hu C (2000) FinFET–a self-aligned double gate MOSFET scalable to 20nm. IEEE Trans Electron Devices 47:2320–2325
2. Zhang W, Fossum JG, Mathew L, Du Y (2005) Physical insights regarding design and performance of independent-gate FinFETs. IEEE Trans Electron Devices 52(10):2198
3. Endo K, Liu Y, Masahara M, Matsukawa T, O'uchi S, Suzuki E (2007) Fabrication and power-management demonstration of four-terminal FinFEts. ECS Trans 6(4):71
4. Takeda E, Matsuoka H, Igura Y, Asai S (1988) A band to band tunneling MOS device B2TMOSFET. Technical digest IEEE international electron devices meeting, pp 402–405
5. Choi WY, Park BK, Lee JD, Liu TJK (2007) Tunneling field-effect transistors (TFETs) with subthreshold swing (SS) less than 60 mV/dec. IEEE Electron Device Lett. 28(8):743–745
6. Boucart K, Ionescu AM (2007) Double-gate tunnel FET with high-k gate dielectric. IEEE Trans Electron Devices 54(7):1725–1733
7. Fong X, Kim Y, Venkatesan R, Choday SH, Raghunathan A, Roy K (2016) Spin-transfer torque memories: devices, circuits, and systems. Proc IEEE 104(7):1449–1488
8. Wang S, Cai Li, Qi K, Yang X, Feng C, Cui H (2016) Impact of nanomagnets size on switching behaviour of all spin logic devices. Micro Nano Letters 11(9):508–513
9. Chen X, Qi W, Bai F, Tang X, Zhang H, Zhong Z (2016) A methodology to design spin-wave-based logic gates in a single ferromagnetic nanostripe using spin-transfer torque effects. IEEE Trans Magn 52(7):1400104
10. Colinge JP, Baie X, Bayot V, Grivei E (1996) A silicon-on-insulator quantum wire. Solid-State Electron 39:49
11. Lemme MC, Mollenhauer T, Henschel W, Wahlbrink T, Baus M, Winkler O, Granzner R, Schwierz F, Spangenberg B, Kurz H (2004) Subthreshold behavior of triple-gate MOSFETs on SOI material. Solid State Electron 484–529
12. Park JT, Colinge JP (2002) Multiple-gate SOI MOSFETs: device design guidelines. IEEE Trans Electron Devices 49(12):2222–2229
13. Suzuki K, Tanaka T, Tosaka Y, Horie H, Arimoto Y (1993) Scaling theory for double-gate SOI MOSFET's. IEEE Trans Electron Devices 40(12):2326–2329
14. Ijiima S (1991) Helical microtubules of graphitic carbon. Nature 354:56–58
15. Avouris P, Afzali A, Appenzeller J, Chen J, Freitag M, Klinke C, Lin Y-M, Tsang JC (2004) Carbon nanotube electronics and optoelectronics. IEDM technical digest. IEEE international electron devices meeting.
16. Connor IO', Liu J, Gaffiot F, Prégaldiny F, Lallement C, Maneux C, Goguet J, Frégonèse S, Zimmer T, Anghel L, Dang T-T, and Leveugle R (2007) CNTFET modeling and reconfigurable logic-circuit design. IEEE Trans Circuits Syst—I Regular Papers 54(11):2365–2379
17. Nasiri SH, Farshi MKM (2013) Stability analysis in CNTFETs. IEEE Electron Device Lett. 34(2):301–303
18. Lee C-W, Afzalian A, Akhavan ND, Yan R, Ferain I, Colinge J-P (2009) Junctionless multigate field-effect transistors. Appl Phys Lett 94(5):053511
19. Colinge JP, Lee CW, Aryan A et al (2010) Nanowire transistors without junctions. Nat Nanotechnol 5:225–229
20. Akhavan ND, Ferain I, Razavi P, Yu R, Colinge J-P (2011) Improvement of carrier ballisticity in junctionless nanowire transistors. Appl Phys Lett 98(10):103 510-1–103 510-3

21. Rios R, Cappellani A, Armstrong M, Budrevich A, Gomez H, Pai R, Rahhal-orabi N, Kuhn K (2011) Comparison of junctionless and conventional trigate transistors with L g down to 26 nm. IEEE Electron Device Lett 32(9):1170–1172
22. Lin H-C, Lin C-I, Huang T-Y (2012) Characteristics of n-type junctionless poly-Si thin-film transistors with an ultrathin channel. IEEE Electron Device Lett 33(1):53–55
23. Gundapaneni S, Ganguly S, Kottantharayil A (2011) Bulk planar junctionless transistor (BPJLT): an attractive device alternative for scaling. IEEE Electron Device Lett 32(3):261–263
24. Baidya A, Krishnan V, Baishya S, Lenka TR (2015) Effect of thin gate dielectrics and gate materials on simulated device characteristics of 3D double gate JNT. Superlattices Microstruct 77:209–218
25. Baidya A, Lenka TR, Baishya S (2020) Linear distortion analysis of 3D double gate junctionless transistor with high-K dielectrics and gate metals, silicon

Chapter 39
Power- and Area-Efficient FIR Filter for Denoising of Electrooculogram Signal

Kishore Kumar Gundugonti and Balaji Narayanam

1 Introduction

Electrooculography(EOG) is one of the bioindicators that may be measured and monitored. It is a technique of measuring the capacitance of the state of the cornea and retina, which exists between the front and lower parts of the back of the human eye. The subsequent signal is electrooculogram. The EOG signal amplitude is within the range of 50–3500V, and the frequency is within the range of 5–30H z [2]. Movement patterns include physical activity information such as writing [12], reading [4], and driving [7]. Many [15, 19] researchers have analyzed eye movements in applications such as reading, typing environments, and human–computer interfaces for people with disabilities. The multiplier-less hardware architecture is developed for eye blinks and looking up [24]. The reading activity detection is developed in [16]. The novel eye tracking system is developed in [9]. The EOG system is developed to control a serious computer game [13]. The saccade detection system is designed in [14]. The detection of eye closing/opening from EOG is developed in [18]. The saccade detection is used in identifying the epileptic patients in [22].

Several authors proposed methods to reduce partial products using CSA (carry-save architecture) [5, 8, 17, 20, 23]. In conventional shift-add, partial products are added using RCA and this leads to an increase in delay. Complexity reduction of coefficients multiplication has been performed by using diverse method among which the common subexpression elimination (CSE) algorithm is considered to be the most

K. K. Gundugonti (✉)
Department of Electronics and Communication Engineering, V R Siddhartha Engineering College, Vijayawada, India
e-mail: gkishorekumar@vrsiddhartha.ac.in

B. Narayanam
Department of Electronics and Communication Engineering, JNTUK University College of Engineering, Kakinada, India

S. Dawn et al. (eds.), *Smart and Intelligent Systems*, Algorithms for Intelligent Systems,
https://doi.org/10.1007/978-981-16-2109-3_40

famous one. Hartley [6] used the CSE approach with CSD in filters, which result in a decrease in the amount of adders.

The uncooked EOG signals are corrupted with numerous noises. The filtering is the preprocessing for the EOG signal to remove the noise. The mathematical operations in digital filter contains multiplier and adder. The multiplier calculation is substituted with shift-add approach in the implementation of ASIC. The adder counted in shift-add approach is primarily based on the sign-power- of-two (SPT) phrases found in every coefficient. Some types of research were centered at the multiplier-less filter implementation for EOG signal [1, 25]. The implementation in citewu15 is primarily based on variable shifter. The implementation in citeag17 is based totally on distributed arithmetic procedures with excessive throughput. In [11], only speed was concentrated. In [10], differential evolution (DE) algorithm based system is developed to reduce area and power with fewer SPT in each coefficient. In the proposed method, area- and power-efficient FIR filter structure is designed for denoising the EOG signal using CSE with CSD method.

This article is divided into 6 sections. Section 2 discusses the architecture of the FIR filter. In Sect. 3, CSE with CSD is discussed for the FIR filter. In Sect. 4, filter functionality is verified by using the Xilinx-Simulink environment. In Sect. 5, synthesis results are discussed and followed by conclusion in Sect. 6.

2 Finite Impulse Response(FIR) Filter Architecture

The FIR filter equation is given in Eq. (1).

$$\mathrm{y(k)} = \sum_{\mathrm{k}=0}^{\mathrm{N}-1} \mathrm{h(k)} \times \mathrm{x(k} - (\mathrm{N} - 1)) \tag{1}$$

In Eq. (1), x(k) is denoted as filter input and y(k) is denoted as filter output. The h(k) is denoted as filter coefficient and N is denoted as filter order. The basic building blocks of FIR filter architecture are multiplier, adder, and delay. Eq. (1) requires N multipliers, $N - 1$ delays, and $N - 1$ adders.

The FIR filter can be implemented in two ways. First one is direct form approach and second one is transposed direct form approach [21]. These two approaches are shown in Figs. 1 and 2. The choice of filter implementation is clock frequency or maximum combinational path delay. The maximum combinational path delays of the direct form approach and the transposed direct form approaches are given in Eq. (2).

$$\begin{aligned} \mathrm{T}_{\text{direct form}} &= \mathrm{T_M} + \mathrm{T_A} \times (\mathrm{N} - 1) \\ \mathrm{T}_{\text{transposed direct form}} &= \mathrm{T_M} + \mathrm{T_A} \end{aligned} \tag{2}$$

Where TA and TM are adder and multiplier delays. From Eq. (2), the direct shape filter structure includes N-1 adders and one multiplier delay whereas transposed

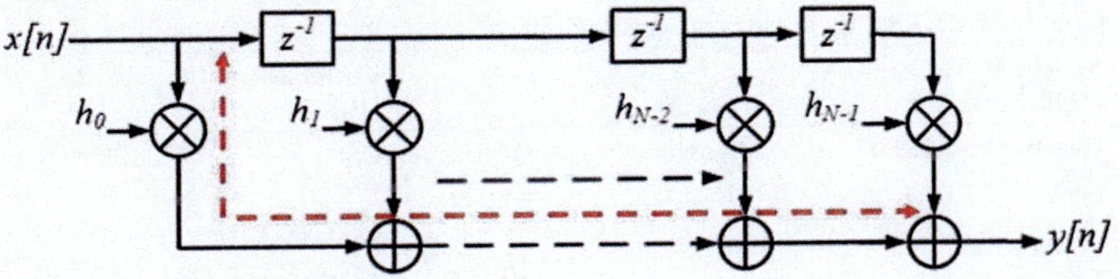

Fig. 1 Direct form FIR filter architecture

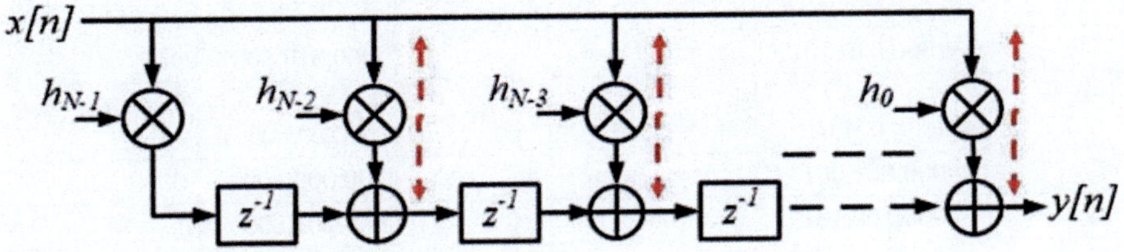

Fig. 2 Transposed direct form FIR filter architecture

direct shape includes adder and multiplier delay only. From the observation of these architectures, transposed direct form architecture is a lesser critical path delay as compared to direct form architecture.

3 Proposed FIR Filter Using Common Subexpression Elimination

The CSD representations of [10] coefficients are shown in Table 1. The input data samples multiplied with filter coefficients are represented in Eq. (3)

$$\begin{aligned}
x \times h_0 &= -(x << 2 + x)\\
x \times h_1 &= -(x << 3 + x << 1)\\
x \times h_2 &= -(x << 4 - x << 2)\\
x \times h_3 &= -(x << 4 - x << 2)\\
x \times h_4 &= -(x << 3)\\
x \times h_5 &= x << 2\\
x \times h_6 &= x << 5 - x << 3\\
x \times h_7 &= x << 6 - x << 4 + x << 2\\
x \times h_8 &= x << 6 + x << 4 + x << 1
\end{aligned}$$

Table 1 MATLAB, [10] filter coefficients, and adders

MATLAB coefficients				Coefficients [10]	
Quantized	CSD	Adders	Quantized	CSD	Adders
−6	−(000,001,010)	1	−5	−(000,000,101)	1
−9	−(000,001,001)	1	−10	−(000,001,010)	1
−11	−(000,010,101)	2	−12	−(000,010,100)	1
−11	−(000,010,101)	2	−12	−(000,010,100)	1
−6	−(000,001,010)	1	−8	−(000,001,000)	0
7	000,001,001	1	4	000,000,100	0
27	000,100,101	2	24	000,101,000	1
53	001,010,101	3	52	001,010,100	2
82	001,010,010	2	82	001,010,010	2
111	010,010,001	2	112	010,010,000	1
133	010,000,101	2	136	010,001,000	1
145	010,010,001	2	148	010,010,100	2
		Total adders = 21			Total adders = 13

$$
\begin{aligned}
x \times h_9 &= x << 7 - x << 4 \\
x \times h_{10} &= x << 7 + x << 3 \\
x \times h_{11} &= x << 7 + x << 4 + x << 2
\end{aligned} \tag{3}
$$

Equation (3) is rewritten as

$$
\begin{aligned}
x \times h_0 &= -(x << 2 + x) \\
x \times h_1 &= -((x << 2 + x) << 1) \\
x \times h_2 &= -(x << 4 - x << 2) \\
x \times h_3 &= -(x << 4 - x << 2) \\
x \times h_4 &= -(x << 3) \\
x \times h_5 &= x << 2 \\
x \times h_6 &= (x << 4 - x << 2) << 1 \\
x \times h_7 &= (x << 4 - x << 2) << 2 + x << 2 \\
x \times h_8 &= (x << 2 + x) << 4 + x << 1 \\
x \times h_9 &= x << 7 - x << 4
\end{aligned}
$$

$$x \times h_{10} = x << 7 + x << 3$$
$$x \times h_{11} = x << 7 + (x << 2 + x) << 2$$

From Eq. (4), the common expressions are (x << 2+x) and (x << 4−x << 2). These common expressions are represented as p1 and p2, respectively. Eq. (4) is rewritten as

$$\begin{aligned}
x \times h_0 &= -p_1 \\
x \times h_1 &= -p_1 << 1 \\
x \times h_2 &= -p_2 \\
x \times h_3 &= -p_2 \\
x \times h_4 &= -(x << 3) \\
x \times h_5 &= x << 2 \\
x \times h_6 &= p_2 << 1 \\
x \times h_7 &= p_2 << 2 + x << 2 \\
x \times h_8 &= p_1 << 4 + x << 1 \\
x \times h_9 &= x << 7 - x << 4 \\
x \times h_{10} &= x << 7 + x << 3 \\
x \times h_{11} &= x << 7 + p_1 << 2
\end{aligned} \tag{4}$$

The FIR filter has symmetrical filter coefficients and 24-tap. The architectural representation of Eq. 4 is shown in Fig. 3. The MATLAB FDATool generated coefficients require 21 adders and DE coefficients require 13 adders. From Fig. 3, common subexpression elimination (CSE) method requires only 7 adders for input data sample multiplied with filter coefficients. So, the CSE-based architecture offers a 53.84% reduction in adders as compared to DE-based architecture.

4 Functional Verification of Proposed Filter

The functionality of the proposed filter has been examined the usage of the Xilinx System Generator (XSG) 2016.2. The digital design diagram for the proposed filter is shown in Fig. 4. Here, the modified filter is a CSE with a CSD-based architecture primarily. The changed filter is carried out with the use of Verilog- HDL. The Verilog-HDL of the proposed filter is imported for the usage of XSG. The filter black box has 16-bit input, reset, and 32-bit output ports. The real-time EOG signal information is considered from [3]. This information is going to the filter unit. The black box is compiled for the usage of XSG after which real-time samples of the EOG signal are passed as input to the filter block. The proposed filter structure is simulated in hardware through uploading samples of the EOG signal from the MATLAB workspace into the XSG-Simulink environment. EOG samples are surpassed to the

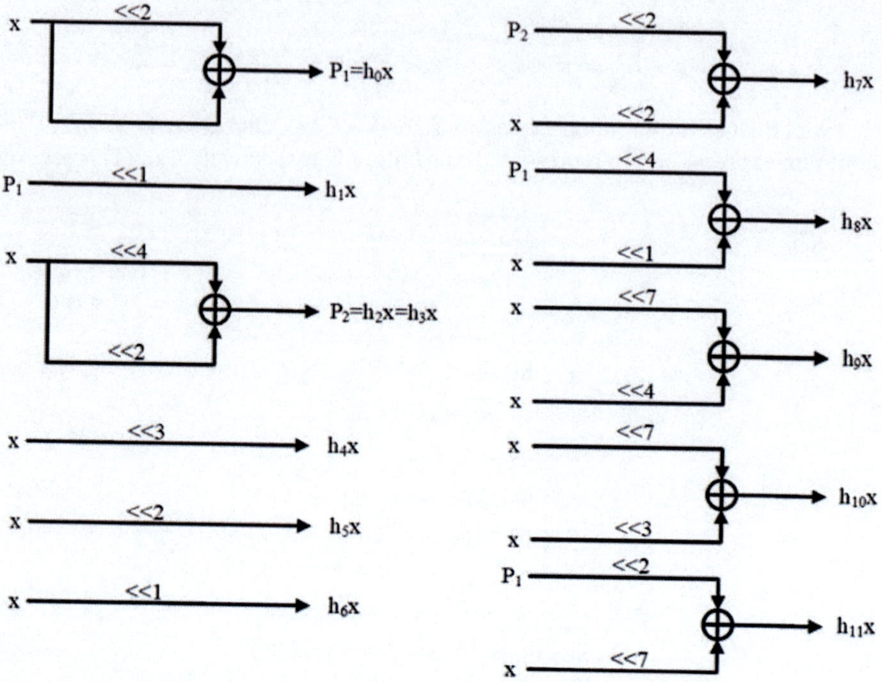

Fig. 3 Proposed architecture for data samples multiplied with filter coefficients

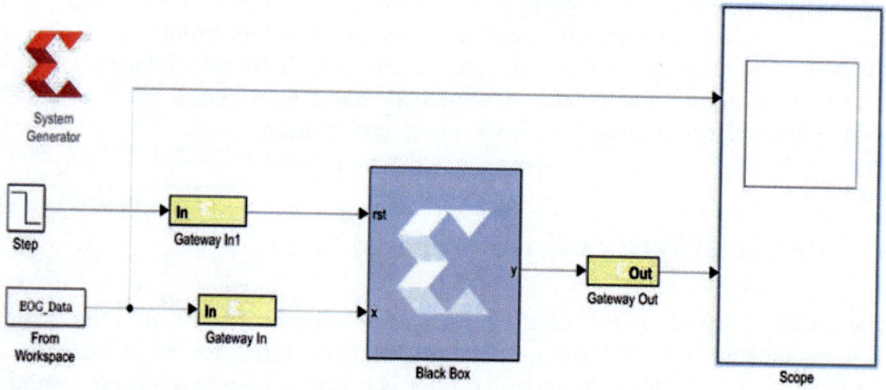

Fig. 4 Digital design setup for the proposed filter

proposed task model via input gateways. The squelch output generated via the developed version is handed over to the MATLAB environment for rendering via the output gateways. The hardware structure of the proposed technique has been tested via joint hardware simulations, and the outcomes are shown in Figs. 5, 6, 7, 8, 9 and 10.

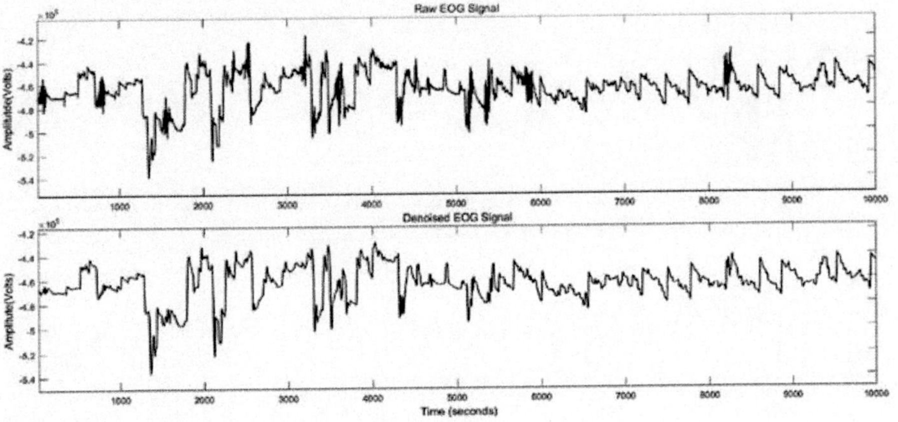

Fig. 5 Horizontal EOG signal for sample 1

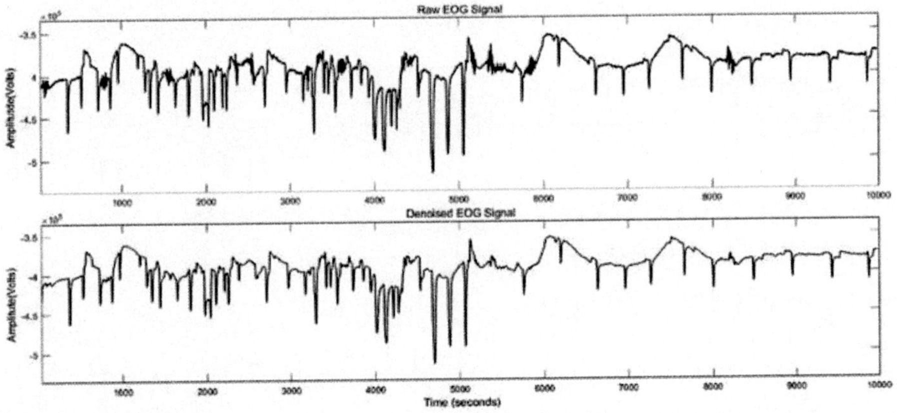

Fig. 6 Vertical EOG signal for sample 1

5 Synthesis Results of Proposed Filters Using Compressors and CSE

The results of the CSE synthesis with the CSD filter are compared with the conventional implementation of the Remez algorithm DE [10, 11] along with the SPT filter implemented in [25]. All filters are implemented using Verilog HDL at the gate level. Here the filter order is 24, and with 16-bit input and 32-bit output. The results of ASIC synthesis using 90 nm UMC technology were obtained using the Cadence RTL compiler. The ASIC synthesis results are shown in Table 2. A common filter in Table 2 uses a radix-4 multiplier to multiply the coefficients. Wu et al. [25] uses a variable shift scheme to multiply the coefficients. The bandwidth of the switch

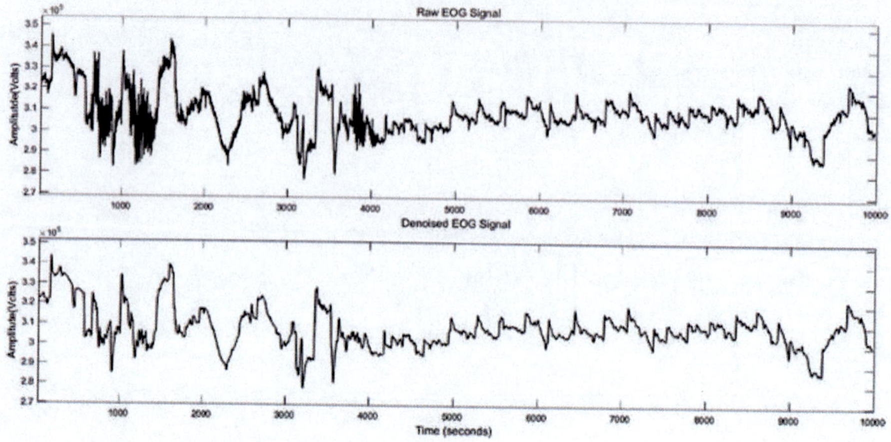

Fig. 7 Horizontal EOG signal for sample 2

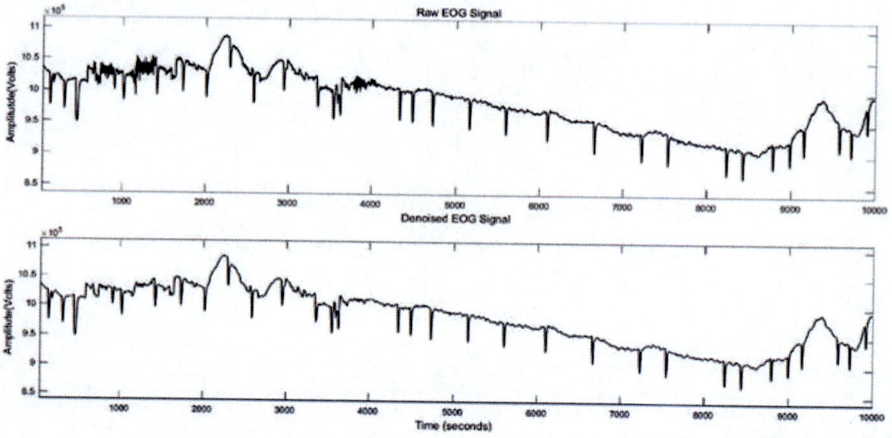

Fig. 8 Vertical EOG signal for sample 2

circuit in [25] is based on the SPT members present in the factor. In Table 2, CSD Rem is a CSD-based offset and added implementation for the Remez coefficient set. CSE with CSD filter uses partial CSD products to define generic subexpressions.

The DE [10] coefficient set calls for 13 adders and Remez coefficient set calls for 21 adders. From Fig. 3, CSE with CSD for coefficient multiplication requires only 7 adders. So, the CSE with CSD-based architecture offers 38.10% and 66.67% reduction in adders as compared to Remez coefficient set and DE [10] coefficient set respirations.

Lowering the number of adders reduces the area, energy, and latency of the CSE with the CSD filter. The ASIC synthesis outcomes in Table 2 displays that CSE with CSD filter attains 114.45% power increase, 148.17% area increase, and PDP increase

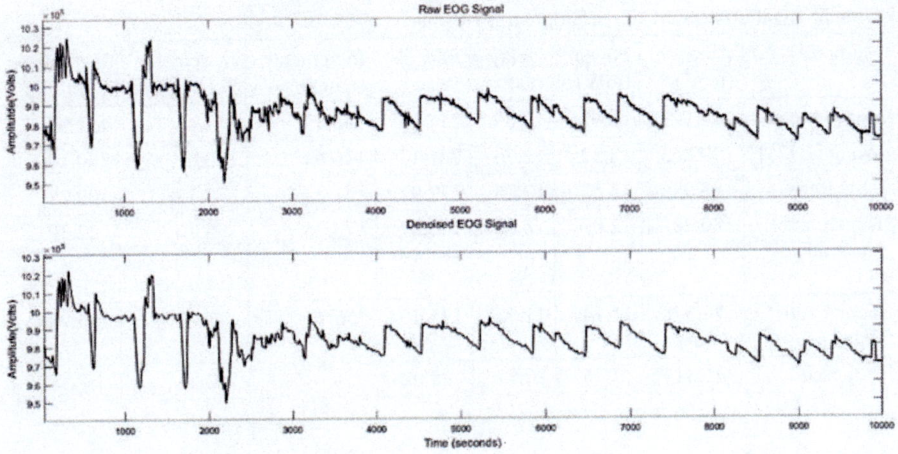

Fig. 9 Horizontal EOG signal for sample 3

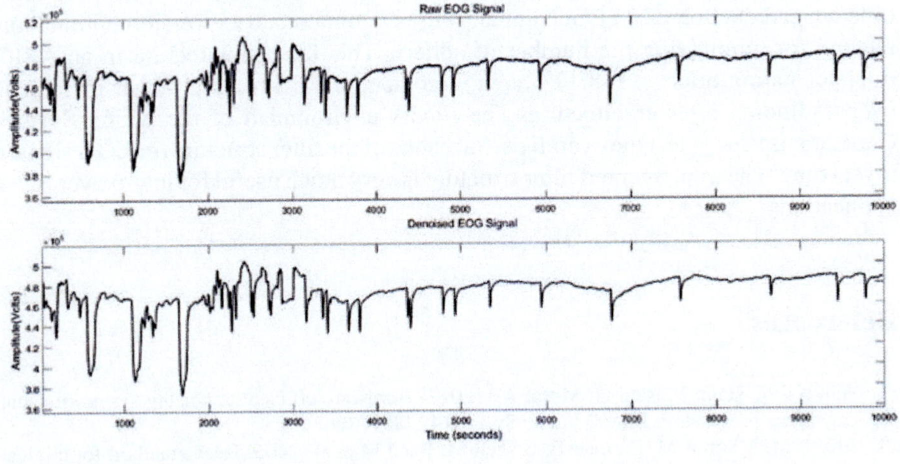

Fig. 10 Vertical EOG signal for sample 3

of 151.99% in comparison to normal filter. The synthesis outcomes also display that CSE with CSD is carried out with a smaller area and less energy in comparison to CSD Remez, [25], CSD with compressors and filter based on DE algorithm [10]. These results are also shown in Table 2.

Table 2 Synthesis results of the proposed FIR filter

Method	Area (μm^2)	Power (mW)	Delay (ns)	PDP %	Incrementr % in area	Incrementr % in power	Incrementr % in PDP
Conventional	114,929	16.62	12.89	214.23	148.17	114.45	151.99
Wu et al. [25]	99,863	17.33	12.76	221.13	115.64	123.61	160.10
CSD Rem	83,926	13.52	13.16	177.92	81.22	74.75	109.28
Kumar and Narayanam [11]	80,287	12.09	8.60	103.97	173.36	56.00	22.30
Kumar and Balaji [10]	72,578	11.14	11.98	133.46	56.72	43.74	56.98
Proposed	46,311	7.75	10.97	85.02	-	-	-

6 Conclusion

This article presents the architecture of an efficient area and power filter for real-time noise reduction of an EOG signal. It uses common sub expression elimination method for minimizing the number of adders. This filter architecture in an ASIC implementation offers a 148.17% area increment and a 114.45% power increment over traditional filter architecture. The FPGA environment of the Xilinx System Generator is used to test the overall performance of the filter structure for EOG signals in real time. The implemented filter structure is very much useful for low-power EOG applications

References

1. Agarwal S, Singh V, Rani A, Mittal AP (2017) Hardware efficient denoising system for real eog signal processing. J Intell Fuzzy Syst 32(4):2857–2862
2. Brown M, Marmor M, Zrenner E, Brigell M, Bach M et al (2006) Iscev standard for clinical electro-oculography (eog) 2006. Doc Ophthalmol 113(3):205–212
3. Bulling A, Ward JA, Gellersen H (2012) Multimodal recognition of reading activity in transit using body-worn sensors. ACM Trans Appl Percept (TAP) 9(1):1–21
4. Bulling A, Ward JA, Gellersen H, Troster G (2011) Eye movement analysis for activity recognition using electrooculography. IEEE Trans Pattern Anal Mach Intell 33(4):741–753
5. Ercegovac MD (2004) Tomas lang. "Digital Arithmetic". Morgan Kaufmann Publishers, by Elsevier Science (USA), vol 1, Ch 2
6. Hartley RI (1996) Subexpression sharing in filters using canonic signed digit multipliers. IEEE Trans Circuits Syst II Analog Digital Signal Process 43(10):677–688
7. Ji Q, Yang X (2002) Real-time eye, gaze, and face pose tracking for monitoring driver vigilance. Real-Time Imaging 8(5):357–377
8. Koren I (2002) Computer arithmetic algorithms. Universities Press
9. Kumar D, Dutta A, Das A, Lahiri U (2016) Smarteye: developing a novel eye tracking system for quantitative assessment of oculomotor abnormalities. IEEE Trans Neural Syst Rehabil Eng 24(10):1051–1059

10. Kumar GK, Balaji N (2020) An area-power efficient denoising hardware architecture for real eog signal. J Circuits Syst Comput
11. Kumar GK, Narayanam B (2020) Low power implementation of fir filter for de-noising the eog signal. Int J Comput Digital Syst 10:1–6
12. Lee K, Chang W, Kim S, Im C (2017) Real-time eye-writing recognition using electrooculogram. IEEE Trans Neural Syst Rehabil Eng 25(1):37–48
13. L´opez A, Fernandez M, Rodr´ıguez H, Ferrero F, Postolache O (2018) Development of an eog-based system to control a serious game. Measurement 127:481–488
14. Lv Z, Wang Y, Zhang C, Gao X, Wu X (2018) An ica-based spatial filtering approach to saccadic eog signal recognition. Biomed Signal Process Control 43:9–17
15. Naga R, Chandralingam S, Anjaneyulu T, Satyanarayana K (2012) Denoising eog signal using stationary wavelet transform. Meas Sci Rev 12(2):46–51
16. Ouyang R, Lv Z, Wu X, Zhang C, Gao X (2017) Design and implementation of a reading auxiliary apparatus based on electrooculography. IEEE Access 5:3841–3847
17. Rabaey J, Chandrakasan A, Nikolic B (2003) Digital integrated circuits, a design perspective. Second
18. Sharma K, Jain N, Pal PK (2020) Detection of eye closing/opening from eog and its application in robotic arm control. Biocybern Biomed Eng 40(1):173–186
19. Soltani S, Mahnam A (2016) A practical efficient human computer interface based on saccadic eye movements for people with disabilities. Comput Biol Med 70:163–173
20. Veeramachaneni S, Krishna KM, Avinash L, Puppala SR, Srinivas M (2007) Novel architectures for high-speed and low-power 3–2, 4–2 and 5–2 compressors. In: 20th international conference on VLSI design held jointly with 6th international conference on embedded systems (VLSID'07). IEEE, pp 324–329
21. Venkataramani B (2002) Digital signal processors. Tata McGraw-Hill Education
22. Vidapanakanti M, Kakarla S, Katukojwala S, Naidu M (2009) Analysis of saccadic eye movements of epileptic patients using indigenously designed and developed saccadic diagnostic system. In: 13th international conference on biomedical engineering. Springer, pp 431–434
23. Wang Z, Jullien GA, Miller WC (1995) A new design technique for column compression multipliers. IEEE Trans Comput 44(8):962–970
24. Wu J, Ang A, Tsui KM, Wu H, Hung YS, Hu Y, Mak J, Chan SC, Zhang Z (2014) Efficient implementation and design of a new single-channel electrooculography-based human–machine interface system. IEEE Trans Circuits Syst II Express Briefs 62(2):179–183
25. Wu JF, Ang AMS, Tsui KM, Wu HC, Hung YS, Hu Y, Mak JNF, Chan SC, Zhang ZG (2015) Efficient implementation and design of a new single- channel electrooculography-based humanmachine interface system. IEEE Trans Circuits Syst II: Express Briefs 62(2):179–183

Chapter 40
Foggy Image Enhancement Using Improved Histogram Equalization and Guided Filter

Shivangi Mishra, Ashish Dwivedi, and Badal Soni

1 Introduction

Nowadays, due to different unwanted particles in the atmosphere like haze, fog, mist, etc., the weather condition is getting degraded. Among these, fog is one of the most common reasons for weather degradation. The visibility of the images taken in such foggy weather conditions gets visually affected, which can cause problems in different fields like analysis of traffic, video surveillance, study of satellite images, etc. In order to remove such a serious problem, defogging the foggy image is necessary [1, 2].

Fog is formed when water vapor condenses and forms tiny water droplets and gets suspended in the air. Fog can be of different types on the basis of their formation processes:

(1) **Radiation Fog**: It is formed due to the cooling of land by thermal radiation. It generally occurs after the sunset.
(2) **Advection Fog**: It is formed when moist air is passed over a cool surface and is gradually cooled. It commonly occurs at the sea, when moist tropical air moves over cold waters.
(3) **Evaporation fog**: This is often caused when cold air gets passed over warm water or over any moist land, which can cause frost or freezing fog.
(4) **Upslope Fog**: It occurs when air goes up the slope of a hill/mountain and condenses to form fog. This kind of fog generally occurs during the winter.

Mathematically, a fog model can be defined as:

S. Mishra (✉) · A. Dwivedi
Department of Computer Science Engineering, United Institute of Technology Allahabad, Allahabad, India

B. Soni
Department of Computer Science Engineering, NIT-Silchar, Assam, India
e-mail: badal@cse.nits.ac.in

S. Dawn et al. (eds.), *Smart and Intelligent Systems*, Algorithms for Intelligent Systems,
https://doi.org/10.1007/978-981-16-2109-3_41

$$I(x) = J(x) \times t(x) + A \times (1 - t(x)) \tag{1}$$

where x is the coordinate of any point on the input image, I(x) is the input foggy image, J(x) is the original color of the image that is to be recovered, A is atmospheric light, t(x) is the transmission rate along the light. J(x) × t(x) together called the attenuation and A × (1 − t(x)) is called the air light.

The paper presented here proposes a method of defogging an image by combining two methods: CLAHE and Guided Filter (GF). The later part of this paper is arranged as follows: Sect. 2 presented the detailed related work. Section 3 addresses the background knowledge. Methodology is described in Sect. 4. In Sect. 5, evaluation measures, experimental results and analysis are given. Finally, the conclusions of the paper are given in Sect. 6.

2 Related Work

In paper [3], a method for calculating the transmission in foggy scenes using a single input image is proposed, which gives good visibility for in-homogeneous fog but fails to give better results when the fog is dense.

Tan et al. in [4] gave a technique for contrast enhancement of the foggy image so as to improve its visibility. It is applicable on both colored as well as gray images. Though the proposed method gives good contrast of the degraded image but can have halo effects and also color may get distorted.

In paper [1], a single-image-based debasing technique is given. The proposed technique is applicable for outdoor images (single color image). This method has good color restoration effect for thin fog but fails to give appropriate results when the fog is dense.

Tarel et al. in [5] proposed a method to restore the visibility of the degraded image using the median filtering approach. The method is quite fast and also has high visibility for thin foggy images but cannot work for images having discontinuous scene depth.

In [6], a Multiscale Depth Fusion procedure is presented to enhance the foggy images with a single input image, which has high visibility for fog-affected images but the limitation lies with a high cost of computation and also have problem with edge and color distortion.

In [7], a new technique to enhance the visibility of stereo images has been proposed by using scattering coefficient and depth. However, fog is not removed from far away areas in case of a small weight coefficient.

Park et al. in [8] proposed a novel approach of defogging a depth-based image by fusing Dark Channel Prior, having distinct patch size to maintain the image details and stereo matching to estimate the transmission map. This method can remove fog without any color distortion.

3 Background Work

In this work, an image defogging method is proposed by combining two methods, i.e., Contrast-Limited AHE and Guided Filter. However, before carrying out the proposed method. Initially, White Balancing is applied to the input foggy image, which is done to modify the original color of the image with respect to white. Applying the white balance function helps in acquiring the color of our resultant image, as accurate as possible.

A. *Contrast Limited Adaptive Histogram Equalization (CLAHE)*

To address the problem of noise over amplification, faced by Adaptive Histogram Equalization (AHE) [9], CLAHE can be used [10]. Contrast-Limited AHE is the modified version of Adaptive Histogram Equalization, whose overall procedure is quite similar to that of AHE, the only difference is in its clip limit.

The method of Contrast-Limited Adaptive Histogram Equalization, requires the image to be divided into different blocks followed by estimating Histogram Equalization (HE) of each block. Finally, a clip limit is calculated to specify the image contrast, ranging between 0 and 1. The gray value distribution of the histogram, for each of the regions, is like that the histogram's height will be in the range of clip limit. However, the portion exceeding the clip limit should not be discarded, rather it should be redistributed evenly to the histogram.

B. *Guided Filter*

Guided Filter (GF) [11] is an edge-preserving smoothing operator, which can filter the input image by using the content of another image called guidance image. This guidance image can be the input image itself or a part of the input image or can be a completely new image. Guided Filter is useful in many ways like it can eliminate false edges and staircase effects from the image, can help in image smoothing and reduction of noise from image, etc.

Mathematically, Guided Filter can be defined as:

$$q_i = a_k I_i + B_k, \forall_i \varepsilon\, \omega_k \tag{2}$$

where I is the guidance image, p is the input image, q be the output image which is a linear transformation of I, ω_k is the window size centering at k pixel, and (a_k, b_k) are the linear coefficients. And the value of a_k and b_k can be estimated as

$$a_k = \frac{\frac{1}{|\omega|}\sum_{i\in\omega_k} I_i P_i - \mu_k \Gamma_k}{\sigma_k^2 + \varepsilon} \tag{3}$$

$$b_k = \Gamma_k - a_k \mu_k \tag{4}$$

where ω is the number of pixels in the window ω_k, μ_k and are, respectively, the mean and variance of I in the window ω_k, ω is used as a regularization parameter to avoid maximizing the value of a_k and p_i is the mean of p in the window ω_k, i.e.,

$$\Gamma_k = \frac{1}{|\omega|} \sum_{i\varepsilon\omega_k} P_i \tag{5}$$

Finally, after estimating the values of a_k and b_k, filtering output q_i can be written as

$$q_i = \frac{1}{|\omega_i|} \sum_{k\in w_i} a_k I_i + b_k \tag{6}$$

$$= \dddot{a}_i I_i + \dddot{b}_i \tag{7}$$

where

$$\dddot{a}_i = \frac{1}{|\omega|} \sum_{i\in\omega_k} a_k \tag{8}$$

and

$$\dddot{b}_i = \frac{1}{|\omega|} \sum_{i\in\omega_k} a_k \tag{9}$$

4 Proposed Methodology

(1) Take a foggy input image.
(2) White Balancing is performed to the input foggy image.
(3) After white balancing is done, the image is transformed into Lab space, which is a three-axis color system, which exhibits all the colors perceived by the eyes of humans and for comparing different color values a standard scale is also provided.
(4) Next, the image (in Lab color space) is divided into several nonoverlapping regions and histogram is calculated for each of the regions.
(5) The calculated histograms may have noise amplification problem. To minimize the noise, a clip limit b is obtained by

$$\beta = \frac{MN}{L}\left(1 + \frac{\alpha}{100}(S_{max} - 1)\right) \tag{10}$$

where M N represents the number of pixels in each region, L denotes the number of grayscales, a indicates the clipping factor which ranges between 0 and100 and Smax is the maximum allowable slope.

(6) Part of the histogram which exceeds the clip limit is not discarded but preferably redistributed so that the height of the histogram should not be greater than the clip limit.

(7) In the next step, Guided Filtering (GF) procedure is applied. For this application, a guidance image (I) is considered. By using the content of this guidance image, the filtration process is carried out on the CLAHE output image (Fig. 1).

5 Experimental Results and Discussion

Image defogging algorithm is used not only for enhancing the image visibility but also for restoring the structure and color of the image. Some parameters by which the image quality can be assessed are Contrast Gain, PSNR, and MSE.

A. ***Contrast Gain***

It is estimated by taking the mean difference between enhanced output image contrast and the input foggy image contrast [12] and can be given by,

$$C_{gain} = C_j - C_i \tag{11}$$

where the mean contrast of the output image is denoted by C_j and the contrast of the input foggy image is denoted by C_i.

B. ***PSNR and MSE***

PSNR [13–16] and MSE are the two commonly used evaluation measures to determine the quality of the output images. PSNR of an image can be achieved by using the following mathematical equation:

$$\mathrm{PSNR} = 10 * \log_{10} \frac{(2^{\beta} - 1)^2}{\mathrm{MSE}} \tag{12}$$

where β denotes the bits per sample and MSE [17] is given by

$$\mathrm{MSE} = \frac{1}{[M \times N]} \sum_{M,N} [I_1(m, n) - I_2(m, n)]^2 \tag{13}$$

where $I_1(m, n)$ is the input foggy image, $I_2(m, n)$ is the recovered image, MN represents the size of the image and m, n is the x, y location of the image pixel. Images

Fig. 1 Flow diagram proposed methodology

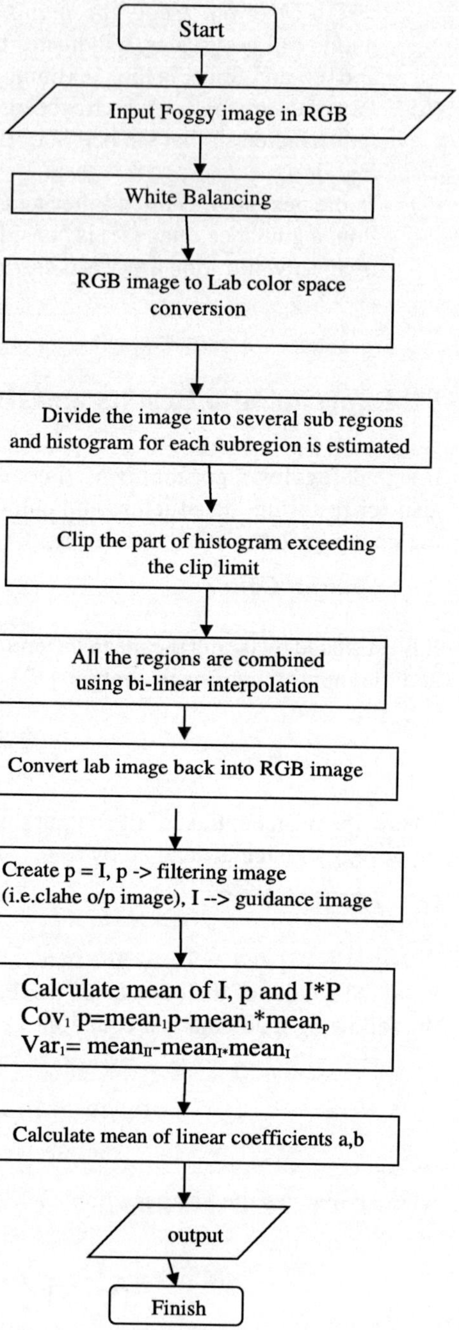

Table 1 Calculated values of the PSNR and MSE

Figure	White balanced		CLACHE		Final output	
	PSNR	MSE	PSNR	MSE	PSNR	MSE
Fig. 2	58.3691	1.4558$_e$–06	28.1276	0.0015	14.8120	0.0330
Fig. 3	47.9876	1.5894$_e$–05	22.1010	0.0062	9.6300	0.1089
Fig. 4	47.5924	1.7409$_e$–05	24.4851	0.0036	8.5865	0.1385
Fig. 5	34.7537	1.3468$_e$–04	19.5059	0.0112	8.4732	0.1421

Fig. 2 Visual analysis on fog image-I

with a high value of PSNR and a low value of MSE value are generally considered the better output images.

Table 1 shows the resultant values of these evaluation measures Figs. 2, 3, 4, and 5.

6 Conclusions and Future Scope

This research work fused two methods to enhance the foggy input images. Initially, CLAHE is applied in the foggy input image after that Guided Filter approach is used to smooth the image for better enhancement. Experimental results show that

Fig. 3 Visual analysis on foggy image-II

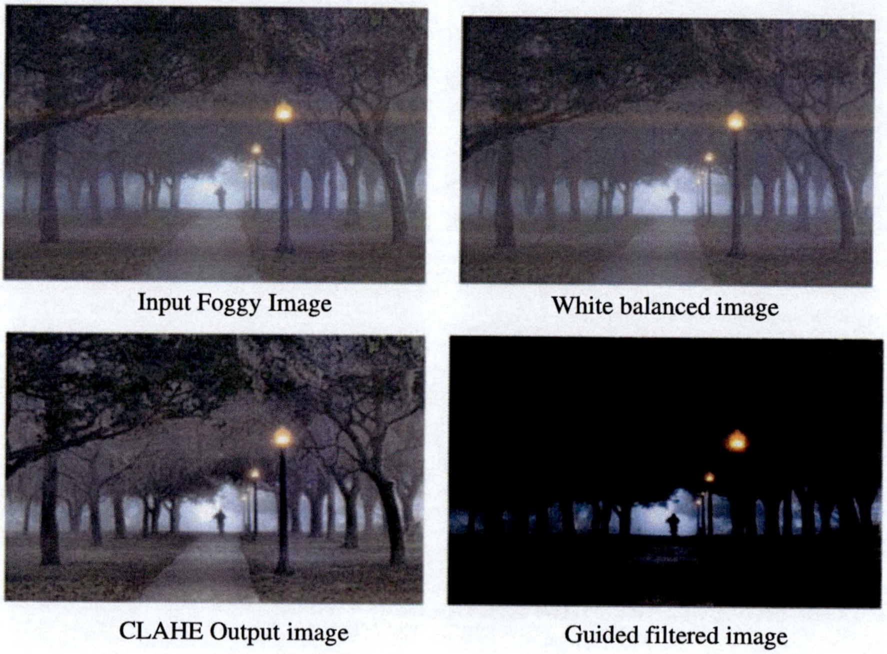

Fig. 4 Visual analysis on foggy image-III

Fig. 5 Visual analysis on foggy image-IV

the proposed system can effectively de-haze the input hazy image. In addition to that PSNR and MSE results as given in Table 1 are also supporting that the proposed system is good for enhancing the input foggy images. The improved version of the proposed method may be used with some more standard datasets. Experimental results compared with existing methods is left as a future work to this research.

References

1. He K, Sun J, Tang X (2011) Single image haze removal using dark channel prior. IEEE Trans Pattern Anal Mach Intell 33:2341–2353
2. Narasimhan SG, Nayar SK (2003) Contrast restoration of weather degraded images. IEEE Trans Pattern Anal Mach Intell 25:713–724
3. Fattal R (2008) Single image dehazing. ACM Trans Graph 27:72:1–72:9
4. Tan RT (2008) Visibility in bad weather from a single image. In: 2008 IEEE conference on computer vision and pattern recognition, pp 1–8
5. Tarel J, Hautiere N (2009) Fast visibility restoration from a single color or gray level image. In: 2009 IEEE 12th international conference on computer vision, pp 2201–2208
6. Wang Y, Fan C (2014) Single image defogging by multiscale depth fusion. IEEE Trans Image Proc 23:4826–4837
7. Lee Y, Gibson KB, Lee Z, Nguyen TQ (2014) Stereo image defogging. In: 2014 IEEE international conference on image processing (ICIP), pp 5427–5431
8. Park H, Park J, Kim H, Paik J (2016) Improved dcp-based image defogging using stereo images. In: 2016 IEEE 6th international conference on consumer electronics-Berlin (ICCE-Berlin), pp 48–49

9. Ketcham DJ (1976) Real-time image enhancement techniques
10. Reza AM (2004) Realization of the contrast limited adaptive histogram equalization (clahe) for real-time image enhancement. VLSI Signal Proc 38:35–44
11. He K, Sun J, Tang X (2013) Guided image filtering. IEEE Trans Pattern Anal Mach Intell 35:1397–1409
12. Economopoulos T, Asvestas P, Matsopoulos G (2010) Contrast enhancement of images using partitioned iterated function systems. Image Vision Comput 28(1):45–54
13. Yu Y-H, Chang C-C, Lin I-C (2007) A new steganographic method for color and grayscale image hiding. Comput Vision Image Underst 107(3):183–194
14. Lin M-H, Hu Y-C, Chang C-C (2002) Both color and gray scale secret images hiding in a color image. Int J Pattern Recog Artif Intell 16(06):697–713
15. Rongrong N, Qiuqi R (2002) Embedding information into color images using wavelet. In: Proceedings 2002 IEEE region 10 conference on computers, communications, control and power engineering, TENCOM '02, vol 1, pp 598–601
16. Liu T, Ding Qiu Z (2002) A dwt-based color image steganography scheme. In: 6th international conference on signal processing, vol 2, pp 1568–1571
17. Hitam MS, Awalludin EA, Yussof WNJHW, Bachok Z (2013) Mixture contrast limited adaptive histogram equalization for underwater image enhancement. In: 2013 international conference on computer applications technology (ICCAT), pp 1–5

Chapter 41
A Compact L-Shape ACS-Fed Printed Monopole Dual-Band Antenna for 3.5 GHz WiMAX and 5.8 GHz WLAN Applications

Sangeeta Sharma, Pramod Kumar Singhal, and Deep Kishore Parsediya

1 Introduction

Nowadays, worldwide interoperability for microwave access (WiMAX) and wireless local area network (WLAN) are the two known wireless transmission bands, and these two wireless communication bands integrate with new 5G bands (i.e., 3.4–3.6 GHz for 3.5 GHz WiMAX and 5.725–5.825 GHz for 5.8 GHz WLAN) having high data rate, high network and spectrum efficiency, high connection density and very high traffic capacity [1–12, 14]. Therefore, to give service of these communication bands in a system or an electronics device, miniaturization of sensor or antenna is a very important and difficult task. A compact antenna gives extra freedom for the user to use more than one antenna in a single device [1–15]. At present, coplanar waveguide (CPW) and asymmetric coplanar strip (ACS) feeding technologies have to turn into the utmost trendy approach for decreasing the size of an antenna alongside wide $S_{11} \leq -10$ dB bandwidth at the desired resonant frequencies [1–11]. But as compared to the CPW-fed antenna, ACS-fed antenna owns a very small area.

We have also studied that by using multiple microstrip radiators, multiple frequency bands can be achieved for various applications [12]. Regular microstrip printed aerial built-in flaw of cross-polar radiation, which is not at all right for monopole aerial [13]. Some linear and slant slots have been occupied with developing a mixed-mode response for specific individual-layer printed aerial; this technique is used for developing broad radiations in entire wireless transmission bands

S. Sharma (✉) · P. K. Singhal · D. K. Parsediya
Department of Electronics, Madhav Institute of Technology and Science, Race Course Road, Gole Ka Mandir, Gwalior 474005, Madhya Pradesh, India

P. K. Singhal
e-mail: pks_65@mitsgwalior.in

D. K. Parsediya
e-mail: parsediyadeep@mitsgwalior.in

S. Dawn et al. (eds.), *Smart and Intelligent Systems*, Algorithms for Intelligent Systems,
https://doi.org/10.1007/978-981-16-2109-3_42

[14]. Defected ground structures (DGS) are integrated with simple printed aerial for achieving good broadside patterns, desirable gain, and compact size of aerials [15].

Recently various multiband antennas with ACS-feeding techniques are reported; some of them are reported in this literature as tabulated in Table 1.

The basic structures of CPW-fed and ACS-fed antennas are displayed in Fig. 1, in which f; F1 and F2 are the coaxial feeding points. From Fig. 1, it is clear that the ACS-fed antenna has smaller size as compared to the CPW-fed antenna. Almost 44% reduction in antenna design is reported.

Table 1 The analogy of planned aerial work alongside past reported ACS-fed aerials

S. num.	Published literature	Aerial dimension (mm2)	Aerial purpose		Aerial class	Average peak gain (dBi)
			WiMAX	WLAN		
1.	[1]	28 × 30	Yes	Yes	Double-band	2.1
2.	[2]	31 × 15	No	Yes	Tri-band	2.6
3.	[3]	21 × 19	No	Yes	Double-band	1.8
4.	[4]	37.5 × 24	No	Yes	Double-band	1.21
5.	[5]	17 × 12	No	Yes	Double-band	2.0
6.	[6]	35 × 19	Yes	Yes	Tri-band	2.2
7.	[7]	13.75 × 26	Yes	Yes	Double-band	3.25
8.	[8]	22 × 20	Yes	Yes	Tri-band	2.0
9.	[9]	14.75 × 26	Yes	Yes	Double-band	1.4
10.	[10]	12.5 × 20	Yes	Yes	Double-band	3.6
11.	Proposed Aerial	12.5 × 20	Yes	Yes	Double-band	~3.8

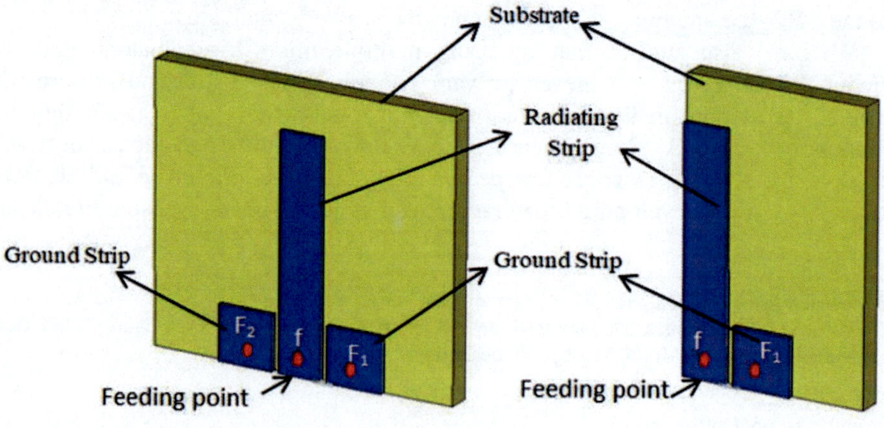

Fig. 1 Diagram of basic CPW and ACS-feeding technique

Here in the research paper, we design a novel antenna including two radiating segments and one uniplanar ground plane alongside ACS-feeding technology in place of CPW-feeding technology; this makes the size of the antenna 43.18% small. This research article further describes the detailed antenna design followed by results and discussions and is the end with its conclusion.

2 Antenna Design

The detailed geometry related to the proposed double-band monopole aerial with the ACS-feeding approach is shown in Fig. 2, where Fig. 2a, b given the front and side aspects of the aerial design, respectively. The optimized parameter of the presented aerial is outlined in Fig. 2 and tabulated in Table 2. The simulation and analysis work is done by CST microwave studio V.2018 software.

The planned configuration is situated on one side of a substrate, that is, FR-4 substrate with 1.6 mm thickness, with a 4.4 dielectric constant and loss tangent 0.02. The required resonant frequencies with 50 Ω impedance bandwidths are achieved by integrating L-profile and rectangular radiating segments with the rectangular ground plane for this ACS-fed double-band monopole aerial design. The detailed description of the optimized parameter for the presented antenna design is tabulated in Table 2, where the red-colored layer depicted the copper and the white-colored layer depicted the FR-4 substrate.

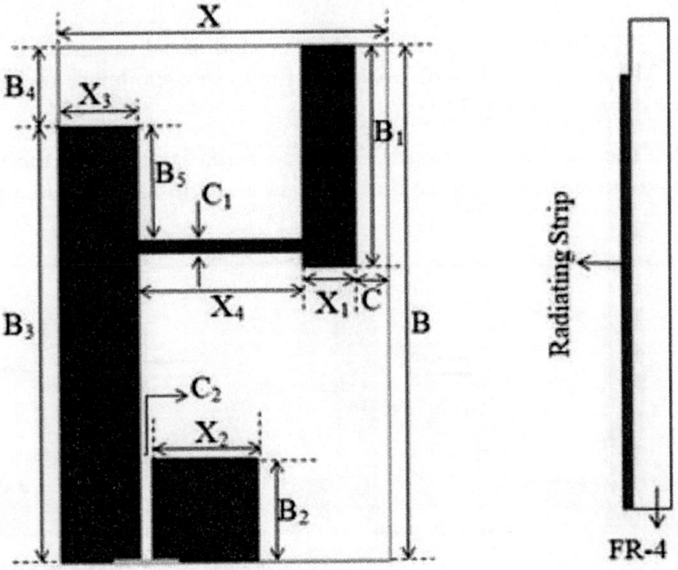

Fig. 2 The precise geometry of the planned aerial **a** Top aspect and **b** Side aspect

Table 2 Upgraded specification of the planned aerial

Upgraded specification	Specification meaning	Value, mm
B	Length substrate	20
X	Width substrate	12.5
B_1	Length of a vertical radiating segment of L-profile aerial	8.6
X_1	Width of a vertical radiating segment of L-shaped aerial	2
B_2	Length of Ground strip	4
X_2	Width of ground strip	4
B_3	Length of rectangular radiating strip	16.9
X_3	Width of rectangular radiating strip	3
B_4	B–B3	3.1
X4	Width of a horizontal radiating segment of L-profile aerial	6.25
B5	Length of the rectangular radiating segment above L-profile aerial	4.4
C	The gap between the L-shaped radiating strip and substrate	1.25
C1	Length of a horizontal radiating strip of L-shaped antenna	0.5
C2	The gap between the ground and a radiating rectangular strip	0.5

The designing of the compact L-shape ACS-fed aerial requires three stages of evolution. These three stages of evolution for the development of the proposed antenna are depicted in Fig. 3.

First stage: The pattern of the basic CPW-fed aerial is shown in the first step as pictured in Fig. 3a by green. This antenna is a single-band antenna resonant at

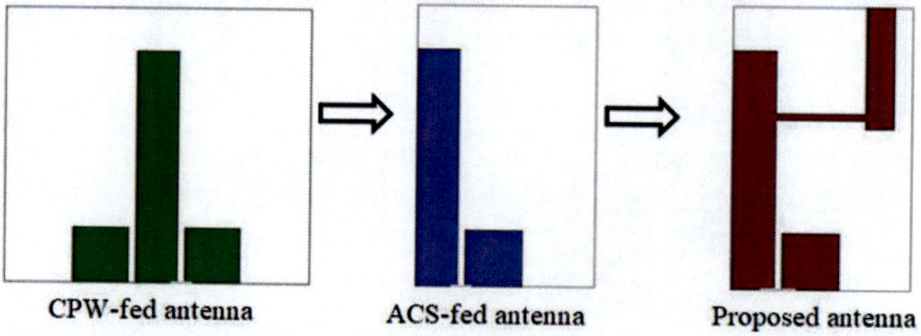

Fig. 3 Three different evolution stages of developing the proposed L-shaped antenna

5.162 GHz with −18.43 dB return loss and having a 1.0116 GHz S11 ≤ − 10 dB impedance bandwidth.

Second stage: The designing of a basic ACS-fed antenna is shown in this stage as shown in Fig. 3b by blue. This stage is developed by abolishing the one-ground face of the CPW-fed aerial that reduces antenna size by 43.18%. This antenna is resonant at 5.426 GHz with −52.027 dB return loss and having 1.578 GHz S11 ≤ − 10 dB impedance bandwidth. Thus we almost achieve one of the required bands, that is, 5.8 GHz WLAN band.

Third stage: This is the final stage of the antenna development, as depicted in Fig. 3c by red. In this stage, we integrate an L-shaped antenna horizontally for achieving another required frequency band, that is, 3.5 GHz WiMAX band. This L-shaped antenna is designed at 3.5 GHz with $\lambda_g/4$ size. Hence we finally design the required ACS-fed double-band aerial.

The explanation of the L-shaped aerial size calculation is done in the following way [11].

The length of the antenna is proportional to the guided wavelength $\lambda_{g,}$ which can be derived by the below formula.

$$\lambda_g = c/[f\sqrt{(\varepsilon_r + 1)/2}] \tag{1}$$

where "c" gives the light's speed and "f" presents an operating frequency.

Now we discuss electromagnetic waves radiating in free space and FR-4 substrate media. If the electromagnetic wave is radiating in the free space, then $\varepsilon_r = 1$. By substituting the $c = 3.8 \times 10^8$ m/s, f = 3.5 GHz in Eq. (1), we get $\lambda_{g1} = 85.7$ mm. Again, if the electromagnetic wave is radiating in the substrate medium, then $\varepsilon_r = 4.4$. By substituting the $c = 3.8 \times 10^8$ m/s, f = 3.5 GHz in Eq. (1), we get $\lambda_{g2} = 66.07$ mm. Thus it is clear that the actual wavelength should be between 85.7 and 66.07 mm at 3.5 GHz.

As we know from our theoretical knowledge, the length of the antenna should be ¼ times of λ_g, hence for the above value, the length of the L-shaped antenna should be between 13.1625 and 21.9 mm. After optimization, we get the total length of the L-shaped antenna is $X_4 + X_1 + B_1 = 16.85$ mm, which is in between the calculated limit.

The comparisons of these three stages are pictured in Fig. 4. Figure 4a, b, and c depict the return loss, impedance, and admittance of these three stages, respectively. The green, blue, and red color lines in graphs represent CPW-fed, ACS-fed, and proposed antenna characteristics.

Figure 4a shows that CPW-fed and ACS-fed antennas are single-band antennas while the proposed antenna is a double-band antenna. Figure 4b represents the impedance characteristics, and it is almost equal to 50 Ω at resonant frequencies which is required for all three-stage antennas. Figure 4c represents the admittance characteristics, and it should be equal to 0.02 mho at the resonating frequency and is almost achieved for all three antenna designs.

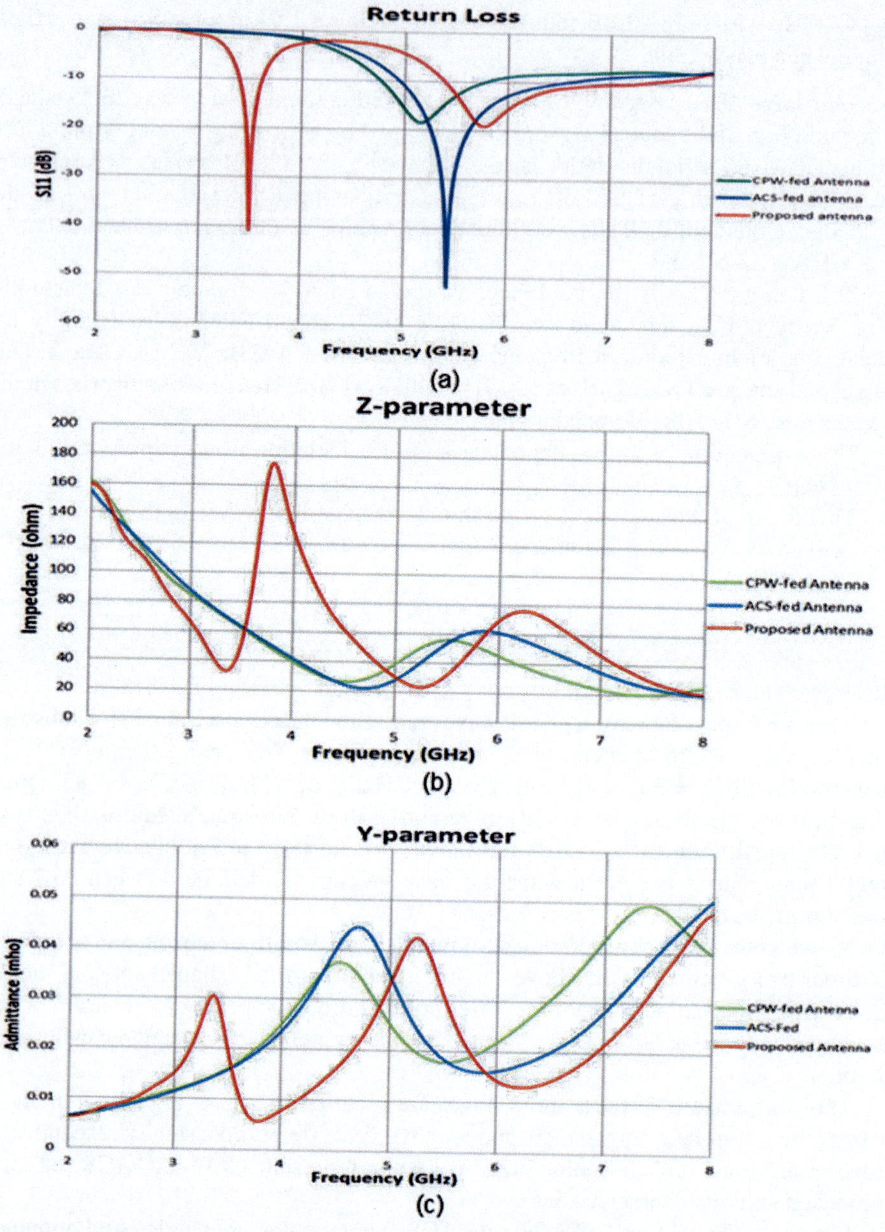

Fig. 4 Comparisons of three stages evolution of proposed antenna in parity of **a** Return loss, **b** Impedance, and **c** Admittance parameters

Further investigation of the planned aerial geometry is achieved by three variables p, q, and r. In Fig. 5, we depict the graphical variation of the performance of the proposed antenna when a variable "p" is changing. The variable "p" is defined as P $= C_1 + p$. From Fig. 5, it is clear that when $p = 0$ mm we get the required dual-bands. Otherwise, when we change p from 0 to 3 the antenna still behaves dual-band, but both resonating frequencies are changing, and the graph is shifted toward the right side.

Now our next target is to achieve the independent frequency-changing variables for both the frequency bands. Figure 6 depicts the working variation of the planned antenna when a variable "q" is changing. The variable q is defined as $B = B_3 + q$. When variable q is changed from 0 to 3 mm the 5.8 GHz WLAN protocol is shifted toward the left side, and the 3.5 GHz WiMAX protocol is not changed. The required dual-bands are achieved at $q = 0$ mm.

Figure 7 shows the working variation of the planned antenna when a variable "r"

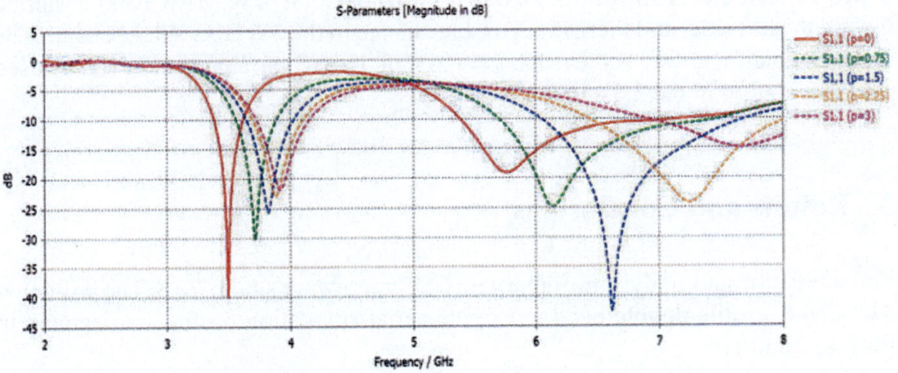

Fig. 5 The working of the proposed antenna when a p-variable is changed

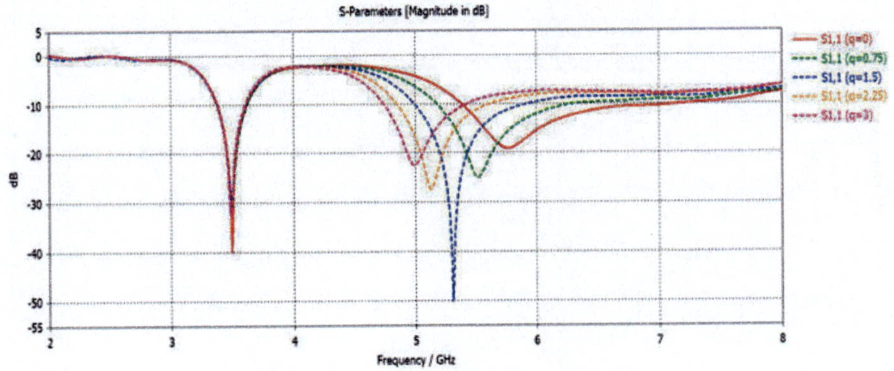

Fig. 6 The working of planned aerial when q-variable is changed

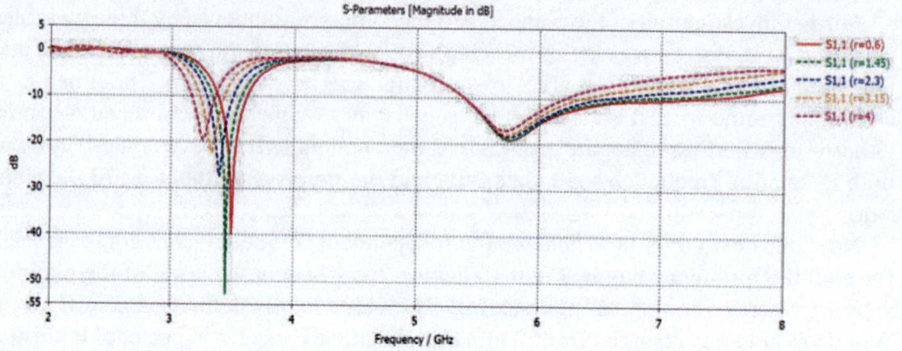

Fig. 7 The working of planned aerial when r- variable is changed

is changing. The variable r is defined as $R = B_1 - r$; hence r is the reduction in the B_1. When variable r is changed from 0.6 to 4 mm the 3.5 GHz WiMAX band is shifted toward the left side, and there is negligible change in the 5.8 GHz WLAN band. The required dual-bands are achieved at r = 0.6 mm. Hence variable q and r can be used for the variation in the 3.5 and 5.8 GHz bands.

3 Results and Conclusions

The simulation and analysis are done with CST microwave studio V.18. The presented ACS-fed L-profile double-band monopole aerial simulation results are depicted in Figs. 8, 9 and 10.

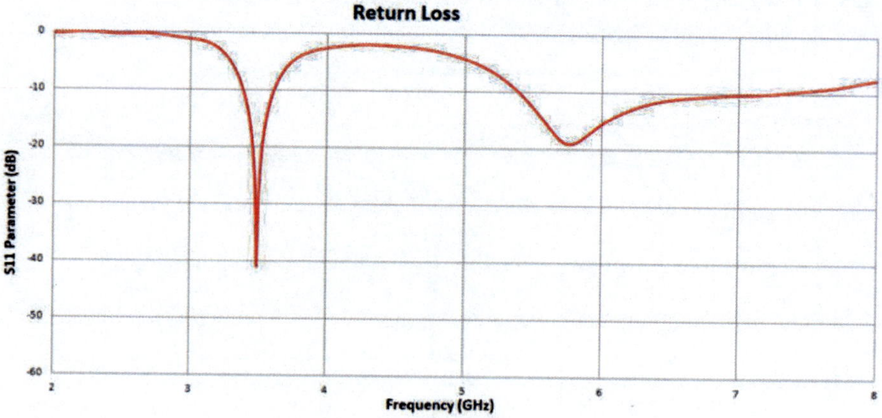

Fig. 8 Return loss chart of the planned aerial

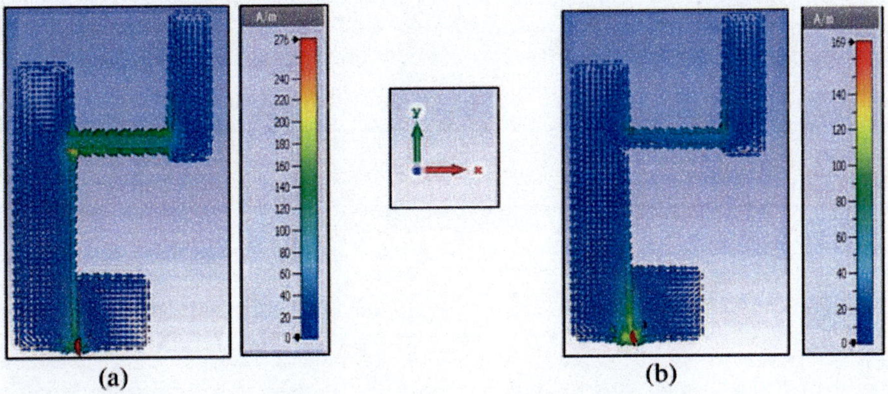

Fig. 9 The surface current of the planned aerial at **a** 3.5 GHz and **b** 5.8 GHz frequencies

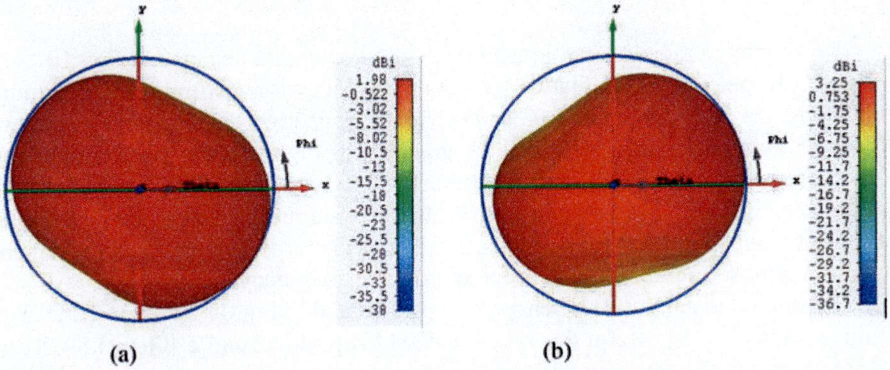

Fig. 10 3-D radiation arrangements at **a** 3.5 GHz and **b** 5.8 GHz

Figure 8 represents the S11-parameter related to the aerial. From Fig. 8, it's clear that this aerial resonates at 3.5002 GHz with −40.765 dB return loss for the 3.5 GHz WiMAX band and 5.7698 GHz with −19.145 dB return loss for the 5.8 GHz WLAN band. The $S_{11} \leq -10$ dB impedance bandwidth is from 3.3925 to 3.6178 GHz, that is, 225.3 MHz for the early band and that is for the later band is from 5.3879 to 7.277 GHz, that is, 1.8891 GHz.

Figure 9 represents the surface current in A/m of the proposed aerial at both resonating frequencies [12]. Figure 9a gives the surface current at 3.5 GHz for the 3.5 GHz WiMAX band with maximum and minimum values of 276 and 0 A/m, respectively. Figure 9b gives the surface current at the 5.8 GHz WLAN band with maximum and minimum values of 169 and 0 A/m. From Fig. 9a, b, it is clear that the utmost of the surface current is riched at the horizontal strip of a proposed L-shaped antenna and the gap between the unipolar ground and the radiating rectangular strip [12].

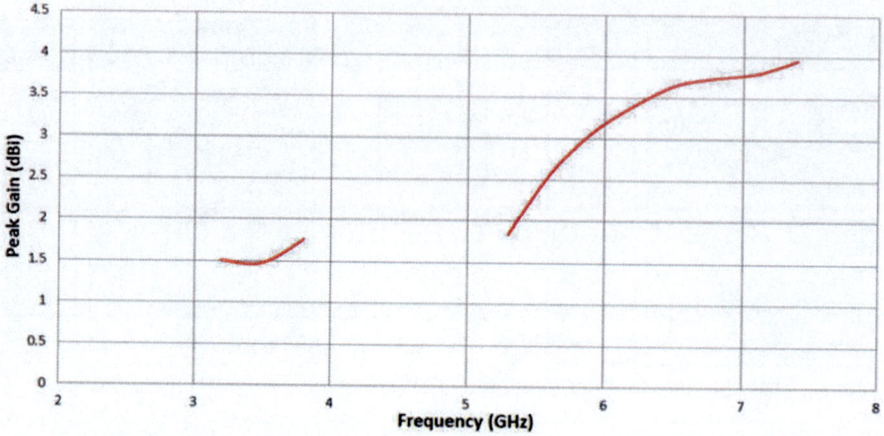

Fig. 11 The peak gain chart planned aerial

The simulated radiation patterns in three-dimension are depicted in Fig. 10.

Figure 10a depicts the radiation pattern at 3.5 GHz with maximum and minimum directivity (or directive gain) 1.98 and −38dBi, respectively. This radiation pattern is almost omnidirectional with a slide angular tilt. Figure 10b gives the radiation arrangement at 5.8 GHz with maximum and minimum directivity 3.25 and −36.7 dBi, respectively. This radiation pattern is also an omnidirectional pattern with a slide angular tilt but opposite to the previous one. Hence the far-field radiation arrangement owns an almost omnidirectional shape in both the frequency bands.

The simulated gain v/s frequency graph is depicted in Fig. 11 by the red line. The gain is 1.456–1.5314 dBi for the 3.5 GHz WiMAX protocol and 2.406 to 3.8918 dBi for the 5.8 GHz WLAN protocol. Hence the average peak gain for the planned aerial design is 3.8 dBi.

The voltage standing wave ratio (VSWR) chart for the planned aerial is depicted in Fig. 12. The ideal value of VSWR is 1, whereas its practical value is 1.5, and also VSWR = 2 can be accepted. From Fig. 12, it is clear that its value is 1.018 and 1.25 at 3.5 and 5.8 GHz frequencies. For the first band, we almost achieve the ideal value of VSWR and for the second band, we achieve lower than the practical value.

4 Conclusion

In this research article, we developed an ACS-fed L-profile double-band monopole aerial design from a basic CPW-fed antenna in three evolution stages for 3.5 GHz WiMAX and 5.8 GHz WLAN functions. The comprehensive area of the proposed antenna is reduced by 43.18% as compared to the CPW-fed antenna design. The aerial is printed on an FR-4 substrate alongside a 1.6 mm thickness. The whole size of the antenna is only $12.5 \times 20\ \text{mm}^2$ which is almost equal to the size of a thumbnail.

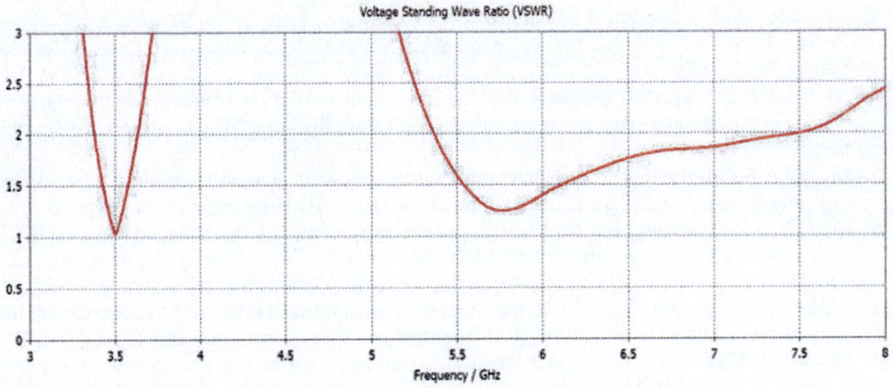

Fig. 12 The VSWR chart for the planned aerial

Table 3 Simulated results of the planner ACS-fed aerial

S. no.	Parameter	3.5 GHz WiMAX band	5.8 GHz WLAN band
1.	Return loss	−40.765 dB	−19.145 dB
2.	Impedance bandwidth	225.13 MHz	1.88891 GHz
3.	Impedance	50 Ω	50 Ω
4.	Peak gain	1.4781 dBi	2.9 dBi
5.	VSWR	1.018	1.25
6.	Radiation pattern	Omnidirectional	Omnidirectional
7.	Directivity	1.98 dBi	3.25 dBi

Due to its small or compact size, it is very easy to integrate this antenna with any electronic device or system. The simulated results for 3.5 GHz WiMAX and 5.8 GHz WLAN bands are achieved by applying CST MWS V.18 and tabulated in Table 3. All the results are tabulated in the table is measured at the resonating frequency. The resonating frequencies for the 3.5 GHz WiMAX band and 5.8 GHz WLAN band are 3.5 GHz and 5.8 GHz, respectively. Next, we fabricate this design, measure all the results, and compare them with the simulated results.

References

1. Deepu V, Raj RK, Joseph M, Suma MN, Mohanan P (2007) Compact asymmetric coplanar strip fed monopole antenna for multiband applications. IEEE Trans Antennas Propagat 55:2351–2357
2. Song Y, Jiao YC, Wang XM, Weng ZB, Zhang FS Compact coplanar slot antenna fed by asymmetric coplanar strip for 2.4/5 GHz WLAN operations. Microwave Optical Technol Lett 50:3080–3083

3. Deepu V, Sujith R, Mridula S, Aanandan CK, Vasudevan K, Mohanan P (2009) ACS fed printed F-shaped uniplanar antenna for dual-band WLAN applications. Microwave Optical Technol Lett 51:1852–1856
4. Ashkarali P, Sreenath SR, Dinesh R, Krishna DD, Aanandan CK (2012) A compact asymmetric coplanar strip fed dual-band antenna for DCS/WLAN applications. Microwave Optical Technol Lett 54:1087–1089
5. Li Y, Li W, Mittra R (2013) Miniaturization of ACS-fed dual-band antenna with loaded capacitance terminations for WLAN applications. IEICE Electron Expr 10
6. Li B, Yan Z-H, Zhang T-L (2013) Triple-band slot antenna with U-shaped open stub fed by asymmetric coplanar strip for WLAN/WiMAX applications. Progr Electromag Res Lett 37:123–131
7. Naidu PV, Malhotra A (2015) Design & analysis of miniaturized asymmetric coplanar strip fed antenna for multi-band WLAN/WiMAX applications. Progr Electromag Res C 57:159–171
8. Naidu PV, Malhotra A (2015) A compact tri-band ACS-fed monopole antenna with mirror L-shape strips for WLAN/WiMAX applications. Int J Microw Optical Technol 10:266–273
9. Naidu PV (2016) Printed V-shape ACS-fed compact dual-band antenna for Bluetooth, LTE, and WLAN/WiMAX applications. Microsyst Technol 23(4):1005–1015
10. Kumar A, Naidu PV (2016) A compact O-shaped printed ACS fed monopole dual-band antenna for 2.4 GHz Bluetooth and 5 GHz WLAN/WiMAX Applications. Progr Electromag Res Symp (PIERS):2004–2008
11. Naidu PV, Malhotra A, Kumar R (2015) A compact ACS-fed dual-band monopole antenna for LTE, WLAN/WiMAX, and public safety applications. Microsyst Technol 22:1021–1028
12. Wu l, Huang J, Yuan N (2016) Compact multi-band monopole antenna for GPS/WLAN/WiMAX application. In: 2016 IEEE international conference on microwave and millimeter wave technology (ICMMT)
13. Chattopadhyay S, Chakraborty S (2018) A physical insight into the influence of dominant mode of rectangular microstrip antenna on its cross-polarization characteristics and its improvement with T-Shaped microstrip antenna. IEEE Access 6:3594–3602
14. Sarkar T, Ghosh A, Chakraborty S, Singh LLK, Chattopadhyay S (2020) Employment of mixed mode in single-layer microstrip antenna for ISM/WiMAX/WLAN/4G/Sub 6 GHz 5G mobile communication. J Electromag Waves Appl 34(7):907–925
15. Sarkar T, Ghosh A, Singh LK, Chattopadhyay S, Sim C-Y-D (2020) DGS integrated air loaded wideband microstrip antenna for X and Ku band. IEEE Antennas Wireless Propag Lett 19(1):114–118

Chapter 42
Performance Investigation of PAPR Mitigation in MIMO-OFDM System Using Unconstrained Global Optimization Base Partial Transmit Sequence

Padarti Vijaya Kumar, V. Siva Reddy, and K. Vara Prasad

1 Introduction

OFDM technology is an attractive system in the field of wireless communication [17]. This technology-based systems are mostly employed for wireless applications such as European DAB, DVB Wireless Land Area Networks (LANs) and Broadband Fixed Wireless Systems (BFWS) because it provides high-spectral efficiency, immunity to the impulse noise, multipath delay spread tolerance and robustness against frequency selective fading channels [8, 10]. The OFDM is a multicarrier modulation method to use the orthogonality of the subcarriers. The OFDM system is applicable for high data rate application based on its flexible inter-symbol interference (ISI) flexible management in high dispersive channels [25]. The OFDM system provides strength not in favor of ISI and multipath fading due to a low-rate subcarrier and increases symbol period, which is robust against narrowband interference. The OFDM system allows Radio Frequency (RF) with the utilization of an adaptive modulation scheme. It enables bandwidth-on-demand technology and privileged spectral efficiency. Hence, the OFDM does not require adjacent bandwidth for Inverse Fast Fourier Transform (IFFT) operation. The OFDM system creates its frequency network, which is suitable for broadcasting applications [6]. In the present day, the OFDM system faces the PAPR problem in multicarrier transmission OFDM. An increase of PAPR in the

P. Vijaya Kumar (✉) · V. S. Reddy · K. Vara Prasad
Department of ECE, Siddhartha Engineering College, VR Siddhartha Engineering College, Vijayawada, India
e-mail: vijayakumar.padarti@vrsiddhartha.ac.in

V. S. Reddy
e-mail: vsivareddy@vrsiddhartha.ac.in

K. Vara Prasad
e-mail: kvprasad.vrsec@gmil.com

S. Dawn et al. (eds.), *Smart and Intelligent Systems*, Algorithms for Intelligent Systems,
https://doi.org/10.1007/978-981-16-2109-3_43

OFDM system leads to different distortions, like inter-symbol interference in the transmitter chain, amplitude-frequency, degradation of the system performance (Bit Error Rate (BER)), high linear power amplifiers that maximize the cost of the system [23]. In recent years, several techniques on PAPR reduction were introduced in the OFDM system, and those techniques are Linear Block Coding (LBC) [21, 22], Peak Insertion (PI), Interleaving technique [3, 14], tone reservation [4, 12], SLM [9, 2], and PTS technique [18] and so on. The conventional PAPR reduction [5, 12, 20] techniques have some drawbacks such as less PAPR reduction in maximum transmit power, reduced data rate, boosted BER, and increased computation complex. The proposed UGO-PTS technique works efficiently in the reduction of PAPR of MIMO-OFDM signal and outperforms the existing algorithms in terms of convergence speed and calculated solution.

This research work is composed as follows; Sect. 2 presents a brief survey of recent papers on PAPR reduction techniques. In Sect. 3, briefly explained the proposed UGO-PTS technique for PAPR reduction. Section 4 described comparative experimental results of a proposed UGO-PTS technique compared to the existing method. Conclusion of this research work is explained in Sect. 5.

2 Literature Survey

The scientists have proposed a few procedures for PAPR decrease in the OFDM framework. This segment exhibited a concise assessment of some critical commitments on the OFDM inquires about.

Hu et al. [7] proposed bedlam based PTS (CTPS) strategy for physical layer security in the OFDM Passive Optical Network (PON). A four-dimensional hyper disorder was utilized to produce the confused parcel information and clamorous stage weighting factors in the PTS, just as disorganized preparing succession for image synchronization. This gave a huge keyspace to improve the physical layer. The proposed strategy improved the exhibition of the OFDM transmission because of the high decrease of the PAPR in the OFDM framework utilizing the confused irregular PTS method. The proposed technique gave a better PAPR decrease, yet that builds calculation multifaceted nature.

Anand [1] introduced the PTS strategy diminished the PAPR esteem in the OFDM framework. In this paper, input information images were characterized into detached sub-squares, and the sub-squares were independently stage turned by an exclusively chosen stage factors alongside the way toward upgrading the exhibition of PAPR in the OFDM framework. The intricacy of the proposed PTS procedure could be decreased by the use of an advanced stage arrangement. The PTS method had the option to achieve huge execution. Be that as it may, this procedure expands the looking through multifaceted nature exponentially with the quantity of sub-square, which is the real disadvantage.

Padarti and Nandhanavanam [19] proposed a PTS strategy called as sequential PTS and parallel PTS to lessen the computational multifaceted nature of the PTS technique dependent on weighting variables browsed the code-wards of the direct square codes. In this paper, stages were haphazardly produced, and the created stages were increased. Henceforth, the information has been changed by utilizing the stage controls when the PAPR was decreased in OFDM. The parallel and sequential PTS procedure had diminished computational multifaceted nature and gave mistake revision ability to weighting factors. The PTS calculation has accomplished a low PAPR and less excess. In any case, the proposed strategy was not contrasted and any current strategies.

Ku et al. [11] proposed another PTS strategy named as Space Frequency Coding (SFBC) in the Multiple-Input and Multiple-Output (MIMO) OFDM frameworks for lessening PAPR. The PTS strategy has utilized the example rundown of forces in the sub-hinders so as to produce cost work for the determination of tests, which is utilized to assess the pinnacle intensity of every single up-and-comer signal that brought about the decrease of the framework multifaceted nature. The reenactment result affirmed that this specific PTS strategy accomplished less PAPR and BER rate with less computational expense. A great deal of calculations was required to pick the best up-and-comer grouping utilizing a problematic calculation that prompts complex issues.

Wang et al. [24] introduced Selective Weighting—PTS (SW-PTS) strategy for diminishing PAPR in the OFDM framework. In the SW-PTS procedure, a specific weighting sub-square arrangement was initially utilized for lessening computational unpredictability. In this work, to amplify the quantity of the OFDM competitor grouping, couple of explicit stage weighting arrangement was created for weighting the underlying sub-square succession. The PC reenactments demonstrated that proposed SW-PTS have accomplished a less computational multifaceted nature and PAPR decrease contrasted with the Original PTS strategy. As this strategy comprises a few cycles, its calculation time is more.

UGO-PTS innovation is executed for decreasing PAPR in the OFDM framework. The guideline of the proposed UGO-PTS procedure is quickly clarified in the underneath area.

3 Proposed UGO-PTS Technique

Multi-receiving wire transmission, called MIMO, is extensively disposed of as the major bent for empowering remote broadband administrations. No matter how you look at it, accomplishment will require multiple times progressed unearthly suitability than existing cell plans multiple times minimal effort per bit. In the different high information rate executions, the OFDM strategy has been maintained as the standard. The utilization of parallel OFDM, known as MIMO-OFDM, can be utilized in the cutting edge correspondence plans, where the PAPR discount is a major framework inconvenience. PAPR reduction procedure depends on the unique fundamentals of the framework. There are various constituents that are to be considered before an assistant PAPR drop-off methodology of the outline, that is, unpredictability, development in diffuse banner influence and augmentation in the bit-botch rate at the recipient, disaster in data rate. In the piece of scrambling upsets to amass the subcarriers, the Partial Transmit Series is a twisting less system. The fundamental idea of the PTS approach is that the information piece is repelled into disconnected sub-squares. The sub-transporters that are diffused by methods for the sub-squares are increased by estimating the estimation of the stage transformation vector expected for those sub-pieces. The stage change vector is proficiently picked to such a degree to the point that the PAPR regard is diminished. PTS is incredibly usable in PAPR decreasing and solid severance use. From another perspective, pivotal computational multifaceted nature is fundamental to look concerning high-dimensional vector space and basic transmission of side data (SI) to the collector trapped for a viable execution. In this article, an effective streamlining strategy is developed inside the PTS intend to demonstrate the stage turn vector with less estimation disarray. The versatile Unconstrained Global Optimization calculation is the improvement system adjusted in our foreseen methodology.

3.1 OFDM System Model

Proposed UGO-PTS is a well-known non-linear PAPR reduction technique with non-linear bandwidth, less computational complexity, and easy implementation without receiver-side cooperation. At the same time, the SLM and PTS technique are known to provide better performance in PAPR reduction without any kind of signal distortion (Fig. 1).

The block diagram of the Proposed UGO-PTS technique consists of the five major blocks in the transmitter side: data block, a modulation block, pilot insertion block, IFFT block, and proposed UGO-PTS block. The receiver side consists of four major sub-blocks that are FFT, pilot inverse insertion, and QPSK. This research used QPSK modulation and higher orders over the Additive White Gaussian Noise (AWGN) channel environment. Here in this system, initially, a specified count of the input data segments are modulated and then mutually related together by applying IFFT at the side of the transmitter. Mathematically, an IFFT associates all the input signals to produce each and every signal of the output OFDM symbol. Mathematically, the below-mentioned equation presents the input frequency domain sequences.

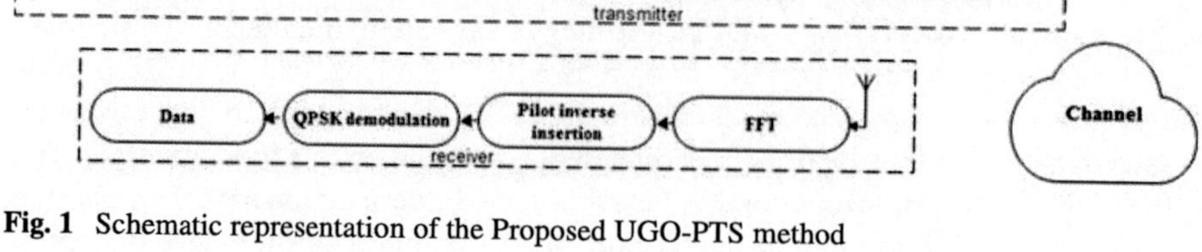

Fig. 1 Schematic representation of the Proposed UGO-PTS method

$$P' = \left[P_0, P_1, \ldots P_{N-1}\right] \tag{1}$$

where N represents the dimension of input data and P denotes input data. In the initial stage, the ordered OFDM input frequency samples modulate into time-domain sequences. The subcarrier sequence selects orthogonally to support several transmissions. Equation (2) presents time-domain modulated signals.

$$p_n = \frac{1}{\sqrt{N}} \sum_{k=0}^{n-1} P_n W_N^{kn} \tag{2}$$

where P_n represents the signal further processed by IFFT, kn represents the orthogonal value of the subcarriers. At the transmitter, the integration of signals may direct to high PAPR. PAPR is denoted as the ratio between maximum power and average power of the modulated signal.

$$\text{PAPR} = 10\log_{10} \frac{\max\left\{|p_n|^2\right\}}{E\left\{|S_n|^2\right\}} \tag{3}$$

where $E\left\{|S_n|^2\right\}$ represented as the average power of p_n.

3.2 UGO-PTS Method

The PTS technique is one of the familiar methods used for PAPR mitigation in OFDM systems. The block diagram of the PTS method is shown in Fig. 2. The applied input data block is first to split into multiple numbers of sub-blocks. After that, each of the

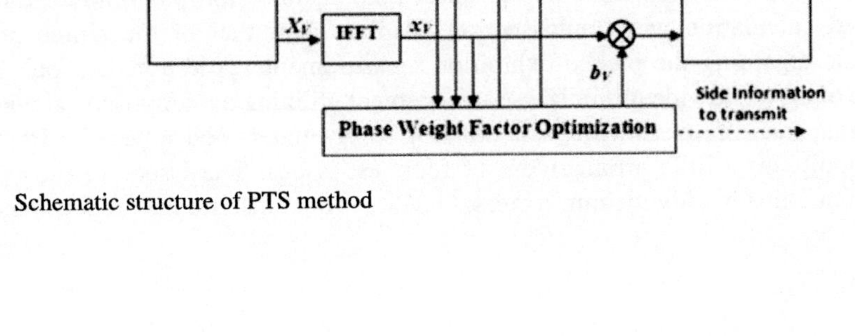

Fig. 2 Schematic structure of PTS method

sub-blocks is appended with zeros in such a way to make them identical in size and whirled by complex phase factors.

The data vector $X = [X_0, X_1, \ldots X_{N-1}]^T$ is separated in V disarticulated sets, f g $X_V = 1, 2, 3, \ldots$ v using identical size groups, and the frequency domain representation is specified by:

$$X' = \sum_{V=1}^{V} X_V b_V \tag{4}$$

where $b_v = e^{j\O v}$ are the phase factors and $\O v = \frac{2\pi i}{\omega}$, $i = 0, 1, ..w - 1$ W represents the phase weight for the phase factors. In time domain X_v(IFFTofX_v) is called partial transmit sequence. The optimum phase factor is chosen from all the allowable phase factors such that PAPR of candidate signal X' is minimum:

$$X' = \sum_{V=1}^{V} \tilde{b}_V X_V \tag{5}$$

For W allowable phase weights and V sub-blocks, phase weight for the first block is usually chosen as $b1 = 1$, then to find the optimum phase factors overall $W_v - 1$ phase combinations are to be analyzed, which leads to minimum PAPR. But at the same time to decode the OFDM signal at the receiver, precise knowledge of optimal phase factor is necessary, which is transmitted in the form of side information. This side information requires log $2W_v$ bits to be transmitted separately more than existing BW.

3.3 *Unconstrained Global Optimization*

Worldwide improvement is the investigation of how to locate the best (ideal) answer for an issue. For the estimate of multivariate polynomials, the B-spline structure is utilized. First proposed an essential calculation for worldwide streamlining that uses a few quickening calculations, for example, lapse test and monotonicity test. The proposed Unconstrained Global Optimization [16, 15] fundamental and improved calculations are tried and looked at on 16 test issues under a set of double test conditions. The consequence of the quest demonstrates the predominance of the bettered calculation with multi-section B-spline over that of the single portion B-spline regarding the picked exhibition measurements. Additionally, look at the nature of all worldwide minimizers' arrangement utilizing the proposed calculations (essential and improved) with those utilizing surely understood solvers BARON and Gloptipoly, on a little arrangement of four test issues. The issues in the last set have numerous worldwide minimizers. Outcomes demonstrate the predominance of targeted calculations, though Gloptipoly and BARON neglect to do so in a few or most of the test issues individually [24].

Input:
Degree N of each variable occurring in the objective function polynomial, the coefficients matrix A of the objective function polynomial in the power form, the initial box x, the tolerance limit $\in$ and number of B-spline segments K.
Output:
Global minimum $\hat{p}$ to the specified tolerance, $\in$ and all the global solution(s) $\hat{b}$ in the box x.
Start Algorithm 1
{Enumerate the B-spline segments}
if $K = \emptyset$
then compute
$K = N + 2$
else go to 2 end 2 {Calculate B-spline factors}
$D(x) = B - spline\ coefficient\ matrix\ (A, N, K, x)$
{Set current lowest estimate}
Prepare presnt lowest estimate $\hat{p}$ =max D(x)
{Initialize lists}
$Set\ L \leftarrow \{x, D(x)\}, Lsol \leftarrow \{\}$
$\{Sort\ the\ list\ L\}$
Systemize list L in ascending order of (min D(x)).
$\{Start\ iteration\}$
if $L = \emptyset$ then
go to 12
else pick the last item from L, denote it as $\{b, D(b)\}$, and delete this item entry from L
end
{Test if vertex condition is met within $\in$ }
(See Definition 1)
For the item $\{b, D(b)\}$
if min $D(b)$ satisfies vertex condition within $\in$ then
enter the item {b, min D(b)} in list L^{sol} and return to 6
end

4 Results and Discussion

The Proposed UGO-PTS procedure is executed in MATLAB (R2018a). The Proposed UGO-PTS method is connected to the sub-squares of un-coded data, and that data is regulated by utilizing QPSK adjustment. Stage revolution variables are transmitted straightforwardly to the collector through the sub-obstruct in the OFDM framework. The presentation estimation was dissected as far as the Complementary Cumulative Distribution Function (CCDF) and reenactment specifications considered for execution investigation are IFFT length$-$1024, bearer count$-$32 Bit Per Symbol = 16, Symbol per Carrier$-$8. This examination work considered 1000 examples for one example timeframe indicated in time and recurrence space. These examples are additionally prepared for square isolating, parallel to sequential transformation and transmission. Reenactment parameters have appeared in Table 1.

Table 1 Simulation Parameters

Simulation parameters	Types/value
OFDM blocks	10000
Total sub-carriers	1024, 512, 256 and 128
Totality of sub-blocks	4
Over sampling factor	4,8
Modulation	QPSK

The SLM requires the transmission of many side data bits for each datum square, and it brings about certain information rate misfortune. Consequently, the cutting method is utilized in the OFDM framework for PAPR decrease, which does not consider the side data in the exploration work. Figures 2, 3, 4 and 5 demonstrate the examination of the current and Proposed UGO-PTS system for various IFFT lengths:

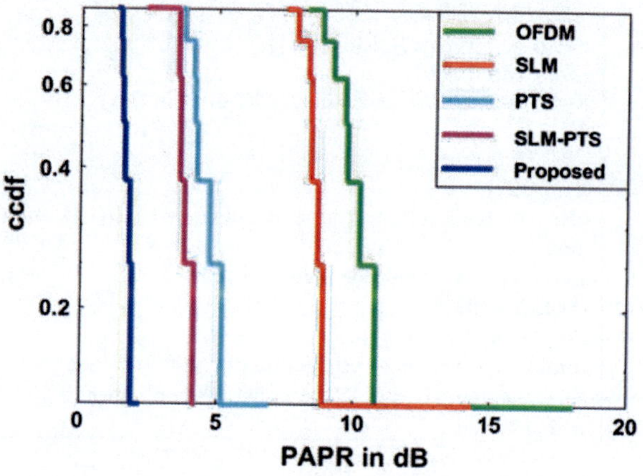

Fig. 3 Performance comparison of proposed UGO-PTS with IFFT length 1024

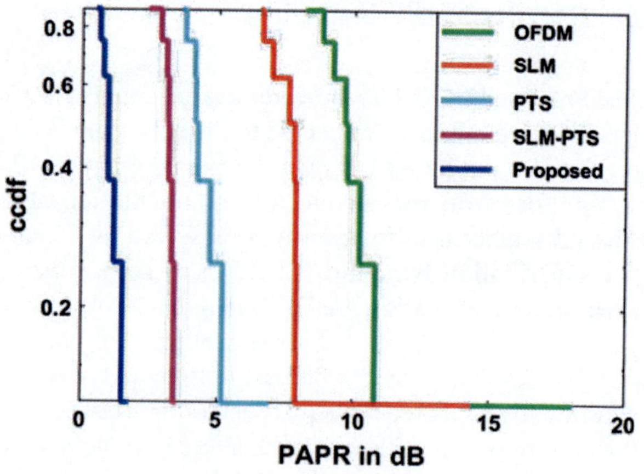

Fig. 4 Performance comparison of proposed UGO-PTS with IFFT length 512

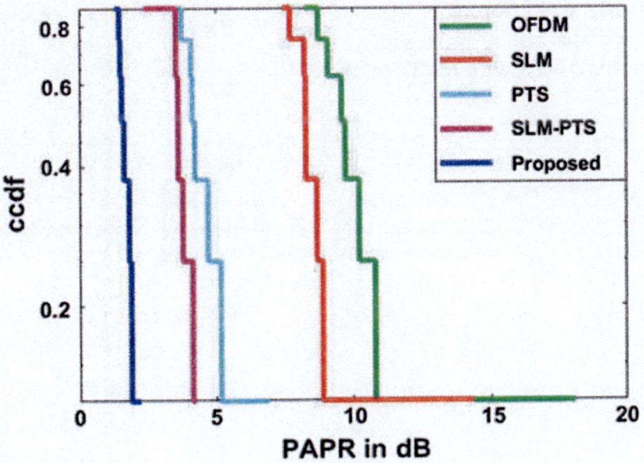

Fig. 5 Performance comparison of proposed UGO-PTS with IFFT length 256

1024, 512, 256, and 128 bits. On the off chance that the length of IFFT has changed to various sizes of the examples, such as 512, 256, and 128 bits. Figure 2 speaks to the correlation of PAPR decrease for existing and Proposed UGO-PTS strategy. From Fig. 2, unmistakably Proposed UGO-PTS strategy accomplished 1.9 dB of PAPR decrease and existing strategies, for example, SLM, PTS, SLM-PTS systems accomplished 14.01, 7.51, 3.85 dB separately. It plainly demonstrates that the Proposed UGO-PTS system performed all around contrasted with the current techniques. Here, the most extreme cycle of the PTS and SLM are 4 and 8 individually. Besides, the Proposed UGO-PTS method accomplishes 1.17 dB for 512 IFFT lengths, 1.95 dB for 256 IFFT lengths, and 1.16 dB for 128 IFFT lengths, which is appeared in Figs. 3, 4, and 5. From Figs. 2, 3, 4, and 5, it is reasoned that the Proposed UGO-PTS technique altogether diminished the PAPR worth contrasted with the current strategies. These figures portrayed that the PAPR worth was decreased in the wake of applying less IFFT length. The 128 IFFT lengths give 1.16 dB in the PAPR decrease, which is the best in the Proposed UGO-PTS method. Here, the inspecting commotion and the quantity of the sign diminished by utilizing a Proposed UGO-PTS strategy. Though, the computational multifaceted nature consequently decreased in the Proposed UGO-PTS system. The outcomes affirmed that the IFFT length of 512 and 128 demonstrates a critical PAPR decrease of 1.17 and 1.16 dB (Fig. 6).

Figure 7 portrays the BER versus SNR curves for the proposed system. For 20 iterations, BER is measured for distinct IFFT sizes of the OFDM system. Also from the graph, it clearly indicates that 1024 IFFT signal length outperforms when compared with other FFT length. Clearly, the proposed techniques achieved less BER.

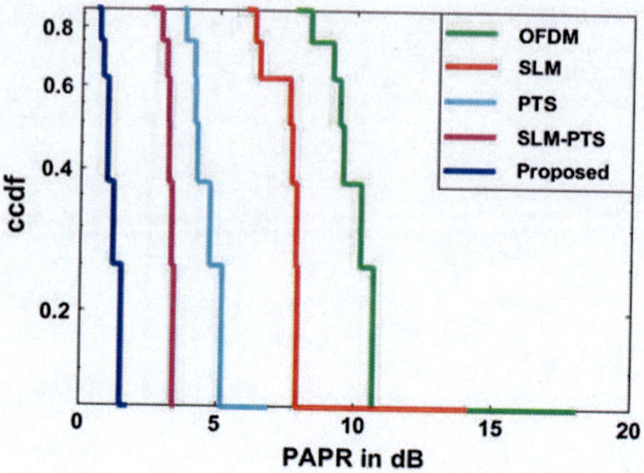

Fig. 6 Performance comparison of proposed UGO-PTS with IFFT length 128

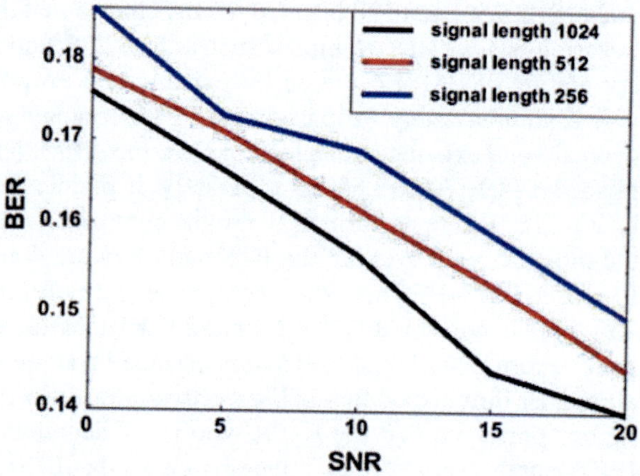

Fig. 7 BER versus SNR with various IFFT signal lengths

5 Conclusion

This paper utilized EM-based strategy to get the ideal stage factor for the PTS method to decrease computational complexity and for PAPR mitigation. We planned the stage factor search of the PTS system as a worldwide enhancement issue with bound imperatives. We, at that point, connected the EM-based strategy to look for the ideal stage factor. The PC reenactment results demonstrated that contrasted and the different stochastic pursuit strategies grew already, the proposed EM technique got the attractive PAPR decrease with low computational complexity. Results have been validated through substantial simulation tests. The proposed technique PAPR performance with different parameters is also verified through extensive computer simulations. The proposed technique offers significant PAPR reduction compared to conventional PAPR reduction methods.

References

1. Anand KP (2017) PAPR reduction technique: partial transmit sequence (PTS). Int Res J Eng Technol (IRJET) 04(03):2395–0072. www.irjet.net. ISSN: 2395–0056
2. Bahra N, Abed Hodtani G (2018) A modified SLM technique for PAPR reduction in OFDM systems by using a novel phase sequence. Int J Electron Lett 6(1):90–97
3. Bai Jurong, Cao Chenfei, Yang Yi, Zhao Feng, Xin Xiangjun, Soliman Abdel-Hamid, Gong Jiamin (2018) Peak-to-average power ratio reduction for DCO-OFDM underwater optical wireless communication system based on an interleaving technique. Optical Eng 57(8):
4. Bo L, Hu L, Yang F, Ding L, Song T (2018) Tone reservation ratio optimization for PAPR reduction in OFDM systems. In: Wireless communications and networking conference (WCNC), pp 1–6. IEEE
5. Gawali DD, Zidna A, Nataraj PSV (2017) Algorithms for unconstrained global optimization of nonlinear (polynomial) programming problems: the single and multi-segment polynomial B-spline approach. Comput Operat Res 87:205–220
6. Gayathri R, Sangeetha V, Prabha S, Meenakshi D, Raajan NR (2013) PAPR reduction in OFDM using Partial Transmit Sequence (PTS). Int J Eng Technol (IJET) 5(2):1428–1431
7. Hu Xiaonan, Yang Xuelin, Shen Zanwei, He Hao, Weisheng Hu, Bai Chenglin (2015) Chaos-based partial transmit sequence technique for physical layer security in OFDM-PON. IEEE Photon Technol Lett 27(23):2429–2432
8. Ibraheem ZT, Md Rahman M, Fazea Y, Ahmed KK (2018) PAPR reduction in OFDM signal by incorporating mu-law companding approach into enhanced pts scheme. J Optical Commun. https://doi.org/10.1515/joc-2017-0215
9. Jo H, Lee Y, Jeong S, Choi J (2018) A technique to reduce PAPR for OFDM-IM using multiple mapping rules for IM. In: 2018 IEEE 87th vehicular technology conference (VTC Spring), pp 1–5. IEEE
10. Kim K-H (2016) On the shift value set of cyclic shifted sequences for PAPR reduction in OFDM systems. IEEE Trans Broadcast TBC 62(2):496–500
11. Ku Sheng-Ju (2014) Low-complexity PTS-based schemes for PAPR reduction in SFBC MIMO-OFDM systems. IEEE Trans Broadcast 60(4):650–658
12. Kumar PK, Venkateswara Rao N (2018) A novel method for ICI cancellation in OFDM systems. In: 2018 third international conference on electrical, electronics, communication, computer technologies and optimization techniques (ICEECCOT), pp 14–15
13. Kumar PV, Nandhanavanam VR (2018) A novel method for joint- PAPR mitigation in OFDM based massive MIMO downlink systems. Int J Eng Technol 7(3):1185–1188
14. Mathew ME, Jeevitha J, Reethu K, Chacko A (2014) Performance improvement in OFDM system with interleaving and blanking over impulse noise environment. Int J Eng Trends Technol 9
15. Murthy NS (2020) Improved isolation metamaterial inspired MM-wave MIMO dielectric resonator antenna for 5G application. Progr Electromag Res C 100:247–261. https://doi.org/10.2528/PIERC19112603
16. Nimmagadda SM (2020) Enhancement of efficiency and performance gain of massive MIMO system using trial-based rider optimization algorithm. Wireless Pers Commun. https://doi.org/10.1007/s11277-020-07921-y
17. Ogunkoya F, Popoola WO, Shahrabi A, Sinanovic S (2015) Performance evaluation of pilot-assisted PAPR reduction technique in optical OFDM systems. IEEE Photon Technol Lett 27(10):1088–1091
18. Padarti VK, Nandhanavanam VR (2019) Performance evaluation of coexistence of Wi-Fi and LTE licensed–assisted access to unlicensed spectrum using markov chain analytical model. Int J Eng Adv Technol 8(6):2032–2037
19. Padarti VK, Nandhanavanam VR (2021) An improved ASOICF algorithm for PAPR reduction in OFDM systems. Int J Intell Eng Syst 14(2):352–360. ISSN No: 2185–3118
20. Padarti VK, Rao NV (2020) Adaptive SOICAF algorithm for PAPR mitigation in OFDM systems. Wireless Pers Commun 113:927–943. https://doi.org/10.1007/s11277-020-07260-y

21. Padarti VK, Nandanavanam VR (2019) Performance analysis of OFDM-based massive MIMO downlink system. Int J Recent Technol Eng 8(3):7161–7165. ISSN No: 2277–3878
22. Shiragapur BK, Wali U (2016) Peak-to-average power ratio reduction using coding and hybrid techniques for OFDM system. ICTACT J Commun Technol 7(01)
23. Vijaya Kumar P, Venkateswara Rao N (2019) A survey on mitigation techniques of PAPR in OFDM MIMO and massive MIMO systems. Int J Current Eng Scient Res 6(6):48–55. ISSN No: 2393–8374
24. Wang L-Y, Yuan H, Liu L-G (2016) Selective weighting PTS PAPR reduction scheme in OFDM systems. Electro Commun Netw V 343–349. Springer, Singapore
25. Xiaoying Z, Zhang L, Xiao P, Ma D, Wei J, Xin Y (2018) Mixed numerologies interference analysis and inter-numerology interference cancellation for windowed OFDM systems. IEEE Trans Veh Technol

Chapter 43
Novel Construction of Eight Transmit Antennas by Varying the Receiving Antennas Using Maximum Likelihood Decoding Algorithm

Satyanarayana Murthy Nimmagadda, Venkata Subbaiah Gandham, and Khaleel Ahmed Shaik

1 Introduction

In recent years, the demand for mobile communication systems with high data rates and high data reliability has increased. Due to this, the proposing methods should satisfy these wireless communication demands, which are limited bandwidth and transmitted power. That means we can send huge data through the channel with high data rates and with high reliability. By achieving these requirements, here we proposed technique of STBC for the MIMO system in wireless communication. This MIMO system allows multiple antennas at both transmitter and receiver. By this, we can achieve limited bandwidth and less transmitter power. We can say that the data among those antennas will split and then be transmitted to the multiple receiving antennas for multiple antennas. So, by this, we can send the huge data, and also by replicas, the reliability of the data also increases. This STBC technique gives a high data rate, full diversity. Wireless communication, the data is transfer without using the cables. With this communication system, we can transfer the data to large distances without cables. But the main drawback of this wireless system is this system easily affected by fading. Fading is caused by the multi-paths. So, the data was contained the noise and path gains, so the reliability of data degrades. We used the STBC technique to achieve high reliability of the data and the channel capacity [27]. By using the ML decoding for neglecting the noise and desired data estimation.

S. M. Nimmagadda (✉) · V. S. Gandham · K. A. Shaik
ECE Department, Velagapudi Ramakrishna Siddhartha Engineering College, Vijayawada 520007, India

S. Dawn et al. (eds.), *Smart and Intelligent Systems*, Algorithms for Intelligent Systems,
https://doi.org/10.1007/978-981-16-2109-3_44

2 Related Work

Papadias et al. [28] proposed a transmit diversity scheme for the downlink communication system employing four transmit antennas. This technique requires open-loop Shannon capacity, which means it allows single users. In Shannon capacity approaching, it uses the log delta formula means this log delta formula does not provide feedback to the transmitter. In order to achieve the high data rates, the Space-Time Coding (STC) technique was used. The STC code technique Bell Labs Layered Space-Time (BLAST) is used. By this technique, the complexity at the receiver will decrease.

Zhang [31] proposed quasi-orthogonal space-time block code (QOSTBC) orthogonal frequency-division multiplexing (OFDM) system is an application in frequency selective scheme. By using this method in broadband, that can increase the data rates. The decoding complexity increases with modulation level M^2, but the reliability of data increases in this scheme. To decreasing the decoding complexity, a newly proposed method is to be used. To further reducing the decoding complexity, a new approach is used, which combines the decoding with array processing which results in the complexity of $2\sqrt{M}$, which is considerably lower than the previous method. Here, they used the sample by sample method to calculate the null values because they used scalars and vectors to calculate the null values in decomposition method.

Shr et al. [29] described the Space-Time Trellis Coding STTC widely used to get high coding and diversity gains than STC. By this system can achieve low computational complexity. They used a diversity technique here is STTC (Space-Time Trellis Code). This diversity technique is resisting the channel impairments and also improves the spectral efficiency and coding gain. The basic almouti code is extended to STBC to provide high diversity gain and coding gains. STTC has high capability of resisting the channel impairments than STBC. But STTC decoder uses the viterbi algorithm, which results in high decoding complexity.

Erotokritou et al. [4] explored and shows the gain of the framework in what capacity can be accomplished by expanding the receiving antennas on remote correspondence framework. It gives the unique group of the symmetrical square of codes for different transmitting antennas. A bit error rates equal to $10 - 6$, the plan with five transmit antennas accomplishes more than 2-dB increase over the plan with four transmits antennas. In addition, the performance of four transmit antennas accomplish 5-dB increase over the three transmit antennas, and in conclusion, the three transmit antennas more than 5-dB gain over the two transmit receiving antennas.

Chaudhary et al. [3] proposed the space-time block codes with lower modulation always give low bit error rates and low modulation errors. When compared to the higher-order modulation, the lower modulation technique gives low bit error rates.

Berenguer et al. [2] discussed include the development of efficient coding and modulation signal processing techniques to improve the quality and spectral efficiency of wireless systems.

Aktas et al. [1] presented a simple generalization to the space-time block codes system for the Maximum ratio Combining (MRC) principle. And also allows the significant reduction in the ML decoding. Especially for large antennas at the receiver system gives decoding complexity. This is the basic principle of the MRC system. Furthermore, the decoder implementation for more receiving antennas gives high computational complexity. Finally, the canonical space-time code construction with optimized construction gives different values.

3 Proposed Method

The main objective of this paper is to get high data rates, reliability of the data, and increase the system capacity. These can achieve by using the STBC technique in the MIMO system. The BER of the (8, 8) system outperforms the BER of (4, 4) system with less complexity and high data rates, and also reliability of the data [28]. Using the STBC technique with the MIMO system improves the system performance in terms of the high performance of the BER. And reliability of the data and system capacity was also increased by the STBC technique. The 8×8 system outperforms the 4×4 system with low decoding complexity.

3.1 System Model

Here the information source gives binary data to the STBC encoder. The STBC encoder encoded the given binary data to encoded format (s_1 to s_8) and then mapped it to the BPSK constellation. After that, it gives the data in the block of codes. The rows and columns are time slots and antennas, respectively.

Those signals are received at the receiving antennas. Antennas at the receiver after that those signals are combined and decoded by the maximum likely hood decoder and then retrieve the original signal. The 8×8 system here is shown in above Fig. 1. This fig contributes to how the system performs and how to increase the reliability of the data. The above system model describes that first, we generate the random integers for the system (8, 8). Then modulate the given random integers by using BPSK modulation. Using the STBC encoder encoded the given random integers to the symbols after that by using the diversity technique (STBC) to generate the symbols to coded format then transmit the symbols signals. The transmitted signals are passing through the white Gaussian channel hence the noise will add. And then received at the receiver antennas and then decoded by using the STBC decoder and then combined the signal. Finally detect the original data at the receiver output.

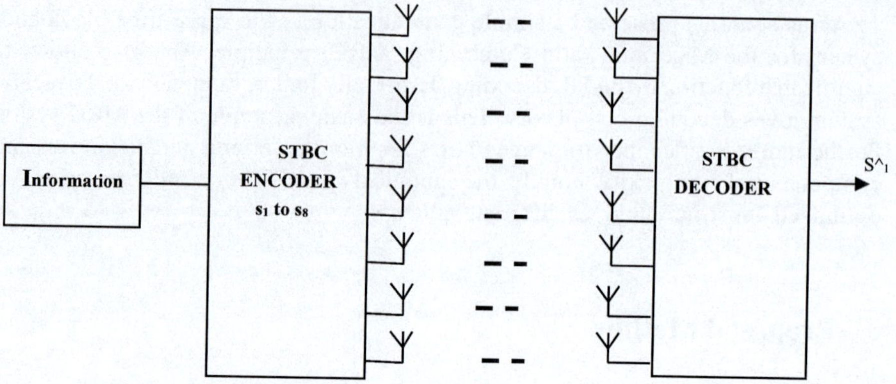

Fig. 1 8 × 8 system model OSTBC of proposed system

4 OSTBC (8 * 8) Scheme

The 8 * 8 system consists of eight transmitting antennas and eight receiving antennas. The increasing of antennas at both sides is done by the STBC technique of the MIMO [30, 26] system. This system has spatial multiplexing means splitting among the eight transmit antennas [8–20]. So the data is demultiplexed among the eight transmit antennas and then dent to the receiving antennas. The size of the message is 2 × b, where b is the transmitted bits. The transmitted encode symbols are s(t) [31]. The s (t) is demultiplexed into eight substreams as s_i (t), where i = 1 to 8. At the first time slot t the first data is transmitted. And at the second time slot t + T, the next data will transmit. The transmitted symbol matrix is Eq. 1

$$s = \begin{bmatrix} s_1 & s_2 & s_3 & s_4 & s_5 & s_6 & s_7 & s_8 \\ -s_2^* & s_1^* & -s_4^* & s_3^* & -s_6^* & s_5^* & -s_8^* & s_7^* \\ -s_3^* & -s_4^* & s_1^* & s_2^* & -s_7^* & -s_8^* & s_5^* & s_6^* \\ s_4 & -s_3 & -s_2 & s_1 & s_8 & -s_7 & -s_6 & -s_5 \\ -s_5^* & -s_6^* & -s_7^* & -s_8^* & s_1^* & s_2^* & s_3^* & s_4^* \\ s_6 & -s_5 & s_8 & -s_7 & -s_2 & s_1 & -s_4 & s_3 \\ s_7 & s_8 & -s_5 & -s_6 & -s_3 & -s_4 & s_1 & s_2 \\ -s_8^* & s_7^* & s_6^* & -s_5^* & s_4^* & -s_3^* & -s_2^* & s_1^* \end{bmatrix} \quad (1)$$

In the above matrix S, the rows represent the number of antennas, and columns represent the time slots. Where * indicates that the conjugate of the transmitted symbol, this matrix is designed based on the almouti code [29]. The encoded symbols are transmitted from the eight antennas through the Rayleigh channel to eight receiving antennas [5, 6]. When the signal is transmitting through the channel, the signal is added with path gains and noise. Here the path gain for each path is constant because the channel used here is quasi-static. And the noise is eliminated by

ML detection, and then the signal is decoded by the STBC decoder and then combined combiner as the transmitted symbols. The path gain matrix is shown below. H is the path gain matrix; for each antenna, the path gain is h_m, where $m = 1$ to 8 is expressed as Eq. 2. For the system 8×8 [7, 21–25], the r and **n** are the received signal and noise. The obtained receiving signal at the receiver is shown Eq. 3. The signal's received signal is multiplied with path gain and added with the noise, then the received signal is [4].

$$H = \begin{bmatrix} h_1 & h_2 & h_3 & h_4 & h_5 & h_6 & h_7 & h_8 \\ -h_2^* & h_1^* & -h_4^* & h_3^* & -h_6^* & h_5^* & -h_8^* & h_7^* \\ -h_3^* & -h_4^* & h_1^* & h_2^* & -h_7^* & -h_8^* & h_5^* & h_6^* \\ h_4 & -h_3 & -h_2 & h_1 & h_8 & -h_7 & -h_6 & -h_5 \\ -h_5^* & -h_6^* & -h_7^* & -h_8^* & h_1^* & h_2^* & h_3^* & h_4^* \\ h_6 & -h_5 & h_8 & -h_7 & -h_2 & h_1 & -h_4 & h_3 \\ h_7 & h_8 & -h_5 & -h_6 & -h_3 & -h_4 & h_1 & h_2 \\ -h_8^* & h_7^* & h_6^* & -h_5^* & h_4^* & -h_3^* & -h_2^* & h_1^* \end{bmatrix} \tag{2}$$

$$r = s * \begin{bmatrix} h_1 \\ h_2 \\ h_3 \\ h_4 \\ h_5 \\ h_6 \\ h_7 \\ h_8 \end{bmatrix} + n = sh + n \tag{3}$$

4.1 Decoding of Eight Transmits Antennas

At the receiver side, we have eight receiving antennas to collect the replicas of each antenna from the transmitter side. We have an STBC decoder [2] for minimizing the noise and collect all received symbols; among those symbols, we accept only the desired signal by nullifying the remaining symbols by using the maximum likelihood decoding, then combined and finally got at the receiver with high reliability [3]. The received signals at the receiver are Eq. 4. The received signal is the product of the transmitted symbols multiplied by the path gains. The noise in the transmitted signals is neglected by the Rayleigh fading channel. And the path gains are constant for each path because we used the quasi-static channel that's reason for keeping the path gains constantly. After receiving the signals undergoing the STBC decoding where Rayleigh fading the noise was neglected by using an ML detector; the desired symbols were got the remaining signals will be nullified. After receiving the data, the noise contained by the signals is reduced and then combined. By using the maximum likely hood decoding [1], the transmitted symbols are retrieved at the receiver by nullifying the independent terms of the received symbols. Finally, the results are given below Eq. 5.

5 Results and Discussions

The performance and BER of the MIMO system with the STBC technique are observed. That is MIMO system of STBC technique means simply it is an 8×8 system. It resembles that multiple transmit antennas and multiple receiving antennas. The number of antennas increased at the receiver side while keeping the eight transmitter antennas constant at the transmitter side. And then investigate how the performance of BER and data rates of the system increases. This flow chart, Fig. 2, describes that we generate random integers for the system (8, 8). Then modulate the given random integers by using BPSK modulation. Using the STBC encoder encoded the given random integers to the symbols after that using diversity technique (STBC) to generate the symbols to coded format then transmit the signals (Fig. 3).

$$
\begin{aligned}
r_1 &= s_1 * h_1 + s_2 * h_2 + s_3 * h_3 + s_4 * h_4 + s_5 * h_5 + s_6 * h_6 + s_7 * h_7 + s_8 * h_8 \\
r_2 &= -s_2^* * h_1 + s_1^* * h_2 - s_4^* * h_3 + s_3^* * h_4 - s_6^* * h_5 + s_5^* * h_6 - s_8^* * h_7 + s_7^* * h_8 \\
r_3 &= -s_3^* * h_1 - s_4^* * h_2 + s_1^* * h_3 + s_2^* * h_4 - s_7^* * h_5 - s_8^* * h_6 + s_5^* * h_7 + s_6^* * h_8 \\
r_4 &= s_3 * h_1 - s_3 * h_2 - s_2 * h_3 + s_1 * h_4 + s_8 * h_5 - s_7 * h_6 - s_6 * h_7 - s_5 * h_8 \\
r_5 &= -s_5^* * h_1 - s_6^* * h_2 - s_7^* * h_3 - s_8^* * h_4 + s_1^* * h_5 + s_2^* * h_6 + s_3^* * h_7 + s_4^* * h_8 \\
r_6 &= s_6 * h_1 - s_5 * h_2 + s_8 * h_3 - s_7 * h_4 - s_2 * h_5 + s_1 * h_6 - s_4 * h_7 + s_3 * h_8 \\
r_7 &= s_7 * h_1 + s_8 * h_2 - s_5 * h_3 - s_6 * h_4 - s_3 * h_5 - s_4 * h_6 + s_1 * h_7 + s_2 * h_8 \\
r_8 &= -s_8^* * h_1 + s_7^* * h_2 + s_6^* * h_3 - s_5^* * h_4 + s_4^* * h_5 - s_3^* * h_6 - s_2^* * h_7 + s_1^* * h_8
\end{aligned}
\tag{4}
$$

$$
\begin{aligned}
s_1 &= 2 * r_8 * s_1 * h_8^* + 2 * r_7 * s_1^* * h_7^* + 2 * r_6 * s_1^* * h_6^* + 2 * r_5 * s_1 * h_5^* - 2 * r_4 \\
&\quad * s_1^* * h_4^* + 2 * r_3 * s_1 * h_3^* + 2 * r_2 * s_1 * h_2^* + 2 * r_1 * s_1^* * h_1^* + |s_1|^2 * \left[\sum_{j=1}^{8} |h_j|^2 \right] \\
s_2 &= 2 * r_1^* * s_2 * h_2 - 2 * r_2^* * s_2^* * h_1 + 2 * r_3^* * s_2^* * h_4 - 2 * r_4^* * s_2 * h_3 + 2 * r_5^* \\
&\quad * s_2^* * h_6 - 2 * r_6^* * s_2 * h_5 + 2 * r_7^* * s_2 * h_8 - 2 * r_8^* * s_2^* * h_7 + |s_2|^2 * \left[\sum_{j=1}^{8} |h_j|^2 \right] \\
s_3 &= 2 * r_1^* * s_3 * h_3 + 2 * r_2^* * s_3^* * h_4 - 2 * r_3 * s_3 * h_1^* - 2 * r_4^* * s_3 * h_2 + 2 * r_5^* \\
&\quad * s_3^* * h_7 + 2 * r_6^* * s_3 * h_8 - 2 * r_7^* * s_3 * h_5 - 2 * r_8^* * s_2^* * h_7 + |s_3|^2 * \left[\sum_{j=1}^{8} |h_j|^2 \right] \\
s_4 &= 2 * r_1^* * s_4 * h_4 - 2 * r_2^* * s_4^* * h_3 - 2 * r_3^* * s_4^* * h_2 - 2 * r_4^* * s_4 * h_8 + 2 * r_5^*
\end{aligned}
$$

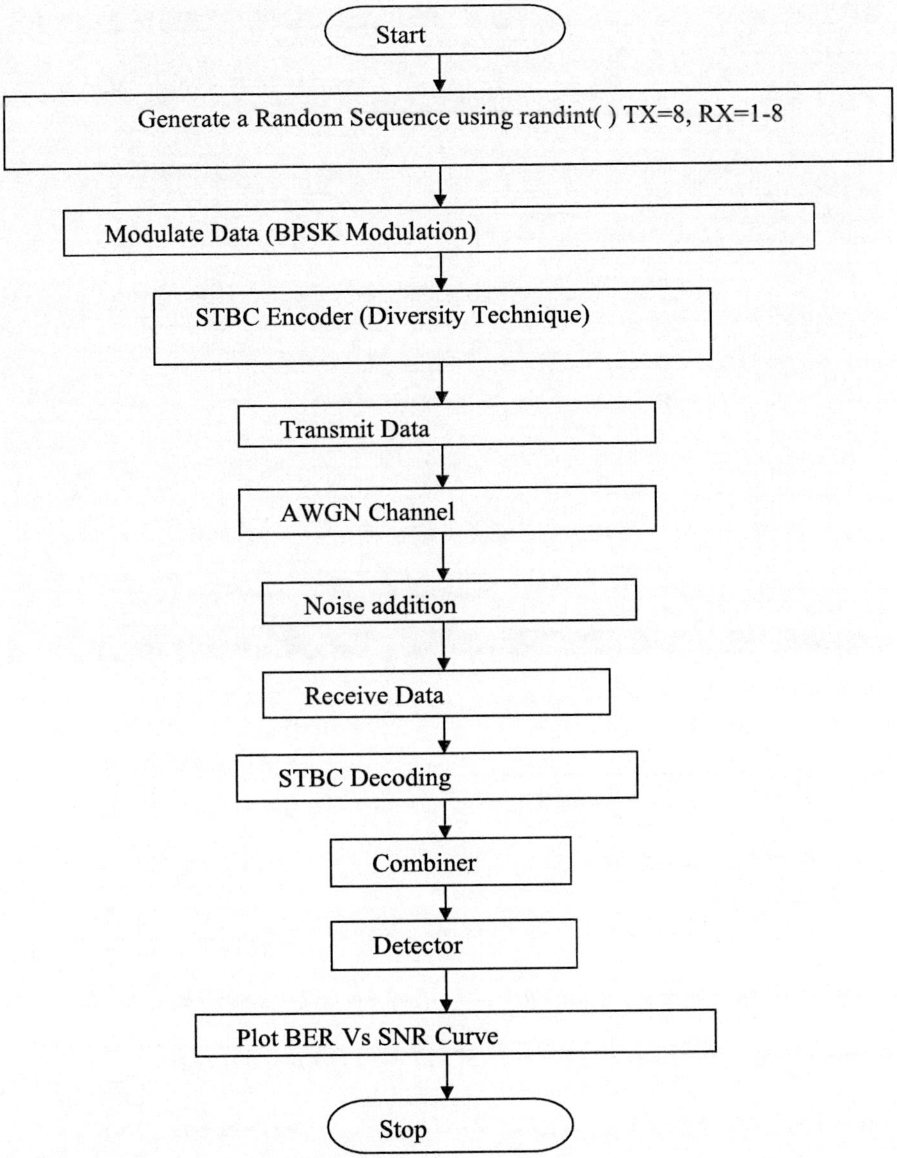

Fig. 2 Flow chart of proposed simulation system

$$* s_4^* * h_8 - 2 * r_6^* * s_4 * h_7 - 2 * r_7^* * s_4 * h_6 + 2 * r_8^* * s_4^* * h_5 + |s_4|^2 * \left[\sum_{j=1}^{8} |h_j|^2 \right]$$

$$s_5 = 2 * r_1^* * s_5 * h_5 + 2 * r_2^* * s_5 * h_6 + 2 * r_3^* * s_5^* * h_7 - 2 * r_4^* * s_5 * h_8 - 2 * r_5^*$$

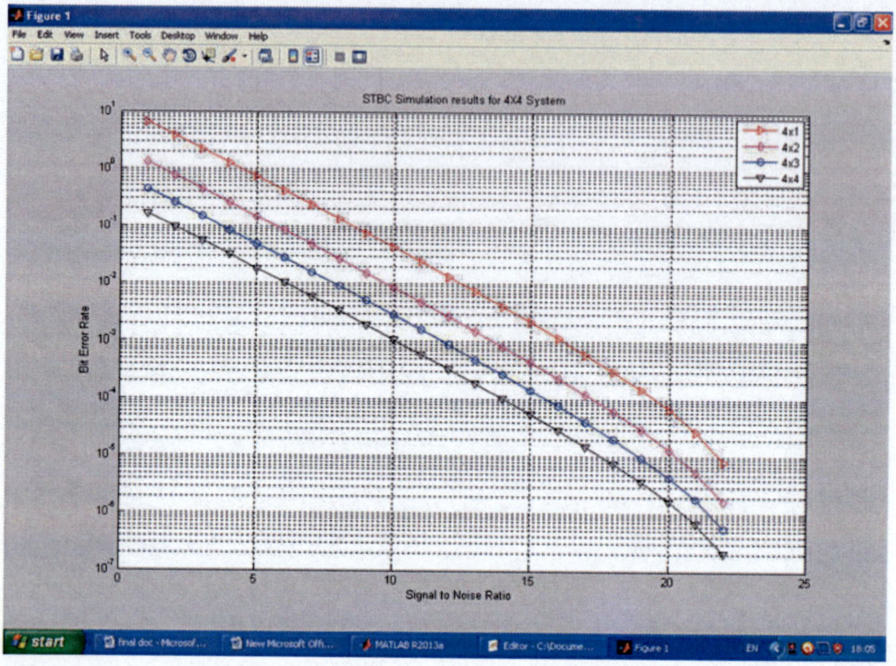

Fig. 3 Simulation results for 4 × 4 systems

$$* s_5^* * h_1 - 2 * r_6^* * s_5 * h_2 - 2 * r_7^* * s_5 * h_3 - 2 * r_8^* * s_5^* * h_4 + |s_5|^2 * \left[\sum_{j=1}^{8} |h_j|^2 \right]$$

$$s_6 = 2 * r_1^* * s_6 * h_6 - 2 * r_2^* * s_6^* * h_5 + 2 * r_3^* * s_6^* * h_8 - 2 * r_4^* * s_6 * h_7 - 2 * r_5^*$$
$$* s_6^* * h_2 + 2 * r_6^* * s_6 * h_1 - 2 * r_7^* * s_6 * h_4 + 2 * r_8^* * s_6^* * h_5 + |s_6|^2 * \left[\sum_{j=1}^{8} |h_j|^2 \right]$$

$$s_7 = 2 * r_1^* * s_7 * h_7 + 2 * r_2^* * s_7^* * h_8 - 2 * r_3^* * s_7^* * h_5 - 2 * r_4^* * s_7 * h_6 - 2 * r_5^*$$
$$* s_7^* * h_3 - 2 * r_6^* * s_7 * h_4 + 2 * r_7^* * s_7 * h_1 + 2 * r_8^* * s_7 * h_2 + |s_7|^2 * \left[\sum_{j=1}^{8} |h_j|^2 \right]$$

$$s_8 = 2 * r_1^* * s_8 * h_8 - 2 * r_7^* * s_4^* * h_3 - 2 * r_8^* * s_8 * h_8 - 2 * r_2^* * s_8^* * h_7 - 2 * r_3^*$$
$$* s_8^* * h_6 + 2 * r_4^* * s_8 * h_5 - 2 * r_5^* * s_8^* * h_4 + 2 * r_6^* * s_8 * h_3 + |s_8|^2 * \left[\sum_{j=1}^{8} |h_j|^2 \right] \quad (5)$$

The above simulation results Fig. 4 shows how the BER performance increases when the number of antennas increases. When the system has fewer antennas, the receiver gives less system performance, less BER performance, and less system capacity. So if we increase more antennas, that means we can achieve high data

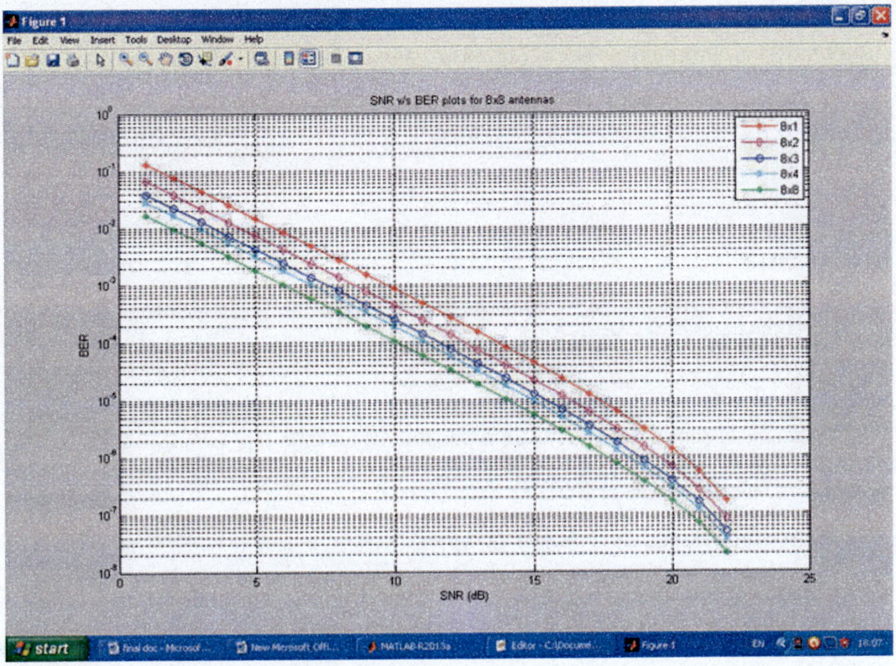

Fig. 4 Simulation results for the 8 × 8 system

rates and high BER performance. Here we said that the reliability of data increases means the data sending through eight transmit antennas through the Rayleigh fading channel to the eight receiving antennas. So from the first transmitting antenna, the replicas of transmitted data arrived at all the receiving antennas. The BER is at for the 8 × 1 is 10^0. Similarly, by increasing the antennas at the receiver, the BER performance increases concerning the number of antennas. Finally, the 8 × 8 system the BER is at 10^ − 2 that while increasing antennas, the data rates and reliability of data increased.

6 Conclusion

This paper has proposed a MIMO system with space-time block codes consisting of eight transmitter antennas and eight receiving antennas. This new design of (8, 8) system has linear processing at the receiver side. The maximum likelihood decoding is the linear processing technique. Here we choose the ML decoding for 8 × 8 antennas system to make the decoding simpler at the receiver. If we increase the number of antennas at the receiver gives high computational decoding complexity. So here we just examine how the BER increases with increasing antennas. The simulation results for 8 × 8 system give high BER than the existed system. That means this 8 × 8 system outperforms the 4 × 4 system without increasing the computational complexity. The other advantage of increasing the antennas is higher data rates by sending more data through the multiple antennas and increasing the receiving antennas with low computational complexity.

References

1. Aktas E, Gama HE, Fitz MP (2003) On the design and maximum-likelihood decoding of space–time trellis codes. IEEE Areas Commun 51(6)
2. Berenguer I, Wang X (2003) Space-time coding and signal processing for MIMO communications. J Comput Sci Technol 18(6):689–702
3. Chaudhary S, Patil AJ (2012) Performance analysis of MIMO-space time block coding with different modulation techniques. IEEE Areas Tele Commun 3
4. Erotokritou ID (2006) Space-time block coding for multiple transmit antennas over time-selective fading channels. IEEE Trans Signal Proc
5. Khaleelahmed S, Venkateswara Rao N (2020) Energy efficient power allocation using salp particle swarm optimization model in MIMO–NOMA systems. Wireless Pers Commun 111(02):1235–1254. https://doi.org/10.1007/s11277-019-06911-z. ISSN: 1572–834X
6. Khaleelahmed SK, Venkateswara Rao N (2019) Priority based scheduling for energy efficient power allocation in MIMO-NOMA system with multiple users. Int J Intell Eng Syst 12(4):348–357. https://doi.org/10.22266/ijies2019.0831.32. ISSN: 2185–3118
7. Murthy NS (2020) Improved isolation metamaterial inspired mm-wave MIMO dielectric resonator antenna for 5G application. Progr Electromag Res C 100:247–261. https://doi.org/10.2528/PIERC19112603
8. Murthy NS, Sri Gowri S, Prabhakara Rao B (2015) Space time block codes using multiple transmitting antennas. J Wireless Netw Commun 5(1):9–18. https://doi.org/10.5923/j.jwnc.20150501.02
9. Murthy NS, Sri Gowri S, Prabhakara Rao B (2014) Orthogonal Stbcs for five, six and seven transmit antennas. Recent Adv Commun Netw Technol 3(2):93–96. ISSN: 2215–082X (Online), ISSN: 2215–2811
10. Murthy NS, Sri Gowri S, Prabhakara Rao B (2014) Non orthogonal quasi-orthogonal space-time block codes based on circulant matrix for eight transmit antennas. Int J Appl Eng Res 9(21):9341–9351. ISSN 0973–4562 (Scopus index)
11. Murthy NS, Sri Gowri S, Prabhakara Rao B (2013) Low delay general complex orthogonal space-time block code for seven and eight transmit antenna. Int J Eng Res Technol (IJERT) 2(6). ISSN: 2278–0181
12. Murthy NS, Sri Gowri S, Prabhakara Rao B (2013) Orthogonal space-time block codes for four, five and transmit antennas with 4/7 and 5/11 code rates. Int J Eng Res Appl (IJERA) 3(3):1380–1382. ISSN: 2248–9622
13. Murthy NS, Sri Gowri S, Prabhakara Rao B (2012) Orthogonal space-time block codes for 6 transmit antennas", Int J Eng Res Appl (IJERA) 2(4):362–364. ISSN: 2248–9622
14. Murthy NS, Sri Gowri S, Prabhakara Rao B (2012) Two new general complex orthogonal space-time block codes for 6 and 16 transmit antenna. Int J Electron Commun Eng Technol 3(2). ISSN: 0976–6464
15. Murthy NS, Sri Gowri S (2012) Full rate general complex orthogonal space-time block code for 8-transmit antenna. Elsevier, Sciverse Science Direct Procedia Engineering in IWIEE 2012, China, Jan. 9–10. H Index: 11, IF: 0.61 (Scopus index). ISSN: 1877–7058
16. Murthy NS, Sri Gowri S, Prabhakara Rao B (2014) New high rates non-orthogonal STBCs for seven and eight transmit antennas. IETE Natl J Innov Res II(II), pp 9–12. ISSN: 2320–8961

17. Murthy NS, Sri Gowri S, Prabhakara Rao B (2014) Quasi-Orthogonal space-time block codes based on circulant matrix for eight transmit antennas. In: IEEE International conference on communication and signal processing- ICCSP 14, Melmaruvathur, 3–5th April 2014
18. Murthy NS, Sri Gowri S, Prabhakara Rao B (2014) New orthogonal space-time block code for eight transmit antenna. In: Third international conference on recent trends in engineering & technology (ICRTET'2014), 28–30th. Elsevier Publications, Chandwad, Nasik
19. Murthy NS, Sri Gowri S, Prabhakara Rao B (2013) Orthogonal space-time block code for six transmit antenna", international conference on communications, networking and signal processing 19–21 September 2013, School of Electronics Engineering, VIT University, Vellore, Tamilnadu, India
20. Murthy NS, Sri Gowri S, Prabhakara Rao B (2013) Orthogonal Space-Time Block Code for maximal rate Nine Transmit antenna. In: Second international conference on communication and signal processing on April 2013, QIS engineering college, Ongole
21. Murthy NS, Sri Gowri S (2012) Complex orthogonal space-time block codes rates 3/7,4/8 and 6/8for 3,4 and 4transmit antennas in space time coding. In: International conference on devices, circuits and systems–ICDCS 2012, Mar 15–16, 2012, Karunya University, Coimbatore, IEEE Explore
22. Murthy NS, Sri Gowri S, Satyanarayana P (2011) Complex orthogonal space-time block codes rates 5/13 and 6/14 for 5 and 6 transmit antennas. Wireless Communications, Networking and Mobile Computing (WiCOM). In: 2011 7th International conference Sept 23–25 at Wuhan, China, IEEE Explore Digital Object Identifier. https://doi.org/10.1109/wicom.2011.6040107
23. Murthy NS, Praneeth KJ, Praneeth Sai P (2020) Spectral efficiency of the linear, precoded zero force and mrc using massive mimo. Int J Adv Sci Technol 29(3):6618–6622
24. Nimmagadda SM (2020) Enhancement of efficiency and performance gain of massive MIMO system using trial-based rider optimization algorithm. Wireless Pers Commun. https://doi.org/10.1007/s11277-020-07921-y
25. Nimmagadda SM (2020) Optimal spectral and energy efficiency trade-off for massive MIMO technology: analysis on modified lion and grey wolf optimization. Soft Comput 24:12523–12539. https://doi.org/10.1007/s00500-020-04690-5
26. Padarti VK, Nandhanavanam VR (2018) A novel method for joint-PAPR mitigation in OFDM-based massive MIMO downlink systems. Int J Eng Technol 7(3):1185–1188. ISSN No: 2227-524X
27. Papadias CB, Foschini GJ (2003) Capacity-approaching space–time codes for systems employing four transmitter antennas. IEEE Trans Inform Theory 49(3):726–733
28. Papadias CB, Foschini GJ (2001) A space-time coding approach for systems employing four transmit antennas. ICASSP 4(7):2481–2484
29. Shr K-T, Chen H-D, Huang Y-H (2010) A low- complexity viterbi decoder for space-time trellis codes. IEEE Trans Circuits Syst Regul Papers 57(4):873–855
30. Vijaya Kumar P, Venkateswara Rao N (2020) Adaptive SOICAF algorithm for PAPR mitigation in OFDM systems. Wireless Pers Commun 113(2):927–943. ISSN No: 0929-6212. Springer
31. Zhang H, Yuan D, Chen H-H (2010) On array- processing-based quasi-orthogonal space-time block-coded OFDM systems. IEEE Trans Veh Technol 59(1):508–513

Chapter 44
Effect of Channel Doping Variation on Electrostatic Characteristics of 3D Double Gate Junctionless Transistor

Achinta Baidya, Rajesh Saha, Jayendra Kumar, Sadhan Gope, and Chaitali Koley

1 Introduction

Incessant development in complementary metal-oxide semiconductor (CMOS) technology is required to meet the present-day requirements of high-speed and high-performance applications. Moore's Law guided us to the extreme scaling process of the transistor to increase the packing density of integrated circuits. Scaling of the devices is aimed to improve functionality, low power dissipation, and high speed. In the process of scaling, novel structures and new materials are introduced to fulfill the requirement and overcome the shortcomings of conventional CMOS. To achieve these goals, researchers are required to carry forward their work to new heights. The microelectronics industry significantly invested their knowledge, time, and infrastructure to invent novel devices and new technology for extremely low dimension fabrication. Short channel effects (SCEs) are the main drawback of devices in the

A. Baidya (✉)
Department of Electronics and Communication Engineering, Mizoram University, Aizawl 796004, India

R. Saha
Department of Electronics and Communication Engineering, Malaviya National Institute of Technology Jaipur, Jaipur 302017, India

J. Kumar
Department of Electronics and Communication Engineering, Vellore Institute of Technology, Amaravati 522 237, Andhra Pradesh, India

S. Gope
Department of Electrical Engineering, Mizoram University, Aizawl 796004, India

C. Koley
Department of Electronics and Communication Engineering, National Institute of Technology Mizoram, Aizawl 796012, India
e-mail: chaitali.ece@nitmz.ac.in

S. Dawn et al. (eds.), *Smart and Intelligent Systems*, Algorithms for Intelligent Systems,
https://doi.org/10.1007/978-981-16-2109-3_45

nanometer regime [1]. Emerging novel structures like Fin-FET [2–4], Silicon-on-Insulator (SOI) [5–8], Tunnel field-effect transistor (TFET) [9–11], Carbon nanotube field-effect transistor (CNTFET) [12–15], spin-based device [16–18], ferromagnetic logic [19–21], junctionless FET [22–24] were introduced to overcome the limitations of conventional MOSFETs.

Novel structure—"junctionless transistor" with a junction-free exceptional architecture was proposed in 2009 [22] and later fabricated in 2010 [23]. To know the effect of aggressive scaling of junctionless transistor, a comprehensive study is necessary. With the introduction of the new semiconductor device, it is important to understand the device working properly. Although junctionless transistors have technological advantages over inversion mode MOSFETs, they are not fully free from scaling challenges. Double gate silicon-based junctionless transistor (JLT) with 20 nm gate length was proposed [24]. The same structure is considered for the present investigation. In order to address the challenges in the new device structure of the junctionless transistor, tunability of device doping profile is tested in the present paper.

2 Device Dimension and Simulation Environment

A thin body junctionless transistor with a double gate of 20 nm length is considered for the present investigation. A silicon bar of 10 nm thickness and 12 nm width is used for the body of JLT. Polysilicon is used for gate metal, and SiO_2 of 1 nm is used as a gate oxide for both gates. The body acts as a channel of the JLT that is doped with n-type doping, Arsenic. The concentration of the channel doping is varied from 1.5×10^{20} to 1.5×10^{15} atoms/cm^3. The body thickness is kept minimum to ensure the full depletion of the body for the switch-off condition. The three-dimensional simulation structure of the double-gated JLT is shown in Fig. 1, and the dimension details are mentioned in Table 1.

The device is simulated with the help of Synopsys Technology Computer-Aided Design (TCAD) tool where several tools like 3D device structure editor, Sdevice and Svisual are used for device structure building, device simulation based on given model, and performance evaluation, respectively. Suitable models like the drift–diffusion model, High Field Saturation model, Auger, Shockley–Read–Hall (SRH), and avalanche model are used in Sdevice.

3 Result and Discussion

This section focuses on the three-dimensional simulation study of double-gated JLT for different electrostatic characteristics. JLT channel doping concentration is varied, and the effect of the variation is evaluated. The Fig. 2 shows the transfer characteristics of JLT for different doping concentration with $V_{DS} = 0.05$ V. JLTs follows the characteristics similar to the normal MOSFET. But with the change in doping

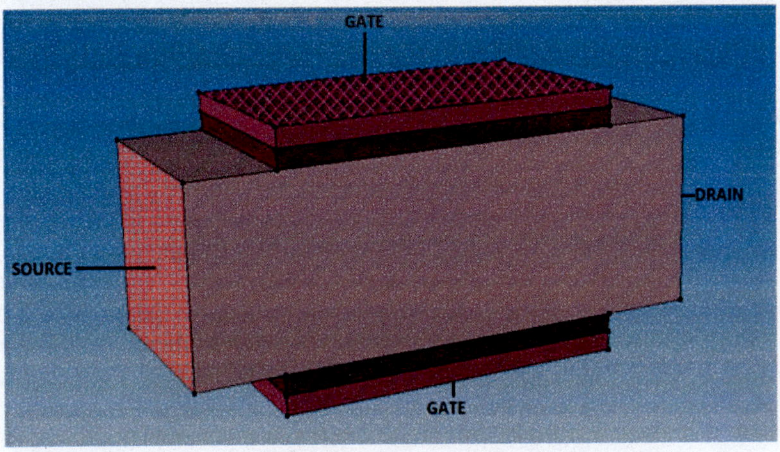

Fig. 1 Three-dimensional simulated structure of double-gated junctionless transistor [24]

Table 1 Device structural factors of the double gate junctionless transistor

Structural factors	Values
Gate length (L_g)	20 nm
Channel width (W)	12 nm
Channel thickness (T_{si})	10 nm
Doping concentration at channel (N_D)	Variable
Doping material	Arsenic
Gate oxide thickness	1 nm

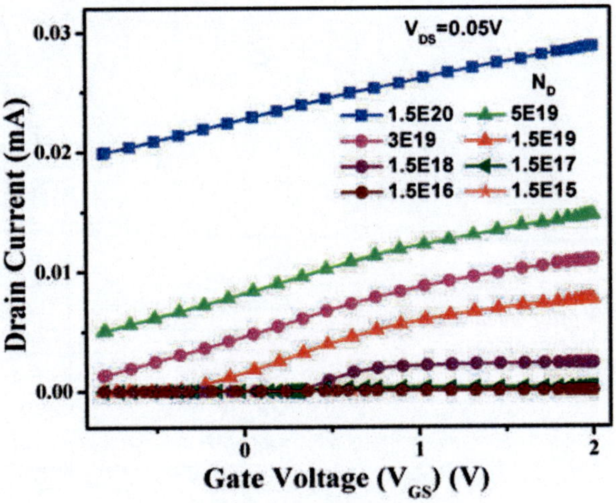

Fig. 2 Transfer characteristics (in linear scale) of JLT for different doping concentrations with $V_{DS} = 0.05$ V

concentration, the threshold voltage and the current level change. High on-current is observed for the highest doping profile, and the lowest drain current is observed with the lowest doping of 1.5×10^{15} atoms/cm^3. Though the 1.5×10^{20} atoms/cm^3 doping profile gives the highest current, it is observed that JLT will require a very high negative gate voltage (V_{GS}) to deplete the channel and switch-off the device. So, too high doping in the channel makes the device unrealistic from the circuit application point of view.

Figure 3 shows the transfer characteristics of JLT for different doping concentrations on a log scale. From this figure, it is very clear that doping concentration from 1.5×10^{19} cm^{-3} to 1.5×10^{15} cm^{-3} gives a suitable on/off current ratio greater than 10^3. In contrast, doping concentration higher than these do not show a good on/off current ratio for the given range of V_{GS} (-1 to 2 V). On current and threshold voltage of the JLTs with different channel doping concentrations are detailed in Table 2. Threshold voltage decreases with the increase in channel doping. That means the device gate will require more negative voltage to occur the depletion in the channel.

Figures 4 and 5 show the transconductance and transconductance generation factor

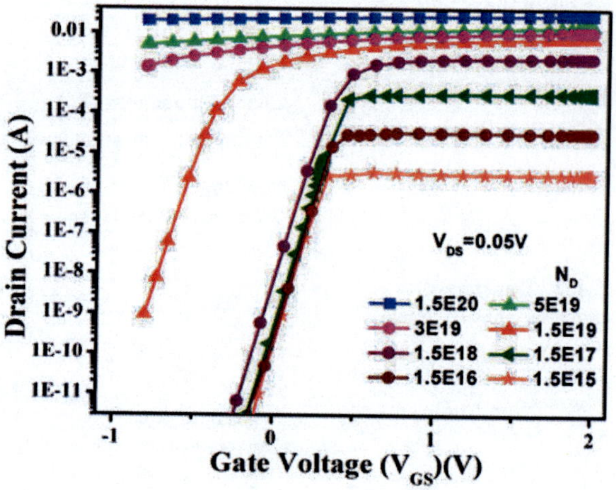

Fig. 3 Transfer characteristics (in log scale) of JLT for different doping concentrations with $V_{DS} = 0.05$ V

Table 2 Performance of JLT for different channel doping concentrations

Channel doping (N_D) (Atoms/cm^3)	Threshold voltage (V_{th}) (V)	'On' current at $V_{GS} = 2$ V (μA)
1.5×10^{20}	< -1	28.8
5×10^{19}	<-1	14.7
3×10^{19}	-1.0	10.9
1.5×10^{19}	-0.304	7.62
1.5×10^{18}	0.289	2.34
1.5×10^{17}	0.309	0.31
1.5×10^{16}	0.333	0.032

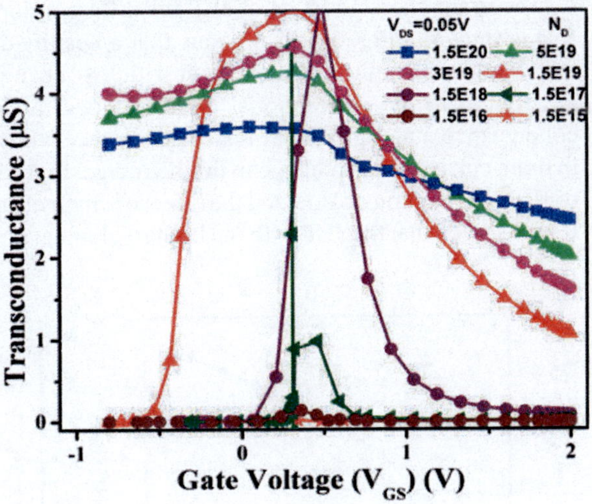

Fig. 4 The transconductance of JLT with channel doping variation at $V_{DS} = 0.05$ V

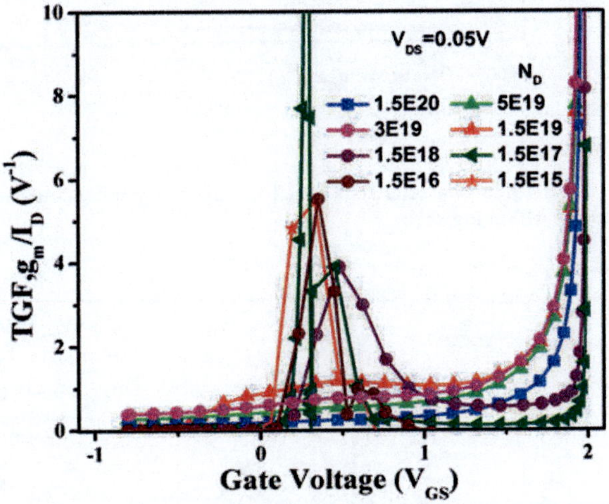

Fig. 5 Transconductance generation factor (TGF) of JLT for different channel doping with $V_{DS} = 0.05$ V

of JLT with different channel doping. These two parameters are very important for device performance. It is observed that heavy doping in the JLT channel gives high transconductance compare to a very low doping profile. Among the heavy doing profiles, the concentration of 1.5×10^{19} cm^{-3}gives a better transconductance value for the subthreshold region. The highest peak is also observed in the case of channel doping of 1.5×10^{19} cm^{-3}. Transconductance generation factor (TGF) gives the effectiveness of transconductance generation of a specific device. JLT devices with moderately high channel doping show higher TGF in most of the regions of operation. Very heavily doped channels do not show very good TGF performance due to their high drain current in the region of operation.

The drain characteristics of JLT for different doping concentration with $V_{GS} =$ 1 V is shown in Fig. 6. It is obvious that a heavily doped channel (with 1.5×10^{20} cm^{-3}) gives the highest drain current with V_{DS} change. The JLT drain characteristics are similar to the conventional MOS devices. A device should show low output conductance and high output resistance in the saturation region to avoid the change in drain current with a change in drain voltage. Figure 7 shows the drain conductance of JLT as a function of $V_{DS,}$ and the effect of channel doping variation is also observed. It is observed that the JLT with the highest channel doping shows higher drain/output

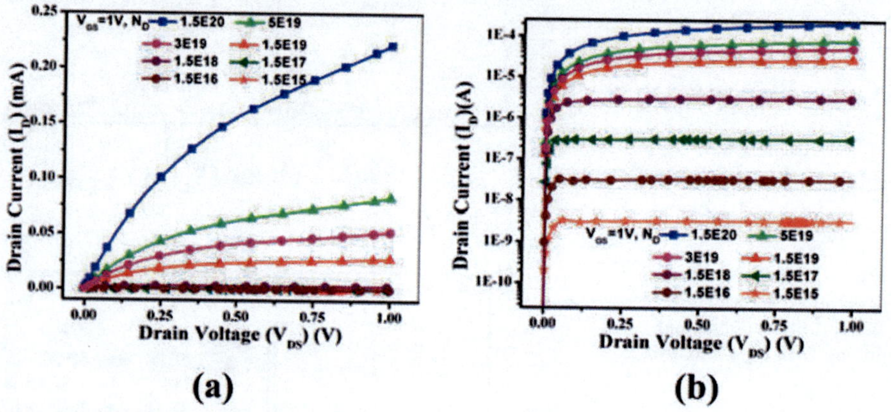

Fig. 6 Drain characteristics of JLT for different doping concentrations with $V_{GS} = 1$ V (**a**) in linear scale, (**b**) in log scale

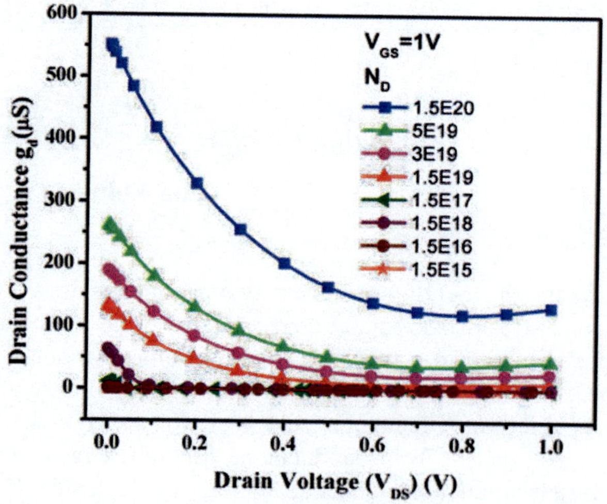

Fig. 7 Drain conductance of JLT for different channel doping with $V_{GS} = 1$ V

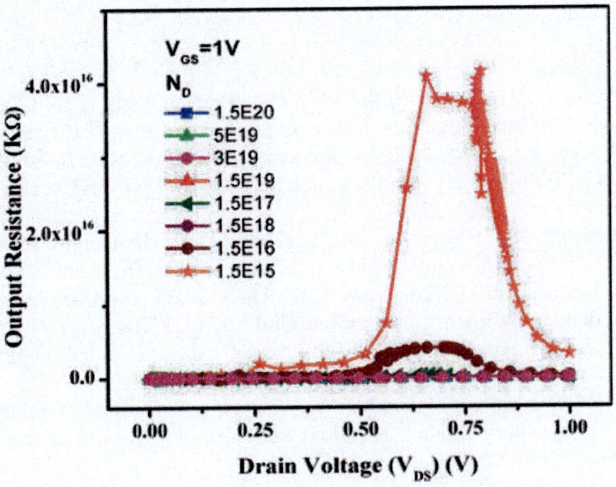

Fig. 8 Output resistance of JLT for different channel doping with $V_{GS} = 1$ V

conductance for all values of drain voltages. So it will be more prone to channel length modulation.

Figure 8 shows the output resistance of JLT as a function of V_{DS} for different doping concentrations. It is quite clear that the JLT with low doping concentration showed the highest output resistance compare to all. But JLT with a very low doping concentration is not suitable for a very low drain current. So a moderately high doping concentration would be suitable in all respect. In the present investigation, JLT with channel doping of 1.5×10^{19} cm^{-3} gives overall better performance w.r.t very high and very low doping concentrations.

4 Conclusion

The investigation focuses on the variation of the doping concentration of the JLT channel to understand the behavior of scaled JLT device. The electrostatic characteristics of JLT are evaluated for different channel doping concentrations, and the effect of doping variation is analyzed. Detailed analyses of transconductance, TGF, output resistance, and output conductance are performed for the analysis. It is observed that compare to very high and low doping concentrations, a moderate channel doping of 1.5×10^{19} atoms/cm^3 gave better or optimized device performance in all respect.

References

1. Frank DJ, Dennard RH, Nowak E, Solomon PM, Taur Y, Wong H-SP (2001) Device scaling limits of Si MOSFETs and their application dependencies. Proc IEEE 89(3):259–288
2. Hisamoto D, Lee W-C, Kedzierski J, Takeuchi H, Asano K, Kuo C, Anderson E, King T-J, Bokor J, Hu C (2000) FinFET–a self-aligned double gate MOSFET ccalable to 20 nm. IEEE Trans Electron Dev 47:2320–2325
3. Zhang W, Fossum JG, Mathew L, Du Y (2005) Physical insights regarding design and performance of independent-gate FinFETs. IEEE Trans Electron Dev 52–10:2198
4. Endo K, Liu Y, Masahara M, Matsukawa T, O'uchi S, Suzuki E (2007) Fabrication and power-management demonstration of four-terminal FinFEts. ECS Trans 6–4:71
5. Colinge JP, Baie X, Bayot V, Grivei E (1996) A silicon-on-insulator quantum wire. Solid State Electro 39:49
6. Lemme MC, Mollenhauer T, Henschel W, Wahlbrink T, Baus M, Winkler O, Granzner R, Schwierz F, Spangenberg B, Kurz H (2004) Subthreshold behavior of triple-gate MOSFETs on SOI Material. Solid State Electron 484–529
7. Park JT, Colinge JP (2002) Multiple-gate SOI MOSFETs: device design guidelines. IEEE Trans Electron Dev 49(12):2222–2229
8. Suzuki K, Tanaka T, Tosaka Y, Horie H, Arimoto Y (1993) Scaling theory for double-gate SOI MOSFET's. IEEE Trans Electron Dev 40(12):2326–2329
9. Takeda E, Matsuoka H, Igura Y, Asai S (1988) A band to band tunneling MOS device B2TMOSFET. Tech Digest IEEE Int Electron Dev Meet 402–405
10. Choi WY, Park BK, Lee JD, Liu TJK (2007) Tunneling field-effect transistors (TFETs) With subthreshold swing (SS) less than 60 mV/dec. IEEE Electron Dev Lett 28(8):743–745
11. Boucart K, Ionescu AM (2007) Double-gate tunnel FET with high-k gate dielectric. IEEE Trans Electron Dev 54(7):1725–1733
12. Ijiima S (1991) Helical microtubules of graphitic carbon. Nature 354:56–58
13. Avouris P, Afzali A, Appenzeller J, Chen J, Freitag M, Klinke C, Lin Y-M, Tsang JC (2004) Carbon nanotube electronics and optoelectronics. IEDM technical digest. IEEE international electron devices meeting
14. O'Connor I, Liu J, Gaffiot F, Prégaldiny F, Lallement C, Maneux C, Goguet J, Frégonèse S, Zimmer T, Anghel L, Dang T-T, Leveugle R (2007) CNTFET modeling and reconfigurable logic-circuit design. IEEE Trans Circuits Syst I Regular Papers 54(11):2365–2379
15. Nasiri SH, Farshi MKM (2013) Stability analysis in CNTFETs. IEEE Electron Dev Lett 34(2):301–303
16. Fong X, Kim Y, Venkatesan R, Choday SH, Raghunathan A, Roy K (2013) Spin-transfer torque memories: devices, circuits, and systems. Proc IEEE 104(7):1449–1488
17. Wang S, Cai L, Qi K, Yang X, Feng C, Cui H (2016) Impact of nanomagnets size on switching behaviour of all spin logic devices. Micro Nano Lett 11(9):508–513
18. Chen X, Wang Qi, Bai F, Tang X, Zhang H, Zhong Z (2016) A methodology to design spin-wave-based logic gates in a single ferromagnetic nanostripe using spin-transfer torque effects. IEEE Trans Magn 52(7):1400104
19. Hu XK, Dey H, Liebing N, Csaba G, Orlov A, Bernstein GH, Porod W, Krzysteczko P, Sievers S, Schumacher HW (2015) Edge-mode resonance-assisted switching of nanomagnet logic elements. IEEE Trans Magn 51(11):3401004
20. Behin-Aein B, Salahuddin S, Datta S (2009) Switching energy of ferromagnetic logic bits. IEEE Trans Nanotechnol 8(4):505–514
21. Ju X, Wartenburg S, Rezgani J, Becherer M, Kiermaier J, Breitkreutz S, Schmitt-Landsiedel D, Porod W, Lugli P, Csaba G (2012) Nanomagnet logic from partially irradiated Co/Pt nanomagnets. IEEE Trans Nanotechnol 11(1):97–104
22. Lee C-W, Afzalian A, Akhavan ND, Yan R, Ferain I, Colinge J-P (2009) Junctionless multigate field-effect transistors. Appl Phys Lett 94(5):053–511

23. Colinge JP, Lee CW, Aryan A et al (2010) Nanowire transistors without junctions. Nat Nanotechnol 5:225–229
24. A. Baidya, V. Krishnan, S. Baishya, T.R. Lenka.: Effect of thin gate dielectrics and gate materials on simulated device characteristics of 3D double gate JNT. Superlattices and Microstructures,vol. 77, (2015) 209–218.

Chapter 45
Design of Multi-Stage Dodecapole Electrical Propelling System (DEPS) and Its Possible Use in the Hyperloop Transportation

Srichandan Kondamudi and Sandhya Thotakura

1 Introduction

Being fastened in traffic gives adequate time to conjure up innovative thoughts for reforming how we travel. Over the most recent couple of years, there is a fast advancement of innovation in transport frameworks. The famous engineer, inventor, and businessman-billionaire Elon Musk famously thought up the hyperloop as the best alternative to rapid transportation [1].

Dudnikov has explained the merits of the hyperloop and mentioned points suggested by hyperloop designers. Some of the critical issues are:

1. The system can transport people and cargo at a very high speed.
2. The system conquers the confinement on the rate of land transportation.
3. The system can ensure low expenses in the plan and execution of the vehicle framework.
4. The system gains autonomy from climate conditions [2].

Ahmed et al. have proposed an approach to levitate, propel, and control a pod using permanent magnets and electromagnets. The electrodynamic suspension system is a small pod attached with permanent magnets from the bottom, which resembles a short-rotor linear synchronous motor (LSM) [3]. Flank et al. have proposed a simple, reliable analytical model for the electrodynamics suspension (EDS) [4]. Linear motors have a few shortcomings, such as motor cogging, air-gap alignment, complicated installation, flux utilization, and heat dissipation.

S. Kondamudi (✉) · S. Thotakura
Department of Electrical, Electronics and Communication Engineering, Gitam Institute od Technology, GITAM University, Visakhapatnam 530045, India
e-mail: skondamu@gitam.edu

S. Thotakura
e-mail: sthotaku@gitam.edu

S. Dawn et al. (eds.), *Smart and Intelligent Systems*, Algorithms for Intelligent Systems,
https://doi.org/10.1007/978-981-16-2109-3_46

Based on a brief literature survey on the Hyperloop [1–4], every existing strategy or design is suffering from insufficiencies such as air-gap alignment, complicated installation, flux utilization, and heat dissipation.

Multipole Field Electromagnetic driving framework appears to have a fantastic guarantee as an idea for the future; its practical realization anticipates the advancement of an outline that is sufficiently solid to have the capacity to convey the payloads.

Zhu et al. proposed the basic design and analysis of Multipole Field Electromagnetic Propeller [5]. The author has indicated that the Multipole field electromagnetic launcher has a great axial component of acceleration force and high muzzle speed. From the author's observations, the total acceleration time is 0.327 ms, the maximum acceleration force is 266.006 kN, and the muzzle speed of the projectile is 232.478 m/s. As per the qualities, the Multipole Field Electromagnetic Propeller can be considered as one of a contrasting option to the present accessible plan.

Yingwei et al. presented theoretical analysis and numerical evaluation of a three-stage twisty octapole field electromagnetic launcher. The researcher has suggested that it be applied in the evacuated tube vehicle, tunnel boring machine, and spacecraft propulsion launcher [6].

Antonino et al. has derived an analytic solution when the system is excited by a sinusoidal current flowing in a saddle coil moving in the axial direction through traveling wave configuration. The author has suggested a model that analyzes the induction launcher called the multipolar electromagnetic field launcher in its moving wave configuration [7].

Wenbo et al. has proposed four different driving coil connection patterns of sextupole field electromagnetic launcher [8]. The author has submitted that the overall propulsion efficiency will increase rapidly as the capacitor's initial voltage increases.

Concerning the Multiple Electromagnetic propeller's literature surveys, the multipole field electromagnetic propelling system can be an exciting alternative for the existing drive system in high-speed transportation.

The available literature on the Multiple Field Electromagnetic System built on assumptions such as.

1. Uniform magnetic field throughout the system
2. Estimated inductance and mutual inductance values

Therefore, this paper's contribution is toward investigating the achievability and confinements of using Dodecapole Field Electrical Propelling System (DEPS) in high-speed transportation and analyzing the Electromagnetic propulsion by minimizing the assumptions.

Detailed modeling of the Dodecapole Field Electrical Propelling System (DEPS) using Flux distribution method (FDM) in Sect. 1. The Lorentz force equation is derived analytically considering the secondary position (based on the reluctance principle) in Sect. 2. FEM simulations are to validate the derived analytical equation in Sect 3. Next, a generalized algorithm for the design of DEPS in Sect. 4. Afterward, the application example specifications from the Hyperloop student competition organized by the Space Exploration Technologies Corporation [9–12]. The results and simulations in Sect. 5.

2 Magnetic Circuit of Dodecapole Electrical Propelling Systems (DEPS)

The critical components of DEPS are the electrically driven coils (primary) and conductive body (secondary) [13–15]. Primary coils provide a magnetic field around the secondary, which induces eddy currents in the secondary surface—this reaction between the eddy current and main field results in linear motion (Fig. 1).

Figure 2 exhibits the dimensions and directions of the forces. Flux paths of the primary coil and secondary surface are estimated. Later the analysis is scaled up to all the coils. The flux lines in a single-coil are observed with FEMM software's guidance, as shown in Fig. 3.

The Reluctance of Flux path 1:

Reluctance in coil

$$\mathcal{R}_{c1} = \frac{5l_c}{\mu_o \mu_r w_c (h_c - t)} \tag{1}$$

Reluctance in Air-gap

$$\mathcal{R}_{g1} = \frac{40 l_g}{\mu_o w_p (4h_c + 5l_p - 9t)} \tag{2}$$

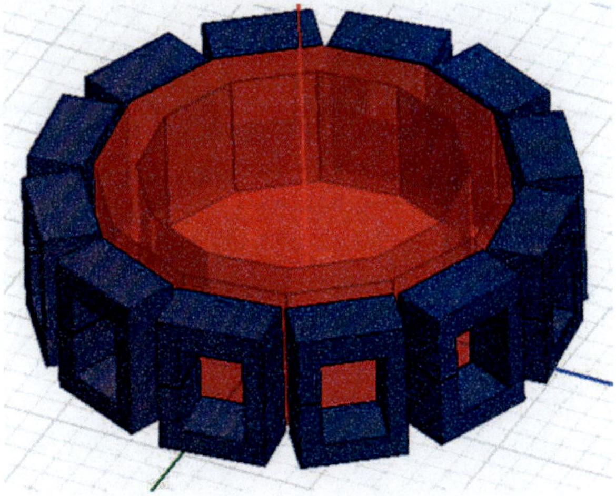

Fig. 1 Arrangement of single-stage DEPS

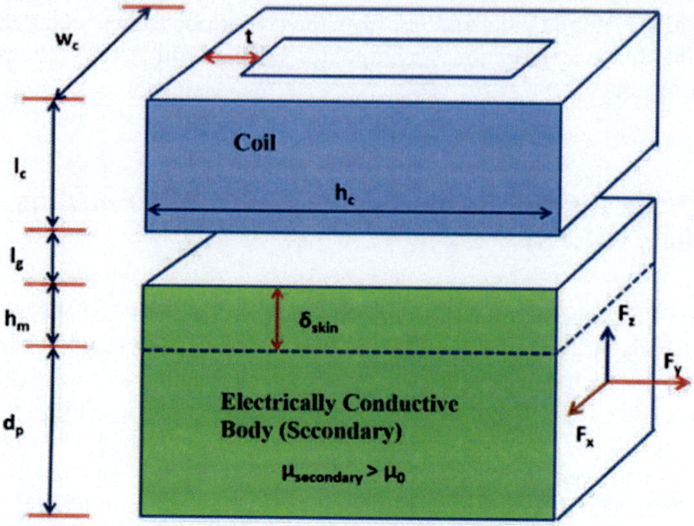

Fig. 2 Illustration of the Forces in DEPS (for single coil), which also expose the system parameters

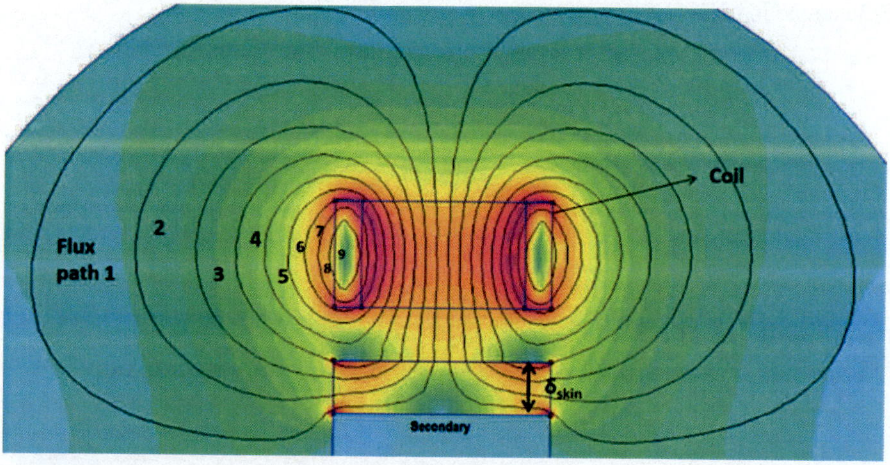

Fig. 3 Illustration of flux distribution in DEPS using FEMM software (side view)

Reluctance in Secondary

$$\mathrm{R}_{p1} = \frac{\pi}{4\mu_o \mu_{rp} h_m w_p} \left(\frac{\sqrt{l_p^2 + h_m^2}}{2l_p} - 3\frac{\sqrt{\left(\frac{3l_p+t}{4}\right)^2 + \left(\frac{2h_m}{3}\right)^2}}{(3l_p+t)} \right) \times \sqrt{\left(\frac{7l_p+t}{8}\right)^2 + \left(\frac{5h_m}{6}\right)^2} \quad (3)$$

Reluctance in the air-gap to reach another side of the coil

$$\mathcal{R}_{og1} = \frac{4\pi(7h_m+6l_g+12l_c)}{3\mu_o w_p h_m} \quad (4)$$

Equivalent Reluctance of Flux path 1

$$\mathcal{R}_{Fluxpath1}(x) = \mathcal{R}_{c1} + \mathcal{R}_{g1} + \mathcal{R}_{p1} + \mathcal{R}_{og1} \quad (5)$$

The Reluctance of Flux path 2:

Reluctance in coil

$$\mathcal{R}_{c2} = \frac{5l_c}{\mu_o \mu_r w_c (h_c - t)} \quad (6)$$

Reluctance in Air-gap

$$\mathcal{R}_{g2} = \frac{40 l_g}{\mu_o w_p (4h_c + 5l_p - 9t)} \quad (7)$$

Reluctance in Secondary

$$\mathcal{R}_{p2} = \frac{3\pi}{4\mu_o \mu_{rp} h_m w_p}\left(\frac{\sqrt{\left(\frac{3l_p+t}{4}\right)^2+\left(\frac{2h_m}{3}\right)^2}}{(3l_p+t)} - \frac{\sqrt{\left(\frac{l_p+t}{2}\right)^2+\left(\frac{h_m}{3}\right)^2}}{(l_p+t)}\right) \times \sqrt{\left(\frac{5l_p+3t}{8}\right)^2+\left(\frac{h_m}{8}\right)^2} \quad (8)$$

Reluctance in air-gap to reach other side of the coil

$$\mathcal{R}_{og2} = \frac{2\pi(h_m+3l_g+6l_c)}{\mu_o w_p h_m} \quad (9)$$

Equivalent Reluctance of Flux path 2

$$\mathcal{R}_{Fluxpath2}(x) = \mathcal{R}_{c2} + \mathcal{R}_{g2} + \mathcal{R}_{p2} + \mathcal{R}_{og2} \quad (10)$$

l_c primary coil length in meters
l_g air-gap length in meters
h_c primary coil height in meters
l_p secondary surface length in meters
h_p secondary surface height in meters
h_m secondary surface's metallic sheath height in meters

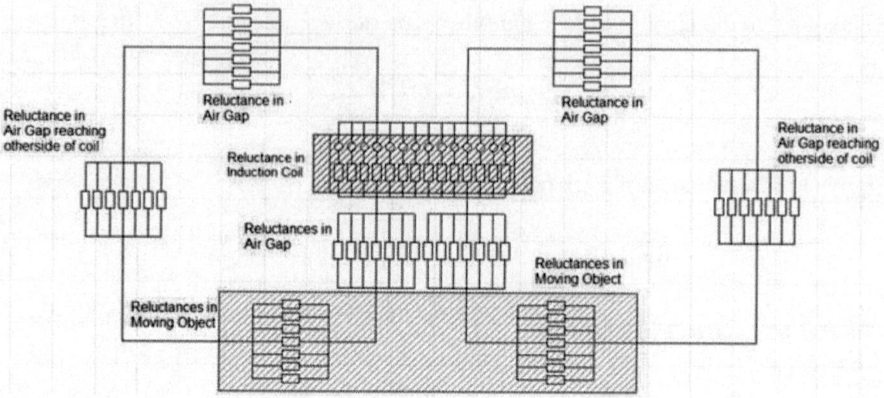

Fig.4 Sectionized magnetic network model of DEPS for single coil

w_p secondary surface width in meters
w_c primary coil width in meters
t thickness of the coil winding in meters
μ_0 permeability of the air
μ_{rp} relative permeability of the projecting object
x position of the secondary

Identically, all the flux paths reluctances are derived. And a sectionized magnetic circuit is developed, as shown in Fig. 4.

Generalized equation of reluctance of single-coil is

$$\begin{aligned}\mathcal{R}_{Total}(x) = {} & \mathcal{R}_{Fluxpath1}(x) + \mathcal{R}_{Fluxpath2}(x) + \mathcal{R}_{Fluxpath3}(x) + \mathcal{R}_{Fluxpath4}(x) \\ & + \mathcal{R}_{Fluxpath5}(x) + \mathcal{R}_{Fluxpath6}(x) + \mathcal{R}_{Fluxpath7}(x) + \mathcal{R}_{Fluxpath8}(x) \\ & + \mathcal{R}_{Fluxpath9}(x)\end{aligned} \tag{11}$$

And the same is applied for finding the reluctance equations in x, y, and z-plane. The total inductance of the DEPS of the secondary in the z-plane is

$$L_z(x) = \frac{N^2}{\mathcal{R}_{total}(x)} \tag{12}$$

where N is the number of turns in the coil.

3 Verification: Derived Equation Versus FEM Analysis

3D FEM simulations using ANSYS were used for validation of the derived inductance equation shown in Fig. 5.

In comparison, the maximum and minimum inductances of the DEPS are 22.097 and 0.319 mH (for derived equations) 21.81 and 0.28 mH (for FEM simulations) shown in Fig. 6. With the error of 1 mH, the derived equation and 3D FEM results agree well.

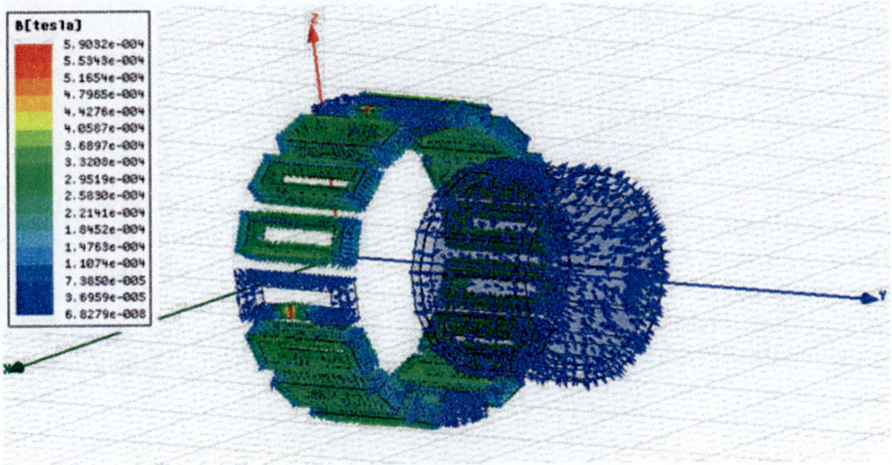

Fig. 5. 3D FEM simulations of DEPS for single stage

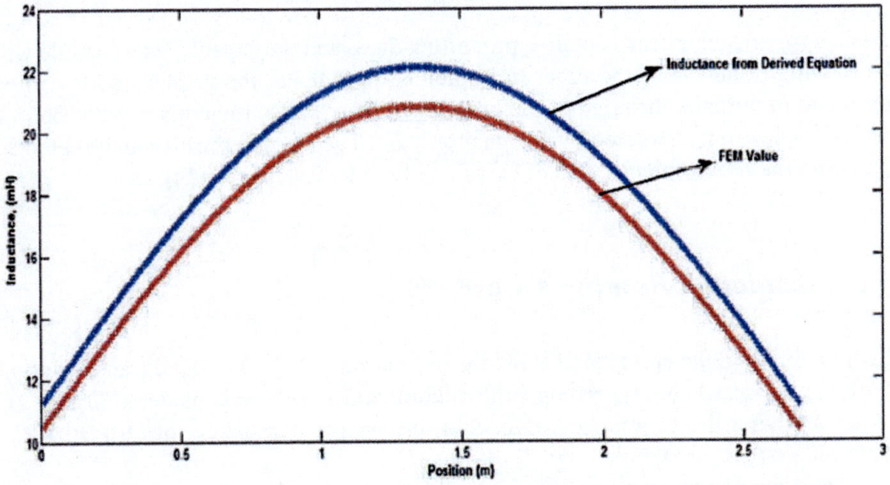

Fig. 6 Position versus inductance

Table 1 Force values in XYZ plane

Symbol	Description	Value	Units
F_x	Drag force	1*10^3	(N)
F_y	Lateral force	24*10^2	(N)
F_z	Lift force	57.2 *10^3	(N)

Moreover, the most significant force value obtained from the derived equation is 57.2 kN, and the force value from FEM is 50 kN. This exhibits that the analytically derived equations are substantial for the underlying plan of functional frameworks (Table 1).

4 Design Algorithm

The primary target is to execute the Dodecapole Electrical Propulsion System (DEPS) for quick transportation. Rapid transportation, as it has numerous applications, requires distinctive velocities. So a generalized plan is necessary to make it fit for every application.

This multi-constraint algorithm expands the flexibility to the consumer in designing the final dimensions of the DEPS based on the requirements, shown in Fig. 7.

5 Implementation of DEPS for Hyperloop Transportation: A Case Study

Basic hyperloop systems contain a propelling drive and the capsule. The capsule will be moving in an evacuated tube. In implementing DEPS, the primary coils can be arranged in the tube throughout the length. At every stage, the coils need to be in a series connection, energized by a capacitor bank. The voltage profile can be selected based on the requirements. The 3D estimated model is shown in Fig. 8.

5.1 *Simulation Results: A Case Study*

Keeping in mind the end goal to show the utilization of DEPS in rapid transportation further, case details of Hyperloop student competition [9] are considered (Table 2).

By applying the generalized algorithm, the design parameters of DEPS for the example systems are tabulated in Table 3.

According to the results depicted in Table 3, a force obtained is 120 MN and the speed obtained is 363 m/sec (approximately 1300 km/h). The results turn for

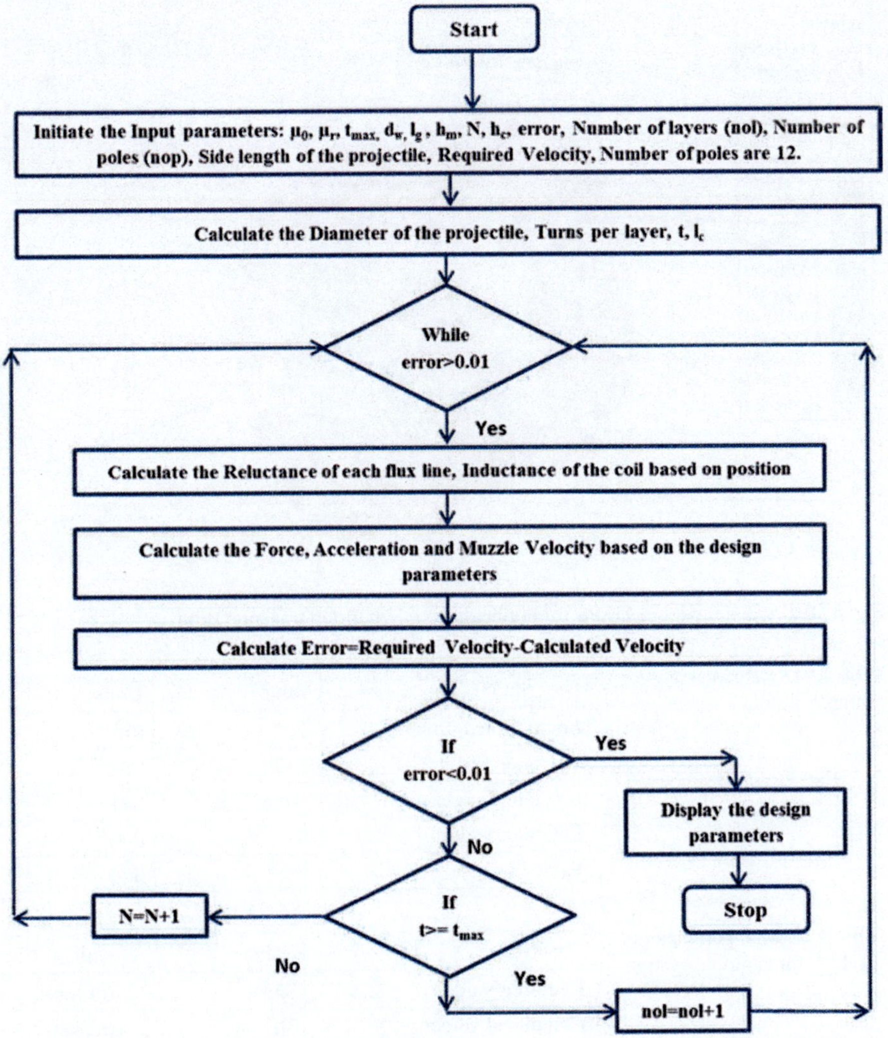

Fig. 7 Multi-constraint algorithm for design of DEPS

Electromagnetic propulsion vehicle when it is contrasted with cutting-edge with existing rapid transportation frameworks.

According to the results depicted in Figs. 9, 10 and 11, a maximum force obtained is 200 MN, the speed obtained is 390 m/sec (approximately 1300 km/h), and the current taken by the system is 800 amps. In Table 4, the switching times and operating durations of the different stages are tabulated. Hence, it can be concluded that because of the switch of the coils based on the position of the secondary, the amount of power consumption can be significantly reduced to cruise at speeds exceeding 250 m/sec compared to the existing maglev machines (Fig. 12).

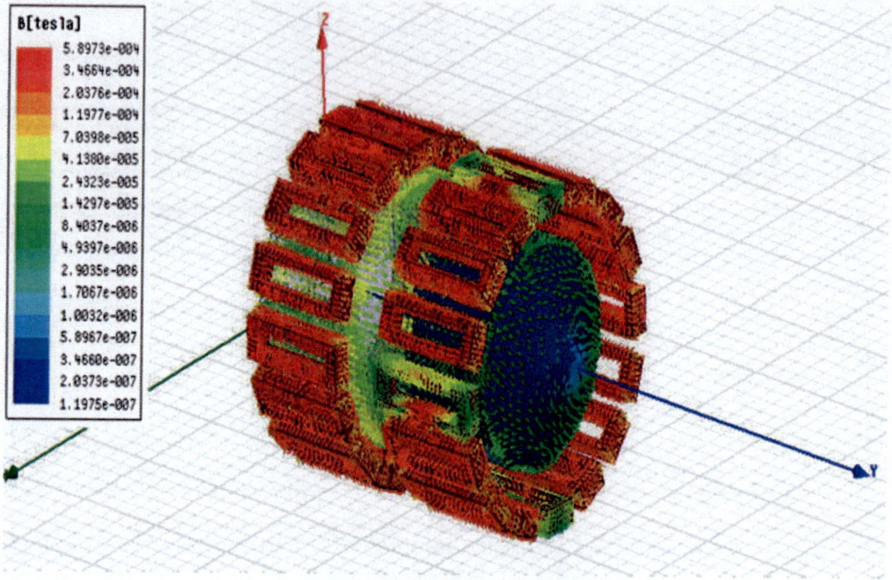

Fig. 8 Estimated model of DEPS for application of hyperloop transportation

Table 2 Dimensions of example system

Description	Value	Units
Levitator length	2	(m)
Levitator width	0.12	(m)
Levitator height	0.016	(m)
Levitator weight	30	(kg)
Payload	250	(kg)

Table 3 Design parameters of DEPS for example system

Description	Value	Units
Required velocity	333	(m/sec)
Calculated velocity	363	(m/sec)
Force obtained	120.22*10^6	(N)
Number of turns	18,906	–
Number of layers	137	–
Diameter of secondary	4.2501	(m)
Height of the coil	0.688921	(m)
Thickness of the coil	0.797500	(m)
Length of the secondary	2	(m)

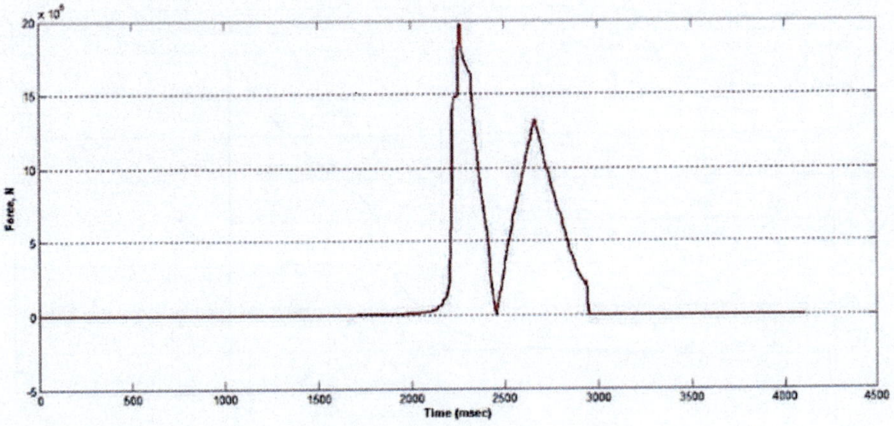

Fig. 9 Simulation results for force generated by DEPS

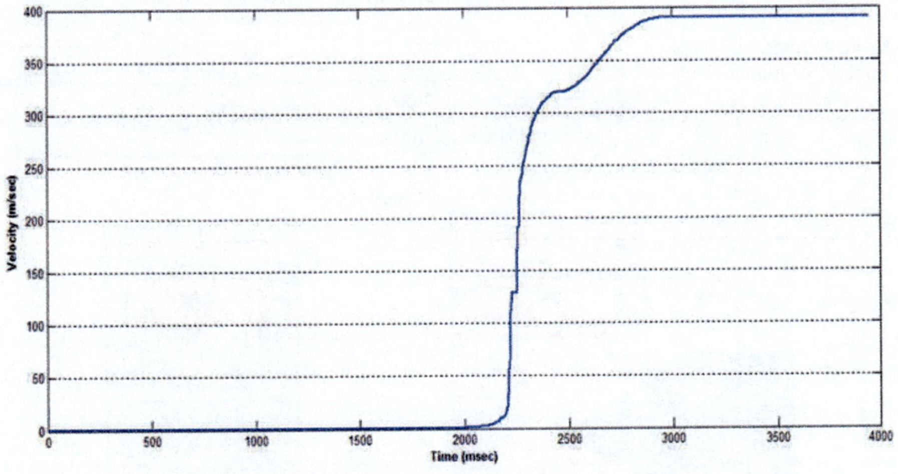

Fig. 10 Simulation results for velocity obtained using DEPS

6 Conclusions

A simple yet accurate analytical model for a Dodecapole Electrical Propelling System (DEPS) has been developed in this paper. Reluctances are calculated for each flux path, giving insight into building up the equation for Lorentz force. The magnetic equivalent circuit's detailed modeling is developed by omitting the assumptions such as uniform magnetic field, approximated inductance and mutual inductance values, and negligible current sheet. This empowers the utilization of the present technique to assess the different electromagnetic propulsion framework with various designs. 3D FEM models are used to approve the analytical model. Simulations verify the

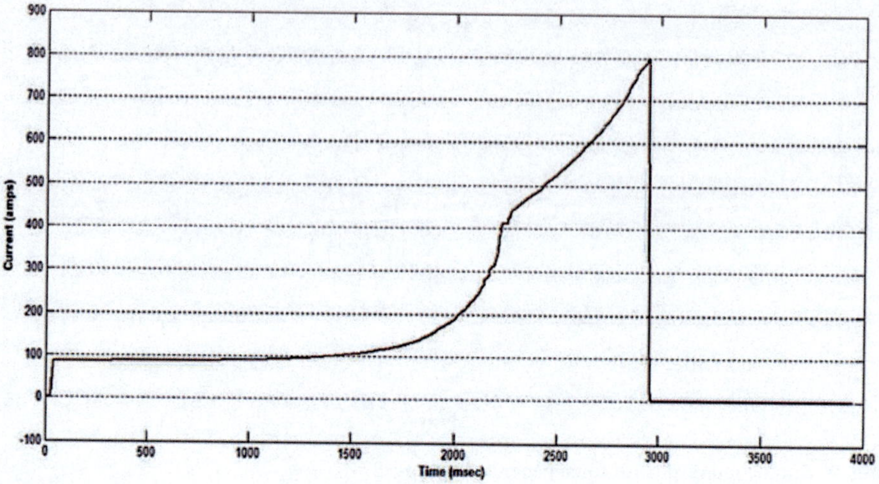

Fig. 11 Simulation results of current characteristics of DEPS

Table 4 Switching and operating duration

	Current value (A)	Operating duration (s)
First stage	400	2
Second stage	400	1

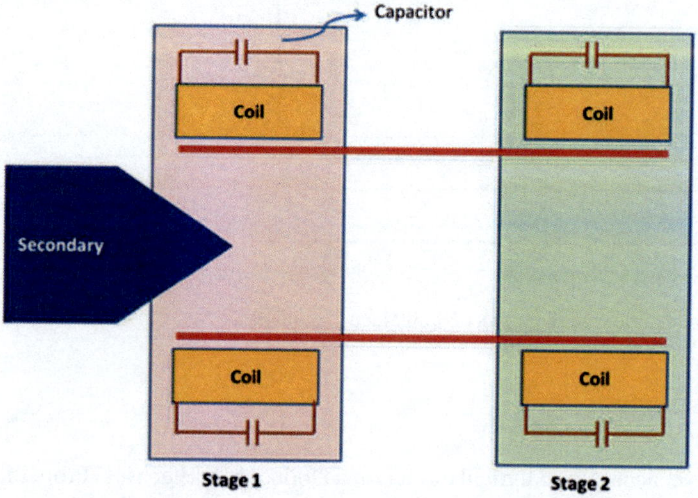

Fig. 12 View of multi-stage DEPS

outcomes acquired from the calculations. Finally, looking at the force profile of the DEPS framework to other types of cutting-edge fast ground and air transportation presents an intriguing picture. Over the cruising velocities of present-day subsonic aircraft, DEPS, in Hyperloop, can significantly expand the cruising-speed energy efficiency.

References

1. Elon Musk, Hyperloop Alpha, SpaceX (2013). https://www.spacex.com/sites/spacex/files/hyperloop_alpha-20130812.pdf
2. Dudnikov EE (2010) Advantages of a new hyperloop transport technology. In: Tenth international conference management of large-scale system development
3. Abdelrahman AS, Sayeed J, Youssef MZ (2018) Hyperloop transportation system: analysis, design, control and implementation. IEEE Trans Ind Electro 7427–7436
4. Flank M, Wellerdieck T, Tüysüz A, Kolar JW (2018) Scaling laws for electrodynamic suspension in high-speed transportation. IET Electric Power Appl 357–364
5. Zhu Y, Wang Y, Yan Z, Dong L, Xie X, Li H (2010) Multipole field electromagnetic launcher. IEEE Trans Mag 2622–2627
6. Yingwei Z, Yu W, Weirong C, Zhongming Y, Haitao L (2012) Analysis and evaluation of three-stage twisty octapole field electromagnetic launcher. IEEE Trans Plasma Sci 1399–1406
7. Musolino A, Rizzo R, Tripodi E (2013) Travelling wave multipole field electromagnetic launcher: an SOVP analytical model. IEEE Trans Plasma Sci 1201–1208
8. Luo W, Wang Y, Gui Z, Yan Z, Chen W (2013) Connection pattern research and experimental realization of single stage multipole field electromagnetic launcher. IEEE Trans Plasma Sci 3173–3179
9. SpaceX. Spacex hyperloop pod competition ii. http://www.spacex.com/hyperloop. Accessed 17 Oct 2017
10. Fair HD (2009) Advances in electromagnetic launch science and technology and its applications. IEEE Trans Mag 225–230
11. McNab IR (2009) Progress on hypervelocity railgun research for launch to space. IEEE Trans Mag 381–388
12. Becherini G, Tellini B (2003) Helicoidal electromagnetic field for coilgun armature stabilization. IEEE Trans Mag 108–111
13. Kondamudi SC, Thotakura S, Pasumarthi MR, Reddy GR, Sathyaraj SM, Jiang XX (2019) A novel type coil-multipole field hybrid electromagnetic launching system. Results Phys 15:102786
14. Kondamudi S, Pasumarthi MR (2019) Computations of magnetic forces in multipole field electromagnetic launcher. Int J Mathemat Eng Manag Sci 4(3):761–774
15. Kondamudi S, Thotakura S (2020) Constraint based design of multi-stage core type multipole field electromagnetic launching system (CMFELS) and its possible use in the catapult system. Lecture Notes Electrical Eng 630:499–510

Chapter 46
A Miniaturized Multi-Band Antenna Operating in C, X, and Ku Bands

G. Ramya Sri, P. Srujana, A. Jhansi Rani, V. Saritha, and S. Tanmayi

1 Introduction

A dual-band operating patch antenna was designed and presented for C and X band applications at a resonated frequency of 5.9 GHz and 8.8 GHz by using Defected Ground Structures (DGS). The antenna is a modified quadrangular shaped patch that has 4 U- shaped slots and 1 I- shaped slot. It has H-shaped DGS. It gives 10 dB impedance bandwidth with 140 MHz operating in 5.85–6 GHz and 10 dB impedance bandwidth with 1.21 GHz operating in 7.87–9 GHz [1]. Rectangular patch antenna achieved dual-band operating at 6 and 9 GHz that find applications intended for satellite communication & radio navigation at small power levels and good bidirectional radiation pattern. It is appropriate for incorporation into portable devices [2]. A Novel printed monopole antenna consisting of the radiator of square shape with a couple of slots is fed by a coplanar waveguide (CPW). Intended to have double band operation, the slots are positioned lateral to the patch; added resonances are observed leading to a wider bandwidth. Also achieved radiation pattern with realistic at the operated bands [3]. A novel Gemini-shaped patch antenna with dual-band features was proposed for applications of C and X band at a resonated frequency of 6.53 GHz and 8.62 GHz. And also observed good impedance matching and high gain performance [4]. A wideband antenna operating in C & X bands that has impedance bandwidth of 400 MHz at 7.47 GHz, 790 MHz at 11.01 GHz. It is also observed as of small size, has high gain [5]. A simple single-layer antenna with a T-shaped radiator is designed to improve cross-polarization suppression of the antenna [6]. A novel antenna operating at multiple bands from 1.5 to 6 GHz is presented for IOT applications [7]. A key-shaped patch antenna exhibited a wide band that operates at frequencies intended for applications over C/X bands. It has a rectangular-shaped

G. Ramya Sri · P. Srujana · A. Jhansi Rani · V. Saritha (✉) · S. Tanmayi
Electronics and Communication Engineering, Velagapudi Ramakrishna Siddhartha Engineering College, Vijayawada, India
e-mail: vsaritha@vrsiddhartha.ac.in

S. Dawn et al. (eds.), *Smart and Intelligent Systems*, Algorithms for Intelligent Systems,
https://doi.org/10.1007/978-981-16-2109-3_47

defected ground structure. Impedance bandwidth of the antenna with the frequency range 7.02–8.45 GHz is 18.48% and observed the peak gain [8]. A planar rectangular patch antenna is designed for applications in X and Ku bands with an air gap between ground and substrate [9]. A compact planar antenna using a reversible F-shaped structured slot is designed to operate in ultra-wideband spectrum. It has resonances at 5.4 GHz and 8.75 GHz [10]. An antenna radiator using F-shape radiated patch that is fed by a microstrip line is designed. Using the square strips and rectangular stub that is positioned in the ground achieved three bands with circular polarization—this covered frequency of the C band and X band destined for satellite applications [11]. A quad-band antenna is presented for sub 6 GHz–5 G applications [12]. This paper proposes a very compact structure to operate in multi-bands from C to Ku bands.

2 Antenna Design

The proposed structure of a very compact antenna with dimensions is depicted in Fig. 1. It comprises a modified rectangular patch with square cuts on four corners of it with a U slot cut. It is fed by a microstrip line with a quarter-wave transformer for proper impedance matching. Two U-shaped resonators are placed in the vicinity of the microstrip line. The ground plane consists of a discrete rectangle and a square with a square cut in its middle. It has dimensions of $16 \times 15 \times 1.6\ mm^3$. The values of dimensions are given in Table 1.

Progressive stages of Evolution: The progressive stages for achieving multibands are depicted in Fig. 2

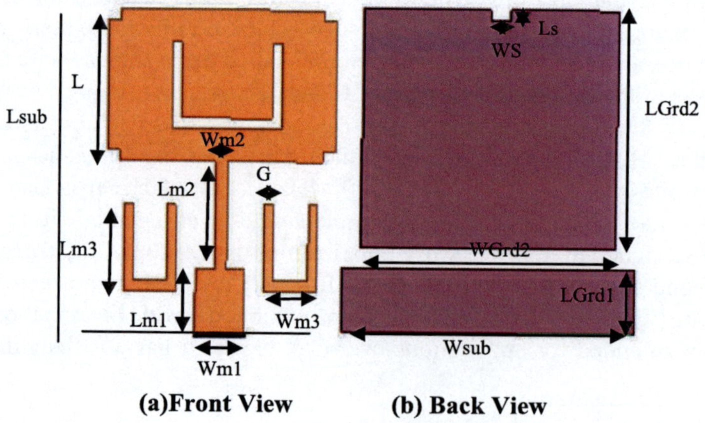

Fig. 1 Proposed antenna

Table 1 Parametric dimensions of the proposed antenna design

Basic configurations	Parameters	Values(mm)
Substrate (FR4-epoxy)	WSub	15
	LSub	16
	Thickness	1.6
Antenna Patch	W	13
	L	7
	Wm2	0.7
	Lm2	5
	Wm1	2.8
	Lm1	3
Ground Plane	WGrd2	13
	LGrd2	11
	LGrd1	3
	Is	0.5
	Ws	1
U-Shaped slot	Lm3	3.5
	Wm3	2.5
	G	0.5

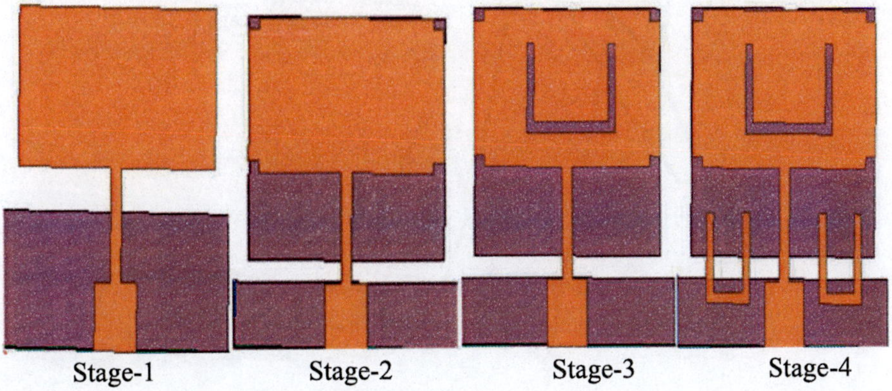

Fig. 2 Evolution of the proposed antenna structure

Stage-1: Stage-1 consists of a rectangular patch fed by a microstrip line which is included with a quarter-wave transformer. The ground plane is a defected with a rectangular shape. This stage not resonated at any frequency. Its return loss and VSWR are displayed in Figs. 3 and 4.

Stage-2: Stage-2 consists of a modified quadrangular patch with square cuts at its four corners. A microstrip line which is included with a quarter-wave transformer

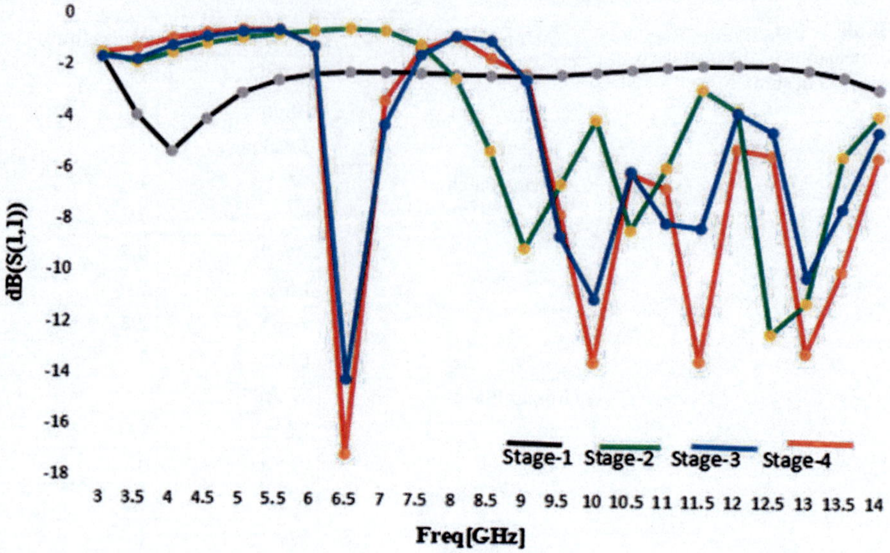

Fig. 3 Return loss plots for four stages in the evolution

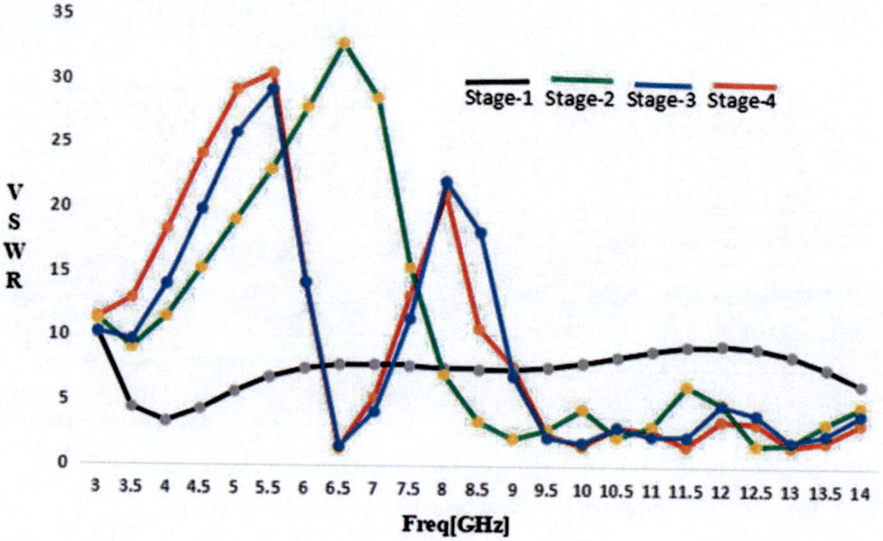

Fig. 4 VSWR plots for four stages in the evolution

feeds the antenna. The ground plane is a defected with a discrete rectangle and square. This structure resonated at 9 and 12.5 GHz. Its return loss and VSWR are shown in Figs. 3 and 4.

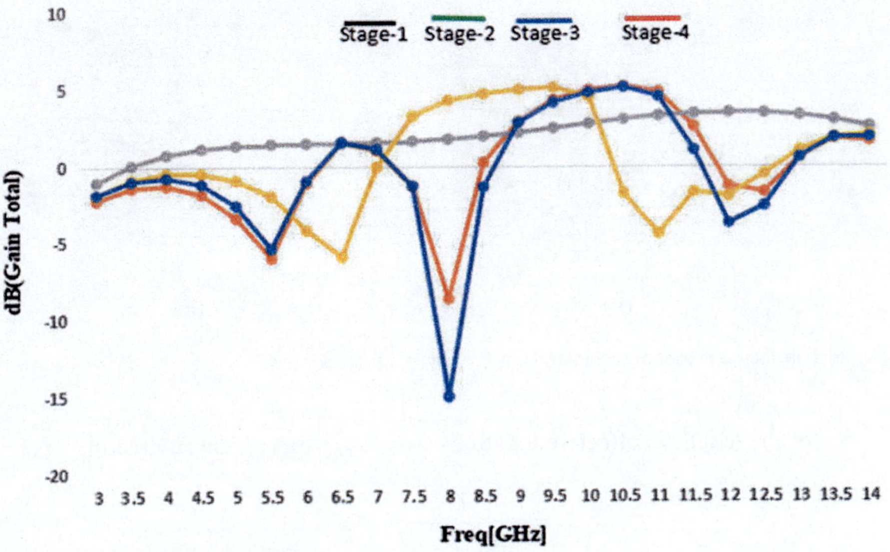

Fig. 5 Comparison of gain plots for four stages in the evolution

Stage-3: In addition to structure in stage-2, stage-3 consists of a U slot in the rectangular patch to resonate at 10 GHz. This structure resonated at 6.5 GHz,10 GHz, and 13 GHz. Its return loss and VSWR are shown in Figs. 3 and 4.

Stage-4: In addition to structure in stage-3, stage-4 consists of two U shape resonators in the vicinity of the feed line. This structure resonated at 6.5 GHz, 10 GHz, 11.5 GHz, and at 13 GHz. Its return loss and VSWR are shown in Figs. 3 and 4.

The gain plot at four stages is plotted in Fig. 5.

The dimensions of the proposed antenna are designed using the equations given from (1) to (5):

$$\varepsilon_{reff} = \frac{\varepsilon_{r+1}}{2} + \frac{\varepsilon_{r-1}}{2}\left(1 + 12\frac{h}{W}\right)^{\frac{1}{2}} \frac{W}{h} >> 1 \tag{1}$$

$$\varepsilon_{reff} = \frac{\varepsilon_{r+1}}{2} + \frac{\varepsilon_{r-1}}{2}\left(1 + 12\frac{h}{W}\right)^{\frac{1}{2}} \frac{W}{h} >> 1 + 0.041\left[1 - \sqrt{\frac{W}{h}}\right], \frac{W}{h} << 1 \tag{2}$$

The fringing effect influences the calculation of L and W as follows:

$$W = \frac{C}{2fr\sqrt{\frac{\varepsilon_r+1}{2}}} \tag{3}$$

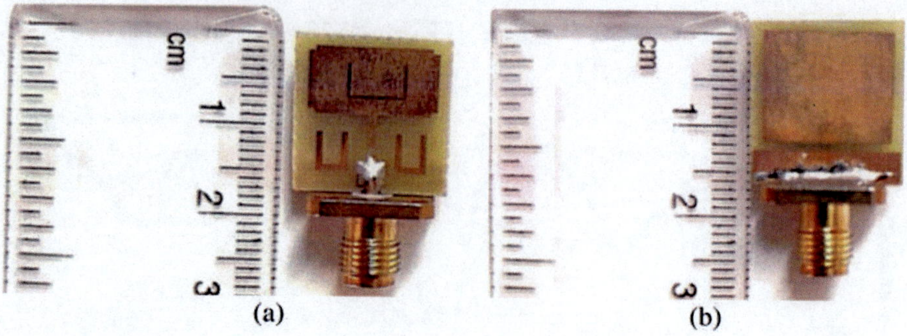

Fig. 6 Fabricated model of the antenna. **a** Front view. **b** Back view

As for the length L, effective length by using ε_{reff} in (1) and subtracting $2\Delta L$:

$$L = \frac{C}{2fr\sqrt{\varepsilon_{reff}}} - 2\Delta L \tag{4}$$

ΔL is called the normalized length extension and is given by:

$$\Delta L = 0.412h\frac{(\varepsilon_r + 0.3)\left(\frac{W}{h} + 0.364\right)}{(\varepsilon_r + 0.258)\left(\frac{W}{h} + 0.8\right)} \tag{5}$$

The length of U slot in rectangular patch antenna is calculated using (6)

$$L = \frac{C}{2f_r\sqrt{\varepsilon_r}} \tag{6}$$

where f_r is the resonating frequency.

The printed model is shown in Fig. 6. The connector used is an SMA connector. The measurements are done by means of vector network analyzer of model N5242A (product by Agilent PNA X series).

The comparison of return loss and VSWR between the simulated and measured results is depicted in Fig. 7.

Radiation Patterns: The plots of co-polarization and cross-polarization are depicted in Fig. 8. In the E-plane patterns, the simulated co-polarization and cross-polarization are displayed in red color and blue color while the measured co-polarization and cross-polarization are shown as dotted in red color and blue color as shown in Fig. 8a at 6.5, 10, 11.5, and 13 GHz. In the H-plane patterns, the simulated co-polarization and cross-polarization are shown in red color and blue color while the measured co-polarization and cross-polarization are shown as dotted in red color and blue color as shown in Fig. 8b at 6.5, 10, 11.5, and 13 GHz.

The 3D polar plots at 6.5, 10, 11.5, and 13 GHz are shown in Fig. 9.

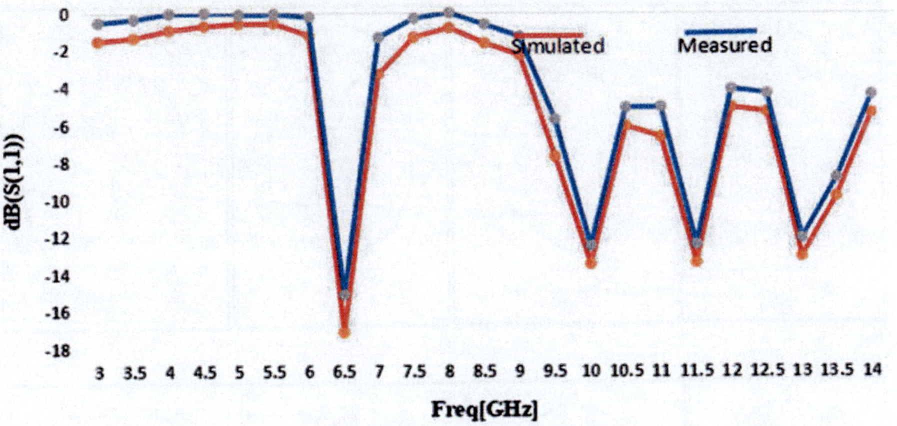

(a) Frequency Vs Return Loss

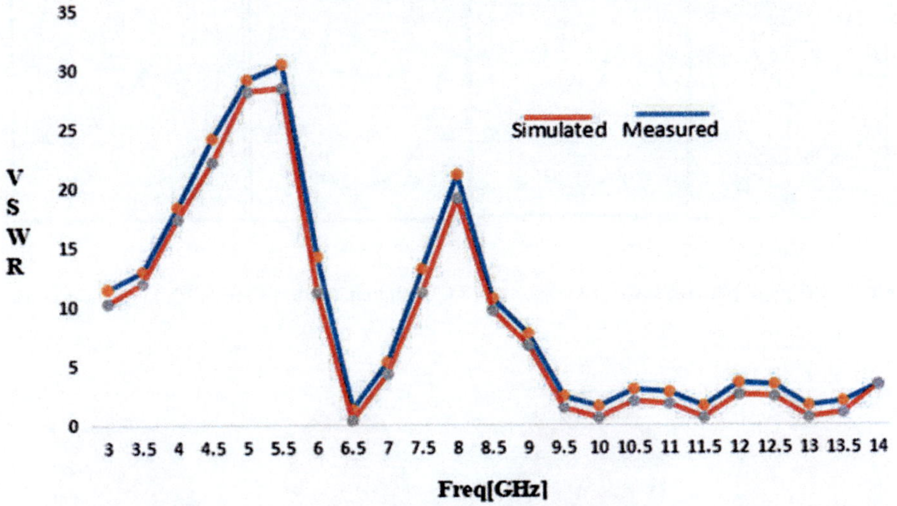

(b) Frequency Vs VSWR

Fig. 7 VSWR and return loss plot for proposed antenna design

Table 2 compares antennas in terms of size, bands of operation, and gain cited in the references.

The proposed design is very compact with reasonable gain.

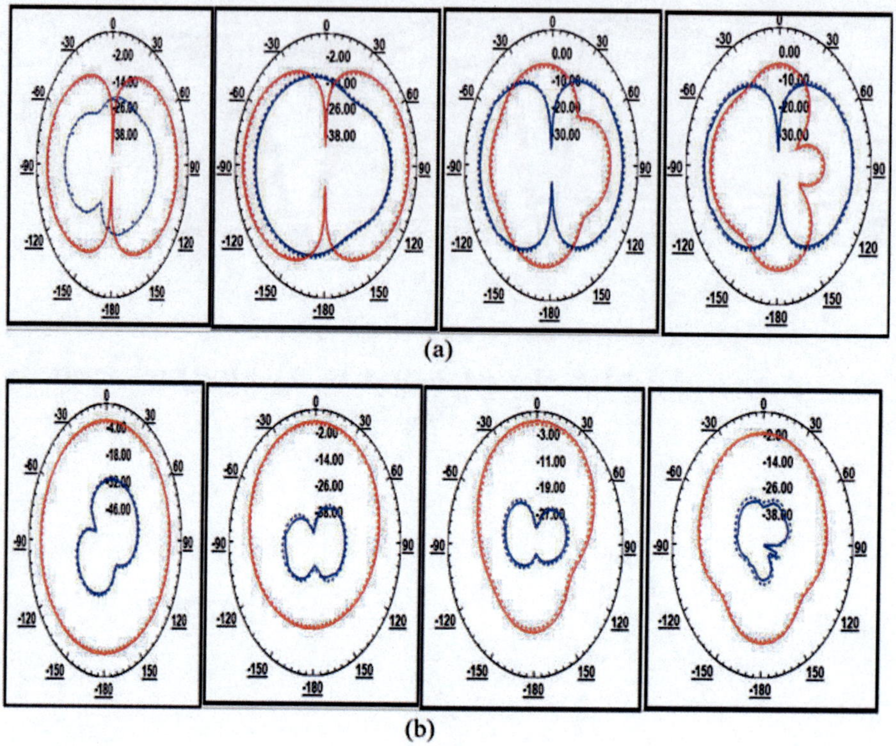

Fig. 8 **a** E-plane patterns at 6.5, 10, 11.5, and 13 GHz, **b** Plane patterns at 6.5, 10, 11.5, and 13 GHz.

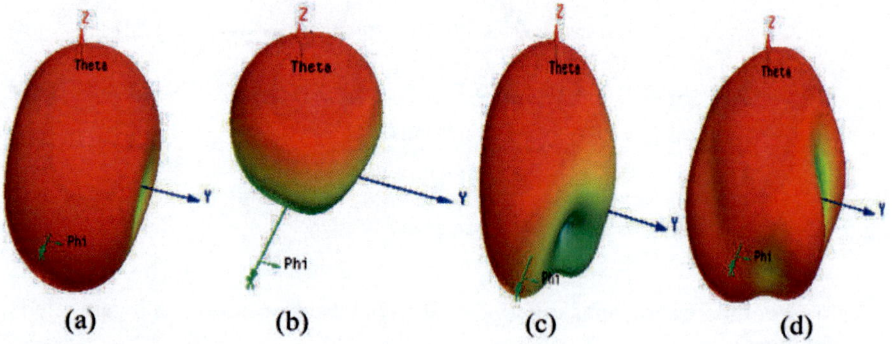

Fig. 9 3D polar plots at 6.5, 10, 11.5, and 13 GHz

3 Conclusion

A new miniaturized rectangular patch antenna is designed, which resonates at multiple frequencies, with dimensions (16 × 15 × 1.6) mm^3. The patch antenna

Table 2 Comparison of dimension of antenna cited in the references

Ref. no	Sizes (mm^3)	Bands	Frequency (GHz)	Gain (dBi)
1	25 × 23 × 1.5	C and X	5.9, 8.8	1.2, 4.4
2	65 × 58 × 1.6	C and X	6.07, 9.15	5.53,7.96
3	26 × 22 × 0.8	C and X	4.4, 8.5	2.4, 4.4
4	25 × 25 × 0.76	C and X	6.53, 8.62	6.032, 9.556
5	16 × 18 × 1.6	C and X	7.47, 11.01	5.36, 4.65
6	27.5 × 8 × 1.6	C and X	7.5, 11.2	2.25, 3.47
7	34 × 36 × 1.6	C and X	5.4, 8.95	2.63
8	60 × 60 × 1.575	X and Ku	8.30–15.20	5, 6.5
9	32 × 32 × 1.6	C and X	5, 7.3, 11	6.23, 2.07,2.11
11	70 × 70 × 1.575	Sub 6 GHz	2.5, 3.5, 5.2, 5.8	4–6
Proposed design	15 × 16 × 1.6	C,X and Ku	6.5,10, 11.5 and 13	3.2, 5.1 and 6.2

is designed using the slot technique and partial ground plane technique, which help in reducing the size of the antenna. The simulated rectangular patch antenna resonates at 6.5, 10, 11.5 GHz, and also at 13 GHz with a gain of 1.3706, 4.7263, 2.9413, and 1.2754 dB, respectively, and with a bandwidth of 257.3, 560, 451.7, and 390 MHz, respectively. The antenna is simulated using ANSYS High-Frequency Structure Simulator (HFSS). From the results, it is concluded that the antenna can operate in C and X bands which is suitable for numerous wireless applications intended for radar and satellite communications.

References

1. Gupta S, Dhillon SS, Khera P, Marwaha A (2013) Dual band U-slotted microstrip patch antenna for C band and X band radar applications. In: International conference on computational intelligence and communication networks
2. Sharma R, Kumar M (2013) Dual band microstrip antenna for C- and X band wireless applications. In: IMPACT
3. Rajkumar R, Thiruvalarselvan P, Elavarasi C, Banu RJ (2015) Cpw-fed square patch compact antenna for c and x band applications. In: International conference on information communication and embedded systems (ICICES2014). Accessed 09 Feb 2015
4. Djengomemgoto G, Altunok R, Karabacak C, UImeci T, Durak T (2017) Dual-band gemini-shaped microstrip patch antenna for C-band and X-band applications. In: International applied computational electromagnetics society symposium, Italy (ACES)
5. Soufian L (2018) Design of dual and wideband rectangular patch antenna for C and X band applications. Adv Electromag 7(5)
6. Chattopadhyay S, Chakraborty S (2018) A physical insight into the influence of dominant mode of rectangular microstrip antenna on its cross-polarization characteristics and its improvement with Tshaped microstrip antenna. IEEE

7. Injarapu P, Sanka H, Manasa NS, Chowdary V, Vanka S (2019) Analysis and design of a low profile multiband antenna for IOT applications. Int J Innov Technol Explor Eng 8(4). ISSN: 2278–3075
8. Mishra PK, Yadav S, Jaiswal K, Singh R (2019) Dual band key-shaped microstrip patch antenna for C/X band applications. In: Proceedings of the international conference on advances in electronic electrical & computational intelligence (ICAEEC)
9. Sarkar T, Ghosh A, Singh LLK, Chattopadhyay S, Sim C-Y-D (2019) DGS integrated air loaded wideband microstrip antenna for X and Ku band. IEEE Antennas Wireless Propagat Lett
10. Prabhu T, Suganya E, Pandian SC (2020) Design of an UWB antenna for microwave C and X band applications. In: International conference on advanced computing and communication systems (ICACCS). Accessed 23 April 2020
11. Srivastava K, Mishra B, Pate AK, Singh R (2020) Circularly polarized defected ground stub-matched triple band microstrip antenna for C- and X-band applications. Microwave Optical Technol Lett. Accessed 22 May 2020
12. Sarkar T, Ghosh A, Chakraborty S, Singh LLK, Chattopadhyay S (2020) Employment of mixed mode in single-layer microstrip antenna for ISM/WiMAX/WLAN/4G/Sub 6 GHz 5G mobile communication. J Electromag Waves Appl

Chapter 47
Maiden Application of Seagull Optimization Algorithm for the Study of Load Frequency Control

CH. Naga Sai Kalyan and Chintalapudi V. Suresh

1 Introduction

One of the preliminary issues that the modern electric system facing is reducing the active power generation and demand mismatch; this leads to the generators running in synchronism in a complex network may get disturbed. Moreover, this unwanted non-synchronism will be imparted within the system due to load perturbations also. This adverse situation is mainly addressed by the primary regulator by changing the fuel input to the generator thereby the generation valve set-point operation gets altered. However, this operation will hold good only under small perturbing conditions and may not be applicable for the cases of medium and large perturbations. In the event of large perturbations, the dynamical behavior of the electric system can be handled by the secondary regulator. Designing the secondary controller for the LFC problem is one of the most recent trends in power system control and automation study.

It is confessed from the literature survey that the study on LFC was first carried out in [1], since then, more researchers are contributed to this study. The control approaches that are available in literature so far are primitive approach of genetic algorithm (GA)/gray wolf optimizer (GWO) tuned PI [2]/PID [3], firefly algorithm (FA) [4]/cuckoo search optimizer (CSO) [5] tuned classical PI/PID, differential evolution (DE)[6]/pattern search (PS) [7] approach of PI/PID, hFA-PS [8], HAEFA [9] approach of PID, etc. Apart from the classical controllers, other fractional order (FO) [10] and fuzzy-based approaches have also been implemented extensively for LFC study. Some cascade FO and fuzzy [11] are also reported. Even though many intelligent and other intelligent cascade controllers are available in the literature, still most of the researchers are more concentrated on the establishment of soft computing algorithms based on the traditional controller of PI/PID/PIDN [6] and the extended classical controllers of PIDD [12]. Because the fuzzy logic systems necessitates much approximations in framing the fuzzy rules results in ambiguity of robustness.

CH. N. S. Kalyan (✉) · C. V. Suresh
EEE Department, Vasireddy Venkatadri Institute of Technology, Guntur, India

S. Dawn et al. (eds.), *Smart and Intelligent Systems*, Algorithms for Intelligent Systems,
https://doi.org/10.1007/978-981-16-2109-3_48

So, in this work, a novel implementation of SOA-based PIDD controller is executed for the first time in LFC study, and no literature is available till now.

On the other side, the territorial control strategies of implementing HVDC line [13], incorporating energy storage devices (ESDs) [14], and FACTs [15] in the multi-area systems for performance enhancement are also reported. In this work, the HVDC line is erected in parallel with the AC line that already existed in the system for improvement in system dynamical behavior.

The contributions of this work are:

(a) SOA algorithm is successfully implemented for the solution of the LFC problem.
(b) The efficacy of the SOA tuned PIDD control technique is established by validating with other control approaches reported in the literature.
(c) Additional tie-line of DC is incorporated to dampen system variations further.
(d) Robustness of controlling technique is demonstrated.

2 Investigated Model

A widely accepted model of dual area hydro-thermal system is deliberated for analyzing the performance in the situation of load disturbances. The system model rendered in Fig. 1, is taken from [2, 8] and is explicated in the platform of MATLAB/SIMULINK (R2016a) version. Due to the additional benefits provided by the double derivative (DD) gain in dampening the system deviations, it is chosen to be employed as a regulator. Moreover, the design of PIDD almost resembles that of traditional PID, which acquires simplicity in design and efficiency in operation motivates to carry out in this work. However, the parameters of PIDD are constituted with an algorithm derived from natural inspired biodiversity of seagulls named as seagull optimization (SOA) algorithm subjected to the chasteness of the absolute error multiplied with the time over the integral (ITAE) function given in Eq. (1) (Fig. 2).

$$J_{ITAE} = \int_{0}^{T_{sim}} t * (\Delta F_1 + \Delta P_{tie12} + \Delta F_2)dt \tag{1}$$

Further, the investigative conventional system is equipped with a DC tie-line with an already existing AC line to enable the bulk power transfer under sudden large load disturbances so that the system stability can be sustained. The modeling of the DC line is considered from [13] and is provided in Eq. (2).

$$G_{DC} = \frac{K_{DC}}{1 + ST_{DC}} \tag{2}$$

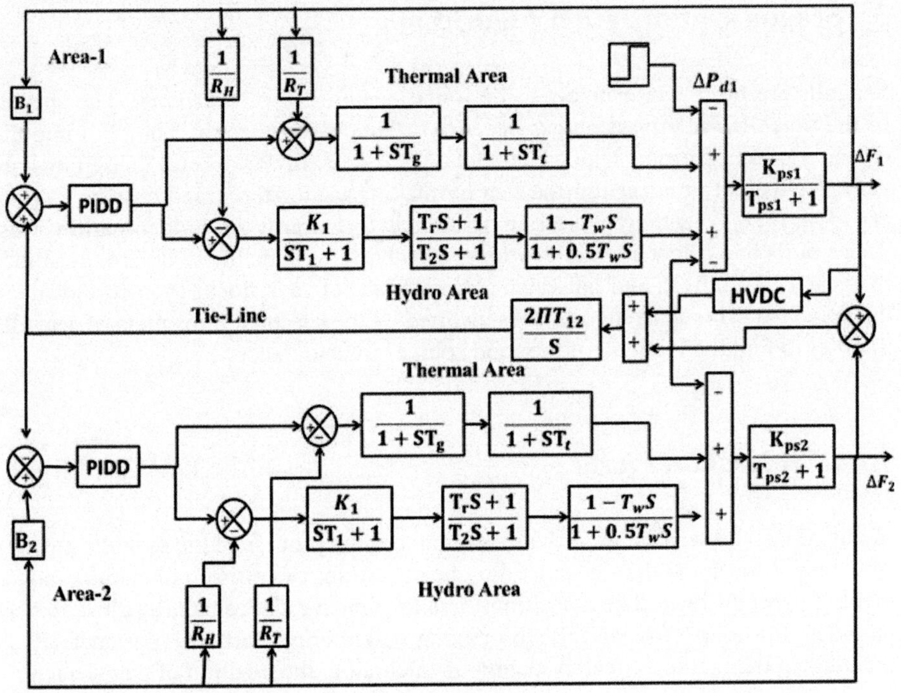

Fig. 1 Dual area conventional system with AC-DC links

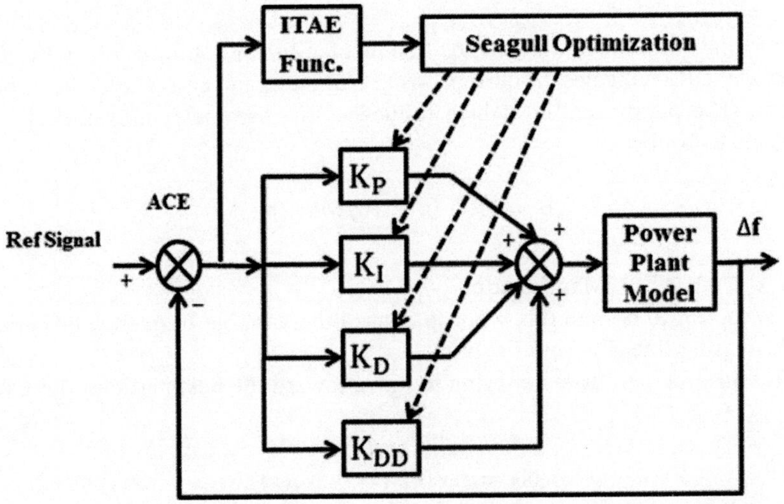

Fig. 2 PIDD controller optimization

3 Seagull Optimization Algorithm

Seagulls are birds which likely to be found at the base of seashores. The species of the seagulls are in vast ranges because of length and masses. They fall into the category of omnivorous, which can feed on fishes, earthworms, etc. These birds are very intellectual in attracting the fish by making use of the bread crumbs as trap, and with their feet, they make raining sounds to ambush the hidden earthworms. These birds had a special quality where the other bird's unable to do are the ability to consume both fresh and saltwater. Normally, they live in flocks to trace and attack the prey [16]. This SOA algorithm is inspired by the seasonal movement of seagulls in order to find the location of the food sources in abundance.

3.1 Exploration Phase

In this phase, the searching mechanism simulates the process of the seagulls moving from one position to other by satisfying the conditions of collision obviation, movement of particles toward best positions, and the tendency of remaining close to best particles. In order to avoid collision among neighboring particles, parameter B is included in the mathematical modeling of calculating the position of a new agent:

$$\vec{P}_S = B\,\chi\,\vec{K}_S(x) \tag{3}$$

where $\vec{P}_S$ indicates the new agent position that facilitates not to collide with a neighboring agent, $\vec{K}_S$ indicates the current position of the agent, and 'x' represents present iteration. The parameter 'B', which influences the parameter movements, can be calculated as follows:

$$B = F_c - (x * (F_c/\text{Max.Iter})) \tag{4}$$

where x = 0,1,2.......Max Iteration.

F_c is chosen to be 2 in this work to control the variable frequency of parameter 'B', decreasing linearly from Fc to 0.

After this, the particles are trying to move toward the best particles direction by using

$$\vec{N}_S = A\chi(\vec{P}_{bs}(x) - \vec{K}_S(x)) \tag{5}$$

The position of the search agent on the way toward the best particle position $\vec{P}_{bs}$ is represented with $\vec{N}_S$. Where the parameter 'A' is randomized by implementing the balance between the phases of exploration and exploitation can be calculated as

$$A = 2 * B^2 * rand() \tag{6}$$

At last, the update in the position of the agents with regards to the best agent position can be modeled as

$$\vec{R}_s = \left|\vec{P}_s + \vec{N}_s\right| \tag{7}$$

3.2 *Exploitation Phase*

The seagulls can continuously alter the attacking angle and speed based on their experience during migration. The behavior under three-dimensional plane can be modeled as

$$U' = r * Cos(k) \tag{8}$$

$$V' = r * Sin(k) \tag{9}$$

$$W' = r * k \tag{10}$$

'r' indicates the radius of the spiral in each turn, and 'k' is a random number from [0 2Π].

The best solution be saved, and other agents position will be updated as

$$\vec{K}_S(x) = (\vec{N}_S * U' * V' * W') + \vec{P}_{bs}(x) \tag{11}$$

The better searching and migration mechanism that is involved in the SOA algorithm motivates to implement for the study of the LFC problem. Moreover, this SOA approach is already implemented by the researchers to image processing problems as proved as the best.

4 Simulation Results

4.1 *Case-1: System Performance with Various Control Approaches*

The conventional dual area system of hydro-thermal generations is analyzed under the supervision of the PIDD regulator fine-tuned with an SOA algorithm for a load

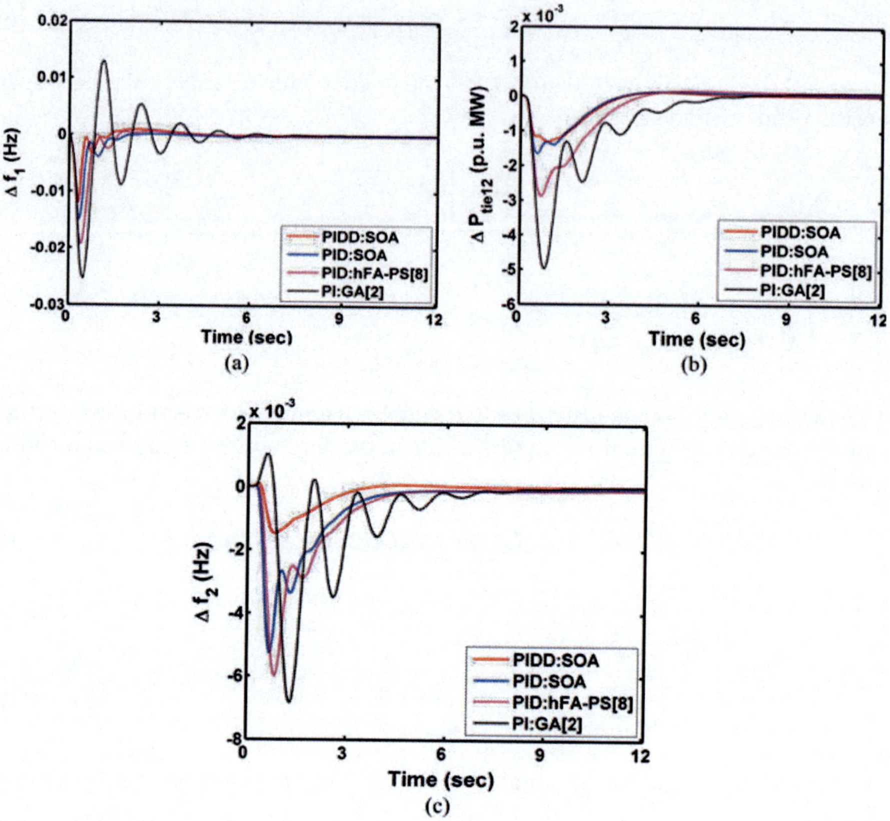

Fig. 3 Responses for case-1. **a** Δf_1, **b** ΔP_{tie12}, **c** Δf_2

change on area-1 of 1%SLP. To get crystal clear visualization with regards to the presented SOA-based PIDD control approach, system performance is also assessed with other controlling approaches such as GA tuned PI and hFA-PS tuned PID for the same loading conditions and the responses are compared in Fig. 3, and the respective gains are given in Table 1. The settling time of Fig. 3 responses is provided in Table II. Interpreting the results provided in Table 2 and noticing Fig. 3, it is absolved that SOA tuned PIDD regulator is much superior in palliating the deviations compared to GA tuned PI [2] and hFA-PS tuned PID [8] approaches available in the literature.

4.2 Case-2: System Performance with AC-DC Lines

An additional tie-line of DC is placed with the system to enable the ease of transferring bulk power. With this, the deviations that are going to have occurred in the system dynamical behavior during large load fluctuations is reduced, and stability be attained.

Table 1 Optimal parameters of controllers employed for case-1 and case-2

Parameter		K_P	K_I	K_D	K_{DD}
Area-1	PI:GA	0.9190	0.606	–	–
	PID:hFA-PS	1.1930	0.332	0.7262	–
	PID:SOA	1.0880	0.520	0.9072	–
	PIDD:SOA	1.1930	0.4603	0.7184	0.2862
	PIDD:SOA with AC-DC lines	1.099	0.781	0.990	0.1343
Area-2	PI:GA	0.8990	0.515	–	–
	PID:hFA-PS	0.9921	0.298	0.555	–
	PID:SOA	1.3010	0.641	0.744	–
	PIDD:SOA	1.0269	0.753	0.668	0.3250
	PIDD:SOA with AC-DC lines	1.2475	0.7673	0.8634	0.3815

Table 2 Settling time of system responses

Control strategy	Settling time (Sec)			ITAE
	ΔF1	ΔPtie,12	ΔF2	
PI:GA [2]	11.06	10.16	10.61	$7.56*10^{-3}$
PID:hFA-PS [8]	6.817	9.529	9.950	$1.26*10^{-3}$
PID:SOA	5.290	7.123	8.223	$9.32*10^{-4}$
PIDD:SOA	5.862	8.714	7.14	$3.41*10^{-5}$
PIDD with AC-DC Lines	3.738	5.49	4.373	$2.15*10^{-5}$

The SOA tuned PIDD approach is continued for this case as it revealed its superiority corresponding to others in the above case. The responses are rendered in Fig. 4. By the arrangement of AC-DC lines, the behavior of the system is enhanced by far.

4.3 *Case-3: Robustness Test*

The presented secondary and territorial control approach robustness is validated by subjecting the system with the variations in the load of 1, 3, and 5% SLP in only area-1 to get a comparative approach, and the corresponding responses are portrayed in Fig. 5. Even after the system is loaded with much higher loading than the load considered for controller optimization, the proposed technique, both secondary (SOA-PIDD)and territorial (AC-DC lines), is much more efficient enough to restore system stability.

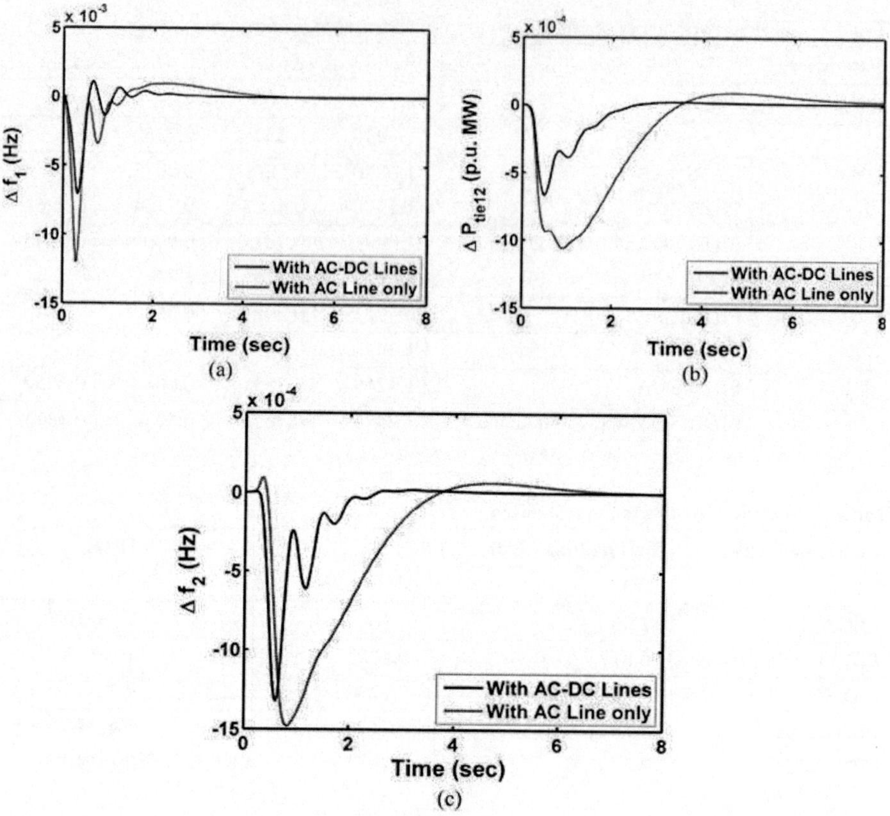

Fig. 4 Responses for case-2. **a** Δf1, **b** ΔPtie12, **c** Δf2

5 Conclusion

The seagull optimization algorithm is successfully implemented in this paper for the optimal solution of the LFC study. For this, the PIDD controller is selected and fine-tuned with SOA technique and proves to best regulator compared to GA-based PI and hFA-PS based PID techniques available in the literature. Moreover, an additional tie-line of DC is incorporated in alignment with the AC line to ease transferring bulk power. Finally, the system is loaded with various loadings to validate the presented coordinated secondary (SOA-PIDD) and territory (AC-DC lines) approach robustness.

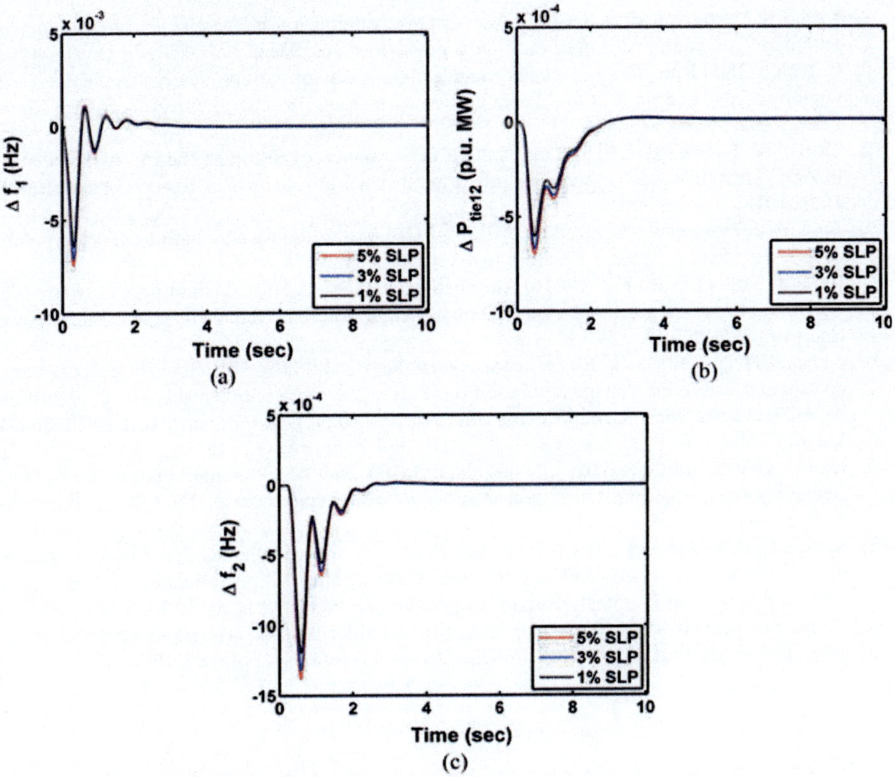

Fig. 5 Responses for case-3. **a** Δf1, **b** ΔPtie12, **c** Δf2

References

1. Fosha CE, Elgerd OI (1970) The megawatt- frequency control problem: a new approach via optimal control theory. IEEE Trans Power Appar Syst 89(04):563–577
2. Chandrakal KRMV, Balamurugan S, Sankaranarayanan K (2013) Variable structure fuzzy gain scheduling based load frequency controller for multi-source multi area hydro thermal system. Electr Power Energy Syst 53:375–381
3. Guha D, Roy PK, Banerjee S (2016) Load frequency control of interconnected power system using grey wolf optimization. Swarm Evol Comput 27:97–115
4. Pradhan PC, Sahu RK, Panda S (2016) Firefly algorithm optimized fuzzy PID controller for AGC of multi-area multi-source power systems with UPFC and SMES. Eng Sci Technol Int J 19(01):338–354
5. Abdelaziz AY, Ali ES (2015) Cuckoo search algorithm based load frequency controller design for nonlinear interconnected power system. Int J Electr Power Energy Syst 73:632–643
6. Kalyan CNS, Rao GS (2021) Stabilizing frequency and voltage in combined LFC and AVR system with coordinated performance of SMES and TCSC. In: Singh AK, Tripathy M (eds) Control applications in modern power system. lecture notes in electrical engineering, vol 710. Springer, Singapore. https://doi.org/10.1007/978-981-15-8815-0_4
7. Sahu RK, Panda S, Padhan S (2015) A novel hybrid gravitational search and pattern search algorithm for load frequency control of nonlinear power system. Appl Soft Comput 29:310–327

8. Sahu RK, Panda S, Padhan S (2015) A hybrid firefly algorithm and pattern search technique for automatic generation control of multi area power systems. Electr Power Energy Syst 64:9–23
9. Kalyan CHNS, Rao GS (2020) Performance comparison of various energy storage devices in combined LFC and AVR of multi area system with renewable energy integration. Int J Renew Energy Res 10(02):933–944
10. Tasnin W, Saikia LC (2018) Comparative performance of different energy storage devices in AGC of multi-source system including geothermal power plant. J Renew Sustain Energy 10:024101
11. Arya Y (2019) A new optimized fuzzy FOPI-FOPD controller for automatic generation control of electric power systems. J Franklin Inst 356:5611–5629
12. Raju M, Saikia LC, Sinha N (2016) Automatic generation control of a multi-area system using ant lion optimizer algorithm based PID plus second order derivative controller. Electr Power Energy Syst 80:52–63
13. Kalyan CHNS, Rao GS (2020) Frequency and voltage stabilization in combined load frequency control and automatic voltage regulation of multi area system with hybrid generation utilities by AC/DC links. Int J Sustain Energy 39(10):1009–1029. https://doi.org/10.1080/14786451.2020.1797740
14. Kalyan CHNS, Rao GS (2020) Coordinated SMES and TCSC damping controller for load frequency control of multi area power system with diverse sources. Int J Electr Eng Infor 12(04):747–769
15. Kalyan CHNS, Rao GS (2020) Combined frequency and voltage stabilization of multi-area multisource system by DE-AEFA optimized PID controller with coordinated performance of IPFC and RFBs. Int J Ambient Energy. https://doi.org/10.1080/01430750.2020.1860130
16. Dhiman G, Kumar V (2019) Seagull optimization algorithm: theory and its applications for large- scale industrial engineering problems. Knowl Based Syst 165:169–196

Chapter 48
Single and Multi-objective Optimal Generation Expansion Planning with HVDC Systems by Using HSDE Algorithm

Kumari Maddipati Veera and K. Vaisakh

1 Introduction

The optimal generation expansion planning (Optimal Generation Expansion Planning) problem plays a vital role in power system planning to meet the forecasted load demand. During OGEP, outage of existing generators, transformers and transmission lines need to be considered. For planning several constraints like construction constraints, reliability constraints, power flow constraints and security constraints are considered. In OGEP, the objective is to expand the existing generators to serve the growing load demand in the future by satisfying the reliability criteria.

1.1 Literature Survey

The OGEP determines size, place, technology and the time of installing new generating plants to meet the forecasted load demand. As OGEP is a highly constrained, nonlinear, discrete [1, 2] optimization problem, it is a highly challenging problem for the decision-makers. The solution for this GEP problem is obtained by complete enumeration of each possible combination [3–8] in the entire planning horizon. Since 1950, many optimization techniques such as linear programming, integer programming and dynamic programming (DP) are applied to solve these combinatorial optimization problems [9–13]. Habib et al. [14] analyzed utilization of renewable energy with domestic coal in place of import coal and oil in Bangladesh Power System from 2010 to 2030. Also, the modelling of long-term renewable and domestic coal utilization is examined by expansion cost and environmental impacts.

K. Maddipati Veera (✉)
EEE Department, Sir CRR College of Engineering, Eluru, Andhra Pradesh, India

K. Vaisakh
EEE Department, Vishakapatnam, Andhra Pradesh, India

S. Dawn et al. (eds.), *Smart and Intelligent Systems*, Algorithms for Intelligent Systems,
https://doi.org/10.1007/978-981-16-2109-3_49

Jadidoleslam and Ebrahimi [15] introduced a modified shuffled frog leaping algorithm (MSFLA) to solve the reliability constrained generation expansion planning (GEP) problem. To solve the GEP problem by MSFLA, a new frog leaping rule, associated with a new strategy for frog distribution into memeplexes are proposed to improve the local exploration and performance of the SFLA.

Sardou et al. [16] presents an application of genetic algorithm (GA) to TC-GEP problem for simultaneously determination of new generation site, capacity and fuel type for a multi-period generation expansion plan. Abdi et al. [17] reviewed the metaheuristic techniques applied for solving the problem of generation expansion planning, applications, description of algorithms and how they operate. Jadidoleslam et al. [18] formulated GEP problem with an optimization technique SFLA which determines the optimal investment plan for adding power plants over a planning horizon to meet the demand criteria, fuel mix ratio and the reliability criteria. Park et al. [19] defined that Least-cost GEP problem is a highly constrained nonlinear discrete dynamic optimization one that can be fully solved by complete enumerations, which leads to computational explosion in a real-world GEP problem. Kannan et al. [20] described the GEP with Power flow Constraints in deregulated environment. Murugan et al. [21, 22] applied NSGA-II to GEP as single objective and Multi-objective optimization. SADE, its Control parameters and Mutation Strategies are implemented to various problems [23–29]. VSC HVDC inclusion and its optimal power flow solution were discussed in [30–33] for IEEE standard systems.

Wang and Zhang [34] proposed multi-criteria decision-making methods based on Fuzzy Sets with Certain Information on Weights. MCDM techniques applied to various decision making problems [35–37]. Contingency assessment under uncertainty for voltage collapse and its application in risk-based contingency ranking were implemented for different problems [38–42].

In this paper, HSDE algorithm is implemented to solve the OGEP problem for two test systems like IEEE 30 bus test system and Indian 82 bus system. This paper is organized as follows: Sect. 2 describes the Problem formulation & Sect. 3 discusses about the methodology used in the paper; Sect. 4 gives the test systems and results obtained; Sect. 5 concludes the paper.

2 OGEP Problem Formulation

The main objective of OGEP is to minimize the total investment and operating costs associated with the addition of new units and to satisfy the reliability, fuel mix and the demand criterion. An innovative Multi-Criteria Decision-Making methods like AHP (Analytical Hierarchical Process), SAW(Simple additive Weighing) and WPM (Weighing Product Method) were implemented using the Weighted Cost, FLCC, Power loss, L index in all the three stages as the criteria.

OGEP is formulated as two single objectives

(i) Minimizing Total Cost
(ii) Minimizing FLCC

Multi-objective is

(iii) Minimizing Weighed Sum of Total Cost and FLCC
(iv) Outage Cost

2.1 *First Objective*

First objective is the total cost function (objective function) which is represented by the following expression:

$$\min CC = InvC(N_x) + MaintC(Y_x) + OutC(Y_x) - SalvC(N_x) \quad (1)$$

$$InvC(N_x) = (1 + D)^{-2s} \sum_{x=1}^{N} [CapInv_x N_{x,s}] \quad (2)$$

$$MaintC(Y_x) = \sum_{t'=0}^{l} (1 + D)^{-(1.5+s+s')} \left[\sum [(Y_s * FuelC) + OperC]\right] \quad (3)$$

$$OutC(Y_x) = OC * \sum_{t'=0}^{l} (1 + D)^{-(1.5+s+s')} \quad (4)$$

$$SalvC(N_x) = (1 + D)^{-T'} \sum_{x=1}^{N} [CapInv_x \delta_x N_{x,s}] \quad (5)$$

Upper Construction limit

If N_x be the units to be committed, in the expansion plan at stage 's', N_x must satisfy the maximum construction capacity of the units to be committed.

$$0 \le N_x \le N_{\max,x} \quad (6)$$

where, $N_{max,x}$ is the maximum construction capacity of the units in stage 's'

Reliability criterion

The selected units along with the existing units must satisfy the reliability criterion, Loss of Load Probability (LOLP).

Where,

$$LOLP\,(X_t) \leq \varepsilon \tag{7}$$

ε is the Reliability criterion expressed in LOLP.

Reserve margin

The selected units must satisfy the minimum and maximum reserve margin.

$$(1 + R_{\min}) * D_s \leq \sum_{x=1}^{N} Y_{x,s} \leq (1 + R_{\max}) * D_s \tag{8}$$

where,

R_{min}: Minimum reserve margin in MW
R_{max}: Maximum reserve margin in MW
D_s: Demand at stage 's', in MW
$Y_{x,s}$: Cumulative capacity of unit i at stage t in MW

Power flow constraints

The operation and maintenance cost is calculated using the OPF algorithm. The OPF (NR) algorithm basically conducts the economic dispatch by satisfying all the power flow constraints (MVA limits, voltage, etc.) to find the online generating capacity of each unit.

Second objective: Second objective function is FLCC (Fuzzy Logic Composite Criteria) which is a combination of line loading NLL, voltage profiles VP, square of voltage stability index VSI and reactive power indices QP as given in Eq. (9).

$$FLCC = NLL + VP + VSI + QP \tag{9}$$

2.2 *Multi-objective*

The Weighed sum of Total Cost and the FLCC objective was considered as third objective.

$$WCCFLCC = w1 * CC + w2 * FLCC \tag{10}$$

where CC is Total Cost objective and FLCC is Fuzzy logic Composite Criteria objective and $w1$, $w2$ are weights.

2.3 VSC-HVDC System

The VSC-HVDC system is represented by two shunt-connected voltage sources linked together by an active power constraint equation. Each voltage source is connected to the AC system by means of its transformer reactance. Suitable expressions for the three phase voltage sources and the linking power equation are:

$$E^{\rho}_{vR1} = V^{\rho}_{vR1}(\cos\delta^{\rho}_{vR1} + j\sin\delta^{\rho}_{vR1}) \tag{11}$$

$$E^{\rho}_{vR2} = V^{\rho}_{vR2}(\cos\delta^{\rho}_{vR2} + j\sin\delta^{\rho}_{vR2}) \tag{12}$$

$$\mathrm{Re}\{-E^{\rho}_{vR1}I^{*\rho}_{vR1} + E^{\rho}_{vR2}I^{*\rho}_{vRm}\} = 0 \tag{13}$$

where ρ indicates phase quantities, a, b and c.

3 Algorithm

The proposed HSDE algorithm for OGEP problem is as given below:

Step 1: Parameter setup and declaration of constants
The parameters of the algorithm are Dimension D, number of population (NP), for HSDE are chosen and declared.

Step 2: Reading data
System Bus data, generator data, transmission line data are declared.

Step 3: Initialization
Generate initial population of individuals with size (NP x D) randomly distributed across the domain of the problem and store them in archive X.

$$X = [x_1, x_2, \ldots x_i, \ldots x_{NP}]^T \tag{14}$$

where $X_{i,0} = [P_{G,0}\ V_{G,0}\,tap\ Q_{G,0}]_{1xN}$
Control parameters are randomly generated within the feasible range.

$$F_{i,G+1} = \begin{cases} Fl + rand_1 * F_u\, if(f(u_{i,G}) > f(X_{i,G}) \\ F_{i,G}\ otherwise \end{cases} \tag{15}$$

$$CR_{i,G+1} = \begin{cases} rand_2\ \ if(f(u_{i,G}) > f(X_{i,G}) \\ CR_{i,G}\ otherwise \end{cases} \tag{16}$$

Step 4: Run the Newton Raphson load flow with individuals of the population to check the feasibility of the solution and satisfy all the equality and inequality constraints, Calculate the optimum fuel cost which is considered in maintenance cost calculation of OGEP. Also calculate power flows, transmission losses, bus voltages, voltage stability index.

Step 5: *0000nnk [[[ppoo]Evaluate objective function values for each individual in the population*
Total Cost CC Objective evaluated
FLCC Objective evaluated
Weighed sum of CC and FLCC objective evaluated.

Step 6: Determine the fitness function fit = 1/ (Objective function +K*penalty function)

Step 7: Sort out the local best and global best solutions and best fitness of the population.

Step 8: Select five random individuals which are mutually different from each other and
Calculate Hybrid mutation operator.

$$V_{i,G} = \begin{cases} X_{r3,G} + F_{i,G}*(X_{r4,G} - X_{r5,G}) \\ if(f(X_{i,G}) > f(X_{r1,G})\, and\, f(X_{i,G}) > f(X_{r2,G}) \\ X_{i,G} + F_{i,G}*(X_{Best,G} - X_{i,G}) + F*(X_{r4,G} - X_{r5,G}),\ else \end{cases} \tag{17}$$

Step 9: Crossover

$$U_{ji,G} = \begin{cases} V_{ji,G}\, if(rand\,[0,\ 1] \leq CR)\, or\, (j = jrand) \\ X_{ji,G},\ otherwise \end{cases} \tag{18}$$

Selection

$$X_{i,G+1} = \begin{cases} U_{ji,G}\, iff(U_{ji,G}) \leq f\,(X_{i,G}) \\ X_{i,G}\, else \end{cases} \tag{19}$$

Step 10: Check for limit violations ($P_G V_G\, Tap\, Q_c$)

Step 11: Check for stopping criteria, if the iterations reach maximum, go to next step else go to step 4. Repeat the steps 5–10 again.

Step 12: Obtain the global best solution.

4 Results and Discussion

The best parameters of HSDE are given in Table 1. The test system results for three objectives are given in Tables 2, 3 and 4. To meet the forecasted load demand, OGEP is essential, either for expansion of existing units or addition of new generator units

Table 1 Optimal control parameters for three objectives under normal operation and network contingency in Stage 1 (Addition of one 40 MW generator at bus7 to meet the demand 341 MW) with HVDC

Control variables	Total cost objective		FLCC objective		Weighed cost objective	
	Normal	NtwConting	Normal	NtwConting	Normal	NtwConting
P_{G1}(MW)	199.08	199.08	143.73	105.73	174.2	127.95
P_{G2}(MW)	80	80	65.12	80	52.12	59.35
P_{G3}(MW)	10	10	35	31	20.73	24.69
P_{G4}(MW)	10	10	16.98	27.77	15.99	30
P_{G5}(MW)	30.76	35.67	47.13	47.75	24.55	39.28
P_{G6}(MW)	12	12	27.32	31.35	34.9	35.5
P_{G7}(MW)	10	10	13.57	24.8	30.41	34.44
V_1(p.u)	1.1	1.1	1.0578	1.0368	0.9926	1.0434
V_2(p.u)	1.0893	1.1	1.0171	1.0237	0.9701	1.0088
V_3(p.u)	1.0641	1.0651	0.991	1.0177	0.9736	0.9875
V_4(p.u)	1.1	1.1	1.0638	1.062	1.0379	1.0378
V_5(p.u)	1.0594	1.0695	0.95	0.9975	0.9317	0.9488
V_6(p.u)	1.1	1.1	1.0381	1.0382	1.0514	1.0531
V_7(p.u)	1.1	1.1	1.0186	1.0387	1.0602	1.0349
T1(p.u)	1.0634	1.0857	0.9781	1.0017	1.0486	1.026
T2(p.u)	0.9281	0.9	1.0178	1.1	1.0165	0.9872
T3(p.u)	0.9667	0.966	0.9628	1.0415	0.9628	0.9488
T4(p.u)	0.9573	0.9586	0.993	1.0447	1.0422	1.0327
Q_{c1}(p.u)	0.2	0.1568	0.0557	0.0588	0.1245	0.0578
Q_{c2}(p.u)	0.2	0.2	0.0419	0.0859	0.1295	0.1252
Q_{c3}(p.u)	0.0645	0.0594	0.0901	0.0336	0.1663	0.1957
Q_{c4}(p.u)	0.0758	0.076	0	0.0861	0.0662	0.0623
Q_{c5}(p.u)	0.0457	0.044	0.0236	0.0837	0.0907	0.1291
Q_{c6}(p.u)	0.1253	0.1176	0	0.0919	0.0171	0.1171
Q_{c7}(p.u)	0.0131	0.016	0.0753	0.0847	0.0872	0.1457
Q_{c8}(p.u)	0.0299	0.0271	0.0885	0.1049	0.1072	0.0975
Q_{c9}(p.u)	0	0.0099	0.1133	0.1027	0.034	0.1394
Cost($)	7.3266e + 9	7.4689e + 9	8.5028e + 9	9.2087e + 9	8.5561e + 9	9.3039e + 9
FLCC index	3.3934e + 3	3.6632e + 3	950.7511	982.0011	2.2551e + 3	2.5712e + 3
Loss	0.1176	0.1667	0.0878	0.0833	0.1283	0.1115
L-index	0.0834	0.0843	0.1072	0.1114	0.1247	0.126

(continued)

Table 1 (continued)

Control variables	Total cost objective		FLCC objective		Weighed cost objective	
	Normal	NtwConting	Normal	NtwConting	Normal	NtwConting
ENS	8.5525e + 4	8.5525e + 4	8.5525e + 4	8.5525e + 4	8.5525e + 4	8.5525e + 4
LOLP	0.005	0.005	0.005	0.005	0.005	0.005
VD	0.1985	0.206	0.0267	0.0287	0.0488	0.0833
Sum P_g	352.76	357.67	349.78	349.33	353.83	352.15
Vmin(pu)	1.05	1.05	0.9689	1.0023	0.9532	0.9689
Vmax(pu)	1.1118	1.1113	1.0538	1.0541	1.06	1.0863
No:of Iter	150	150	150	150	150	150
Time	544.532	481.386	455.822	400.562	536.766	429.906

without expansion of transmission network with the inclusion of VSC-HVDC in between the two buses (namely 9–10). This algorithm is implemented for modified IEEE 30 bus system with 6 years planning horizon. Figure 1 shows the IEEE 30 bus system with inclusion of HVDC. This algorithm is also implemented for Indian 82 bus system with 6 years planning horizon. Table 1 shows the Optimal control parameters for three objectives under normal operation and network contingency in Stage 1 (Addition of one 40 MW generator at bus7 to meet the demand 341 MW) with HVDC.

As shown in Tables 2, 3 and 4 the ranks are assigned to the alternatives in all the three stages using all the three methods of MCDM for IEEE 30 bus system and Indian 82 Bus system. Figures 1, 2, 3 and 4 shows the convergence characteristics of IEEE 30 bus system in different stages with the objective functions.

AHP, WPM SAW have matched for the normal case whereas top rank for alternative is same in all the three methods and three stages, however ranking order is slightly changed in network Contingency of Stage 1 unlike the remaining stages of ranking.

5 Conclusion

Though there are several optimization methods implemented (PSO, DE, SFLA, TS, ES, EA) for GEP, Hybrid optimization methods are not implemented in literature for OGEP. In this paper, the methodology of HSDE algorithm implementation for the GEP is illustrated. In each stage (2 years = 1stage), the three objectives, i.e., total cost (investment cost, maintenance cost, outage cost, salvage cost), FLCC objective and weighed sum of total cost and FLCC for both cases (normal operation, network contingency) are evaluated. The total cost in the three objectives is less in normal operation case when compared to the network contingency case. Comparing the real power loss in the Total cost objective, there is an increase in the network Contingency

Table 2 Comparison of ranking of 5 best run solutions by Multi-criteria Decision making methods in Stages 1,2 and 3 with HVDC

Stage	Run	Weighted cost	FLCC	Power loss	Lj index	AHP rank	SAW rank	WPM rank
I	*Normal operation*							
	1	8.5561e + 9	2.2551e + 3	0.1283	0.1247	1	1	1
	2	8.3538e + 9	1.9037e + 3	0.1059	0.1030	2	2	2
	3	8.3133e + 9	1.4255e + 3	0.0972	0.1112	4	4	4
	4	8.3822e + 9	1.5044e + 3	0.1273	0.1305	5	5	5
	5	8.3712e + 9	1.4667e + 3	0.1268	0.1167	3	3	3
	Network contingency							
	1	8.7893e + 9	1.9914e + 3	0.1171	0.1264	4	4	4
	2	9.3717e + 9	1.6053e + 3	0.1170	0.1266	3	3	3
	3	9.0839e + 9	1.9842e + 3	0.1351	0.1292	1	1	1
	4	9.3039e + 9	2.5712e + 3	0.1115	0.1260	5	5	5
	5	8.9364e + 9	1.8322e + 3	0.1205	0.1251	2	2	2
II	*Normal operation*							
	1	8.5425e + 9	1.5042e + 3	0.1355	0.1473	3	3	3
	2	8.3919e + 9	2.1190e + 3	0.1740	0.1550	2	2	2
	3	8.6725e + 9	2.6927e + 3	0.1652	0.1514	4	4	4
	4	8.4763e + 9	1.4792e + 3	0.1523	0.1549	5	5	5
	5	8.2252e + 9	1.3820e + 3	0.1483	0.1634	1	1	1
	Network contingency							
	1	9.3284e + 9	2.0502e + 3	0.1252	0.1432	2	2	2
	2	9.2780e + 9	2.8777e + 3	0.1253	0.1353	4	4	4
	3	9.3456e + 9	1.4195e + 3	0.1417	0.1571	1	1	1
	4	9.2171e + 9	2.5219e + 3	0.1103	0.1408	3	3	3
	5	9.2951e + 9	1.4350e + 3	0.1285	0.1608	5	5	5
III	*Normal operation*							
	1	8.3198e + 9	1.7265e + 3	0.1707	0.1968	5	5	2
	2	8.2971e + 9	2.4607e + 3	0.1636	0.1888	2	2	5
	3	8.3250e + 9	1.4945e + 3	0.1723	0.2274	3	1	1
	4	8.2987e + 9	1.4945e + 3	0.1633	0.1969	1	3	3
	5	8.3941e + 9	2.2426e + 3	0.1790	0.1944	4	4	4
	Network contingency							
	1	9.2482e + 9	2.1379e + 3	0.2002	0.2012	2	2	2
	2	9.2710e + 9	2.0682e + 3	0.2016	0.2101	3	1	3

(continued)

Table 2 (continued)

Stage	Run	Weighted cost	FLCC	Power loss	Lj index	AHP rank	SAW rank	WPM rank
	3	9.2685e + 9	2.0682e + 3	0.2014	0.2059	5	5	5
	4	9.2685e + 9	2.0682e + 3	0.2014	0.2021	1	3	1
	5	9.2683e + 9	2.0857e + 3	0.2014	0.2031	4	4	4

Table 3 Stage wise comparison of cost and power loss for three objectives of IEEE 30-bus system with HVDC

Stage	Objective	% load	Normal cost($)	Network contingency cost($)	Power loss	
					Normal case	NWCong case
I	CC	120	7.3266e + 9	7.4689e + 9	0.1176	0.1667
	FLCC		8.5028e + 9	9.2087e + 9	0.0878	0.0833
	CC + FLCC		8.5561e + 9	9.3039e + 9	0.1283	0.1115
II	CC	140	7.8143e + 9	7.9536e + 9	0.1213	0.1785
	FLCC		8.4886e + 9	9.1860e + 9	0.1279	0.1199
	CC + FLCC		8.6725e + 9	9.2780e + 9	0.1652	0.1253
III	CC	160	8.1627e + 9	8.2878e + 9	0.1393	0.1821
	FLCC		8.2288e + 9	9.2479e + 9	0.1440	0.2148
	CC + FLCC		8.3941e + 9	9.2710e + 9	0.1790	0.2016

Table 4 Stage wise comparison of cost and power loss for three objectives of 82-bus system with HVDC

Stage	Objective	% load	Normal cost($)	Network contingency cost($)	Power loss	
					Normal case	NWCong case
I	CC	120	2.4707e + 12	2.4920e + 12	1.8414	1.9126
	FLCC		3.1479e + 12	2.7666e + 12	1.6843	1.9164
	CC + FLCC		2.8453e + 12	2.8702e + 12	1.8414	1.9126
II	CC	140	1.9632e + 12	1.9746e + 12	1.5309	1.6250
	FLCC		2.4080e + 12	2.6873e + 12	1.7703	1.9213
	CC + FLCC		2.4977e + 12	2.5527e + 12	1.6587	1.8231
III	CC	160	1.6535e + 12	1.6522e + 12	1.6276	1.6185
	FLCC		2.3398e + 12	2.0954e + 12	1.7221	1.8039
	CC + FLCC		2.2508e + 12	2.0842e + 12	1.7047	1.7320

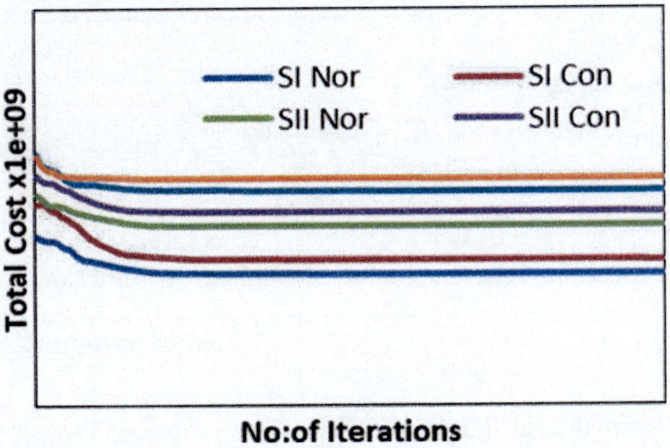

Fig. 1 Convergence characteristics of total cost in three stages during minimization of CC objective with HVDC

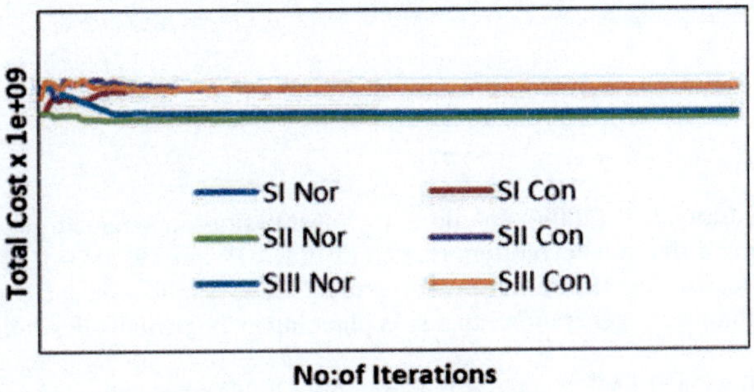

Fig. 2 Convergence characteristics of Total Cost in three stages during minimization of CC&FLCC objective with HVDC

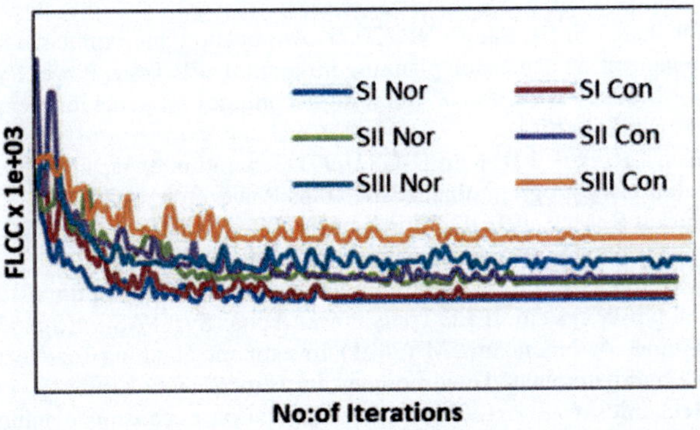

Fig. 3 Convergence characteristics of FLCC value in three stages during minimization of FLCC objective with HVDC

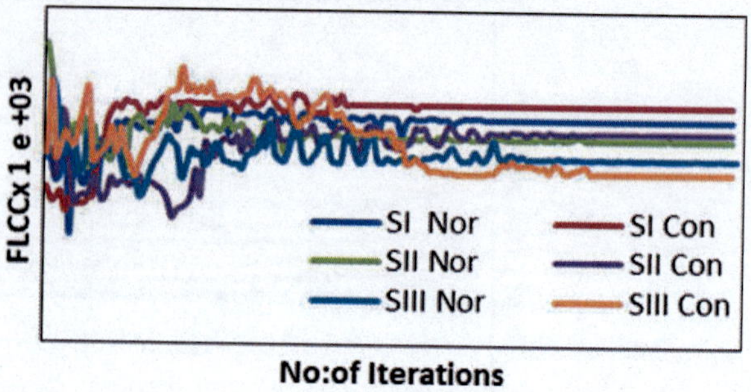

Fig. 4 Convergence characteristics of FLCC value in three stages during minimization of FLCC objective with HVDC

Case than the normal case. The feasibility of the proposed approach was tested on Modified IEEE 30 bus system with three different objective functions. Several cases were investigated to test and validate the robustness of the proposed method in finding the optimal solution.

References

1. Kannan S, Murugan P (2009) Solutions to transmission constrained generation expansion planning using differential evolution. Eur Trans Electr Power 19:1033–1039
2. Kannan S, Slochanal SMR, Baskar S, Murugan P (2006) Application and comparison of metaheuristic techniques to generation expansion planning in the partially deregulated environment. IJEPS
3. Kannan S, Slochanal SMR Subbaraj P, Padhy NP (2004) Application of particle swarm optimization technique and its variants to generation expansion planning problem. Electric Power Syst Res 70:203–210
4. Park J-B, Park Y-M, Won JR, Lee KY (2000) An improved genetic algorithm for generation expansion planning. IEEE Trans Power Syst 15(3)
5. Kannan S, Slochanal SMR, Padhy NP (2005) Application and comparison of metaheuristic techniques to generation expansion planning problem. IEEE Trans Power Syst 20(1):466–475
6. Zhu J, Chow M (1997) A review of emerging techniques on generation expansion planning. IEEE Trans Power Syst 12(4)
7. Park Y-M, Won J-R, Park J-B, Kim D-G (1999) Generation expansion planning based on an advanced evolutionary programming. IEEE Trans Power Syst 14(1)
8. Kannan S, Baskar S, McCalley JD, Murugan P (2009) Application of NSGA-II algorithm to generation expansion planning. IEEE Trans Power Syst 24(1):454–462
9. Beglari F, Laughton MA (1975) The combined costs method for optimal economic planning of an electrical power system. IEEE Trans Power Appar Syst PAS-94(6):1935–1942
10. Kenfack F, Guinet A, Ngundam JM (2001) Investment planning for electricity generation expansion in a hydro dominated environment. Int J Energy Res 25:927–937
11. Sirikum J, Techanitisawad A (2005) Power generation expansion planning with emission control: a nonlinear model and a GA-based heuristic approach. Int J Energy Res Wiley Inter Sci 81–99
12. Sirikum J, Techanitisawad A, Kachitvichyanukul V, A new efficient GA-benders' decomposition method for power generation expansion planning with emission controls. IEEE Trans Power Syst 22(3):1092–1100
13. Park JB, Won JR (1998) A hybrid genetic algorithm/dynamic programming approach to optimal long-term generation expansion planning. Electr Power Energy Syst 20(4):295–303

14. Habib MA, Chungpaibulpatana S (2014) Electricity generation expansion planning with environmental impact abatement: case study of Bangladesh, international conference on alternative energy in developing countries and emerging economies, Energy Proced 52:410–420
15. Jadidoleslam M, Ebrahimi A (2015) Reliability constrained generation expansion planning by a modified shuffled frog leaping algorithm. Electr Power Energy Syst 64:743–751
16. Sardou IG, Ameli MT, Sepasian MS, Ahmadian M (2014) A novel genetic-based optimization for transmission constrained generation expansion planning. Int J Intell Syst Appl 01:73–83
17. Abdi H, Ramzanpour M, Abednejad B, Moghadam S, Habibi Y (2013) Application of Metaheuristic techniques to generation expansion planning problem. Tech J Eng Appl Sci 3(18):2252–2257
18. Jadidoleslam M, Bijami E, Amiri N, Ebrahimi A, Askari J (2012) Application of shuffled frog leaping algorithm to long term generation expansion planning. Int J Comput Electr Eng 4(2)
19. Kim J-H, Park J-B, Parkand J-K, Joo S-K (2005) A market-based analysis on the generation expansion planning strategies. In: 13th International conference on intelligent systems application to power systems
20. Kannan S, Jawahar S, Slochanal SMR, Padhy NP (2003) Generation expansion planning with power flow constraints evolutionary strategy approach. Water Energy Int J 60(3):37–44. Central Board of Irrigation and Power, New Delhi, India
21. Murugan P, Kannan S, Baskar S (2009) Application of NSGA-II algorithm to single-objective transmission constrained generation expansion planning. IEEE Trans Power Syst 24(4):1790–1798
22. Murugan P, Kannan S, Baskar S (2009) NSGA-II algorithm for multi-objective generation expansion planning problem. Electr Power Syst Res 79(4):622–628
23. Zheng F, Zecchin AC, Simpson AR (2013) Self–adaptive differential evolution algorithm applied to water distribution system optimization. J Comput Civil Eng
24. Karthikeyan K, Kannan S, Bhaskar S, Thangaraj C (2013) Application of self adaptive differential evolution algorithm to generation expansion planning problem. J Electr Syst 9–2:203–211
25. Yan X (2015) Self–adaptive differential evolution algorithm with zoning evolution of control parameters and adaptive mutation strategies. IEEE Trans Cyber
26. Maucec MS (2011) Self–adaptive differential evolution algorithm using population size reduction and three strategies. J Soft Comput
27. Yan X (2015) Self –adaptive differential evolution algorithm with discrete mutation control parameters. J Expert Syst Appl
28. Sathish kumar M, Nirmal Kumar A, Lakshminarasimman L, Thiruvenkadam S (2012) A self adaptive hybrid differential evolution algorithm for phase balancing of unbalanced distribution system. Electr Power Energy Syst 42:91–97
29. Yi W, Gao L, Li X, Zhou Y (2014) A new differential evolution algorithm with a hybrid mutation operator and self- adapting control parameters for global optimization problems. Springer Appl Intell J
30. Wei Y, He Q, Sun Y, Sun Y, Ji C (2013) Improved power flow algorithm for VSC-HVDC system based on high-order newton-type method. Mathemat Probl Eng 235316
31. Wiget R, Andersson G (2012) Optimal power flow for combined AC and multi-terminal HVDC grids based onVSC converters. IEEE Trans High Voltage Eng
32. Feng W, Tuan L, Tjernberg L, Mannikoff A, Bergman A (2014) A new approach for benefit evaluation of multiterminal VSC–HVDC using a proposed mixed AC/DC optimal power flow. IEEE Trans Power Delivery 29(1)
33. Karthik M, Arul P (2013) Optimal power flow control using FACTS devices. Int J Emer Sci Eng 1(12):2319–6378
34. Wang J-Q, Zhang H-Y (2013) Multicriteria decision-making approach based on atanassov's intuitionistic fuzzy sets with incomplete certain information on weights. IEEE Trans Fuzzy Syst 21(3)
35. Berrie S, Kockar I, Galloway S (2011) Multi criteria decision analysis as a tool in planning and development of future power systems infrastructure. In: UPEC, 46th international Universities' power engineering conference, 5–8th September 2011

36. Jalao ER, Wu T, Shunk D (2014) A stochastic AHP decision making methodology for imprecise preferences. Inform Sci 270:192–203
37. Demirel T, Tüzün S (2011) Multi criteria evaluation of the methods for preventing soil erosion using fuzzy ANP: the case of Turkey. In: Proceedings of the world congress on engineering. vol II, WCE 2011, July 6–8, London, U.K
38. Abdelaziz AY, Taha ATM, Mostafa MA, Hassan AM (2013) Fuzzy logic based power system contingency ranking. Int J Intell Syst Appl 03:1–12
39. Matarucco RR, BoniniNeto A, Alves DA (2014) Assessment of branch outage contingencies using the continuation method. Electr Power Energy Syst 55:74–81
40. Verma K, Niazi KR (2012) Supervised learning approach to online contingency screening and ranking in power systems. Electr Power Energy Syst 38:97–104
41. Kamarposhti MA, Lesani H (2010) Contingencies ranking for voltage stability analysis using continuation power flow method. J Electr Electron Eng 3(99)
42. Krishnan V, McCalley JD (2012) Contingency assessment under uncertainty for voltage collapse and its application in risk based contingency ranking. Electr Power Energy Syst 43:1025–1033

Chapter 49
Design and Implementation of Thirty One Level Multilevel Inverter

K. Aravinda Shilpa, M. Revathi, and K. Rama Sudha

1 Introduction

The idea of multilevel inverter (MLI) is to give many levels of output voltages with less power switching loss and less harmonic distortion. They are being utilized generally in sustainable energy sources, engine drives applications, adaptable AC transmission frameworks and power quality improvement gadgets. The main features of multilevel inverters are:

1. The output voltage can generate with low distortion and dv/dt stress.
2. low distortion of input current is drawn.
3. The common mode voltage is very small.
4. It operates low switching frequency.

The MLIs include DC voltage sources, power semiconductors, control units and a load. The input to MLI a DC voltage and the output obtained is AC stepped voltage waveform. Cascaded H-Bridge (CHB) MLI is most common when compared with the other two which are Diode clamped MLI and Flying capacitor MLI because the output voltage has low harmonics.

By placing the power semiconductor switches and the DC sources in the right place one can obtain the required number of output voltage levels. When the number of levels at the output is increased one can obtain nearly a sinusoidal voltage. This increase causes a decrease in the total harmonic content (THD) of the output voltage waveform. If output levels increase the number of DC sources and switches also increase. This results in complexity of the circuit thereby increasing the switching losses. The reduction in the number of switches is significant and leads to easy controlling, less power consumption, smaller volume, simple circuit and cost reduction.

K. A. Shilpa (✉) · M. Revathi · K. R. Sudha
Department of Electrical and Electronics Engineering, Andhra University College of Engineering, Visakhapatnam, India

S. Dawn et al. (eds.), *Smart and Intelligent Systems*, Algorithms for Intelligent Systems,
https://doi.org/10.1007/978-981-16-2109-3_50

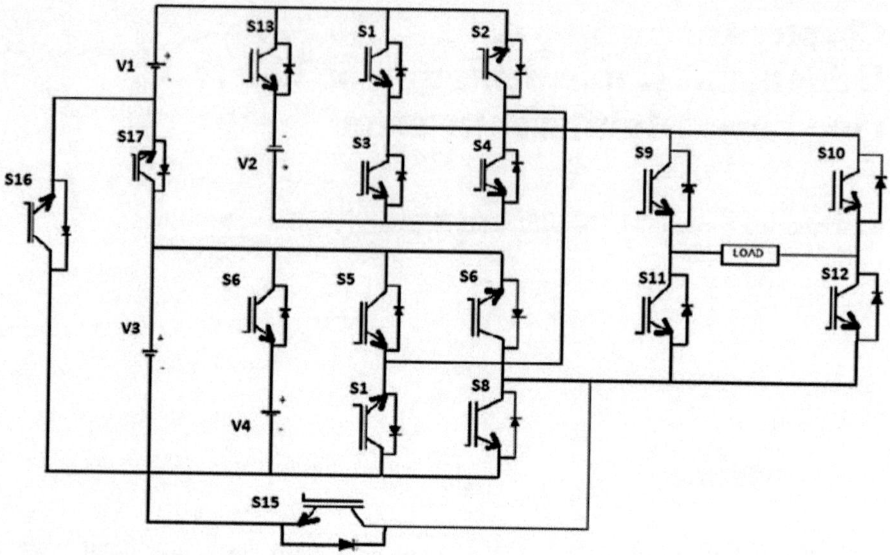

Fig. 1 Prososed topology

The topology is proposed with four DC voltage sources, seventeen power switches and three cascaded H-bridges to get thirty-one level as output level. The power switches used are unidirectional. The high quality of output waveforms with reduced switching devices is achieved. The THD is reduced Simulation Results on MATLAB/SIMULINK have been carried out.

2 Proposed Topology

The proposed topology is for implementing the thirty-one level asymmetric inverter model was shown in Fig. 1. It has three H-bridges connected in cascaded topologies. These connections are powered by four DC sources and seventeen switches.

3 Switching Table

The switching conditions for 31 level MLI is as follows:

V1 = Vdc, V2 = 2Vdc, V3 = 3Vdc, V4 = 4Vdc; Switches position: 1 = ON, 0 = OFF.

Using suitable switching combinations the desired output level of the 31 level MLI is obtained. In these 31 levels, 15 are positive DC levels and 15 are negative DC levels and one zero level.

Table 1 Switching table consdered

O/P Voltage	S1	S2	S3	S4	S5	S6	S7	S8	S9	S10	S11	S12	S13	S14	S15	S16	S17
V_{dc}	1	0	0	0	0	0	0	1	1	0	0	1	0	0	0	1	0
2 V_{dc}	0	1	1	0	1	1	0	0	1	0	0	1	1	0	0	0	0
3 V_{dc}	0	0	1	0	0	0	0	1	1	0	0	1	1	0	0	1	0
4 V_{dc}	0	0	1	1	1	0	0	0	1	0	0	1	0	0	1	0	0
5 V_{dc}	1	0	0	0	0	0	0	0	1	0	0	1	0	0	1	0	1
6 V_{dc}	0	1	1	0	1	0	0	0	1	0	0	1	1	0	1	0	0
7 V_{dc}	0	0	1	0	0	0	0	0	1	0	0	1	1	0	1	0	1
8 V_{dc}	0	0	1	1	0	1	1	0	1	0	0	1	0	1	0	0	0
9 V_{dc}	1	0	0	0	0	1	0	0	1	0	0	1	0	1	0	1	0
10 V_{dc}	0	1	1	0	0	1	1	0	1	0	0	1	1	1	0	0	0
11 V_{dc}	0	0	1	0	0	1	0	0	1	0	0	1	1	1	0	1	0
12 V_{dc}	0	0	1	1	0	0	1	0	1	0	0	1	0	1	1	0	0
13 V_{dc}	1	0	0	0	0	0	0	0	1	0	0	1	0	1	1	1	0
14 Vdc	0	1	1	0	0	0	1	0	1	0	0	1	1	1	1	0	0
15 Vdc	0	0	1	0	0	0	0	0	1	0	0	1	1	1	1	1	0
0	0	0	0	0	0	0	0	0	1	1	1	1	0	0	0	0	0

In this circuit, it contains three cascaded H-bridges where the first bridge consists of the switches S1, S2, S3, S4 and the second bridge consists of S5, S6, S7, S8, and the third bridge consists of S9, S10, S11, S12 and all the switches on each bridge forms an H-bridge. All the power switches are unidirectional which clearly states that the current passes through every switch in only one direction. The semiconductor device which is used in this circuit that acts as a switch is IGBT. Gate pulses are given to the switches according to the switching table given below (Table 1).

4 Mathematical Formulations

The number of inverter levels that can be obtained for this topology is as follows: Number of inverter levels (N) = 2^(K + 1) + 1.

Number of switches = 4K + 1.

where K = number of voltage sources.

4.1 Voltage Stress

Voltage stress can be defined as the ratio of change in voltage levels to the change in time interval.

It is represented by dv/dt.

Voltage stress of switch Sj = p*2f*N Where,

p = number of switching states of Sj. f = input frequency.

N = number of inverter levels.

4.2 Conduction Losses

The losses that occur during conduction are known as conduction losses Pc = $\sum$I2j R.

where

I = current flowing through switch Sj and R = internal resistance of the switch Sj.

5 Pulse Width Modulation

Multicarrier PULSE Width Modulation is carried out to turn on the switches, carrier signals are compared with the reference signal and the pulses so obtained are used for switching the devices corresponding to the respective voltage levels. The Fig. 2

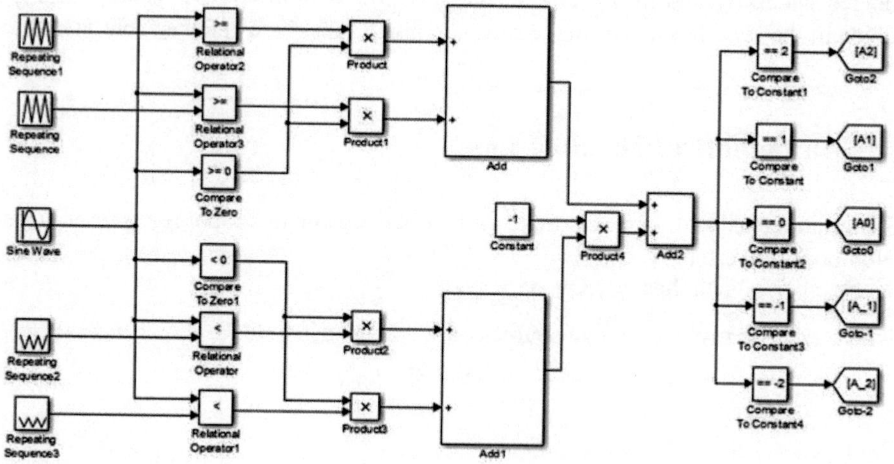

Fig. 2 Control scheme

shows modulation scheme of this MLI when carrier is less than reference signal, the comparator gives '1'. when carrier is more than reference signal, the comparator gives −1, if carrier is equal to reference, the comparator gives '0'. The signals so obtained are synthesized to get an aggregated signal'. This signal has same number of levels as desired in the output waveform. The modulation process employed produces pulses that are sent to every switch. This enables sequential switching. For this topology, a single sinusoidal reference waveform and thirty carrier waveforms which are triangular are used (Figs. 3 and 4).

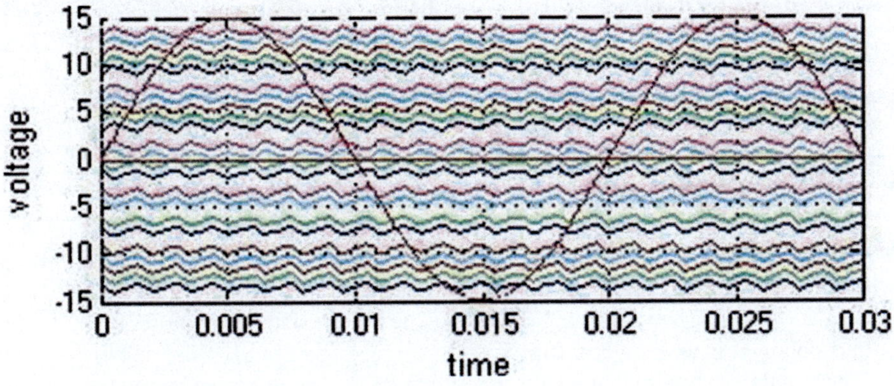

Fig. 3 Reference versus carrier

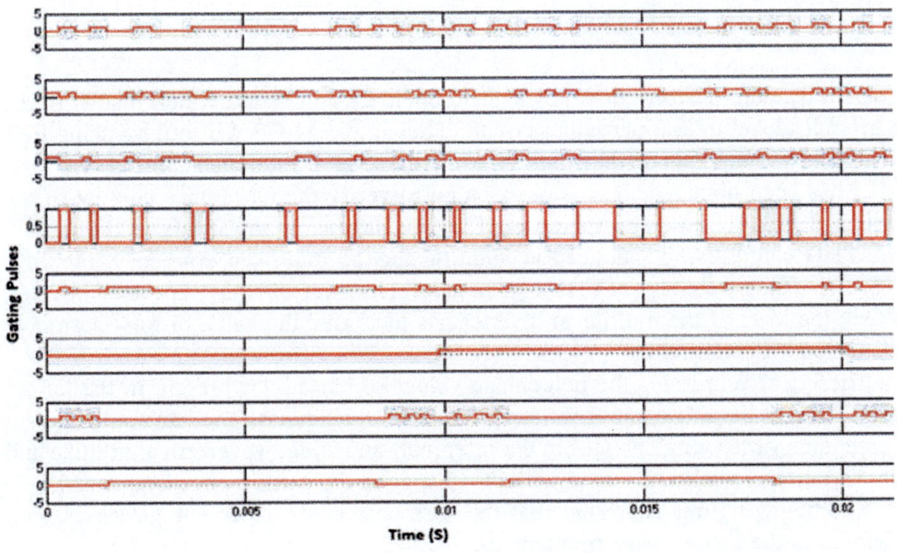

Fig. 4 Gating pulses to switches

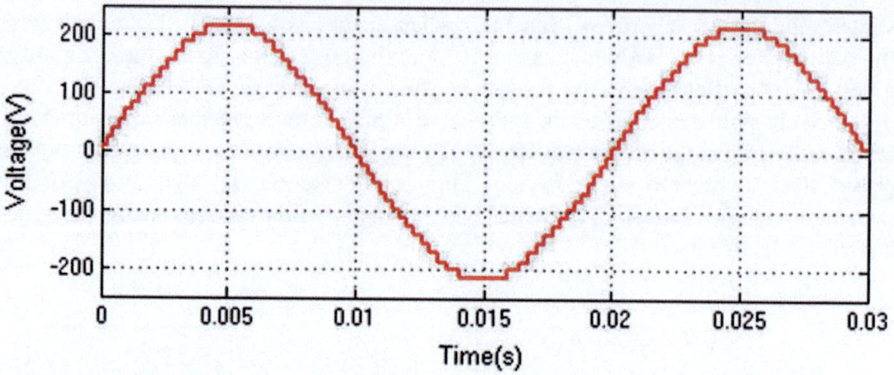

Fig. 5 Output waveform across load

6 Simulation Output

When the switches in the proposed inverter are turned ON at their respective timing with the input to the multilevel inverter to get 230 V AC voltage waveform, V1 = 20 V, V2 = 40 V, V3 = 80 V, V4 = 160 V. The load resistance should be 100 Ω the output obtained is as shown in Fig. 5.

The simulation model that results in a symmetric thirty one level output waveform.

7 Total Harmonic Distortion

Total Harmonic Distortion of any system could be calculated which results in the harmonics level in that particular circuit. Thus in the MATLAB tool by using FFT analysis percentage of THD in this simulation model of symmetric thirty one level MLI has been obtained. It is very essential to verify the harmonics content of the circuit to find its efficient operation. Thus by increasing the levels of the output, the THD could also be reduced accounting for the improved efficiency. Figure 6 represents the percentage of THD of a three level inverter and thirty one level inverter. When the level of the multilevel inverter has increased the value of total harmonic distortion is decreased.

The Fig. 7 represents the percentage values of harmonics present in the output wave form at the DC component and at various frequencies. The harmonic content percentage varies with changes in the reference sinusoidal waveform amplitude and the variations are also observed with the change in the carrier wave frequency.

The below figures represent that the harmonic content percentage varies with changes in the carrier wave frequencies.

Comparisons of the harmonic profiles obtained using carrier waves of frequencies 100 and 200 Hz show a considerable difference in their harmonics (Figs. 8 and 9).

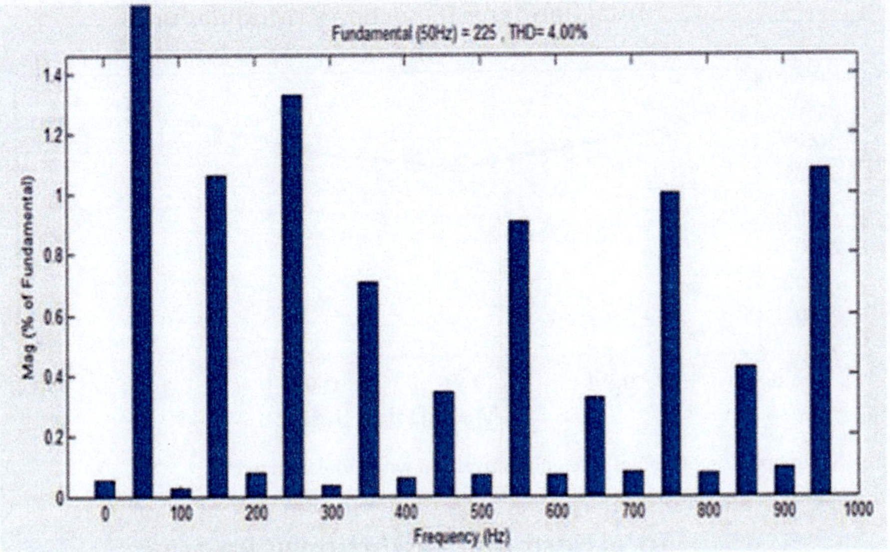

Fig. 6 THD Analysis of 31 level multi level inverter

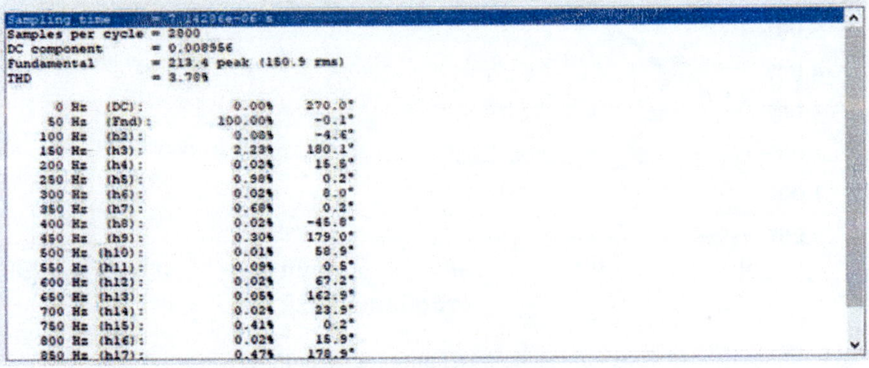

```
Sampling time      = 7.14286e-06 s
Samples per cycle  = 2800
DC component       = 0.008556
Fundamental        = 213.4 peak (150.9 rms)
THD                = 3.78%

    0 Hz   (DC):      0.00%   270.0°
   50 Hz  (Fnd):    100.00%    -0.1°
  100 Hz   (h2):      0.08%    -4.6°
  150 Hz   (h3):      2.23%   180.1°
  200 Hz   (h4):      0.02%    15.6°
  250 Hz   (h5):      0.98%     0.2°
  300 Hz   (h6):      0.02%     8.0°
  350 Hz   (h7):      0.68%     0.2°
  400 Hz   (h8):      0.02%   -45.8°
  450 Hz   (h9):      0.30%   179.0°
  500 Hz  (h10):      0.01%    58.9°
  550 Hz  (h11):      0.09%     0.5°
  600 Hz  (h12):      0.02%    67.2°
  650 Hz  (h13):      0.05%   162.9°
  700 Hz  (h14):      0.02%    23.9°
  750 Hz  (h15):      0.41%     0.2°
  800 Hz  (h16):      0.02%    15.9°
  850 Hz  (h17):      0.47%   178.9°
```

Fig. 7 Percentage value of the harmonics present in the output

8 Conclusion

A novel topology of MLI with reduced device count and harmonic reduction is proposed in this paper. An overview of multicarrier pulse width modulation is also given here. The device count is significant when compared with other topologies with same level. The proposed concept is investigated through simulations. The design and implementation of cascaded thirty one level inverter is done and simulation output has been verified in MATLAB/SIMULINK and THD obtained 4%.Analysis of THD wrt to modulation index and frequency.

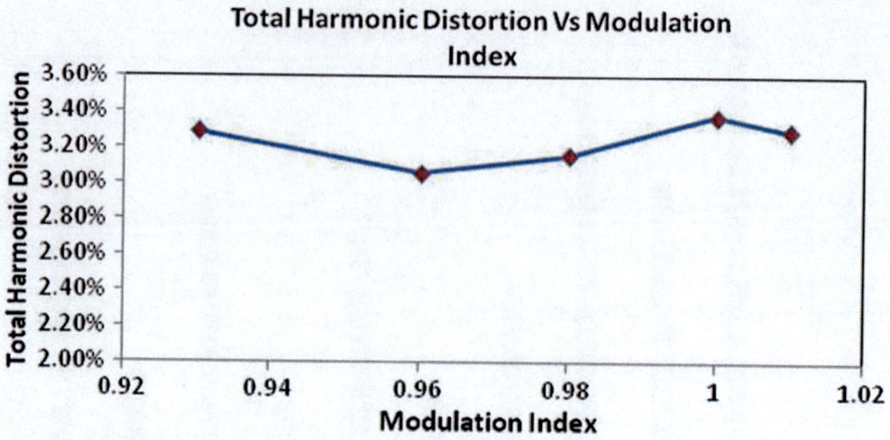

Fig. 8 Comparison of Thd of triangular waveform wrt modulation index

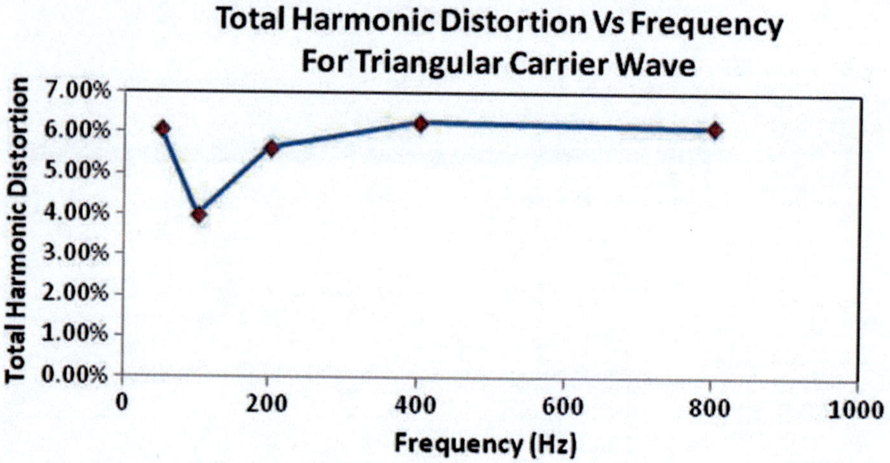

Fig. 9 Comparison of Thd of triangular waveform at different frequencies

References

1. Ozdemir S, Tolbert LM (2009) Fundamental frequency modulated six-level diode-clamped multilevel inverter for three-phase stand-alone photo voltaic system. IEEETrans Ind Electr 56(11):4407–4415
2. Klabunde M, Zhao Y, Lipo TA (1994) Current control of a 3 level rectifier/inverter drive system. In: Conference record 1994 IEEEIAS annual meeting, pp 2348–2356
3. Menzies W, Steimer P, Steinke JK (1994) Five-level GTO inverters motor drives. IEEE Trans Industry Appl 30(4):938–944
4. Sinha G, Lipo TA (1996) A four level rectifier-inverter system for drive application. In: Record IEEE IAS annual meeting, pp 980–987

5. Steinke JK (1988) Control strategy for a three phase AC traction drive with three level IGTO PWM inverter. In: Power electronic specialist—conference(PESC), pp 431–438
6. Zhang J (1995) High performance control of a three level IGBT inverter fed AC drive. In: Conferences Record IEEEIAS annual meeting, p. 22
7. Monopoli (1990) Design of H-bridge multilevel active rectifier for traction systems." IEEE Trans Ind Appl 39(5):1541–1550
8. Villanueva E, Correa P, Rodriguez J (2008) Control of a single phase H-bridge multilevel inverter for grid- connected PV applications. In: Proceedings power electronics motion control conferences, pp 451–455
9. Gonzalez SA, Verne SA, Valla MI (2013) Multilevel converters for industrial applications. CRC Press
10. Marquardt R, Lesnicar A (2003) A new modular voltage source inverter topology. Eur Power Electron Conf 1–10
11. Liu Y, Hong H, Huang AQ (2009) Real-time calculation of switching angles minimising THD for multilevel inverters with steep modulation. IEEE Trans Ind Electron 56(2):285–293
12. Jaqmal A, Weindl C, Herold G (2005) An optimized snubber design for three level inverter systems. EPE2005 Dresden Ger., pp 683–688
13. Al-Nasseir J, Weindl C, Heroid G, Flotthmesch J (2006) A dual-use snubber design for multi-level inverter systems. In: 12th International power electron, motion control Conferences, pp 638–688
14. Ikeda Y, Itsumi J, Funato H (1988) The power loss of the PWM voltage-fed inverter. In: Power electronic specialist conference 1988. PESC '88 Record, 19th Annual IEEE, vol 1, pp 277–283

Chapter 50
A Comparative Study of LCLC Type PV_DSTATCOM Using Improved ALST Control Algorithms

Soumya Mishra, R. Sreejith, M. Pavan Kalyan, M. L. Spoorthi, and J. G. Hemalatha

1 Introduction

Using clean, economical energy resources as an alternative to fossil fuels has been an essential need for the most recent decade due to the shortage of resources. Over this period, 70% of the total energy is supplied by petroleum products. With the increased energy consumption, the necessity of external green energy sources, and their integration into the grid has increased. Out of all green energies, the solar photovoltaic system is widely accepted for both home and commercial uses. It is clean, inexhaustible, needs less maintenance as no moving parts, modular type for mass production, and conversion efficiency is size-independent [1, 2].

Grid connection of PV system is advised for better infliction of energy and power reaped. The liquidation of batteries' usage has made the system cost-effective and with less maintenance [3]. The solar photovoltaic (PV) cells have a non-linear relationship between voltage and current and only give maximum power for one set of values. The particular point where maximum power is measured is called Maximum Power Point (MPP) [4]. As a result of non-linearity, an arrangement must track the MPP; hence the Maximum Power Point Tracking (MPPT) is used. Various MPP algorithms are implemented to track maximum power [5–9], out of which incremental conductance is implemented in this paper as the complexity is medium and does not need periodic tuning.

An inverter is the main component of a grid-connected PV system. The three main Inverter topologies are String or multi-string, Module integrated microinverter and Central inverter. The central inverter is preferred for an extensive generation because of the inverter which is centralized and coarse MPPT for PV array. A string inverter is a cutdown kind of the Central inverter topology, all the modules are connected in series like a string with every string having its own MPPT, a multi-string is an

S. Mishra (✉) · R. Sreejith · M. P. Kalyan · M. L. Spoorthi · J. G. Hemalatha
MVJ College of Engineering, Bangalore, Karnataka, India
e-mail: soumya.mishra@mvjce.edu.in

S. Dawn et al. (eds.), *Smart and Intelligent Systems*, Algorithms for Intelligent Systems,
https://doi.org/10.1007/978-981-16-2109-3_51

improvised version of a string inverter for a more extensive system, and these have their own power electronic chopper. In Module integrated micro converter, each PV module has its sole purpose MPPT and an inverter [10].

These power electronic (PE) converters along with other domestic PE-based loads, inject harmonics into the power system. Since inverter is an expensive and low efficient device it is used as a power quality conditioner to govern the current quality problems like unbalanced load current, reactive power compensation, harmonics and others and at the same time to fulfill the load power requirement by transferring active power from solar system to load. When the solar inverters operate in dual mode (i.e., as power quality conditioner and a device to transfer power from solar system to load), it is called as photovoltaic fed distributed static compensator (PV DSTATCOM) [11–13].

Many control algorithms are implemented to control reference current generation, dc link capacitor voltage and generation of switching pulse [14–17]. With a suitable control algorithm, PV DSTATCOM not only accommodates the purport of active power filtering but additionally provides voltage control and load balancing. Very commonly used control algorithms for PV DSTATCOM are Icos∅ algorithm, Synchronous Reference Frame (SRF) theory and P-Q theory or Instantaneous Reactive Power Theory (IRPT). The performance of IRPT is not satisfactory in the presence of unbalance and voltage distortion and it requires a large filter inductor which increases the system cost [18]. To address the issue with IRP theory, SRFT was developed [19] but the presence of phase-locked loop (PLL) causes the design of control algorithm complex and time-consuming. The Icos∅ algorithm is simple, easy to design and less time-consuming to tune the controller as PLL is absent. It performs better as compared to IRPT and SRF in the presence of voltage disturbance and unbalances. But still, it suffers from the issues that all linear controllers face due to the inherent nonlinearity of the power system network. A modified adaptive linear sinusoidal tracer algorithm (ALST) is proposed in [20] which performs well for adverse conditions like voltage distortion and unbalances.

In the presented work, the improved ALST algorithm is implemented to LCLC type PV DSTATCOM. The performance proposed method is compared with SRFT and Icos∅ algorithm. Section 2 characterizes the system considered for analysis and simulation. The SRFT, Icos∅ and improved ALST algorithm are explained in Sect. 3. The complete system is simulated in MATLAB/Simulink by using Simscape Toolbox and a comparative analysis of the controllers is presented in Sect. 4 with the avail of obtained results. Finally, Sect. 5 concludes the presented work.

2 System Configuration

This system comprises of a solar PV string comprising of series and parallel connections for required output power rating, an interfacing inductor, RC filter and a Voltage Source Converter (VSC). The PV module is affiliated to the inverter/VSC which not only acts as a converter but also performs the role of a compensator for mitigation

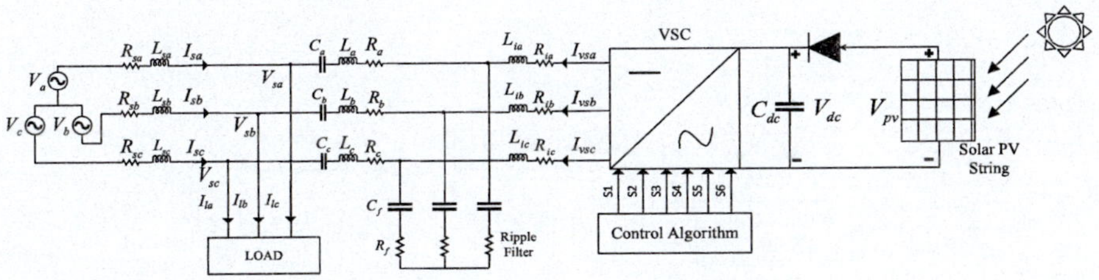

Fig. 1 Network of the proposed model

of potency quality quandaries and as MPPT controller, which then interfaced to the 3-Ø grid system connected with various linear and non-linear load. The network of the initiated system or model is demonstrated out in Fig. 1.

3 Control Algorithm

For power quality amendment and harmonic mitigation, DSTATCOM generates a total power component as required by the load making a source current to be in unity power factor at all the instances of time viewing from the source side.it is clear that the source provides only real power thus the load equilibration is done by making the reference source currents balanced which helps in deciding the switching of the controller of the DSTATCOM network, the fundamental frequency real component is extracted from the load current utilizing sundry algorithms [10].

3.1 IcosØ Algorithm

I_{la}, I_{lb}, I_{lc} are the load currents, bus DC voltage (V_{dc}), supply currents, V_{sa},V_{sb},V_{sc} are the PCC (point of common coupling) voltages which are also the feedback signal sensed by the control methods of DSTATCOM. For phase 'a', the amplitude value of the load current active power component (I_{lpa}) is unearthed from the load current fundamental component (I_{lfa}) (attained after filtering) at the zero crossing of the real unit templates ($\cos\phi_{pa} = U_{sap}$) of the 3-phase Point of Common Coupling voltages.

Figure 2 displays the block diagram for Icos ϕ,the algorithm for generating the supply reference currents.

'sample and hold' and zero-crossing detector circuit is utilized to extract the I_{lpa} [6]. Likewise, component of active power of currents of phase b and c (I_{lpb},I_{lpc}) extracted from the load current at any instant can be written as shown in Eq. (1)

$$I_{la} = \sum_{n=1}^{\infty} I_{lan} \sin(n\omega t - \phi_{an})$$

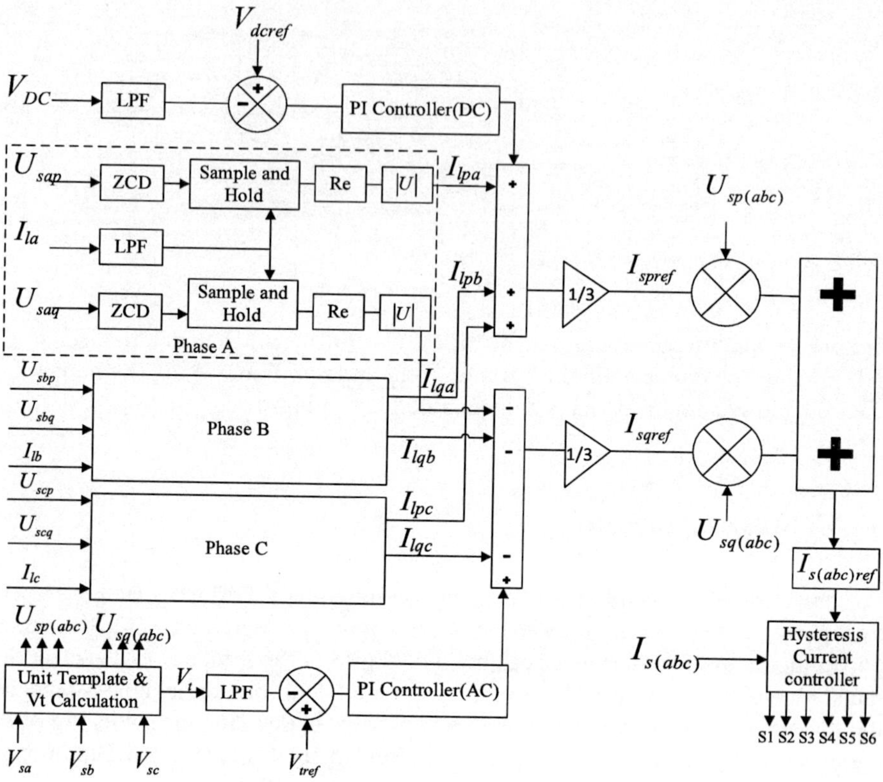

Fig. 2 Block diagram of Icos ∅ algorithm

$$I_{lb} = \sum_{n=1}^{\infty} I_{lbn} \sin(n\omega t - \phi_{bn} - 120^\circ)$$

$$I_{lc} = \sum_{n=1}^{\infty} I_{lcn} \sin(n\omega t - \phi_{cn} - 240^\circ) \quad (1)$$

where $\phi_{(a)n}$, $\phi_{(b)n}$, $\phi_{(c)n}$ is the nth harmonic current phase angles of phase a, b and c respectively. The magnitude of a fundamental load current active component is given as shown in Eq. (2)

$$|\mathrm{Re}(I_{la1})| = |I_{la}|\cos\phi_a; \; |\mathrm{Re}(I_{lb1})| = |I_{lb}|\cos\phi_b; \; |\mathrm{Re}(I_{lc1})| = |I_{lc}|\cos\phi_c; \quad (2)$$

For a stable system, the In-phase component amplitude of load currents is stated in Eq. (3)

$$I_{lp} = (I_{lpa} + I_{lpb} + I_{lpc})/3 \quad (3)$$

The reference currents are used to compare with the supply currents (I_{sa}, I_{sb}, I_{sc}) that are sensed and generated error is used to generate the switching pulse to control the DSTATCOM.

3.2 SRF Theory-Based Control Algorithm

The 3 phase currents at the load are transformed into stationary reference frame $\alpha-\beta$ using Clarke's transformation and later converted into synchronously rotating reference frame d-q employing Park's transformation. A Phase Locked Loop abbreviated as PLL is utilized for producing the voltage templates and then I_d and I_q are calculated which is then circulated through the Low Pass Filter to extract the DC components of I_{1q} and I_{1q}. The direct and quadrature axis currents comprise of harmonic and fundamental components as shown in Eq. (4)

$$I_{ld} = I_{dDC} + I_{dAC}; I_{lq} = I_{qDC} + I_{qAC} \tag{4}$$

This theory gives importance to the transformations of currents in d-q frame or synchronously rotating frame. The block diagram of SRF theory is shown in Fig. 3.

Again, these d-q frame currents are transformed into $\alpha - \beta$ frame by reverse Park's transformation and again transformed into Three phase currents using reverse Clarke's transformation and the reference currents are generated which is compared with sensed currents to produce gate pulses.

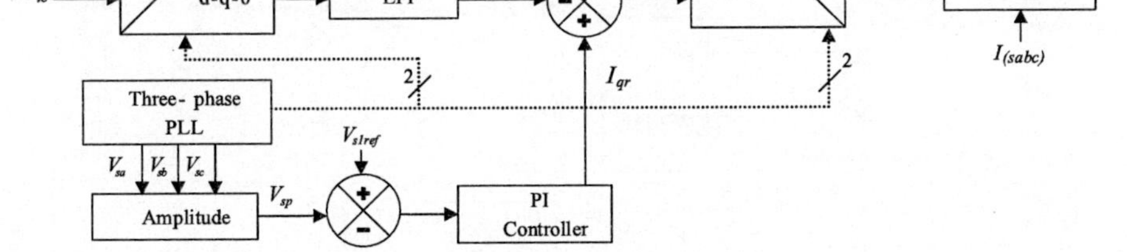

Fig. 3 Schematic diagram of SRF Theory-based control algorithm

3.3 Improved Adaptive Linear Sinusoidal Tracer Algorithm

The PCC line voltages are passed through BPF to eliminate harmonics and switching now and converted to phase voltages and amplitude transformation, the peak voltage is derived in Eq. (5)

$$V_t = \sqrt{\frac{(V_{sc}^2 + V_{sb}^2 + V_{sa}^2)2}{3}} \tag{5}$$

Unit templates in phase with PCC phase voltages are shown in Eq. (6)

$$X_{ap} = \frac{V_{sa}}{V_t}; X_{bp} = \frac{V_{sb}}{V_t}; X_{cp} = \frac{V_{sc}}{V_t} \tag{6}$$

Unit templates with PCC voltages quadrature are shown in Eq. (7)

$$X_{aq} = \frac{-X_{bp} + X_{cp}}{\sqrt{3}}; X_{bq} = \frac{3X_{ap} + X_{bp} - X_{cp}}{2\sqrt{3}}; X_{cq} = \frac{-3X_{ap} + X_{bp} - X_{cp}}{2\sqrt{3}} \tag{7}$$

The load currents contain harmonics, fundamental reactive and active power, dc components. The fundamental load current $I_{lf(a,b,c)}$ extraction is carried out by the ILST algorithm, the 2 parameters $\alpha - \beta$ decide the bandwidth and tuning frequency respectively. Figure 4 illustrates the block diagram for the ILST algorithm.

The transfer function of the improved ALST control algorithm is shown in Eq. (8)

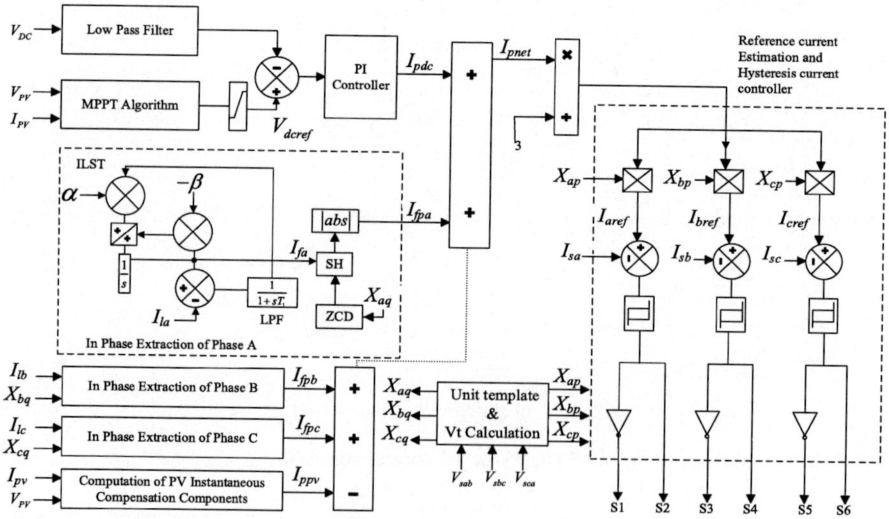

Fig. 4 Schematic block diagram of improved ALST algorithm

$$\frac{I_l(abc)}{I_f(abc)} = \frac{S\alpha}{S^3T_1 + S^2 + (\beta\ T_1 + \alpha)S + \beta} \tag{8}$$

The characteristic equation is given in Eq. (9)

$$S^3T_1 + S^2 + (\beta T_1 + \alpha)S + \beta = 0 \tag{9}$$

The load current active power component I_{fpa} is obtained by samples and the fundamental current output from ILST I_{fa} at every zero-crossing of 90^0 transferred unit templates are given in Eq. (10)

$$\begin{aligned} I_{fa} &= I_m \sin(\omega t - \phi) = 0 \\ I_{fa} &= I_m \cos\omega t \sin\phi + I_m \sin\omega\, t \cos\phi \\ I_{fa} &= (I_m \cos\phi)\cos\omega t = 0 \end{aligned} \tag{10}$$

Similarly, for the other two phases, (b, c) it is derived. The solar PV system the instantaneous compensation is shown in Eq. (11)

$$I_{ppv} = \frac{2P_{pv}}{Vt} \tag{11}$$

To regulate the bus voltage of the DC capacitor of the VSC Proportional Integral controller is implemented. DC bus voltage sensed is compared with the reference DC voltage engendered by the MPPT algorithm and supplied to Proportional Integral controller. I_{pdc} is the output which is the minimum current needed to maintain the bus voltage across the DC capacitor at the reference juncture. The output at discrete instant R is given in Eq. (12)

$$I_{pdc}(R) = K_{i_{loss}} V_{dcerror} + K_{p_{loss}}\{-V_{dcerror}(k-1) + V_{dcerror}(k)\} + I_{pdc}(R-1) \tag{12}$$

I_{pdc} is of Proportional Integral controller output,$K_{p_{loss}}$ is the proportional gain and $K_{i_{loss}}$ is the integral gain. The magnitude of the real power component of the reference source current is shown in Eq. (13)

$$I_{snet} = I_{fp} + I_{pdc} - I_{ppv} \tag{13}$$

4 Results and Analysis

To justify the proposed control technique for the LCLC type shunt active filter fed by a PV system, using MATLAB & SIMULINK and simscape toolbox a 3-Ø 3-wire distribution network connected to a nonlinear load is simulated. The proposed system

Table 1 Parameters of the proposed system

Parameters		Value
Supply Voltage (V_s)		239 V rms
Grid Frequency		50 Hz
DC Bus Voltage (V_{pv})		170 V
DC Bus Capacitor (C_{dc})		3500 μF
Non-linear load- Resistance and Inductance (R_{dc}, L_{dc})		60 Ω, 210 mH
Filter setting, LCL (L_1, L_2, C, C_{se})		1.5, 0.6 mH, 20, 70 μF
Unbalance Load	R_a, R_b, R_c	50, 60, 70 Ω
	L_a, L_b, L_c	0.189, 0.229, 0.139 H

is simulated for different loads to check the efficacy of the implemented controllers. To check the robustness of the improved ALST algorithm, the system performance is compared with SRF theory and Icos∅ algorithm. The total harmonic distortion (THD) is the parameter used for evaluating harmonics for every controller. Table 1 shows the parameters of the proposed power system. Results of the simulation are shown below for several conditions.

4.1 Case-1: Steady-State Performance of the System Connected to Non-Linear Load Under Ideal Grid Condition

The steady-state performance of the system connected to non-linear load is displaced in Fig. 5. For non-linear load, a three phase diode bridge rectifier with R and RL load is used for simulation. Before compensation, the source voltage under ideal conditions is illustrated in Fig. 5a and source current is illustrated in Fig. 5b. After compensation, the source voltage and source current utilizing SRF theory is shown in Fig. 5c, source current utilizing Icos∅ algorithm is shown in Fig. 5d and source current utilizing improved ALST algorithm is shown in Fig. 5e. By means of FFT analysis of Simulink THD is viewed and it is observed that the source current decreased from 30.08 to 2.21% with improved ALST algorithm, 30.08–2.83% using Icos∅ algorithm and 30.08–3.09% with SRF theory, which is properly below the IEEE 519 standard. From the compared result, the source current perhaps is smooth and pure sinusoidal with improved ALST algorithm.

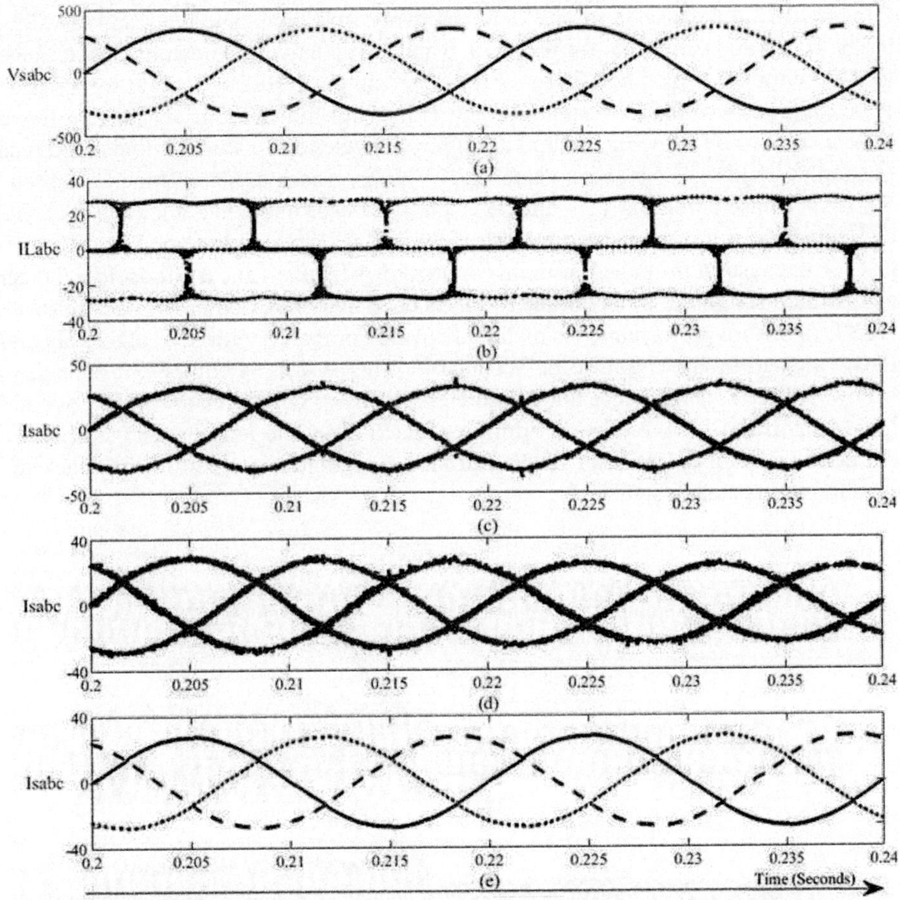

Fig. 5 Steady-state performance of the system proposed in the event of ideal grid condition. **a** Ideal Icos∅ source voltage. **b** Source current simulated before compensation. **c** Source current simulated utilizing SRF theory. **d** Source current simulated utilizing algorithm. **e** Source current simulated utilizing improved adaptive linear sinusoidal tracer algorithm

4.2 Case-2: Transient Performance Under Ideal Grid Condition

To substantiate the performance of system proposed in the event of transient conditions utilizing improved ALST algorithm, load of different types is thought about and used. linear unbalanced loads, 3-Ø rectifier with unbalanced load and 3-Ø rectifier without unbalanced are considered for load dissimilarity. Primarily the set forth system is operated with linear unbalanced load and three phase rectifiers with R-L load. Over this period the value of dc side resistance and inductance is 60 Ω and 210 mH while the linear unbalanced load comprising of resistances(R) are of values 50,

60 and 70 Ω for phases a, b and c, respectively, and inductances(L) are of values 0.189, 0.229 and 0.139 mH for phases a, b and c respectively. The unbalanced linear load is being decoupled at 0.2 secs and the system put forth is operated for a period of 10 cycles only with identical load comprising a non-linear rectifier. The non-linear load is separated following the operation for 10 cycles and a linear unbalanced load is coupled holding resistances values of 5.4 Ω for phase a, 5.6 Ω for phase b and 5.8 Ω for phase c. reiteratively after 0.2 secs the linear unbalanced load is removed and unbalanced linear load along with the non-linear load is linked. The effectiveness of the system proposed utilizing improved Adaptive Linear Sinusoidal Tracer algorithm through the transient operation is been authenticated in Fig. 6. Improved ALST algorithm performance is matched up and compared with SRF algorithm and Icos∅ algorithm and found to be acceptable. This system is competent enough to follow and track its reference and execute efficiently as compared to SRF theory and Icos∅ algorithm for the transient duration. Each time, the load is changed, within the period of 1–2 cycles after modification, the regulator is able to bring back the system into its steady state.

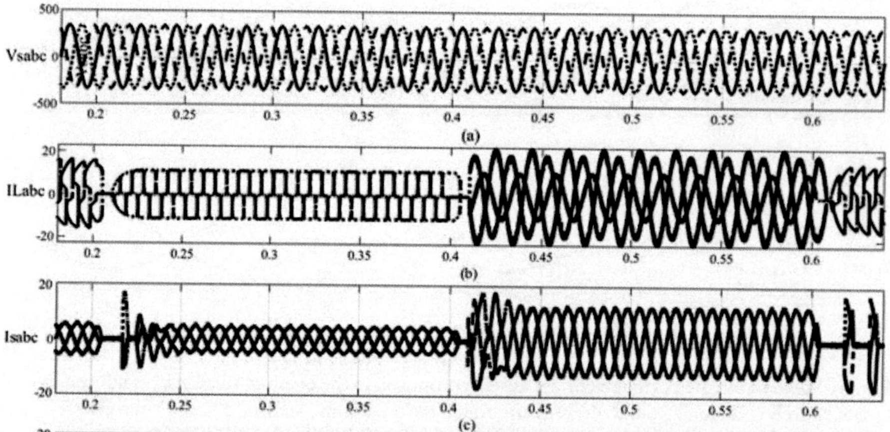

Fig. 6 Performance during Transient period for the system proposed in the event of ideal grid condition. **a** Source voltage. **b** Source current simulated before compensation. **c** Source current simulated utilizing SRF theory. **d** Compensating filter current

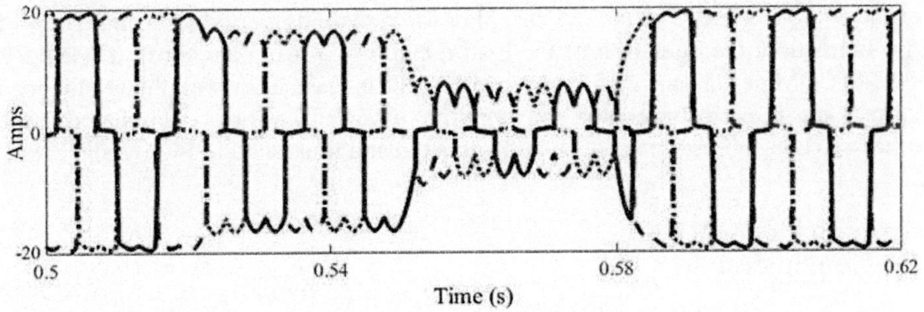

Fig. 7 Source current simulation results before compensation

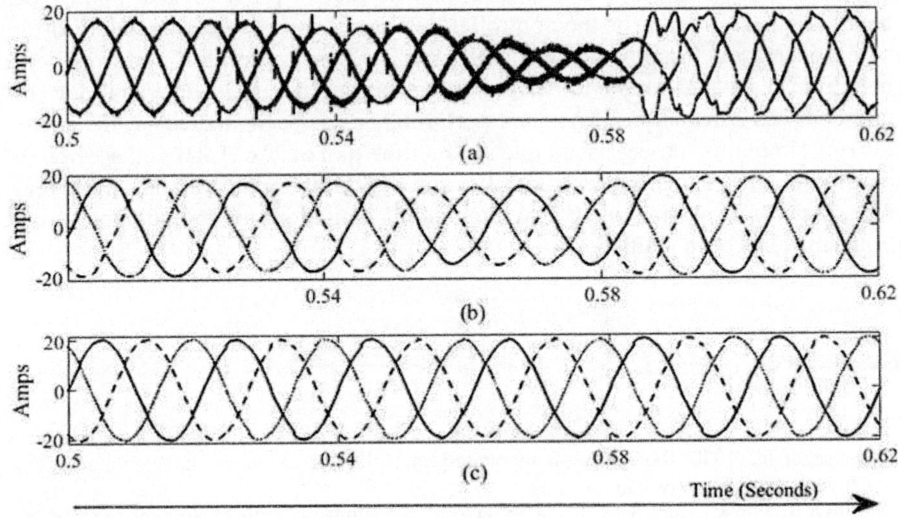

Fig. 8 Icos∅ Source current simulation results after compensation. **a** Utilizing SRF theory. **b** Utilizing algorithm. **c** Using improved ALST algorithm

4.3 Case-3: Dynamic Performance Under Uncertain Load Active Power Variation

Thyristorzed converters in power systems are utilized to regulate the performance of the load according to the necessity. For the production of required output the power electronic rectifier-based loads are often controlled by governing the firing angle. This paper emphasizes on the simulation of a rectifier (Thyristor-based) network with R-L. Taking this into consideration the thyristor firing angle is varied suddenly and abruptly. Keeping the firing angle at 30° the network is simulated till 0.15 secs and then changed to an angle of 60° for 2.5 cycles. Yet again at 0.2 secs the load is simulated with the firing angle of 30° for a period of 0.05 secs and later operated at the firing angle of 90° at 0.25 secs. For the remaining period (after 0.3 secs) of the simulation, the operation of the load remains at a firing angle of 30°. The PV DSTATCOM performance with improved ALST is reasonably well though the load current gets more distorted and consumption of power is varied as compared to the other two controllers, this is been authenticated and illustrated in Figs. 7 and 8.

5 Conclusion

In this paper, a comparison between SRF theory, Icos∅ algorithm and improved ALST algorithm for an enhanced hybrid DSTATCOM supplied by a photovoltaic system has been exemplified. Owing to the existence of phase-locked loop in SRF theory, the performance of the controller deteriorates when the uncertainty of the system increases. In comparison to SRF theory, Icos∅ algorithm performs better though it is easy and simple to design. The response time is also less as it does not have complex calculations. Transient performance has been studied for the system utilizing simulation process available in Simulink tool of MATLAB and ascertained to be reasonable. The proposed controller efficacy for harmonic mitigation and reactive power compensation under different loading conditions are investigated and the proficiency has been studied.

References

1. Singh B, Chandra A, Al-Haddad K (2014) Power quality problems and mitigation techniques
2. Schaefer JC (1990) Review of photovoltaic power plant performance and economics. IEEE Trans Energy Convers 5(2):232–238
3. Singh B, Jain C, Goel S (2014) ILST control algorithm of single-stage dual purpose grid connected solar pv system. IEEE Trans Power Electr 29:5347–5357
4. Singh B, Jayaprakash P, Kothari DP, Chandra A, Haddad KA (2014) Comprehensive Study of DSTATCOM configurations. IEEE Trans Ind Inform 10:854–870
5. Reddy MRK, Veni DKK, Das DT, Pulla DS (2011) Performance analysis of DSTATCOM compensator using control techniques for load compensation. Int J Electr Electron Eng Res 1
6. Subudhi B, Pradhan R (2013) A comparative study on maximum power point tracking techniques for photovoltaic power systems. IEEE Trans Sustain Energy 4(1):89–98
7. Ali A et al (2020) Investigation of MPPT techniques under uniform and non-uniform solar irradiation condition–a retrospection. IEEE Access 8:127368–127392
8. Hanzaei SH, Gorji SA, Ektesabi M (2020) A scheme-based review of MPPT techniques with respect to input variables including solar irradiance and PV arrays' temperature. IEEE Access 8:182229–182239
9. Bollipo RB, Mikkili S, Bonthagorla PK (2020) Critical review on PV MPPT techniques: classical, intelligent and optimisation. IET Renew Power Gener 14(9):1433–1452. Accessed 6 July 2020
10. Esram T, Chapman PL (2007) Comparison of photovoltaic array maximum power point tracking techniques. IEEE Trans Energy Conver 22:439–449

11. Kouro S, Leon JI, Vinnikov D, Franquelo LG (2015) Grid-connected photovoltaic system: an overview of recent research and emerging PV converter technology. IEEE Ind Electron Mag 9:47–61
12. Patel N, Gupta N, Babu BC (2020) Photovoltaic system operation as DSTATCOM for power quality improvement employing active current control. IET Gener Trans Distrib 14(17):3518–3529. Accessed 4 Sept 2020
13. Chen C, Lin C, Hsieh W, Hsu C, Ku T (2013) Enhancement of PV penetration with DSTATCOM in taipower distribution system. IEEE Trans Power Syst 28(2):1560–1567
14. Beniwal N, Hussain I, Singh B (2019) Second-order volterra-filter-based control of a solar PV DSTATCOM system to achieve lyapunov's stability. IEEE Trans Ind Appl 55(1):670–679
15. Bagheri Tolabi H, Ali MH, Rizwan M (2015) Simultaneous reconfiguration, optimal placement of DSTATCOM, and photovoltaic array in a distribution system based on fuzzy-ACO approach. IEEE Trans Sustain Energy 6(1):210–218
16. Modi G, Kumar S, Singh B (2020) Improved widrow–hoff based adaptive control of multiobjective PV DSTATCOM system. IEEE Trans Ind Appl 56(2):1930–1939
17. Mishra S, Ray PK (2016) power quality improvement using photovoltaic fed DSTATCOM based on JAYA optimization. IEEE Trans Sustain Energy 7(4):1672–1680
18. Libo W, Zhengming Z, Jianzheng L (2007) A single-stage three-phase grid-connected photovoltaic system with modified MPPT method and reactive power compensation. IEEE Trans Energy Convers 22(4):881–886
19. Verma AK, Singh B, Sahani DT (2012) Grid interfaced solar photovoltaic power generating system with power quality improvement at AC mains. In: presented at the IEEE 3rd international conferences sustainable energy technology, Kathmandu, Nepal
20. Singh B, Jain C, Goel S (2014) ILST Control algorithm of single-stage dual purpose grid connected solar PV system. IEEE Trans Power Electron 29(10):5347–5357

Chapter 51
QR Code-Based Digital Assistant for Seminar Halls Using Tinker Board and Node-RED

K. V. V. R. S. Vishnu, B. Venkateswara Rao, T. Thirumala Rao, and Y. Jaswanth

1 Introduction

Electricity plays a crucial role in day-to-day life. Therefore human needs increase, technology increases and thus electricity consumption also increases [1]. According to reports, approximately three billion units of power are wasted in one year [2]. It is essential to save electricity for future generations. Adopting automation gives better results in saving electricity. Electricity is wasted due to the negligence of people [3]. The proposed system saves electricity in an efficient way along with providing a rich experience. Node-RED is a programming tool for wiring together hardware devices, APIs and online services in new and interesting ways [4, 5]. It provides a friendly user interface with minimum coding. It has a secured communication system so that only trusted people can access the control [6].

In literature, many authors use Raspberry Pi for Home automation but few of the authors use Raspberry Pi for seminar hall automation [7] but there are some drawbacks, for example, Praveen Kumar et al. [8] propose the Seminar Hall Automation Using Raspberry Pi 3, but their approach needs a smartphone to control the equipment, AC and sound cannot be adjusted, it is not fully automated because electrical items like fans and lights are not controlled through the android app by the user. Hariprasath et al. [9] worked on Intelligent Conference Hall Automation System but in their method counting of visitors is not accurate, Lights and fans control is based on visitors count but not on their location, No GUI, AC and sound cannot be adjusted. Sharma et al. [10] propose interactive projector screen with hand detection using gestures but there are so many compels associated, for example, it only works in a specific region, the handling may be a bit leisurelier than the real touch screen, The user's glove must be dreamily famed from the experience, so it needs high-speed computer, rate of commands must be slow. These entire drawbacks can be reduced by using the proposed device. In this paper, seminar hall automation

K. V. V. R. S. Vishnu · B. Venkateswara Rao (✉) · T. Thirumala Rao · Y. Jaswanth
Department of EEE, V R Siddhartha Engineering College, Kanuru, AP, India

S. Dawn et al. (eds.), *Smart and Intelligent Systems*, Algorithms for Intelligent Systems,
https://doi.org/10.1007/978-981-16-2109-3_52

system can be developed in Tinker Board by using node-red and NodeMCU [11, 12]. This system provides advanced control over the Tinker Board from playing high-resolution videos, to control electrical equipment along with Projector right from the user's device.

2 Mathematical Model of the System

2.1 *Input Set:*

Scans QR code for IP address.

Node-Red User interface opens on Browser.

Tinker Board takes input from the Node-red User Interface running in Mobile Device.

Input = {all inputs from the Node-red User Interface running in Mobile Device}.

2.2 *Output Set:*

The output according to the above input is processed by Tinker Board and responds correspondingly.

Out1 = speech output from mobile device.

Out2 = GUI, Application Response.

Therefore out = out1 + out2.

Output out = Response to the input via Speech, GUI and Application response.

3 Methodology

A QR code scanner installed in a mobile device scans QRcode for the IP address. The IP address points to the graphical user interface in the browser. The main control is hosted on Tinker Board by a programming tool called Node-Red. The inputs are taken from the Browser and sent to the Tinker Board via Wi-Fi. Those inputs are processed by the Node-Red tool and control the functions accordingly. The Tinker Board is connected to Projector via HDMI Cable to control presentations and other Media like videos and photos. The presentations are presented in open source software called LibreOffice. The slides' control action is programmed in Python and triggered with Node-Red.

The Tinker Board is also connected to the Electrical Control Circuit via Wi-Fi. The instructions from the Tinker Board were sent to NodeMCUvia Wi-Fi. The nodeMCU device triggers the BT136 Triac which switches the electrical appliances [13, 14]. The electrical control circuit is designed such that it provides fast, smooth and noiseless operation while switching. This system features a NodeMCU Device, BT136 Triac and MOC3021Optocoupler to control all the electrical equipment. GUI is created by using Node-Red. Decisions taken on the client (mobile device browser) device are processed on a Node-RED server installed on Tinker Board. The processed information triggers the corresponding functions. The Block Diagram of the QR code-based Digital Assistant for Seminar Halls is displayed in Fig. 1. Figure 2 provides the control circuit to control the fans and lights in the seminar hall.

Software design mainly consists of GUI Unit and Processing Unit. Graphical User Interface allows users to interact with the Electronic devices through graphical Items [15]. Node-Red's well-designed GUI provides easy access to all the functions right from the mobile device. Node-RED server initiates the Node-Red Graphical User Interface in client device browser. Configuration is done in node-red server reflected in client device's Browser as shown in Fig. 3.

To create the interface in Node-Red, there are two steps involved. They are Creating Flow and Deploying Flow. The ui_switch from the dashboard module in

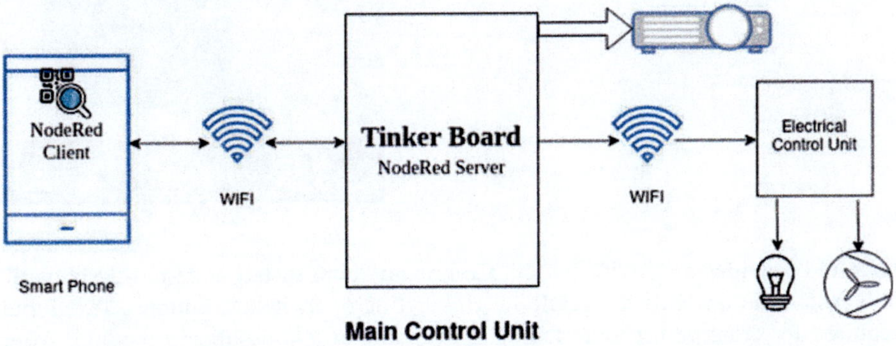

Fig. 1 Block Diagram of QR code-based digital assistant for seminar halls using tinker board and NodeRED

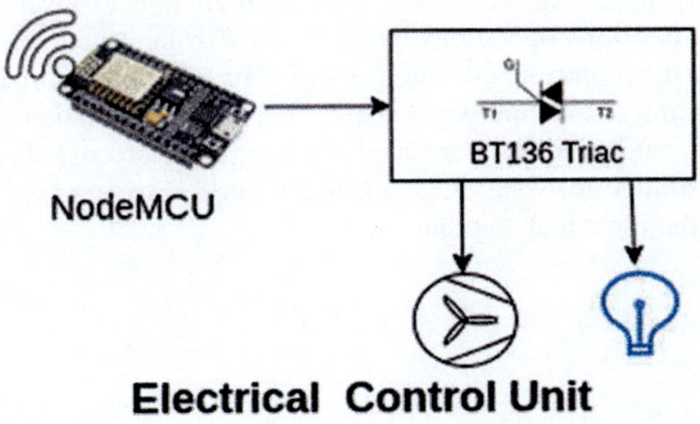

Fig. 2 Electrical appliances control circuit consisting of NodeMCU and triac

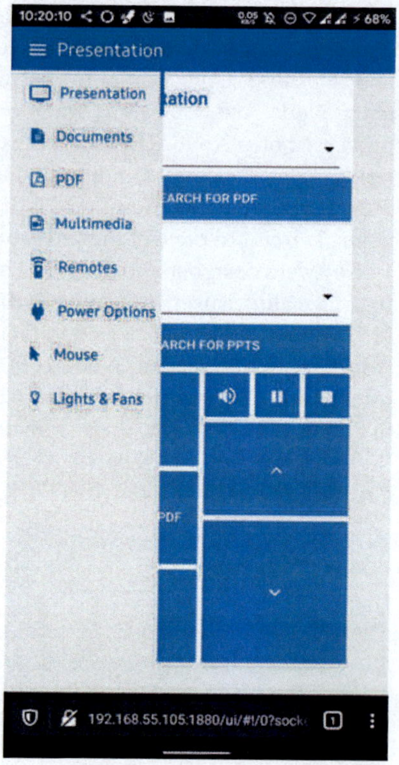

Fig. 3 Various options of UI

node-red provides a UI with switch. Command "npm install node-red-dashboard" will install the dashboard module in Tinker Board. Switches, buttons, Text Input required for creating user interfacing is available at this dashboard module. After the deployment of all flows, the UI developed as shown in Fig. 4. The flow chart to control electrical appliances given in Fig. 5 and GUI consists of different categories including Presentation, Documents, PDF, Multimedia, Lights and Fans, etc. Node Red Flows Showing the control flow of electrical appliances given in Fig. 6. The processing unit performs operations based on input from mobile clients and automatically set up the projector and sound system. For Presentation Inputs, the python code corresponding to the input will be executed. The presentation will be open in fullscreen mode automatically. For Electrical Inputs, the processed instructions are wirelessly transmitted to NodeMCU. Then, NodeMCU triggers the BT136 Triac, which triggers the Electrical Appliances.

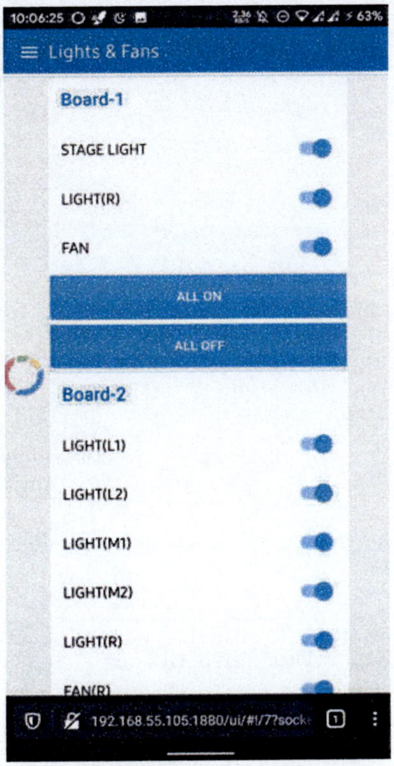

Fig. 4 Electrical appliances control options

4 Results and Discussions

Firstly start the Tinker Board and establish the communication between the Tinker Board and projector is established and it shows the screen as shown in Fig. 7. It shows the QR code and desktop of the Tinker Board. Then connected the mobile to the Wi-Fi network and scanned the QR code using mobile to access the interface of the controls in mobile browser. It is shown in Fig. 8. It consists of various selection parameters like select PPT, open PPT, close PPT, open PDF and close PDF, etc. Communication to Tinker Board and browser web page is established to control the hardware equipment in seminar hall through NodeRED. Figure 8 also shows the digital assistant browser web page to control fan, light and change the powerpoint presentation slides using Tinker Board and NodeRED. Successful in sending the signals to hardware equipment, same are displayed in user interface mobile screen. Figures 9 and 10 show the hardware setup of NodeMCU and Tinker Board, respectively. In these figures, mouse is connected to USB port and projector connected to HDMI port, and various electrical appliances are connected to relays to control. Figure 11 presents the control of presentation using user mobile.

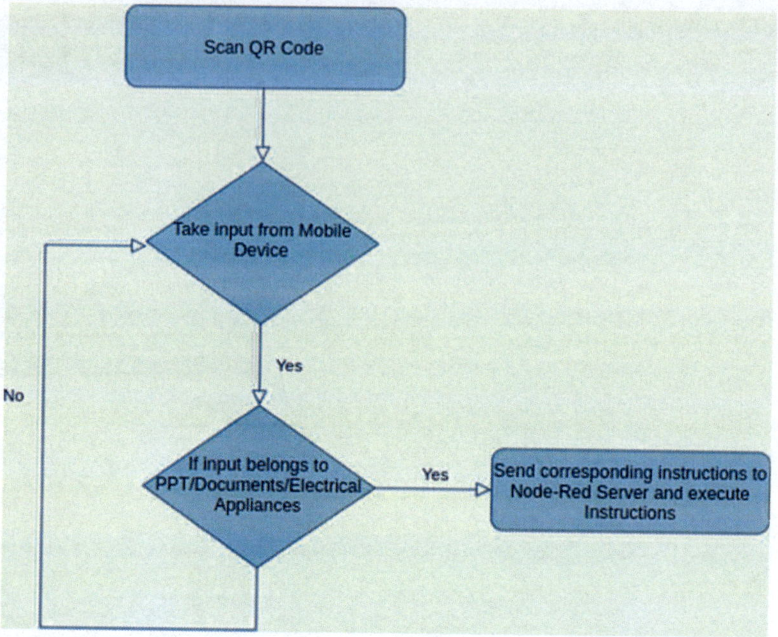

Fig. 5 Flow chart of GUI unit

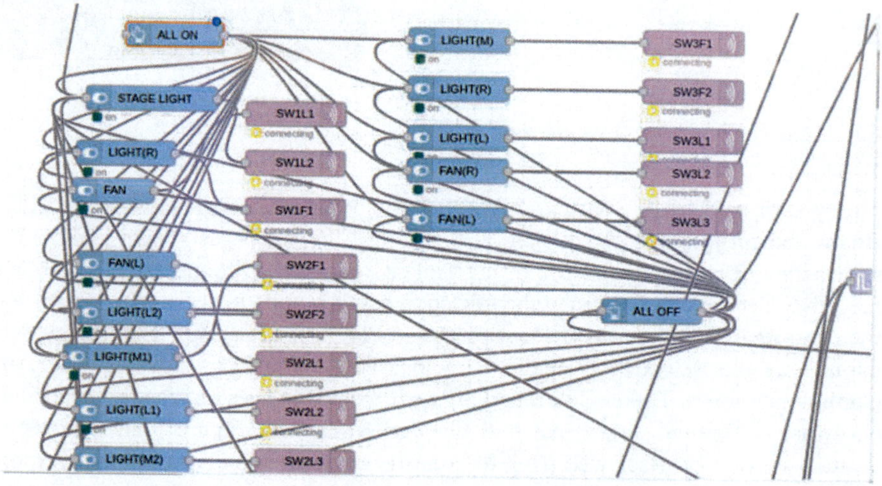

Fig. 6 Node red flows showing the control flow of electrical appliances

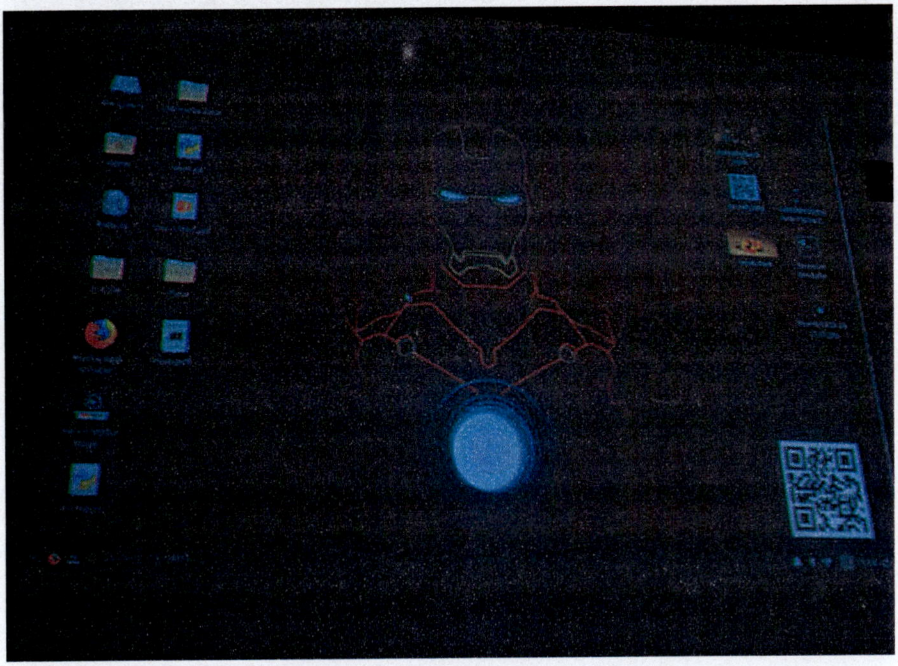

Fig. 7 Tinker board screen with QR code

5 Conclusion

The developed model features the advanced control on Seminal Hall Equipment with Low Cost, Small Size, Reliable and Secured Technology. It can replace Computers and Laptops in seminar halls. It saves a lot of Electrical Energy with flexible controlling. The developed device not only saves energy by adapting automation and can also reduce human errors. It creates a friendly interface for the users. It controls powerpoint presentations and hardware equipment in seminar hall. The device is successful in halting the ppt, pdf, doc files, fans and lights. The device ensures an efficient and reliable operation of seminar halls and is most affordable. The designed seminar hall automation system was tested a number of times. This system had been developed as user friendly, where controlling is done through user mobile phone therefore it can be easily installed in any seminar/conference hall.

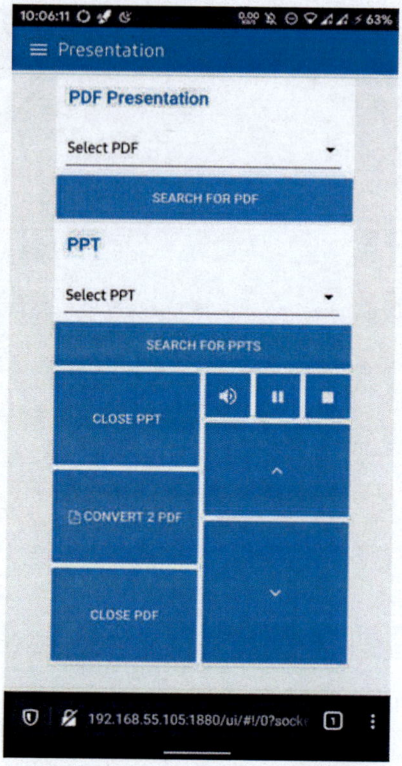

Fig. 8 App on the mobile to control the seminar hall

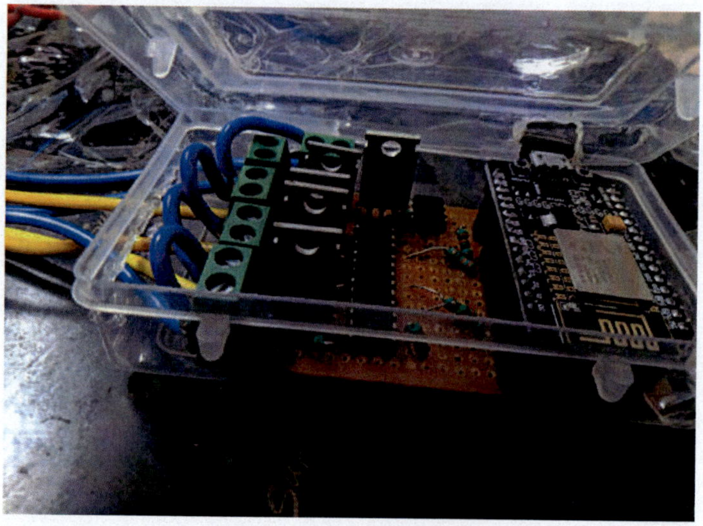

Fig. 9 Connections of electrical appliances with electrical control circuit

Fig. 10 Tinker board connected to projector and mouse

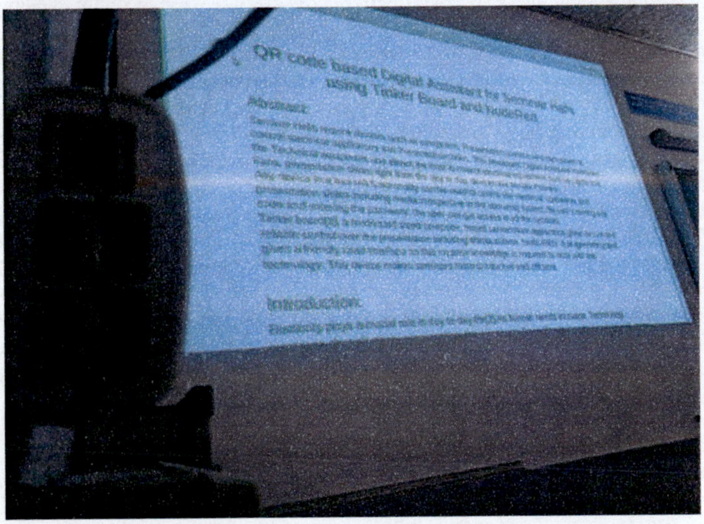

Fig. 11 PPT control through mobile

References

1. Rajesh MV, Venkateswararao B (2019) Conference hall automation system using Python-Kivy application. Int J Innov Technol Explor Eng (IJITEE) 8(6S4):446–449. ISSN: 2278–3075. https://doi.org/10.35940/ijitee.F1092.0486S419
2. Dutta S (2015) Three billion units of power wasted in one year. The Times of India, para. 2, July 5, 2015. https://timesofindia.indiatimes.com/india/Three-billion-units-of-power-wasted-in-one-year/articleshow/47942237.cms. Accessed 30 Dec 2020
3. Chitte PP, Shinde RR, Thosar SV (2015) Smart automated conference room system. Int J Emer Technol Adv Eng 5(3):591–595. ISSN 2250–2459
4. Chaczko Z, Braun R (2017) Learning data engineering: creating IoT apps using the node-RED and the RPI technologies. In: 2017 16th international conference on information technology based higher education and training (ITHET), Ohrid, 2017, pp 1–8. https://doi.org/10.1109/ITHET.2017.8067827.
5. Kodali RK, Anjum A (2018) IoT based HOME AUTOMATION using Node-RED. In: 2018 second international conference on green computing and internet of things (ICGCIoT), Bangalore, India, 2018, pp 386–390. https://doi.org/10.1109/ICGCIoT.2018.8753085
6. Rajesh MV, Venkateswararao B (2017) Artificial intelligence based machine learning assistance for self-driving car using raspberry Pi. Int J Adv Res Sci Eng 6(11):1718–1724
7. Rajesh MV, Venkateswara Rao B, Sai Vamsi Krishna P, Pavan Kumar S (2018) Raspberry PI based digital assistant for seminar halls (D.A.S.H). In: International conference on new trends in engineering & technology (ICNTET)-2018, Organized by GRT institute of engineering and technology, Chennai, during September 07th & 08th 2018
8. Praveen Kumar S, Rajesh K, Subrahmanya S, Kantharaju AG, Moji V (2017) Seminar hall automation using raspberry Pi 3. Int J Technol Res Innov Solut (IJTRIS)
9. Hariprasath M, Karthick PK (2016) Intelligent conference hall automation system. Int J Adv Res Electron Commun Eng (IJARECE) 5(12):2543–2547
10. Sharma R, Shikher R, Bansode NV, Rajarapollu PR (2016) Interactive projector screen with hand detection using gestures. In: International conference on automatic control and dynamic optimization techniques (ICACDOT)
11. Clark L (2019) Project 2: use the tinker board as a media center with Kodi: getting started and building projects with the ASUS single-board computer. https://doi.org/10.1007/978-1-4842-3826-4_8
12. Raju KL, Chandrani V, Begum SS, Devi MP (2019) Home automation and security system with node MCU using internet of things. In: 2019 international conference on vision towards emerging trends in communication and networking (ViTECoN), Vellore, India, 2019, pp 1–5. https://doi.org/10.1109/ViTECoN.2019.8899540
13. Jaswanth Y, Rachana R, Venkateswara Rao B (2019) Protection of microgrids with arduino control scheme. Controller Convert MAT J 4(3):43–49
14. Clougherty H, Brown A, Stonerock M (2017) Home automation and personalization through individual location determination. IEEE 978–1–5386–1848–6/17/$31.00
15. Madupu PK, Karthikeyan B (2018) Automatic service request system for security in smart home using IoT. In: 2018 second international conference on electronics, communication and aerospace technology (ICECA). IEEE

Chapter 52
Transient Response Performance-Based Comparative Studies of TBC Fed UPQC in Multibus-Based System with Conventional Converter

G. V. P. Anjaneyulu

1 Introduction

The main aim of electric power supply at the consumer point having constant magnitude voltage value at ± 5% and also constant frequency of AC supply from source end. At power stations generators produces electrical power and induced e.m.f having almost a wave of sinusoidal nature. The estimation, modeling and functions of all the industrial/agricultural/commercial/noncommercial power system utilities require fairly constant performance values in the evaluation of the starting and future extension. The network stability, reliability, security and capability need to increase with the output and operating necessities.

Always the AC systems are planned and modeled to work with pure magnitudes of sine values of voltages and also currents in practical nature. But all electronic operated loads and also nonlinear loads are deviated from its constant magnitude value of voltages and also currents. Generally, distorted signals are studied by analyzing the harmonic frequency values of components of the signals. These loads harmonics are mainly due to their power electronic and high switching frequency devices. Due to high switching frequency components and also fast active devices in modern equipment more harmonics and distortions induced in the system will also transfer to neighboring systems also [1, 2]. Hence the total system voltage and current magnitudes are deviate from the fundamental values. System frequency also changes. Once the system frequency and voltage values changes the active power and reactive power requirements in the system changes. It turns to introducing compensating equipments in the system [3, 4]. Hence system efficiency reduces and also due to compensating equipment maintenance and cost increases.

G. V. P. Anjaneyulu (✉)
EEE Department, RVR & JC College of Engineering, Guntur, A.P., India

S. Dawn et al. (eds.), *Smart and Intelligent Systems*, Algorithms for Intelligent Systems,
https://doi.org/10.1007/978-981-16-2109-3_53

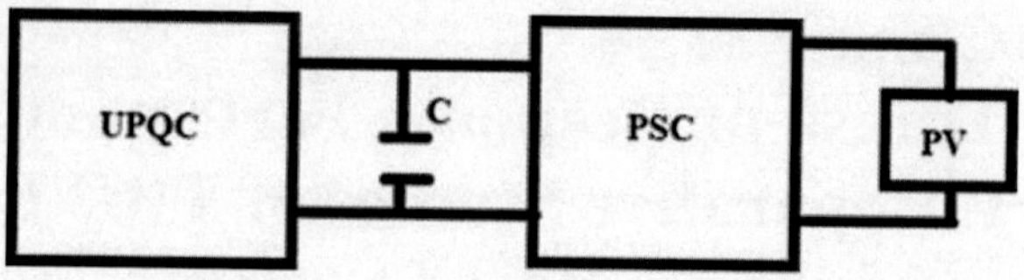

Fig. 1 UPQC configuration with PSC

2 Ease of Use for Open Loop System

In open loop system, the Boost converter can match the magnitude of voltage value of photovoltaic (PV) network. The boost converter output is fed to the DC capacitor. This capacitor acts as link in between peak step-up converter and Unified Power Quality Conditioner (UPQC) [5, 6]. This section deals with the comparison of UPQC systems fed from peak step-up, two inductor boost convertor systems. The comparison is done for open-loop system in terms of input current ripple. The objective of this paper work is selection of desired boost converter from the output of photovoltaic (PV) system to DC capacitor of UPQC system.

2.1 UPQC with Conventional Peak Step-up Converter

Block diagram structure of UPQC system with peak step-up converter [7, 8] is represented in the following Fig. 1. Magnitude of voltage output at PV is boosted using peak step-up converter (PSC). PSC charges the capacitor which is directly connected with UPQC.

2.2 Proposed Two Inductor Boost Converter for UPQC

The block diagram of UPQC system with the proposed boost converter with the two inductors is represented in Fig. 2. Here, the magnitude of output voltage at PV system is boosted using two inductor boost converter (TBC). The TBC output is fed to capacitor C, then it is connected to UPQC system as shown in Fig. 2.

TBC is preferred due to the reduced input current ripple.

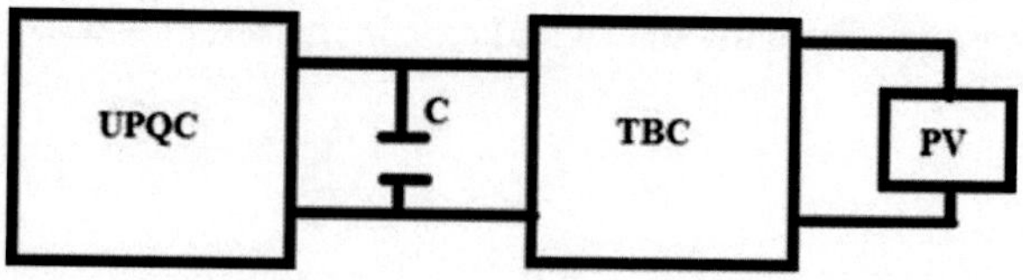

Fig. 2 UPQC configuration with TBC

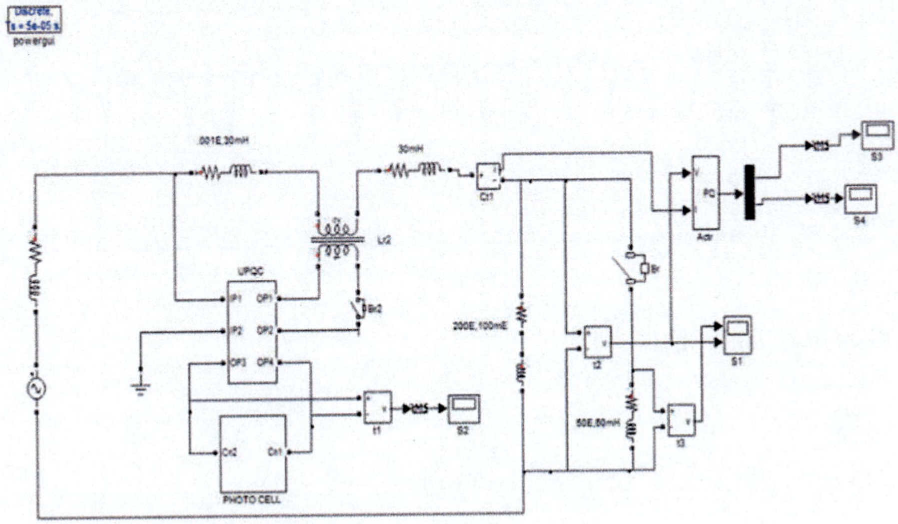

Fig. 3 SIMULINK diagram of PSC based UPQC

3 Implementation for Open Loop System

The following concepts show that PSC and TBC are fed with UPQC system and these models designed are simulated at SIMULINK/MATLAB platform and simulation studies are obtained for analysis.

3.1 *Peak Step up Boost Converter with UPQC System*

The SIMULINK/MATLAB diagram of UPQC-based system of PSC is represented in Fig. 3. The DVR part of the UPQC injects voltage in the middle of the line. The SAF injects current at the sending end. A peak step-up converter is connected between PV source and DC capacitor. Its magnitude current ripple at input is represented in the following Fig. 4. From Fig. 4 it is observed that its peak magnitude value is obtained as 18 A. The zoomed output is shown in Fig. 4.

3.2 *Two Inductor Boost Converter Based UPQC*

The UPQC system with boost converter of two inductors (TBC) is represented in the following Fig. 5. Here, figure shows SIMULINK diagram of two inductor boost converter (TBC) fed UPQC system. The peak step-up converter (PSC) is replaced by two inductor boost converter (TBC). The battery is replaced with PV system.

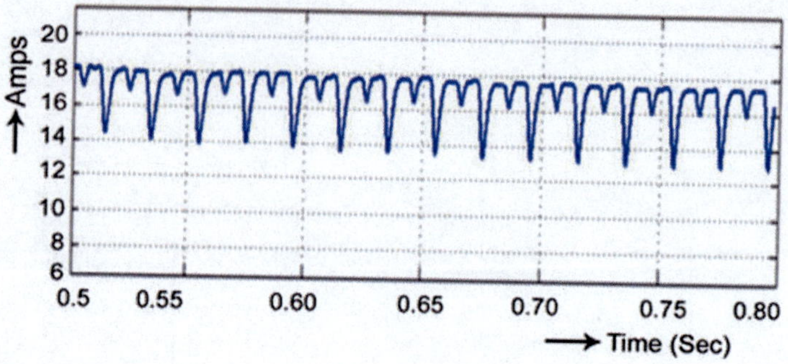

Fig. 4 Input current ripple

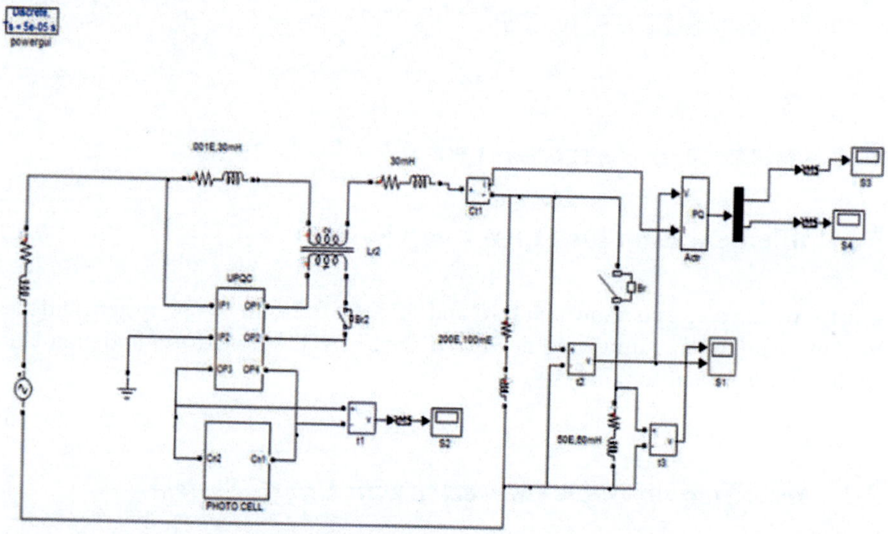

Fig. 5 SIMULINK diagram of PSC-based UPQC

TBC generates the desired value of voltage. With this voltage the DC link capacitor charges. From the Simulation, results are obtained and represented in Fig. 6. Here from Fig. 6 the magnitude of ripple current at input is represented and peak magnitude 19.5 A.

From Figs. 4 and 6 it is clearly observed that two inductor boost converter gives high current with less input ripple. The following table shows quantity of input current ripple of both PSC and TBC-based UPQC system (Table 1).

Hence TBC is utilized in closed-loop system due to its reduced input current ripple.

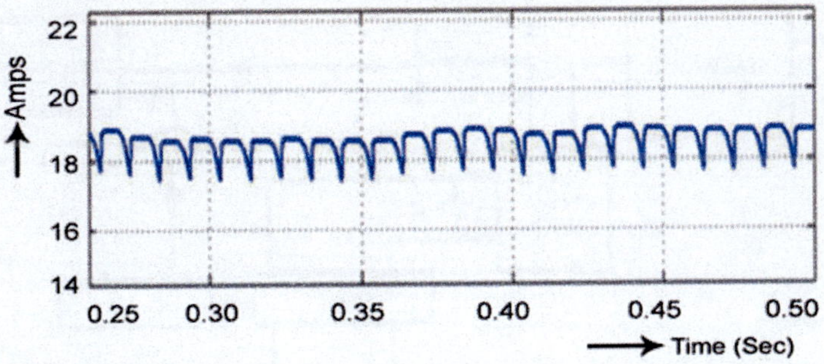

Fig. 6 Input current ripple

Table 1 Input current ripple comparison of PSC, TBC based UPQC system

Model of UPQC system	Input current ripple
PSC	2.15 A
TBC	1.05A

4 Proposed Closed-Loop System

This section deals with the identification of suitable controller for closed-loop operation of two inductor boost converter (TBC)-based UPQC. The analysis and comparison of responses with proportional integral (PI), proportional resonant (PR) are shown here. Simulation results of PI controller with TBC–UPQC are compared and represented also those with PR controller TBC–UPQC system. This concept's intention is how to enhance the time domain characteristics of Two Inductor Boost Converter (TBC) fed UPQC using suitable controller.

The comparison is done in values of their transient values of Rise time and its Peak time also steady state value of its Settling time. Closed-loop control of (TBC) boost converter of two inductor-based UPQC designed and modeled at the platform of SIMULINK. From SIMULINK results indicate superior performance with two inductor-based TBC–UPQC system.

4.1 System Configuration

The two inductor boost converter-based UPQC closed-loop structure diagram is represented in Fig. 7. The capacitor C charged from two inductor boost converter (TBC). The PV system output voltage is enhanced by using two inductor boost converter (TBC). For TBC the modified signals are fed from feedback collected at closed-loop controllers. The feedback is collected from load of receiving end.

TBC is preferred due to reduced input current ripple value. Load voltage at receiving end is measured and it is compared with the specified reference voltage

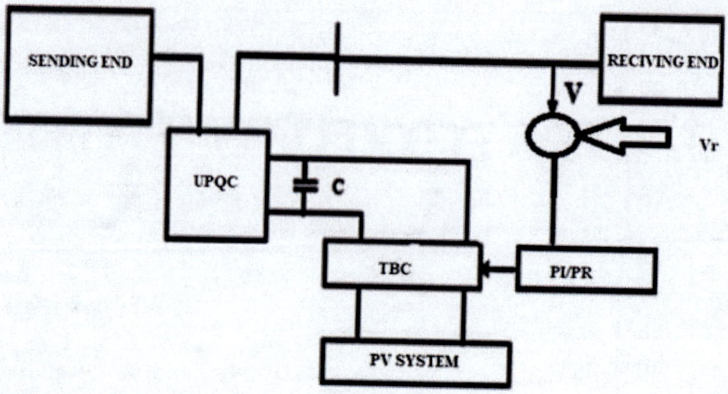

Fig. 7 TBC–UPQC closed-loop Block diagram

value. The resultant is applied to the PI/PR controller. The PI//PR controller updates the pulse width of TBC to regulate the load voltage.

4.2 *PI Controller Based TBC-UPQC Closed Loop System*

PI indicates the combination of Proportional and Integral controllers. The exciting signal is mix of proportional and integral controllers error signal.

$$\mathrm{a}(t) = K_I \int e(t)dt + K_p e(t) \tag{1}$$

The SIMULINK diagram of PI controller-based TBC-UPQC closed-loop system is shown in Fig. 8.

Here, Ki and Kp are Proportional constant and Integral constant, respectively.

Value of receiving end RMS magnitude of voltage is represented in Fig. 9. From this, it is observed that peak value here is 6900 V.

4.3 *Proposed PR Controller Based TBC-UPQC Closed Loop System*

Closed-loop two inductor boost converter (TBC) fed UPQC-based system with the (PR) Proportional Resonant controller is represented in Fig. 10. At closed-loop control system which is shown at Fig. 8 the proportional integral PI controller is replaced with the proportional resonant PR controller. TBC–UPQC with PR controller is modeled and simulated in SIMULINK. The SIMULINK diagram of

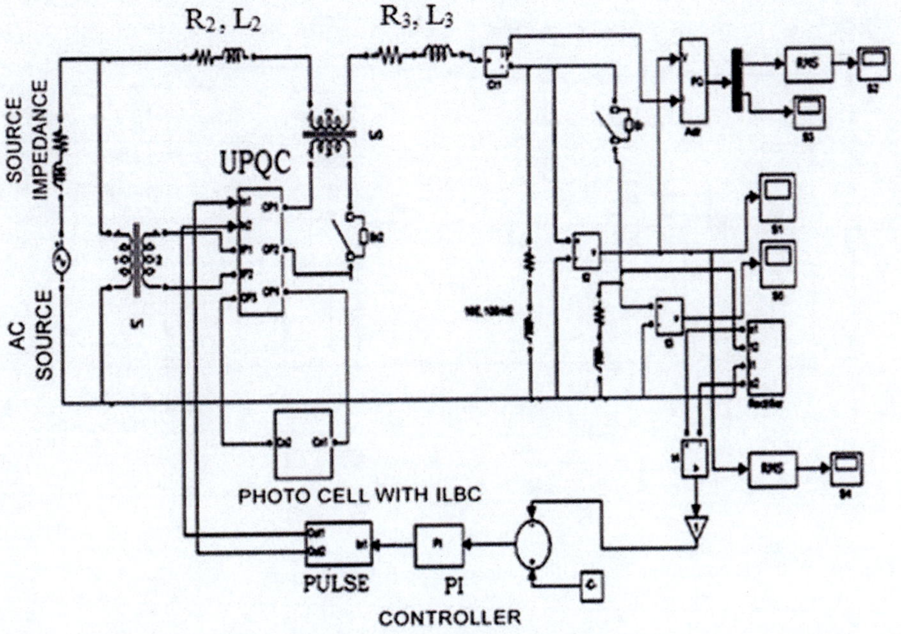

Fig. 8 TBC–UPQC closed-loop Block diagram with PI controller

Fig. 9 Receiving End RMS voltage value o

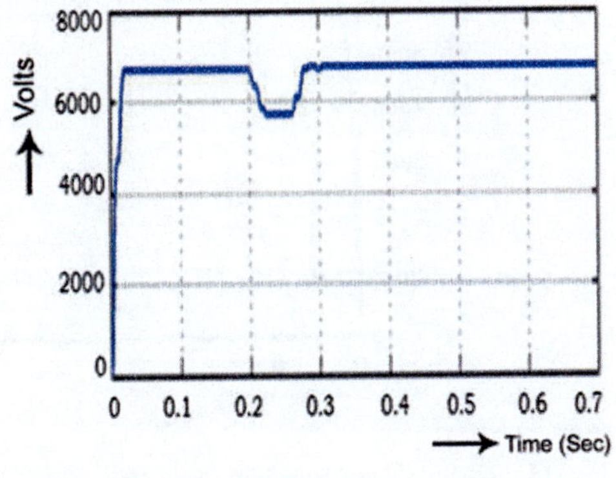

closed-loop TBC–UPQC-based system with the PR controller is represented in Fig. 10.

The magnitude voltage at the output of PR controller is as follows [8, 9].

$$V_O(S) = E(S)\left[K_1 + K_2/(S^2 + W^2)\right] \quad (2)$$

Here, K_1, K_2 are proportional and resonant constants.

Root mean square (RMS) voltage at output is represented in Fig. 11. From this, the magnitude is 6200 V.

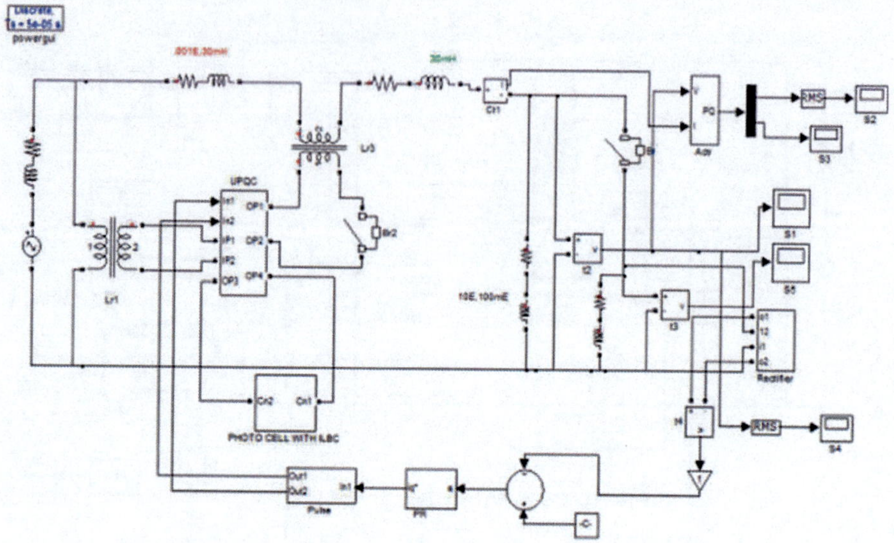

Fig. 10 With PR controller TBC–UPQC closed-loop Block diagram

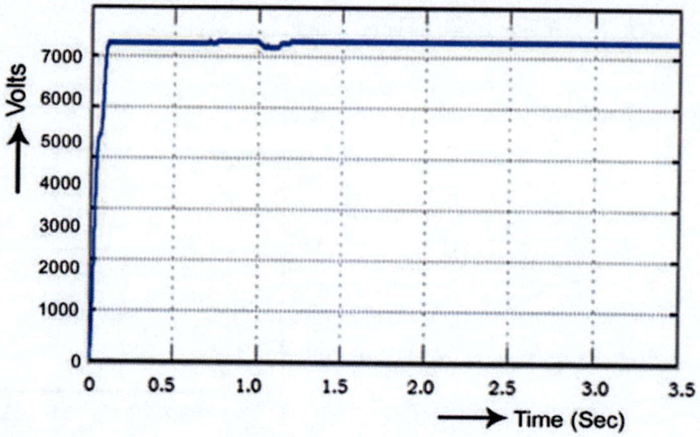

Fig. 11 RMS value of Receiving End Voltage

From the SIMULINK results of Figs. 9 and 10, the identification on their transient values of Rise time and its Peak time also steady state value of its Settling time.

- The Rise time is 0.225 s in that PI-based controller, this is reduced to 0.21 s at PR controller.
- Here, Peak time is 0.25 s in that PI controller, this is reduced to 0.21 s at PR controller.

Table 2 Comparison of time response specifications of PI, PR-based TBC-UPQC system

System	Rise time (T_R)	Peak time (T_P)	Settling time (T_S)
PI-TBC-UPQC	0.225	0.25	0.3
PR-TBC-UPQC	0.21	0.21	0.205

- Here, Settling time is 0.3 s in that PI controller, this is reduced to 0.205 with at PR controller.

Here, comparative analysis of time response parameters values is shown in Table 2. It is observed that TBC–UPQC system with PR controller gives better time response specifications than PI, controlled ILBC–UPQC system.

5 Conclusions

UPQC system is modeled with peak step-up converter and two inductor boost converter in open loop configuration. From open loop configuration results, it is observed that two inductor boost converter is preferred for UPQC. Then two inductor boost converter is implemented in closed, i.e., feedback systems with those PI, PR-based controllers. The PI, PR-based two inductor boost converter-based UPQC systems are modeled in SIMULINK. From the SIMULINK results, comparative time domain specifications are done. It is observed that with PR controller time domain specifications are improved. With PR controller transient values of Rise time and its Peak time also steady state value of its Settling times of two inductor boost converter-based UPQC system are reduced. Hence PR controller is preferred for closed-loop two inductor boost converter for UPQC system.

Acknowledgements Author interested to like thank to the principal, HOD—EEE, Registrar, staff of RVR & JCCE, Chowdavaram, Guntur, A.P., India for giving the accommodation facilities to do simulation worked in the EEE department of college.

References

1. Singh B, Al-Haddad K, Chandra A (1999) A review of active power filters for power quality improvement. IEEE Trans Ind Electron 45(5):960–971
2. Khadkikar V (2012) Enhancing electrical power quality using UPQC: acomprehensive overview. IEEE Trans Power Electron 27(5):2284–2297
3. Khadkikar V, Chandra A (2011) UPQC-S: a novel concept of simultaneous voltage sag/swell and load reactive power compensations utilizing series inverter of UPQC. IEEE Trans Power Electron 26(9):2414–2425
4. Khadkikar V, Chandra A (2008) A new control philosophy for a unified power quality conditioner (UPQC) to coordinate load-reactive power demand between shunt and series inverters. IEEE Trans Power Del 23(4):2522–2534

5. Prasanna Anjaneyulu GV, Sangameswararaju P (2017) Proportional resonant controller based inter leaved boost converter fed unified power quality conditioner system. Int J Eng Technol (IJET) 9(3):2628–2636
6. Yun M, Lee W, Suh I, Hyun D (2004) A new control scheme of unified power quality compensator-Q with minimum power injection. In: Proceeding of IEEE 30th annual industrial electronics social conferences, pp 51–56
7. Kolhatkar YY, Errabelli RR, Das S (2005) A slidingmode controller based optimum UPQC with minimum VA loading. In: Proceeding power engineering social genernation meeting, pp 871–875
8. Kolhatkar YY, Das S (2007) Experimental investigation of a singlephase UPQC with minimum VA loading. IEEE Trans Power Del 22(1):371–380
9. Prasanna Anjaneyulu GV, Sangameswararaju P (2017) Performance of unified power quality conditioner fed from high step up, two inductor and inter leaved boost converters. Int J Sci Technol (IJST) 1(5):1–9

Chapter 53
Smart Wearable Safety Device: A Wearable Anti-Assault and Location Tracking Device

Madhu Agarwal, Simantika Saha, Soumyadeep Pandit, Prasenjit Sarkar, Shreya Shree Das, and Subhojit Dawn

1 Introduction

Assaulting in the streets has become a major threat to the world in this era. Every day women are harassed, people are robbed, and children are kidnapped. Despite the endless protests, measures taken by the police authority, and political intervention the percentage of rapes in the nation has raised over the years. According to the annual data released by the Government of India on January 15, 2020, a rape case is reported every single 15 min [1]. Despite various national-level government initiatives and campaigns like Beti Bachao Beti Padhao (Save the Daughter, Educate the Daughter), the crime rate has increased in 2018 compared to 2017; as the data disclose that crime against women has increased by 0.9%. The National Crime Records Bureau (NCRB) on October 21, 2019, released its report for the year 2017, which stated that 359,849 cases of crime against women were reported in India [2]. One of the most terrifying examples is the rape and murder case of Dr. Priyanka Reddy which took place on November 27, 2019 in Hyderabad, India [3]. There are a large number of such anti-rape/assault products already existing in the market, most of them are made especially for women like the Anti-Rape Condom, Fem Defense Tampon, Hairy Stockings [4], and so on but these are just simple contraptions, designed only for women which may or may not be effective at the time of the assault.

M. Agarwal · S. Saha · S. Pandit · P. Sarkar
Department of Electrical Engineering, Siliguri Institute of Technology, Siliguri, West Bengal, India

S. S. Das (✉)
Department of Electrical Engineering, Mizoram University, Aizawl, India

S. Dawn
Department of Electrical and Electronics Engineering, Velagapudi Ramakrishna Siddhartha Engineering College, Vijaywada, Andhra Pradesh, India

S. Dawn et al. (eds.), *Smart and Intelligent Systems*, Algorithms for Intelligent Systems,
https://doi.org/10.1007/978-981-16-2109-3_54

Whitzman et al. [5] discuss the concept of women's safety in their work. It states that there should be some safety measures that should be taken by women in society. But they had only discussed the security of women, no device has been executed which confines the present work.

After a detailed literature study, it reveals that many authors have raised awareness about women's safety, but they have not proposed any smart device to protect them which is deployed in this work.

2 System Design

This is a device that can be placed in a jacket or a belt for personal safety. This device gets activated by pressing the switch and tracks the location of the victim. In addition to that, it can send an emergency message to family, friends, and police stations as per the contact number installation in the system. It also figures out the attacker through the camera and can give an electric shock to the attacker when they touch the victim. This device increases the chance of avoiding rape or any kind of violence without making a victim more likely to be physically injured. It can store the evidence through the camera for further investigation. It is designed fashionable from outside so that no one would realize the technological power it holds. A device that can be worn on the wrist like a watch or a band and can immobilize the assaulter giving the prospective victim enough time to flee or call for help.

The system works on Raspberry Pi. It consists of a GPS module, camera, and self-defense circuit which is a shock circuit. The switch has been installed in the equipment by which the system can be activated. The Raspberry Pi starts live streaming and GPS tracking. The live location of the victim will be communicated through a link to the police, family, and friends. The camera captures the attacker's identity, saves it, and then uploads it to the memory which gives a plus point to the police in their investigation.

3 Architectural Model

The architectural model of the proposed system is shown below which requires a camera, safety jacket, mobile, contact of police, and friends.

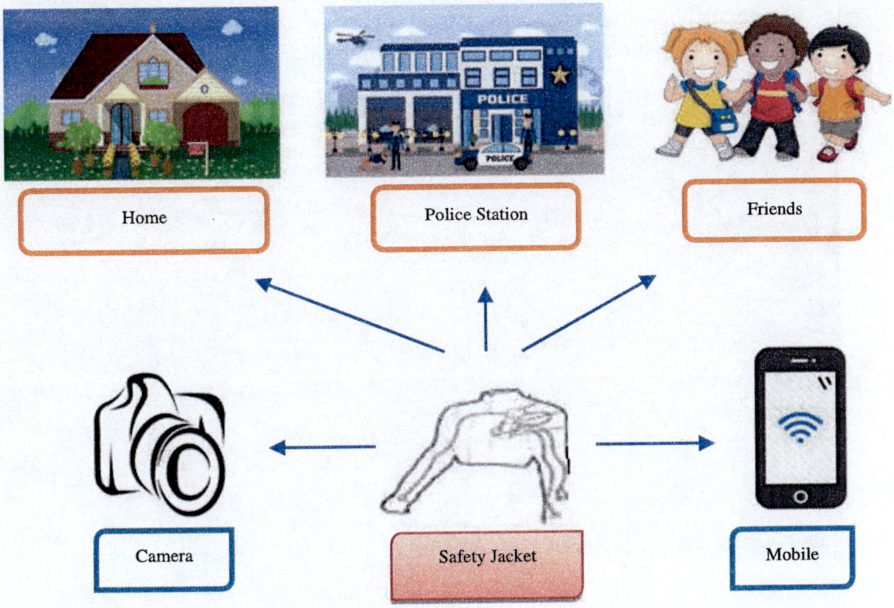

4 Working Principle of the Device

(a) **Triggering mechanism**: Triggering the circuit is the most essential part of the device. The device will get triggered via a switch concealed in the clothing or the wristband. To make it more efficient manual switching is being inserted. The trigger mechanism is such that it is easily accessible by the user and is a fail-proof trigger.

(b) **Raspberry Pi**: It is a microcomputer of the device which starts the live streaming and GPS tracking when the device has been triggered.

(c) **Insulating layer**: The layer provides the user the separation with the conductive shock circuit, which saves the user from the voltage shock. The layer has been so designed that it does not get affected and burnt due to high voltage. Physically, it gives the appearance of normal clothing.

(d) **Conducting layer**: The conductive layer is so designed to give the shock to the attacker when the device is triggered without harming the victim. The layer is covered to the entire surface of the clothing to give maximum effect when it comes to protection. The layer can be attached as straps over the arms, shoulder, and chest to provide cost-effectiveness.

(e) **Battery**: The battery used to power this device should be safe, light, portable, and small enough to be fitted in the clothing or the band. It should have at least 48 h of backup.

(f) **GPS module**: The GPS tracker inside the device provides the exact location and track movement of the user.

(g) **Camera**: The camera captures the attackers' identity, saves it, and uploads it to the memory which gives a plus point to the police in their further investigation.

5 Future Development

As of now, we have introduced only one type of triggering mechanism, but in the future, we would be developing another mechanism to trigger the circuit with more feasible ideas. The defense circuit will be modified.

6 Conclusions

The present invention is a wearable possessing multi-sensory device with personal safety and tracking device. It checks for the input that is when a switch is pressed or any automatic sensing for any kind of sensory inputs which calls for emergency or problematic conditions like sexual assault, kidnapping, human trafficking, accidents, robbery or any kind of violence. Various modules are then triggered which includes activating the defense circuit that may be the shock circuit. Live streaming camera and the link for the continuous live location will be sent through e-mail, SMS, or WhatsApp to the police, family, and friends. The device aims for the safety of people be it either women or children in any kind of situation. It is fashionable and easy to use the device. This device will surely reduce the rate of crimes to a great extent and bring harmony to society.

References

1. A rape in India every 15 minutes: government data, Asia Times, January 2020. https://asiatimes.com/2020/01/a-rape-in-india-every-15-minutes-government-data/
2. Guide prevent rape, think progress. http://thinkprogress.org/health/2014/06/10/3447269/guide-prevent-rape/
3. Hyderabad vet rape-murder case: a timeline since November 27 to 'encounter', The Indian Express, December 2019. https://indianexpress.com/article/india/hyderabad-vet-rape-murder-case-a-timeline-of-events-since-november-27-6153281/
4. The problem with anti-rape underwear. http://www.theguardian.com/lifeandstyle/womens-blog/2013/nov/11/problem-anti-rape-underwear-chastity-belt
5. Witzman C, Shaw M, Andrew C, Travers K (2009) The effectiveness of women safety audits. Secur J 22(3):205–218. https://doi.org/10.1057/sj.2009.1

Author Index

S. Dawn et al. (eds.), *Smart and Intelligent Systems*, Algorithms for Intelligent Systems,
https://doi.org/10.1007/978-981-16-2109-3